T0145267

Smart Innovation, Systems and Technologies

Volume 75

About this Series

The Smart Innovation, Systems and Technologies book series encompasses the topics of knowledge, intelligence, innovation and sustainability. The aim of the series is to make available a platform for the publication of books on all aspects of single and multi-disciplinary research on these themes in order to make the latest results available in a readily-accessible form. Volumes on interdisciplinary research combining two or more of these areas is particularly sought.

The series covers systems and paradigms that employ knowledge and intelligence in a broad sense. Its scope is systems having embedded knowledge and intelligence, which may be applied to the solution of world problems in industry, the environment and the community. It also focusses on the knowledge-transfer methodologies and innovation strategies employed to make this happen effectively. The combination of intelligent systems tools and a broad range of applications introduces a need for a synergy of disciplines from science, technology, business and the humanities. The series will include conference proceedings, edited collections, monographs, handbooks, reference books, and other relevant types of book in areas of science and technology where smart systems and technologies can offer innovative solutions.

High quality content is an essential feature for all book proposals accepted for the series. It is expected that editors of all accepted volumes will ensure that contributions are subjected to an appropriate level of reviewing process and adhere to KES quality principles.

More information about this series at http://www.springer.com/series/8767

Vladimir L. Uskov · Robert J. Howlett
Lakhmi C. Jain
Editors

Smart Education and e-Learning 2017

Springer

Editors
Vladimir L. Uskov
Computer Science and Information Systems
Bradley University
Peoria, IL
USA

Robert J. Howlett
Bournemouth University
Poole
UK

and

KES International
Shoreham-by-Sea
UK

Lakhmi C. Jain
University of Canberra
Canberra, ACT
Australia

and

Bournemouth University
Poole
UK

and

KES International
Shoreham-by-Sea
UK

ISSN 2190-3018 ISSN 2190-3026 (electronic)
Smart Innovation, Systems and Technologies
ISBN 978-3-319-86627-7 ISBN 978-3-319-59451-4 (eBook)
DOI 10.1007/978-3-319-59451-4

Softcover reprint of the hardcover 1st edition 2017

Printed on acid-free paper

This Springer imprint is published by Springer Nature
The registered company is Springer International Publishing AG
The registered company address is: Gewerbestrasse 11, 6330 Cham, Switzerland

Preface

Smart education, smart e-learning, smart classrooms, and smart universities are emerging and rapidly growing areas that represent an innovative and intelligent integration of smart objects and systems, smart technologies, smart environments, smart features or smartness levels, smart pedagogy, smart learning and academic analytics, various branches of computer science and computer engineering, state-of-the-art smart educational software and/or hardware systems. This is the main reason that in June of 2013 a group of enthusiastic and visionary scholars from all over the world arrived with the idea to organize a new professional event that would provide an excellent opportunity for faculty, scholars, Ph.D. students, administrators, and practitioners to meet well-known experts and discuss innovative ideas, findings, and outcomes of research projects, and best practices in Smart Education and Smart e-Learning (SEEL).

The KES International professional association initiated SEEL conference as a major international forum for the presentation of innovative ideas, approaches, technologies, systems, findings, and outcomes of research and design and development projects in the emerging areas of smart education, smart e-learning, smart pedagogy, smart analytics, applications of smart technology and smart systems in education and e-learning, smart classrooms, smart universities, and knowledge-based smart society.

The inaugural international KES conference on Smart Technology-based Education and Training (STET) has been held at Chania, Crete, Greece, on June 18–20, 2014. The 2nd international KES conference on Smart Education and Smart e-Learning took place in Sorrento, Italy, on June 17–19, 2015, and the 3rd SEEL conference—in Puerto de la Cruz, Tenerife, Spain, on June 15–17, 2016.

The main topics of the SEEL international conference are grouped into several clusters and include but are not limited to:

- **Smart Education (SmE cluster):** conceptual frameworks for smart education; innovative smart teaching and learning technologies; best practices and case studies on smart education; smart pedagogy and innovative teaching and learning strategies; smart classroom; smart curriculum and courseware design

and development; smart assessment and testing; smart learning and academic analytics; student/learner modeling; smart faculty modeling, faculty development and instructor's skills for smart education; university-wide smart systems for teaching, learning, research, management, safety, security; smart blended, distance, online and open education; partnerships, national and international initiatives and projects on smart education; economics of smart education;
- **Smart e-Learning (SmL cluster):** smart e-learning: concepts, strategies, and approaches; Massive Open Online Courses (MOOC); Small Personal Online Courses (SPOC); assessment and testing in smart e-learning; serious games-based smart e-learning; smart collaborative e-learning; adaptive e-learning; smart e-learning environments; courseware and open education repositories for smart e-learning; smart e-learning pedagogy, teaching and learning; smart e-learner modeling; smart e-learning management, academic analytics, and quality assurance; faculty development and instructor's skills for smart e-learning; research, design and development projects, best practices and case studies on smart e-learning; standards and policies in smart e-learning; social, cultural, and ethical dimensions of smart e-learning; economics of smart e-learning;
- **Smart Technology, Software and Hardware Systems for Smart Education and e-Learning (SmT cluster):** smart technology-enhanced teaching and learning; adaptation, sensing, inferring, self-learning, anticipation, and self-organization of smart learning environments; Internet of Things (IoT), cloud computing, RFID, ambient intelligence, and mobile wireless sensor networks applications in smart classrooms and smart universities; smart phones and smart devices in education; educational applications of smart technology and smart systems; mobility, security, access, and control in smart learning environments; smart gamification; smart multimedia; smart mobility;
- **"From Smart Education to Smart Society" Continuum (SmS cluster):** smart school; applications of smart toys and games in education; smart university; smart campus; economics of smart universities; smart university's management and administration; smart office; smart company; smart house; smart living; smart health care; smart wealth; smart lifelong learning; smart city; national and international initiatives and projects; smart society.

One of the advantages of the SEEL conference is that it is organized in conjunction with several other Smart Digital Futures (SDF) high-quality conferences, including Intelligent Decision Technologies (IDT), Intelligent Interactive Multimedia Systems and Services (IIMSS), Agent and Multi-agent Systems (AMS), and Innovation in Medicine and Healthcare (IMH). This provides SEEL conference participants with unique opportunities to attend also IDT, AMS, IIMSS, and IMH presentations, and meet and collaborate with subject matter experts in those areas—areas that are conceptually close to SEEL areas.

This book contains the contributions presented at the 4th international KES conference on Smart Education and Smart e-Learning, which took place in

Vilamoura, Algarve, Portugal, on June 21–23, 2017. It contains a total of 48 peer-reviewed book chapters that are grouped into several parts: Part 1—Smart Pedagogy, Part 2—Smart e-Learning, Part 3—Systems and Technology for Smart Education, Part 4—Smart Teaching, and Part 5—Smart Education: National Initiatives and Approaches.

We would like to thank many scholars who dedicated a lot of efforts and time to make SEEL international conference a great success, namely Dr. Luis Anido (Spain), Dr. Claudio da Rocha Brito (Brazil), Dr. Janos Botzheim (Japan), Dr. Dumitru Burdescu (Romania), Dr. Nunzio Casalino (Italy), Prof. Melany Ciampi (Brazil), Mr. Marc Fleetham (U.K.), Dr. Mikhail Fominykh (Norway), Dr. Brian Garner (Australia), Dr. Cristina Gasparri (Italy), Dr. Natalya Gerova (Russia), Dr. Jean-Pierre Gerval (France), Dr. Karsten Henke (Germany), Dr. Alexander Ivannikov (Russia), Dr. Aleksandra Klasnja-Milicevic (Serbia), Dr. Marina Lapenok (Russia), Dr. Andrew Nafalski (Australia), Dr. Zorica Nedic (Australia), Dr. Toshio Okamoto (Japan), Dr. Enn Õunapuu (Estonia), Dr. Mrutyunjaya Panda (India), Dr. Elvira Popescu (Romania), Dr. Valeri Pougatchev (Jamaica), Dr. Ekaterina Prasolova-Førland (Norway), Dr. Danguole Rutkauskiene (Lithuania), Prof. Jerzy Rutkowski (Poland), Dr. Demetrios Sampson (Australia), Dr. Adriana Burlea Schiopoiu (Romania), Dr. Ruxandra Stoean (Romania), Dr. Masanori Takagi (Japan), Dr. Wernhuar Tarng (Taiwan), Dr. Yoshimi Teshigawara (Japan), Dr. Gara Miranda Valladares (Spain), Dr. Heinz-Dietrich Wuttke (Germany), and Dr. Larisa Zaiceva (Latvia).

We also are indebted to international collaborating organizations that made SEEL international conference possible, specifically KES International (UK), InterLabs Research Institute, Bradley University (USA), Science and Education Research Council (COPEC), Institut Superieur de l'Electronique et du Numerique ISEN-Brest (France), Silesian University of Technology (Poland), Multimedia Apps D&R Center, University of Craiova (Romania), and World Council on System Engineering and Information Technology (WCSEIT).

In the near future, we plan to expand the main topics of SEEL conference. Firstly, it can be argued that the primary function of a university is teaching, but other areas of scholarly activity have also always been important. Many modern universities define multiple pillars to characterize their mission, i.e., teaching, research, and engagement with business and the community. Increasingly, students are considering starting their own business instead of getting a job in a company.

Secondly, so far, many researchers still have a vision about a university as a place where a physical location is the hub for learning activity, although some students may be situated remotely. However, in business, the virtual company is becoming increasingly common. Once a company was a building housing the workforce. Now a company may be a collection of people based in geographically dispersed locations, perhaps not even employed, but working together online on some commercial enterprise. In the same way, we should ask if a university needs to be based around a very expensive piece of real estate in the same way it has been in the past. Laboratories need a home, but perhaps the rest can be virtual.

As a result, we plan to add one more important cluster to SEEL main topics—**Smart University: Hub for Students' Engagement into Virtual Business and Entrepreneurship (SmB cluster)**. We plan to introduce it for the 5th SEEL international conference that will be held at Australia's fabulous Gold Coast on June 20–23, 2018.

It is our sincere hope that this book will serve as a useful source of valuable collection of knowledge from various research, design and development projects, useful information about current best practices and case studies, and provide a baseline of further progress and inspiration for research projects and advanced developments in Smart Education and Smart e-Learning areas.

June 2017

Prof. Vladimir L. Uskov, Ph.D. (USA)
Prof. Robert J. Howlett, Ph.D. (UK)
Prof. Lakhmi C. Jain, Ph.D. (Australia)

Contents

Smart Pedagogy

Smart Pedagogy for Smart Universities

Vladimir L. Uskov[1(✉)], Jeffrey P. Bakken[2], Archana Penumatsa[1], Colleen Heinemann[1], and Rama Rachakonda[1]

[1] Department of Computer Science and Information Systems, and InterLabs Research Institute, Bradley University, Peoria, IL, USA
uskov@fsmail.bradley.edu

[2] The Graduate School, Bradley University, Peoria, IL, USA
jbakken@fsmail.bradley.edu

Abstract. The performed analysis of innovative technology-based learning and teaching strategies for smart classrooms clearly shows that in the near future smart pedagogy will be actively deployed by leading academic institutions in the world for teaching of local and remote students in one class. The research project is focused on in-depth analysis of innovative learning strategies, including (1) learning-by-doing, (2) flipped classroom, (3) games-based learning, (4) adaptive teaching, (5) context-based learning, (6) collaborative learning, (7) learning analytics, (8) "bring your own device" (BYOD) strategy, (9) personal enquiry based learning, (10) crossover learning, (11) robotics-based learning, and other advanced technology-based approaches to teaching and learning. The obtained outcomes of this performed research, analysis, and testing of implemented smart pedagogy components undoubtedly prove that those learning and teaching strategies support identified "smartness" levels and smart features of smart classrooms such as (1) adaptivity, (2) sensing, (3) inferring, (4) anticipation, (5) self-learning, and (6) self-organization. The obtained student feedback undoubtedly demonstrates students' strong interest in smart pedagogy – the approach that will be an essential topic of multiple research, design, and development projects in the next 5–10 years.

Keywords: Smart pedagogy · Smart classroom · Teaching strategy · Learning style · Smart education

1 Introduction

Modern sophisticated smart devices, smart systems, and smart technologies create unique and unprecedented opportunities for academic and training organizations in terms of new approaches to education and training, learning and teaching strategies, services to on-campus and remote/online students, IT infrastructure, software and hardware system, technologies, set-ups of modern highly technological lecture halls, classrooms and labs [1].

The performed analysis of 150+ publications and reports relevant to (1) smart classrooms (SmC), (2) smart educational technologies, (3) smart systems, (4) smart universities (SmU), (5) innovative technology-based teaching strategies and learning styles in SmC and SmU – Smart Pedagogy (SmP), (6) smart learning environments

V.L. Uskov et al. (eds.), *Smart Education and e-Learning 2017*, Smart Innovation, Systems and Technologies 75, DOI 10.1007/978-3-319-59451-4_1

(SLE), (7) smart devices, (8) Internet-of-Things technology, (9) cloud computing, and (10) smart educational systems clearly shows that SmP, SmC, and SmU will be essential topics of multiple research, design, and development projects in the next 5–10 years. It is expected that, in the near future, the SmP concept, as well as the necessary supporting hardware/software systems and technologies, will be actively deployed by leading academic intuitions – smart universities - in the world [2–5].

"(1) 76% of teachers say technology allows them to respond to a variety of learning styles, (2) 77% of teachers say technology use in the classroom motivates students to learn, (3) 74% of administrators say digital content in schools increases student engagement, (4) 91% of administrators say effective use of educational technology is critical to their mission of high student achievement" [6].

"Disruptive innovations always start as inferior to incumbent models, but their technological core improves at a steeper curve than that of the incumbent models … Eventually, the disruptive innovations surpass the incumbent models … When traditional faculty teach online, they bring back to their traditional classrooms new pedagogical methods and technologies; online education is thus actually helping to improve traditional delivery models (Fig. 1)" [7].

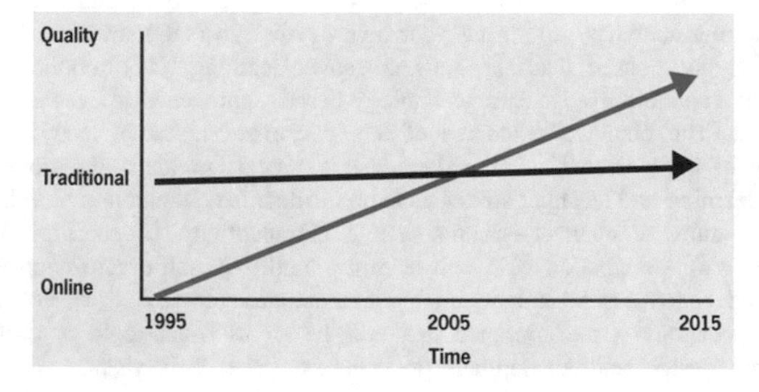

Fig. 1. Quality of online education versus quality of traditional in-classroom education [7]

Multiple publications [8–16] are available on innovative teaching strategies and learning styles. Unfortunately, the aforementioned, as well as multiple additional analyzed publications, do not provide detailed information about SmP that strongly supports SmC concept and emphasizes "smartness" levels and smart features of SmC such as (1) adaptivity, (2) sensing, (3) inferring, (4) anticipation, (5) self-learning, and (6) self-organization [17].

2 Project Goal and Objectives

The overall goal of the on-going research, design, and development project at the InterLabs Research Institute at Bradley University (Peoria, IL, USA) is to use a systematic approach in order to identify, analyze, test, and, eventually, recommend various components of a SmP – teaching strategies and learning styles (or models) that strongly support (a) SmE, SmC, and SmU concepts, and (b) smartness levels and smart features of SmC and SmU. These, therefore, should be actively used in smart classrooms.

In order to achieve this goal, the project team selected the following objectives:

- Analysis of recent (2010–2017) innovative teaching strategies and learning styles;
- Identification of SmP components for SmC – teaching strategies and learning models that strongly support (a) SmE, SmC, and SmU concepts, and (b) smartness levels and smart features of SmC and SmU;
- Qualitative analysis of identified SmP components;
- Formative and summative evaluation of identified SmP components in computer science and information systems (information technology) curriculum.

A summary of up-to-date project findings and outcomes is presented below.

3 Innovative Teaching Strategies Analyzed

In order to arrive with a list of proper teaching and learning strategies for SmC, we analyzed 40+ innovative recent (2010–2017) strategies that include, but are not limited to, the following [8–16]:

- Adaptive Teaching (adapting computer-based teaching to the learner knowledge);
- Game-based Learning (exploiting the power of digital games for learning);
- Learning-by-Doing and Remote Labs (guided experiments on scientific equipment);
- Context-based Learning (SmC context shapes and is shaped by learning);
- Seamless Learning (connecting learning across settings, technologies, and activities);
- e-books pedagogy (innovative ways of teaching/learning with next-generation e-books);
- MOOCs (Massive Open Online Courses);
- SPOC (Small Private Online Courses);
- Crowd Learning (harnessing the local knowledge of many students/learners);
- Badges to Accredit Learning (open framework for gaining recognition of skills and achievements);
- Flipped Classrooms (blending learning inside and outside the classroom);
- Dynamic Assessment (giving the learner personalized assessment to support learning);
- Seamless Learning (connecting learning across settings, technologies and activities);
- Bring Your Own Devices (learners use their personal devices to enhance learning in the classroom);
- Learning Design Informed by Analytics (a productive cycle linking design and analysis of effective learning);
- Learning Analytics and Predictive Analytics (data-driven analysis of learning activities and environments);
- Computational Thinking (solving problems using techniques from computing);
- Stealth Assessment (unobtrusive assessment of learning processes);

- Crossover Learning (connecting formal and informal learning);
- Personal Inquiry Learning (learning through collaborative inquiry and active investigation);
- Analytics of Emotions (responding to the emotional states of students);
- Rhizomatic Learning;
- Learn from Gaming;
- Learning to Learn;
- Learning through Storytelling;
- Learning through Argumentation;
- Incidental Learning;
- Embodied Learning;
- Virtual Assistants;
- Games-Based Learning and Gamification of Learning;
- Learning through Augmented and Virtual Reality;
- Robotics-based Learning.

Based on obtained analysis outcomes and finding of the above-mentioned learning approaches, we arrived with a short list of teaching strategies and learning styles that are focused on active use of advanced technology and, therefore, could support SmP, SmC, SmE, and SmU concepts. Short descriptions of several of these innovative approaches - those that have great prospectives to benefit SmU, SmU, and SmC - are given below.

3.1 Adaptive Teaching

Adaptive Teaching (AT) is based on the idea that computers and corresponding software tools help instructors adapt their classroom activities based on feedback and responses from students/learners. It attempts to adjust to differences in background knowledge and experience, providing ways for learners to cope. AT can also provide just-in time feedback by responding to recent actions, to correct an error or offer a hint [18]. The main features of AT include, but are not limited to:

- Students can choose his/her own sequence in which to study material when reaching portions of the course that students are not interested in;
- Students will receive more hints the more they are stuck on a problem, which is determined at the end of each learning module after the assessment [8, 10];
- Once students understand easier problems, they are given more difficult ones;
- Instructor controls the academic performance of each student and uses feedback from the system to find that a large group of students had trouble with certain topics; in this case, an instructor will (a) re-teach those concepts in class the next day, or (b) split the students into small groups to discuss specific topics.

3.2 Learning Analytics

The purpose of Learning Analytics (LA) is to improve learning and the environment in which learning takes place. Analytics enable visualization and recommendations

designed to influence student behavior while a course is in progress. LA can be used to identify students who are disconnected and are, therefore, at risk. This provides the ability to pick out key information brokers within a class and to find potentially high- and low-performing students so teachers can better plan interventions. For example, the competency map at Capella University helps students by constantly showing them where they are in each course, how much is ahead of them, and where they need to concentrate their efforts to be successful [12].
Instructors can use LA to:

- Monitor the learning process;
- Explore student data;
- Identify problems;
- Discover patterns;
- Find early indicators for success, poor marks or drop-out assess usefulness of learning materials;
- Increase awareness, reflect and self-reflect;
- Increase understanding of learning environments;
- Intervene, supervise, advise and assist;
- Improve teaching, resources and the environment [16].

Students can use LA to:

- Monitor their own learning activities, interactions and learning process, and compare their activities with performance of other students;
- Increase awareness, reflect and self-reflect;
- Improve discussion, participation, learning behavior, and performance.

3.3 Collaborative Learning

Collaborative Learning (CL) is an educational approach to teaching and learning that involves groups of students working together to solve a problem, complete a task, or create a product [13, 15]. The main CL features include (a) active use of online tools and to instruct students, (b) student collaboration (interaction, communication) with those teachers and other students, and (c) team-working approach to problem solving while maintaining individual accountability. Based on published reports, CL (a) develops social interaction skills, and (b) stimulates critical thinking and helps students clarify ideas through discussion and debate [12].

3.4 Context-Based Learning

"Context-based learning is a pedagogical methodology that, in all its disparate forms, centers on the belief that both the social context of the learning environment and the real, concrete context of knowing are pivotal to the acquisition and processing of knowledge. The approach is based on the firm conviction that learning is a social activity that is badly served by most classroom situations due to an inherent misrepresentation of how the mind acquires, processes, and produces knowledge" [19]. Context-Based Learning (CBL) main features include, but are not limited to,

- Use of real-life and fictitious examples in teaching environments in order to learn through the actual practical experience with a subject rather than just its mere theoretical parts;
- Helping students to learn from the world around them and to see how general principles in science and society relate to their everyday lives.

3.5 Games-Based Learning (Gamification of Learning)

Games-Based Learning (GBL) uses a computer or video game for learning. It is aimed at teaching a discrete skill or specific learning outcome, rather than being a complete pedagogical system. GBL is instrumental to [12, 16, 20]:

- Make learning concepts more palatable for students;
- Stimulate creative thinking;
- Encourage student creative behavior;
- Promote divergent ideas, approaches, thoughts;
- Help with fast strategic thinking and problem solving;
- Increase student engagement and motivation;
- Provide immediate feedback on learning assignments.

GBL strongly supports active learning and learning outcomes: analyze, design, create, and evaluate. In accordance with Dales' Cone of Experience (Fig. 2), GBL can help students/learners to remember up to 90% of what they do in the learning process [21].

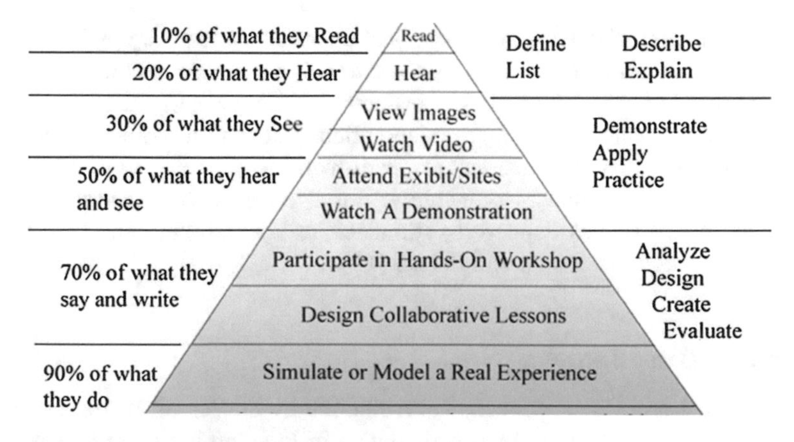

Fig. 2. Dale's cone of experience [21]

3.6 Flipped Classroom

Flipped Classroom (FC) learning is an attempt to get maximum benefits from joint online and face-to-face teaching and learning. In FC concept, direct teaching is taken out of the school or college classroom and put online. Students learn at home first through instructional videos and presentations. The FC-based learning approach has two elements to consider: (1) the "direct instruction" part at home, and (2) the "interactive face-to-face discussion" part in the classroom [22].

The success of FC-based learning significantly depends on how the interactive classroom element is constructed. Some teachers have adopted peer mentoring, where students teach each other; others have used group projects to explore the concepts learned at home. The home element does not have to consist of short videos - textbooks, software or other resources could work as well; however, short video learning modules tend to be the dominant format in FC [22]. Usually, FC teaching and learning helps to promote: (1) creativity; (2) communication; (3) teamwork; (4) leadership; (5) collaboration; (6) initiatives; and (7) problem solving.

4 Research Outcome # 1: Pilot Testing of Smart Pedagogy

Based on the performed research and analysis, we predict that various identified components of SmP will actively and strongly support SmC, SmE, and SmU concepts, as well as "smartness" levels and smart features of SmC.

In 2014–2016, Bradley University created eleven smart classrooms with top-quality multimedia Web-lecturing and capturing equipment. Each classroom is equipped with the SMART Board, HD video cameras, projectors, document camera, instructor's main console, computer systems, microphones, speakers, etc.

The latest smart classroom – Br160 classroom - with the most advanced software and hardware systems and technology was deployed by Bradley University in January of 2017. It is equipped with

- Innovative type of a smart board – the SMART Board 4084 with Interactive Flat Panel; particularly, it may work as a very big tablet with 84-inch touch screen for collaborative learning of in-classroom and online/remote students;
- Twenty-one (21) state-of-the-art DELL 7459 All-In-One tablet/desktop computers; each computers has 24″ touch display and built-in Intel(R) RealSense(R) 3D video camera;
- Three LG 55″ big screen TVs with 1080p Smart HDTV features; these features are very instrumental for a virtual presence of remote/online students during actual classes in a classroom and active communication/collaboration between local/in-classroom and online/remote students (see a simulation on Fig. 3 below);
- BenQ HT2020 ceiling-mounted projector,
- Microphones, speakers;
- 84″ regular screen,
- Various related software systems, technology, and devices.

Fig. 3. 2nd generation Smart Classroom at Bradley University to effectively support Smart Pedagogy

We tested the proposed SmP components in various graduate Computer Science (CS) and Computer Information Systems (CIS) classes against the introduced SmC "smartness" levels and smart features. A list of undergraduate and graduate courses taught includes (1) Software Engineering at undergraduate and graduate levels, (2) Software Project Management, (3) Six Sigma and Total Quality Management, (4), Agile Methodology Applications in Computing, (5) Web-based and Mobile Software, and (6) Computer Information Systems Analysis and Design. A summary of obtained research outcomes is given in Table 1.

5 Research Outcome # 2: Student Feedback Obtained

During the 2016–2017 academic year, we taught multiple CS and CIS undergraduate and graduate courses to master various SmP components, including (1) learning-by-doing, (2) flipped classroom, (3) games-based learning, (4) collaborative learning, (5) learning analytics, and (6) adaptive teaching. For the purpose of this research, a summary of student formative feedback about designated SmP components from 23 undergraduate and 20 graduate students is presented in Tables 2–4 below. The following legend to represent students answers has been used in those tables: 1 – neutral; 2 – somewhat like it; 3 – like it very much; Av. UG – average score by all undergraduate students, Av. GR - average score by all graduate students, Av. UG+GR – average score by both undergraduate and graduate students.

Table 1. SmC "smartness" levels and corresponding SmP components

Smartness levels of SmC	Innovative SmP component to support SmC/SmE/SmU smartness levels (examples)
Adaptation	• GBL allows the learner to control where and when to learn, pausing and reflecting on learning material. • FC enables users to see information about their surroundings displayed in front of them and hence adapt to the surroundings [22]. • AT helps teachers adapt their classroom activities based on responses from students. • Bring-Your-Own-Device (BYOD) provides a convenience because they are familiar and comfortable with using their own technology; as a result, they adapt well to SmC environment and can focus on actual learning process.
Sensing	• BYOD uses cameras and microphones to collect images, video, and audio; location sensing can be used to tag places of interest. Mobile devices with built-in accelerometers, noise, light, humidity, and temperature sensors can be used as science toolkits to collect data and perform experiments [23]. • Robotics teaching strategy helps users to see and sense information about their surroundings displayed in front of them.
Inferring	• LA in SmC or online mode helps a virtual tutor to react based on student facial indications. • AT provides just-in time feedback by responding to recent actions in order to correct an error or offer a hint.
Self-learning	• CBL helps improve learning as it enables people to learn easily and successfully from each other. It also allows students to watch video lectures at home and discuss them in SmC. • CL helps to improve learning by enabling a more social form of study, with a group of students working together on reading, annotating, and comparing one or more texts on the same topic.
Anticipation	• AT provides just in time feedback and hints that can help students to anticipate the answer and keep their attention [13]. • LA helps students know where they are in each course, how much is ahead of them, and where they need to concentrate their efforts to be successful, creating a sense of anticipation of what to do next [24].
Self-organization	• BYOD helps students to be more organized with lecture notes and assignments. • CL is used to improve communication, collaboration, and teamwork within various organized or self-organized teams. • LA helps learners organize their own activities, interactions, and learning process and thereby become better learners.

The summary of undergraduate and graduate student feedback (Tables 2, 3, 4), as well as additional obtained student feedback, clearly shows that innovative teaching strategies are interesting for students. For example, *Learning-by-Doing* is the most popular teaching style among all students in SmC - it has an average score of 2.63 (out of 3.0 points) by both undergraduate and graduate students, followed by *Flipped*

Table 2. *Learning-By-Doing* teaching strategy: a summary of student feedback

How do you like this component of SmP?	1	2	3	Av. UG	1	2	3	Av. GR	Av. UG+ GR
Have in-classroom exercise (s) every class	13	7	3	1.57	11	5	4	1.65	1.60
Have in-classroom exercise (s) one time in 2–3 classes	2	9	12	2.43	1	5	14	2.65	2.53
Have 15-minute long tests/quizzes every other class	19	4	0	1.17	13	5	2	1.45	1.30
Have 15-minute long tests/quizzes 2–3 times per semester	11	8	4	1.70	6	6	8	2.10	1.88
Have homework assignment after every class	18	4	1	1.26	9	7	4	1.75	1.49
Have homework assignment one time in 2–3 classes	1	12	10	2.39	3	5	12	2.45	2.42
Work alone on in-classroom exercise(s)	13	7	3	1.57	5	5	10	2.25	1.88
Work with a classmate (as a team) on in-classroom exercise(s)	4	6	13	2.39	4	7	9	2.25	2.33
Work alone on homework assignment(s)	9	11	3	1.74	1	6	13	2.60	2.14
Work with a classmate (as a team) on homework assignment(s)	2	12	9	2.30	8	4	8	2.00	2.16
Have in-classroom discussions on new topics, i.e. Q&A sessions	3	12	8	2.22	4	4	12	2.40	2.30
In general, what is your opinion about this teaching strategy	0	10	13	**2.57**	1	4	15	**2.70**	**2.63**

Classroom teaching strategy (2.30), *Collaborative Learning* (2.16), *Gamification of Learning* (2.12), *Learning Analytics* (2.10), and *Adaptive Teaching* (2.01). On the other hand, average scores by undergraduate and graduate students are very close to each other for *Learning-by-Doing* teaching strategy – 2.57 and 2.70 accordingly; however, they substantially differ for *Flipped Classroom* strategy (2.09 and 2.55) and *Gamification of Learning* teaching strategy (1.87 and 2.40).

Table 3. *Flipped classroom teaching* strategy: a summary of student feedback

How do you like this component of SmP?	1	2	3	Av. UG	1	2	3	Av. GR	Av. UG + GR
Assignment to learn software systems by yourself before the class	9	9	5	1.83	2	3	15	2.65	2.21
Required use of software systems used in industry to complete lab assignment	3	8	12	2.39	1	3	16	2.75	2.56
Have 1 week or less to complete lab assignment (i.e. to work quickly)	13	9	1	1.48	8	5	7	1.95	1.70
Have 2 weeks or more to complete lab assignment	2	9	12	2.43	4	4	12	2.40	2.42
Work alone on lab assignment with regular set of tasks	9	10	4	1.78	2	5	13	2.55	2.14
Work with a classmate (as a team) on expanded lab assignment (with expanded set of tasks)	3	14	6	2.13	4	7	9	2.25	2.19
Work alone on course-project assignment(s) with regular set of tasks	13	8	2	1.52	10	3	7	1.85	1.67
Work with a classmate (as a team) on course project assignment(s) with expanded set of tasks	2	14	7	2.22	0	6	14	2.70	2.44
Have Web-based discussion forum to discuss progress of course project with a designated student moderator	12	8	3	1.61	4	6	10	2.30	1.93
Instructor should have Skype-based virtual office hours	15	6	2	1.43	6	5	9	2.15	1.77
Have short (5–10 min long) videos by instructor or by external experts on course website	8	10	5	1.87	5	3	12	2.35	2.09
Have a very detailed course website with week-by-week activities, assignments, submission deadlines, access to PPT presentations, etc.	1	7	15	2.61	3	0	17	2.70	2.65
In general, what is your opinion about this teaching strategy	2	17	4	**2.09**	3	3	14	**2.55**	**2.30**

Table 4. *Gamification of learning* teaching strategy: a summary of student feedback

How do you like this component of SmP?	1	2	3	Av. UG	1	2	3	Av. GR	Av. UG + GR
Participate in in-classroom discussions of various topics	4	13	6	2.09	1	7	12	2.55	2.30
Try to solve in-classroom exercise(s) first (i.e. before other students) and present your solution to the class (for possible extra points)	13	5	5	1.65	8	6	6	1.90	1.77
Try to solve required HW assignment as quickly as possible and submit your HW solution before other students	14	6	3	1.52	13	2	5	1.60	1.56
Try to solve optional HW assignment and submit your HW solution to instructor	5	10	8	2.13	6	4	10	2.20	2.16
Know my up-to-date academic performance in the course only 2–3 times per semester	9	7	7	1.91	4	3	13	2.45	2.16
Know my academic performance in the course after each major assignment	4	10	9	2.22	3	5	12	2.45	2.33
Do you want instructor to regularly show the dynamics of getting extra points by all students in the class (even if a student has 0 pts to date)?	11	8	4	1.70	8	4	8	2.00	1.84
Do you want instructor to invite expert(s) from industry to present various topics or systems used in industry?	3	8	12	2.39	4	2	14	2.50	2.44
In general, what is your opinion about this teaching strategy	9	8	6	**1.87**	3	6	11	**2.40**	**2.12**

6 Conclusions. Next Steps

Conclusions. The performed research, analysis, design, and obtained findings and outcomes enabled us to make the following conclusions:

(1) We analyzed 40+ most recent (2010–2016) innovative teaching strategies and learning styles (Sect. 3). This enabled us to identify about 10 learning styles that are most suitable for an implementation and active utilization in SmC (Section 4).

(2) We tested several SmP components against SmC "smartness" levels - the corresponding research outcomes are presented in Table 1. It is clear that the identified SmP components strongly support "smartness" levels of SmC and SmE.
(3) Based on the pilot use of various SmP components in SmC environment, we surveyed 23 undergraduate and 20 graduate students in SmC environment. The summaries of student formative feedback are available in Tables 2, 3, 4. Overall, the obtained data clearly shows student significant interest in and acceptance of various SmP components in Computer Science and Information Systems curricula.

Next Steps. Based on (a) obtained research data, findings, and outcomes, and (b) developed and tested SmP components, the future steps in this research project are:

(1) Implement, test, validate, and analyze additional SmP components, specifically, learning analytics, context-based learning, Personal Inquiry Based Learning (PIBL), and other;
(2) Identify corresponding software and hardware systems and technologies that effectively support various SmP components in SmC environment,
(3) Perform summative and formative evaluations of local and remote students as well as instructors in various courses, and
(4) Gather data on the quality of SmP components in CS and CIS undergraduate and graduate curriculum.

References

1. IBM, Smart Education. https://www.ibm.com/smarterplanet/global/files/au__en_uk__cities__ibm_smartereducation_now.pdf
2. Hwang, G.-J.: Definition, framework, and research issues of smart learning environments – a context-aware ubiquitous learning perspective. Smart Learn. Environ. **1**(4), 1–14 (2014). http://www.slejournal.com/content/1/1/4. a SpringerOpen Journal
3. Coccoli, M., Guercio, A., Maresca, P., Stanganelli, L.: Smarter universities: a vision for the fast changing digital era. J. Vis. Lang. Comput. **25**, 1003–1011 (2014)
4. Tikhomirov, V., Dneprovskaya, N., Yankovskaya, E.: Three dimensions of smart education. In: Uskov, et al. (eds.) Smart Education and Smart e-Learning, pp. 47–56. Springer (2015). 510 p., ISBN 978-3-319-19874-3
5. Shi, Y., Qin, W., Suo, Y., Xiao, X.: Smart classroom: bringing pervasive computing into distance learning. In: Nakashima, H. et al. (eds.), Handbook of Ambient Intelligence and Smart Environments, pp. 881–910. Springer (2010)
6. Innovations at Education. http://d20innovation.d20blogs.org/category/learning-strategies/
7. LeBlanc, P.: Thinking about Accreditation in a Rapidly Changing World, EDUCASE Review, April 2013. http://er.educause.edu/articles/2013/4/
8. van den Broek, G.: "Innovative Research-Based Approaches to Learning and Teaching", OECD Education Working Papers, No. 79, OECD Publishing (2012). http://dx.doi.org/10.1787/5k97f6x1kn0w-en
9. Nair, P.: 30 Strategies for Education Innovation (2015). http://www.fieldingnair.com
10. The NMC Horizon Report: 2016 Higher Education Edition. http://cdn.nmc.org

11. The NMC Horizon Report: 2015 Higher Education Edition. http://cdn.nmc.org
12. The NMC Horizon Report: 2014 Higher Education Edition. http://cdn.nmc.org
13. Innovating Pedagogy: The Open University (UK) (2015). http://proxima.iet.open.ac.uk/public/innovating_pedagogy_2015.pdf
14. Innovating Pedagogy: The Open University (UK) (2014). http://www.openuniversity.edu/sites/www.openuniversity.edu/files/The_Open_University_Innovating_Pedagogy_2014_0.pdf
15. Innovating Pedagogy: The Open University (UK) (2013). http://www.open.ac.uk/iet/main/sites/www.open.ac.uk.iet.main/files/files/ecms/webcontent/Innovating_Pedagogy_report_2013.pdf
16. Innovating Pedagogy: The Open University (2012). http://www.open.ac.uk/iet/main/sites/www.open.ac.uk.iet.main/files/files/ecms/webcontent/Innovating_Pedagogy_report_July_2012.pdf
17. Uskov, V., Bakken, J., Pandey, A.: The ontology of next generation smart classrooms. In: Uskov, et al. (eds.) Smart Education and Smart e-Learning, pp. 3–14. Springer (2015). 510 p., ISBN 978-3-319-19874-3
18. Intelligent Adaptive Learning: An Essential Element of 21st Century Teaching and Learning. http://www.dreambox.com/white-papers/intelligent-adaptive-learning-an-essential-element-of-21st-century-teaching-and-learning#sthash.nkKDQmzA.dpuf
19. Rose, D.: Context-Based Learning (2012). https://nclstage2projects.files.wordpress.com/2012/02/context-based-learning.pdf
20. Games-based learning. http://www.newmedia.org/game-based-learning–what-it-is-why-it-works-and-where-its-going.html
21. Dale, E.: The cone of experience. In: Audio-Visual Methods in Teaching, pp. 37–51. Dryden Press, NY (1946)
22. FLIP Learning. http://flippedlearning.org
23. Zhang, J.: Technology-supported learning innovation in cultural contexts. Educ. Tech. Res. Dev. **58**(2), 229–243 (2010)
24. The Task Force on Innovative Teaching Practices to Promote Deep Learning at the University of Waterloo: Final Report (2011). https://uwaterloo.ca/centre-for-teaching-excellence/

Information Channel Based Measure of Effectiveness of Computer-Assisted Assessment in Flipped Classroom

Jerzy Rutkowski(✉) and Katarzyna Moscinska

Akademicka 16, 44-100 Gliwice, Poland
{jrutkowski,kmoscinska}@polsl.pl

Abstract. Properly designed Assessment Program, both Formative tests and Summative test (Exam), is the basic part of arbitrary engineering course. Today, computerization of Assessment is a norm and in the 2nd generation of Smart Classroom, computerization means first of all active use of mobile technology. Then, measurement of effectiveness of Computer-Assisted Assessment, achievements of learning outcomes, correlation between number of Formative tests solved by the students and the Exam Pass/Fail ratio, is absolutely essential. The proposed measurement method is based on Discrete Memoryless Channel principles and utilizes mutual information as the measure of this correlation. The Case Study is presented and it confirms usefulness of the proposed measure. Some guidelines, good practices in Computer Assisted Assessment design are given.

Keywords: Technology enhanced learning · Computer Assisted Assessment

1 Introduction

In the era of dynamic development of ICT, their use in Higher Education is ubiquitous. The ICT, properly used, may significantly contribute to the quality of education, enable development of new course delivery methods, new methods of assessment. Learning Activities are the core of every engineering course and these Activities can be divided into two components:

- Learning Content (knowledge delivery);
- Assessment Program (knowledge assessment).

These components have to be tightly correlated, regardless the course delivery method, in the Flipped Classroom [1] as well as in the Traditional Classroom. In the Flipped Classroom, the Learning Content is based on e-materials. Then, Assessment Program should be correlated with these e-materials and take into account the Flipped Classroom characteristics, Learning Content modular structure in particular. Today, computerization of Assessment in HE is a norm. Tests (quizzes) are completed by the student at a computer, firstly at home (Formative Assessment), then in a computer lab (Summative Assessment, Exam), without the teacher's intervention. Extensive research in Computer-Assisted Assessment (CAA), also called E-Assessment and Computer-Aided Assessment, has been done so far, its effectiveness has been reported [2–4].

V.L. Uskov et al. (eds.), *Smart Education and e-Learning 2017*, Smart Innovation, Systems and Technologies 75, DOI 10.1007/978-3-319-59451-4_2

However, further studies are necessary to make the Assessment Program fully computer-automated and reliable, such that traditional Assessment can be replaced by CAA. In fact, traditional Assessment has to be practically eliminated from the Flipped Classroom. Organization of Moodle-based CAA in the Flipped Classroom has been discussed in [5]. To make this CAA effective, students have to accept new learning model, teachers have to match their teaching to this model:

- Students have to accept Self-Regulated Learning (SRL) mode of learning, take Formative tests systematically and solve questions with understanding.
- Teachers have to prepare high quality e-materials supporting the SRL. Guidelines how to prepare good e-materials, using different techniques, have been given by many authors, the review can be found in [1]. Then, to ensure high correlation between the number of Formative questions answered and the Exam grade, the teacher has to properly construct and align the Formative tests and the Exam test. Moreover, the teacher has to take into account that, in the 2nd generation of Smart Flipped Classroom, implementations are mainly based on active use of mobile technology and automatic communications in the Smart Classroom environment [6]. Obviously, when designing the CAA, achievements of learning outcomes have to be taken into account as well.

A model of SRL, also called Self-Directed Learning (SDL), correlation between student goals, tactics, strategies and achievements of learning outcomes have been discussed by many authors [7, 8] – this model has been repeated in Fig. 1.

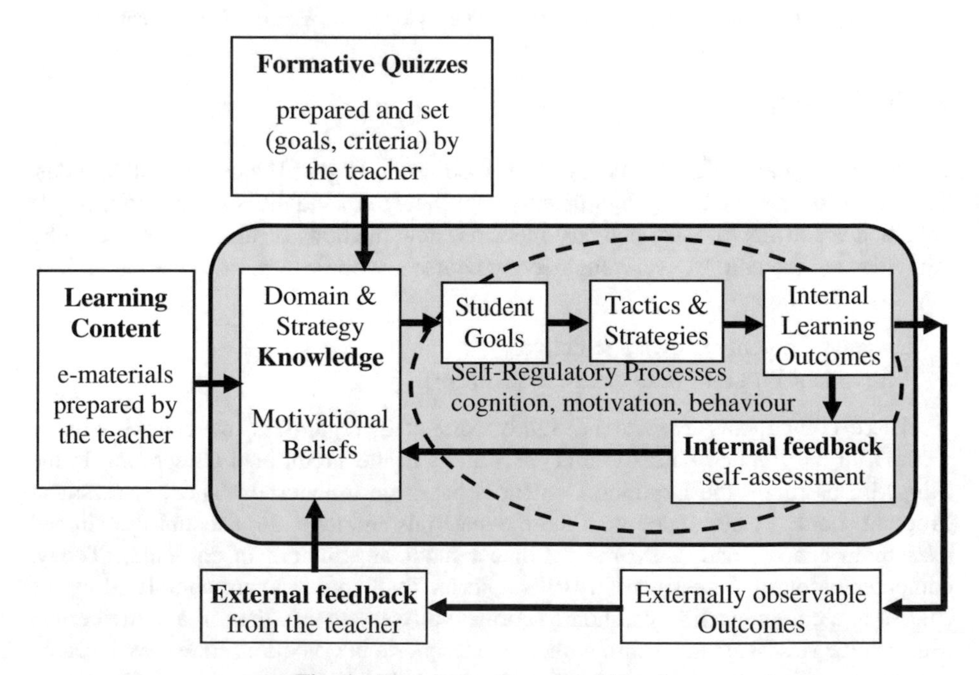

Fig. 1. Model of self-regulated learning

While there would normally be an overlap between the student goals and those of the teacher, the degree of overlap may not be high. The following main barriers in making Formative Quizzes more effective can be enlisted:

1. Students' SDL Readiness (SDLR) [9] is low, freshmen SDLR in particular.
2. If students perceive Formative assessment as primarily examining content knowledge, they will tend to do little more than rote learning, especially when they wish only to pass the Exam [10].
3. Students tend to ignore activities that do not directly contribute to grades and degree class; even though they could see the benefit of developing competencies, they do not take advantage of it [10].

These barriers can be broken by providing students with clear evidence that correlation between Formative quizzes taken and the Exam result is very high. Solving of Formative quizzes is supported by e-materials provided by the teacher and obviously they are only as good as the teacher who prepares them. Then, measurement of correlation between the number of Formative questions solved and the Exam score is essential for both the teacher and the students. A new method for finding this correlation has been proposed in [8]. This method is based in the field of Information Theory [11], developed by C.E. Shannon in the late 40's of the last century. The updated version of this method is presented in Sect. 2. A case study, use of Information Theory to evaluate effectiveness of Formative quizzes for Electric Circuit Analysis course in the academic year 2015/2016, is presented in Sect. 3, some guidelines and conclusions are given in Sect. 4.

2 Evaluation of Correlation Between Formative Quizzes Taken and Exam Results

The relationship between activity in Formative Quizzes and Exam results can be described by means of Discrete Memoryless Information Channel (DMC). Figure 2 presents Source-Channel-Receiver information system [9], where:

$$\mathbf{X} = \{X_1, \ldots, X_M\} \tag{1}$$

is the discrete input source of information, in short the Source, set of samples (ensembles) characterized by the probability assignment

$$\mathbf{P}_\mathbf{X} = \{p\{X_1\}, \ldots, p(X_M)\} \tag{1a}$$

$$\mathbf{Y} = \{Y_1, \ldots, Y_K\} \tag{2}$$

is the channel output source, in short the Receiver, characterized by the probability assignment

$$\mathbf{P}_\mathbf{Y} = \{p\{Y_1\}, \ldots, p(Y_K)\} \tag{2a}$$

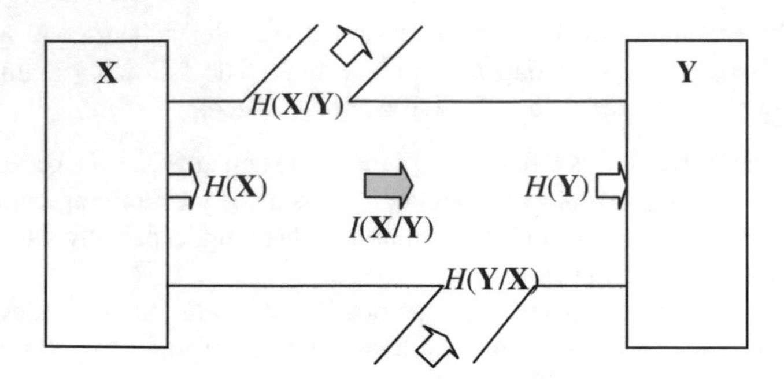

Fig. 2. Information system: source-channel-receiver

and Channel itself is characterized by *MK* transition probabilities that relate ensembles of input and output source:

$$p\left(Y_j/X_i\right); i = 1, \ldots, M, j = 1, \ldots, K \tag{3}$$

Then, for the given probabilistic model of the Source and the Channel, information loss $H(\mathbf{X/Y})$, misinformation $H(\mathbf{Y/X})$ and mutual information $I(\mathbf{X/Y})$ can be defined. Mutual information between events (sources) **X** and **Y** is the information provided about the event **X** by the occurrence of the event **Y**, or vice versa.

$$I(\mathbf{X}/\mathbf{Y}) = H(\mathbf{X}) - H(\mathbf{X}/\mathbf{Y}) = H(\mathbf{Y}) - H(\mathbf{Y}/\mathbf{X}) \tag{4}$$

$$H(\mathbf{X}) = -\sum_{i=1}^{M} p(X_i) \log_2 p(X_i) \tag{5a}$$

$$H(\mathbf{Y}) = -\sum_{j=1}^{K} p(Y_j) \log_2 p(Y_j) \tag{5b}$$

$$H(\mathbf{Y}/\mathbf{X}) = -\sum_{i=1}^{M} \sum_{j=1}^{K} p(X_i, Y_j) \log_2 p(Y_j/X_i) \tag{6}$$

To give a measure, how far the considered channel is from the idealized (target) one, the normalized mutual information is introduced:

$$I_{\mathrm{n}} = I(\mathbf{X}/\mathbf{Y})/I(\mathbf{X}/\mathbf{Y})_{\mathrm{ref}} \tag{7}$$

where $I(\mathbf{X/Y})_{\mathrm{ref}}$ is the mutual information of the reference channel, channel that is considered as the idealized target one.

Exemplary channels: an arbitrary binary channel and the reference 3-input/3-output channel, are presented in Fig. 3.

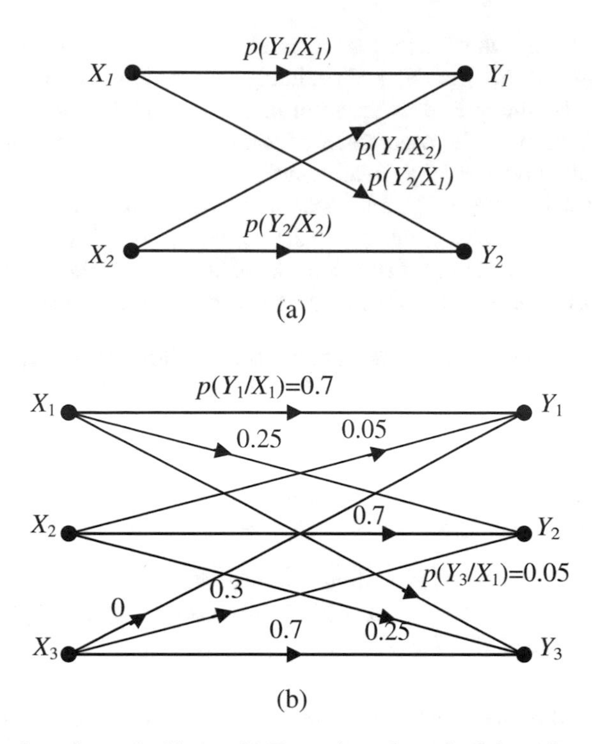

Fig. 3. (a) Exemplary channels: binary (b) Exemplary channels: 3-input/3-output (reference of Case Study)

Exam can be considered as measurement of students' knowledge and then, it can be described by Discrete Memoryless Information Channel (DMC). Students' knowledge, the measured quantity, can be expressed by a number of Formative quizzes taken and, after discretization, it consists the set of samples, the input Source. Set of Exam results consists the output source (Receiver). Exam can be considered as measurement of students' knowledge. It can be assumed that this knowledge is designated by Formative quizzes taken, number of tasks solved by the student. For the binary channel ($M = K = 2$), both sources can be discretized as follows:

- $X_1 = X_D$ Diligent students, D students that solved at least T_D % of tasks,
- $X_2 = X_N$ Negligent students, N students that solved less than T_D % of tasks,
- $Y_1 = Y_P$ P students that Passed Exam,
- $Y_2 = Y_F$ F students that Failed Exam.

It is assumed that numbers D, N, P, F are known and they designate probability assignments $\mathbf{P_X}$ and $\mathbf{P_Y}$, e.g. $p(X_D) = D/M$ is probability of Diligence, $p(Y_F) = F/M$ is probability of Fail, $M = D+N = P+F$. Also, conditional probabilities that relate students' Diligence/Negligence and Exam results: $p(Y_j/X_i)$, $i =$ D,N; $j =$ P,F, are known, e.g. $p(Y_P/X_N) = P_N/N$ is the probability of Passing the Exam by the Negligent student, where P_N is the number of Negligent students that Passed. Then, relationship between

Formative quizzes and the Exam can be modeled by means of binary information channel, as depicted in Fig. 3a, and mutual information can be calculated. This information may be interpreted as the information provided about the measured data (students' knowledge) by the occurrence of measurements (Exam), in other words as the measure of effectiveness of Formative quizzes.

To describe more precisely relationship between students activity in solving quizzes and their performance during the Exam, more complex channel can be considered. The ensemble of Diligent students can be split into Diligent High and Diligent Low, while the ensemble of Negligent students remains unchanged:

- $X_1 = X_{DH}$ D_H Diligent High students that solved more than T_{DH} % of tasks,
- $X_2 = X_{DL}$ D_L Diligent Low students that solved between T_{DL} % and T_{DH} %,
- $X_3 = X_N$ N Negligent students that solved less than T_{DL} % of tasks.

Then, Pass can be split into Pass High and Pass Low while the ensemble of students that Failed remains unchanged:

- $Y_1 = Y_{PH}$ P_H students that Passed with Excellent, Very Good or Good grade,
- $Y_2 = Y_{PL}$ P_L students that Passed with Satisfactory or Sufficient grade,
- $Y_3 = Y_F$ F students that Failed Exam.

The conditional probabilities have to be split accordingly, taking into account how many Diligent High Passed High, etc., e.g. $p(Y_{PH}/X_{DL}) = P_{H,DL}/D_L$ is the probability of Passing High by Diligent Low, where $P_{H,DL}$ is the number of Diligent Low that Passed High. Then, the relationship between Formative quizzes and Exam can be modeled by means of 3-input/3-output information channel. The conditional probabilities can be expressed in the form of table, as presented in Table 1.

Table 1. Conditional probabilities of 3-input/3-output Channel

$p(Y_j/X_i)$	$Y_1 = Y_{PH}$	$Y_2 = Y_{PL}$	$Y_3 = Y_F$
$X_1 = X_{DH}$	$P_{H,DH}/D_H$	$P_{L,DH}/D_H$	F_{DH}/D_H
$X_2 = X_{DL}$	$P_{H,DL}/D_L$	$P_{L,DL}/D_L$	F_{DL}/D_L
$X_3 = X_N$	$P_{H,N}/N$	$P_{L,N}/N$	F_N/N

3 Case Study: Electric Circuit Analysis Course

For the first time, the described methodology of evaluation of relationship between number of Formative quizzes taken and Exam results has been verified in the academic year 2014/2015. Fifty students (all enrolled) of Macro (Electronics + Automatics + Informatics), consisted the test group, results of only the first Exam have been taken into account. Formative quizzes and the Exam have been distributed through Moodle LMS. The obligatory Formative quizzes contained $9 \times 15 = 135$ *Calculated* questions [12] and the following thresholds have been assumed: obligatory minimum $T_N = 30\% = 40$, $T_{DL} = 50\% = 68$, $T_{DH} = 67\% = 90$ questions. The Exam quiz consisted of ten questions: eight *Calculated* questions, marked 0 or 1 and drawn from Formative quizzes, two *Multiple-Choice*, marked 1, 0 (no answer) or −0.5 (wrong answer). The Pass threshold at 3.5 points has been experienced as the most adequate [13]. This case study has been presented in [8], high correlation between students activity in solving Formative quizzes and the Exam results has been confirmed and the following main conclusions have been drawn:

1. Only students that have solved majority of Formative Quizzes (Diligent High students), Pass the Exam with High mark.
2. Minority of Negligent students that have solved only the obligatory number of T_N questions Passed, probability of Passing with a High mark was practically zero.
3. Very small percentage of Diligent High students Failed the Exam.

In the academic year 2015/2016 the study has been repeated, the test group consisted of 276 students (all enrolled), representing three fields of study: Macro (53), Informatics (158) and Teleinformatics (65). The set of Formative questions has been enlarged to $9 \times 21 + 100 = 289$ *Calculated* questions (number of questions per obligatory quiz has been increased from 15 to 21, nonobligatory quiz with 100 questions has been added) and the following new thresholds have been assumed: $T_N = 20\% = 55$, $T_{DL} = 33\% = 96$, $T_{DH} = 50\% = 145$ questions. The Exam organization remained unchanged, i.e. same Pass threshold of 3.5 points has been applied.

The results of the first two Exams have been taken into account, i.e. if the student failed the first Exam, then the resit result has been taken into account. The 3-input/3-output channel provides more information than the binary and only such channel has been considered. This channel, conditional probabilities $\mathbf{P_{Y/X}}$, together with the input probability assignment $\mathbf{P_X}$ and the output probability assignment $\mathbf{P_Y}$ (calculated from $\mathbf{P_X}$ and $\mathbf{P_{Y/X}}$), for each field of study and the aggregate, are presented in Fig. 4. Conditional probabilities for the aggregate channel are repeated in Table 2.

From formulas (4), (5a), (5b) and (6) the mutual information $I(\mathbf{X}/\mathbf{Y})$ can be easily calculated and for the aggregate channel $I(\mathbf{X}/\mathbf{Y}) = 0.40$bit. In an ideal channel (noiseless channel, all crossover probabilities are zero) $I(\mathbf{X}/\mathbf{Y})_{max} = H(\mathbf{X}) = 1.30$bit but such idealization is too rigorous to give the reference channel. To get more realistic reference (target) channel, we may accept that 5% of Diligent High students Fail (explanation will be given in Sect. 4). We may accept even higher percentages for other crossover transitions except one, we may not accept that Negligent students Pass High, i.e. for the reference channel we have to assume $p(Y_{PH}/X_N)_{ref} = 0$. If we assume the reference (target) channel as proposed in Fig. 3b, then the reference mutual information $I(\mathbf{X}/\mathbf{Y})_{ref} = 0.41$bit and consequently $I_n \approx 1$, $I_n\% \approx 100\%$.

This maximum normalized mutual information proves again the extremely high correlation between the Formative quizzes taken and the Exam results.

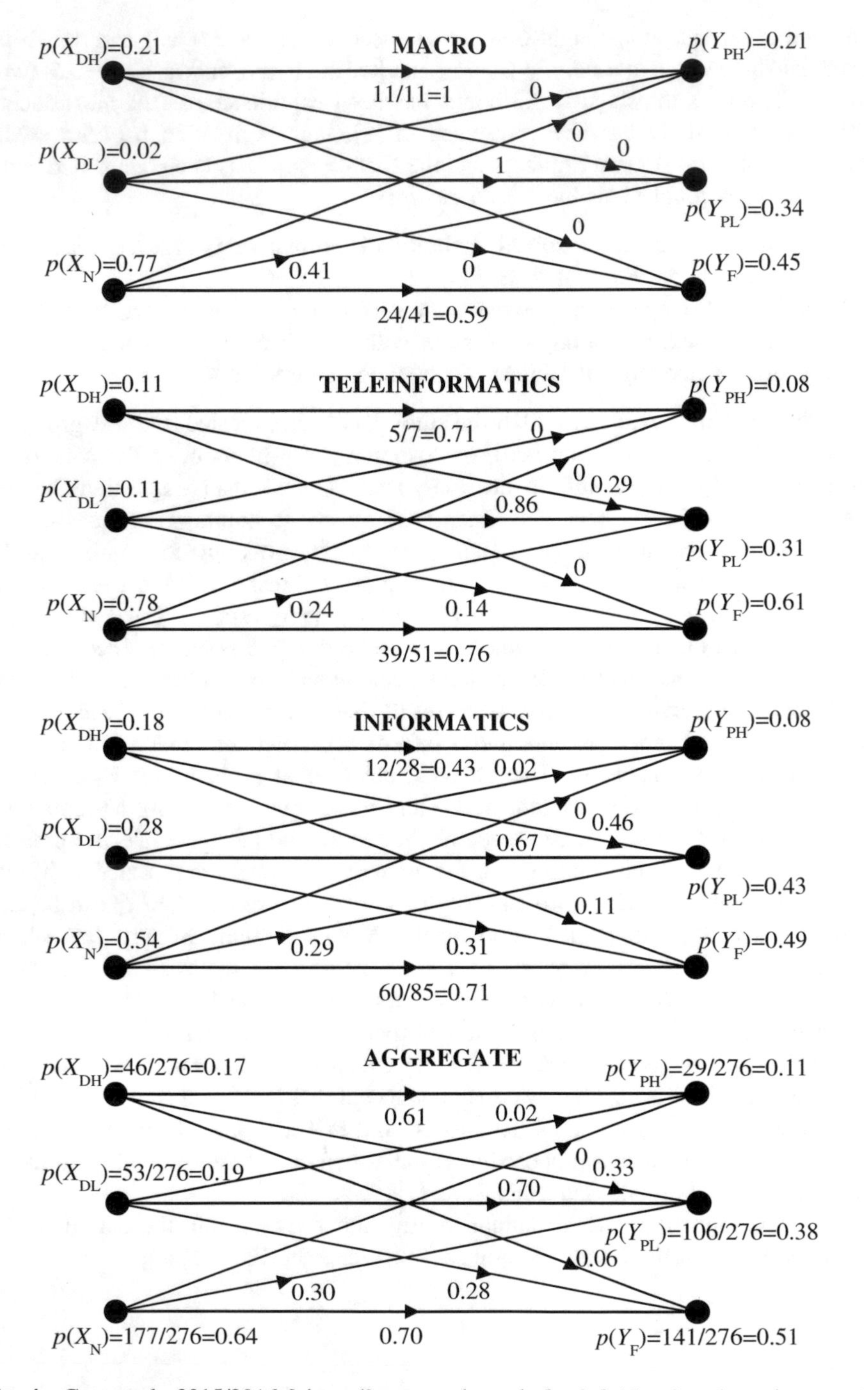

Fig. 4. Case study 2015/2016 3-input/3-output channels for 3 fields of study and aggregate

Table 2. Conditional probabilities of 3-input/3-output aggregate Case Study Channel

$p(Y_j/X_i)$	$Y_1 = Y_{PH}$	$Y_2 = Y_{PL}$	$Y_3 = Y_F$
$X_1 = X_{DH}$	28/46 = 0.61	15/46 = 0.33	3/46 = 0.06
$X_2 = X_{DL}$	1/53 = 0.02	37/53 = 0.70	15/53 = 0.28
$X_3 = X_N$	0/177 = 0	54/177 = 0.30	123/177 = 0.70

4 Final Conclusions and Guidelines

The updated new method that enables quantitative evaluation of effectiveness of Formative tests (quizzes) has been proposed. All three findings of the 2014/2015 case study have been confirmed. Taking also into account 2015/2016 case study, the following final findings can be formulated:

1. Number of Negligent students, students that started Formative quizzes just (one week) before the deadline and solved* only the obligatory minimum of tasks (20%) is high, ranges from 54% to 78%, with 64% (≈2/3) as the average!!!
2. Number of students that started Formative quizzes just before the deadline is even greater, and reaches 75% (11% managed to complete T_{DL} = 33% of tasks during the last week before the deadline to become Diligent Low).
3. None of Negligent students have Passed with High mark.
4. 70% of Negligent students Failed**.
5. Only 6% of Diligent High didn't Pass***.
6. 61% of Diligent High Passed with High mark.
7. Information channels for different fields of study are very similar.

* It can be suspected that some Negligent students cheated when passing obligatory Formative quizzes, used cheat-sheets (repository of answers-formulas) or asked peers to do the job for them.
** Significant percentage (30%) of Negligent students Passed Low only due to low Pass threshold of 3.5 points, for the threshold set on 5 points (50%) only 5% of Negligent students would Pass.
*** It has been observed that some students classified as Diligent High have passed Formative quizzes without understanding, just using cheat-sheets. Then, they have learnt answers-formulas by heart hoping that it will be enough to set the Exam-data to these formulas and Pass without understanding. Unfortunately for them, when preparing the Exam quiz, some minor corrections have been introduced to Formative questions such that the formula learnt by heart didn't give the correct answer.

These findings gave valuable feedback to the teacher:

- Proved high quality of the designed Formative online quizzes, usefulness of e-materials supporting SDL in the Flipped Classroom (video-podcasts, screencasts, e-textbook, e-slides explaining reasoning).
- Proved proper alignment of the Formative online quizzes and the final Exam and consequently its compliance with the teaching goals and learning outcomes.

- Made it clear that there is an urgent need for greater motivation of students to systematic work and solving quizzes with understanding.

To meet this need, some necessary steps have to be undertaken. In the carrot and stick strategy applied more stick has to be added, and the following final and unconfirmed guidelines can be formulated. To improve Pass/Fail ratio, not violating achievements of learning outcomes:

- It is necessary to force students to more systematic work. To reach this goal:
  - All Formative quizzes have to be obligatory, i.e. distribution of questions answered has to be uniform in the set of all quizzes.
  - Common deadline on all quizzes at the end of semester has to be replaced by deadlines on individual quizzes, distributed uniformly during the whole semester, as the lecture goes on.
  - Systematic work, solving quizzes with understanding, has to be verified during classroom tutorials.
  - Top students, some 10% of students that solved systematically the greatest number of tasks may obtain upgrade of the Exam grade or even get the credit, with Excellent grade, based on Formative quizzes alone.
- It is necessary to persuade students that only solving problems with understanding has sense. Completing quizzes using cheat-sheets (repository of formulas) is self-deception and also rote learning of these formulas will not pay. The Exam questions have to differ slightly from Formative questions, such that stored (learnt by heart) formulas are useless.
- The opportunity for cheating, both while solving Formative quizzes and solving the Exam quiz has to be reduced to minimum. What regards Formative quizzes, solving with understanding can be verified during classes, e.g. by short, single question tests. What regards the Exam, some additional technical precautions in a computer lab have to be undertaken, first of all to prevent from using mobile devices.

References

1. Rutkowski, J.: Flip teaching approach in smart education - principles and guidelines. Keynote of the 2nd International KES Conference on Smart Education and E-learning, Sorrento (2015). https://drive.google.com/file/d/0B_esc-ATuPmhX05RXzBKRzZIVTg/view
2. Fielding, A., Bingham, E.: Tools for computer-aided assessment. Learn. Teach. Action **2**(1) (2003). http://www.celt.mmu.ac.uk/ltia/issue4/fieldingbingham.shtml. Learning and Teaching Unit of Manchester Metropolitan University
3. Sim, G., Holifield, P., Brown, M.: Implementation of computer assisted assessment: lessons from the literature. ALT-J Res. Learn. Technol. **12**(3), 215–229 (2004)
4. Computer Assisted Assessment – Research into E-Assessment. In: Proceedings of the International Conference on CAA 2014. Communications in Computer and Information Science Series, Zeist. Springer (2014)
5. Rutkowski, J.: Moodle-based computer-assisted assessment in flipped classroom. In: Smart Education and Smart e-Learning. Smart Innovation, Systems and Technologies, vol. 41, pp. 37–46 (2015)

6. Uskov, V., Bakken, J., Pandley, A.: The ontology of next generation smart classrooms. In: Smart Education and Smart e-Learning. Smart Innovation, Systems and Technologies, vol. 41, pp. 3–14 (2015)
7. Nicol, D.J., MacFarlane-Dick, D.: Formative assessment and self-regulated learning: a model and seven principles of good feedback practice. Stud. High. Educ. **31**(2), 199–218 (2006)
8. Rutkowski, J.: Evaluation of the correlation between formative tests and final exam results – theory of information approach. Int. J. Electron. Telecommun. **62**(1), 55–60 (2016)
9. Guglielmino, L.M.: Reactions to field's investigation into the SDLRS. Adult Educ. Q. **39**(4), 235–240 (1989)
10. Higgins, M., et al.: Effective and Efficient Methods of Formative Assessment, CEBE Innovative Project in Learning & Teaching (2010). www.smu.ca/webfiles/Effective_and_Efficient_Methods_of_Formative_Assessment.pdf
11. Rutkowski, J.: Theory of Information and Coding. The Publishing House of The Silesian University of Technology, Gliwice (2006)
12. Moodle 2.7 Documentation. https://docs.moodle.org/27/en/Main_page
13. Rutkowski, J., Moscinska, K., Chruszczyk, L.: Development and evaluation of computer assisted exam – circuit theory example. In: Proceedings of the 9th IASTED International Conference on CATE, Lima, Peru, pp. 333–338 (2006)

Efforts for Upward Spirals Based on Both Teacher and Student Feedback on Smart Education

Nobuyuki Ogawa[1(✉)] and Akira Shimizu[2]

[1] Department of Architecture, National Institute of Technology, Gifu College, Gifu, Japan
ogawa@gifu-nct.ac.jp

[2] General Education, National Institute of Technology, Gifu College, Gifu, Japan
ashimizu@gifu-nct.ac.jp

Abstract. In order to promote smart education on a collegewide level, it is essential for respective teachers to pursue upward spirals of their own practice of smart education in addition to the introduction of ICT equipment into classrooms, as well as the holding of faculty development sessions related to upward spirals. In this paper, we describe our college's efforts in respect of a "review" of smart education which can be regarded as a future global standard, not only from one's own point of view but also from that of others. We mention the content of upward spirals conducted based on feedback from both teachers and students, which was promoted as efforts by all the faculty members to practice and improve smart education.

Keywords: Smart education · Pedagogy · Collegewide efforts · Upward spirals · Student feedback · Teacher feedback · ICT-driven education · e-Learning

1 Introduction

With the aim of developing educational and research projects working on leading educational reforms more and more, the Ministry of Education, Culture, Sports, Science and Technology (MEXT), Japan is providing subsidies for promoting university reform and other purposes for good projects as competitive funds among Japanese higher education institutions.

For more than fifteen years, the National Institute of Technology (NIT), Gifu College has been practicing various kinds of smart education, such as ICT-driven education, e-Learning, formation of two kinds of distance education-based consortiums using the Internet under the credit transfer agreement. The formation of the consortiums has successfully been continued. One is formed with more than 20 colleges and universities within Gifu prefecture, and the other covers about 30 nationwide higher education institutions including NIT colleges, science and technology colleges, universities, graduate universities and the Open University of Japan.

V.L. Uskov et al. (eds.), *Smart Education and e-Learning 2017*, Smart Innovation, Systems and Technologies 75, DOI 10.1007/978-3-319-59451-4_3

Our leading smart education practiced for the past years was highly evaluated, and it led to our successful acquisition of the "Support Program for Contemporary Educational Needs (GP)" with three-year major financial support from MEXT in the academic year 2004. Also, our application for the "Acceleration Program for Rebuilding of University Education" (AP) was accepted in the academic year 2014 and we started to practice our advanced challenge on smart education, financially supported by MEXT up to the academic year 2019.

Receiving the financial support necessary to fund a project to promote smart education from MEXT, our college is engaged in two collegewide projects, promotion of smart education [1–3] associated with active learning [4–13] using ICT-driven equipment and visualizing education from students' and teachers' point of view. With the view to promote collegewide smart education at an accelerating rate, we started to create an environment for practicing smart education in 2014 by introducing electronic blackboards, tablet computers, notebook personal computers, LMS, wireless LAN devices, software for making teaching materials and other equipment through AP project. The papers shown in the references [14, 15] provide more details on the creation of smart environment. In addition, we are holding faculty development (FD) sessions [16] so that the faculty members can acquire proficiency in the use of the equipment. In order for all the faculty members of our college to promote smart education in his or her own classes, it is essential to make efforts for upward spirals based on both teacher and student feedback in addition to constant holding of FD sessions and the introduction of ICT equipment into all classrooms of all the departments of our college. In this article, we describe efforts for upward spirals based on both teacher and student feedback on smart education out of some aspects of our AP project.

2 Feedback from Students on Smart Education

The following data shows the mean values of class evaluation questionnaire by all the students of all smart education classes in our college, conducted as student feedback on smart education. Regarding the aggregated data as a "review" from students' point of view, we are dealing with it as one of the main themes in FD sessions all the faculty members attend and helping them create an upward spiral of his or her practice. The response alternatives of the questionnaire were on an ascending scale of 1 to 5, 5 being the best rating expressing "Yes, I strongly think so". The questions are shown below. The aggregate results were analyzed in the year our AP project started.

- Question 1 (Enthusiasm) Did you study enthusiastically in smart education classes?
- Question 2 (Understanding) Did you have a good understanding of content in smart education classes?
- Question 3 (Achievement) Did you comprehensively achieve your objectives in smart education classes?
- Question 4 (Satisfaction) Were smart education classes comprehensively satisfactory?
- Question 5 (Adequacy) Did teachers show adequate support for and take care of students in a smart education class?

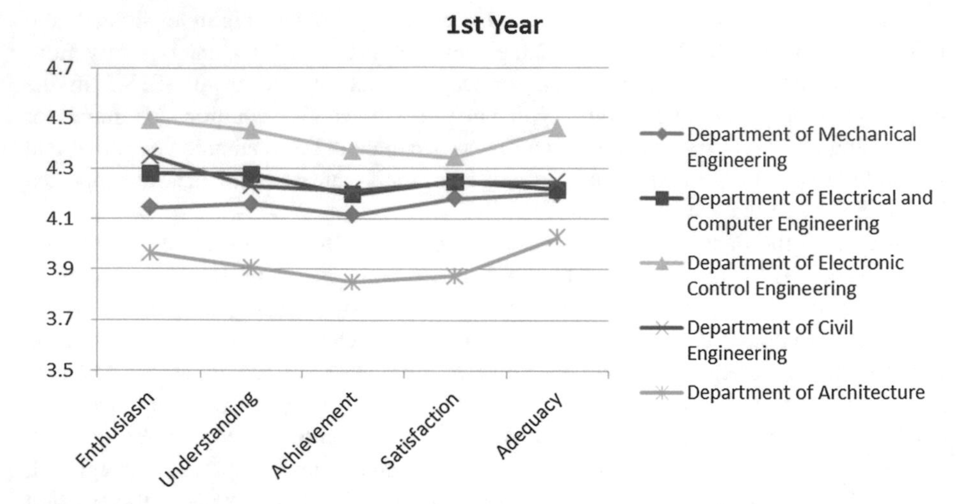

Fig. 1. The aggregate data of the first year students

Figure 1 shows that each of the question items has a very high value as a whole, quite high compared with the values of the other years. Though the values by the students at the Department of Electronic Control Engineering are slightly higher and those by the students at the Department of Architecture are somewhat lower, all the values by the students at the two departments are over 3.7, and those by the students at the other departments are also high, almost at the same level.

In Fig. 2, the low values by the students at the Department of Mechanical Engineering are conspicuous. The values by the students at the other four departments are almost the same. As compared to the values of "Understanding", "Achievement" and "Satisfaction"

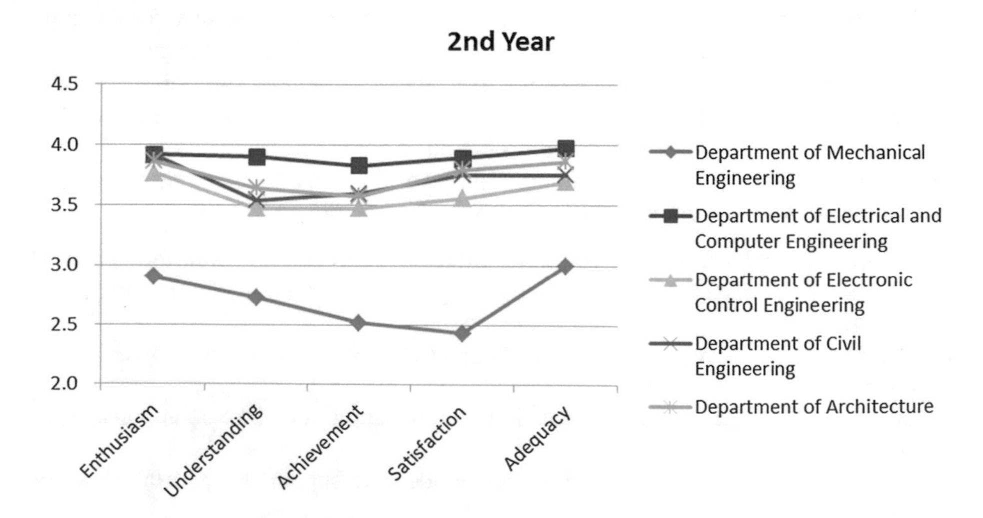

Fig. 2. The aggregate data of the second year students

by the students at the four departments, the values by the students at the Department of Mechanical Engineering have a significant difference among the three items.

Figure 3 shows slightly higher values by the students at the Department of Mechanical Engineering and the Department of Electronic Control Engineering. The data of the latter department indicates that the value of "Satisfaction" is high along with that of "Understanding".

Figure 4 shows that the value of "Understanding" is almost the same with that of "Achievement" in respective departments. The value of "Satisfaction" varies widely

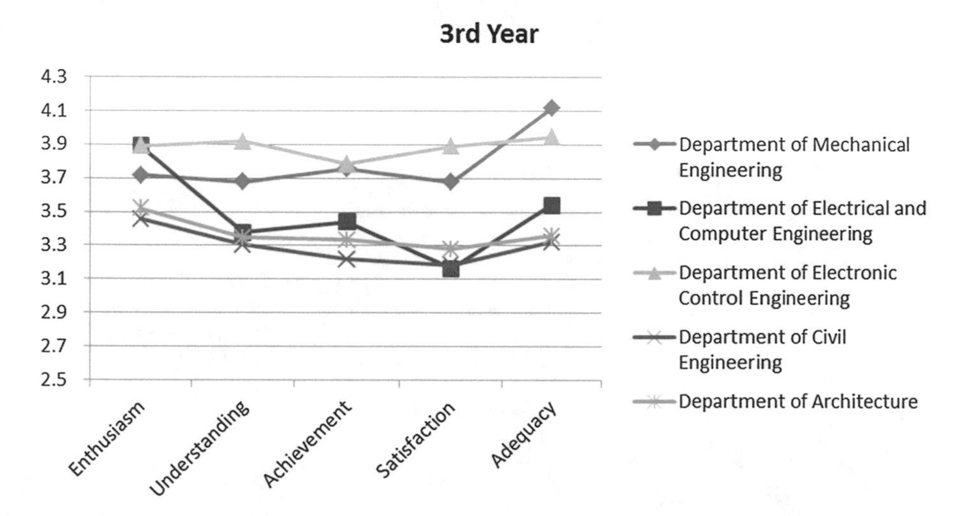

Fig. 3. The aggregate data of the third year students

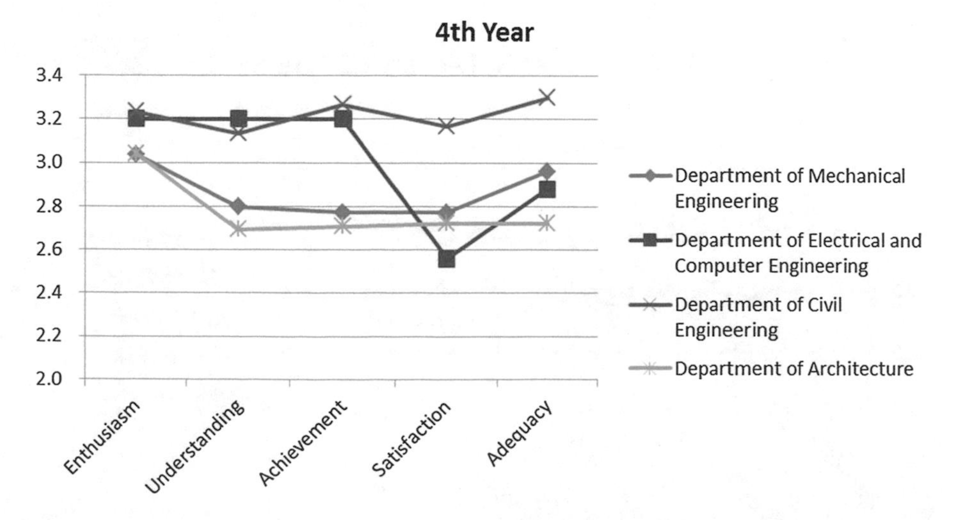

Fig. 4. The aggregate data of the fourth year students (Responses by the students at the Department of Electronic Control Engineering are not shown due to shortage of data)

among the departments. The values by the students at the Department of Electrical and Computer Engineering show that the values of “Understanding” and “Achievement” are high but that the value of “Satisfaction” is significantly low, which might have been caused by the extent to which teachers had supported students.

Figure 5 shows that each of the values of “Enthusiasm”, “Understanding”, “Achievement” and “Satisfaction” are about 3.

Figure 6 shows that the values decrease as their year in college progresses. The values by the first, second and third year students show a tendency that the values of “Enthusiasm” and “Adequacy” are high.

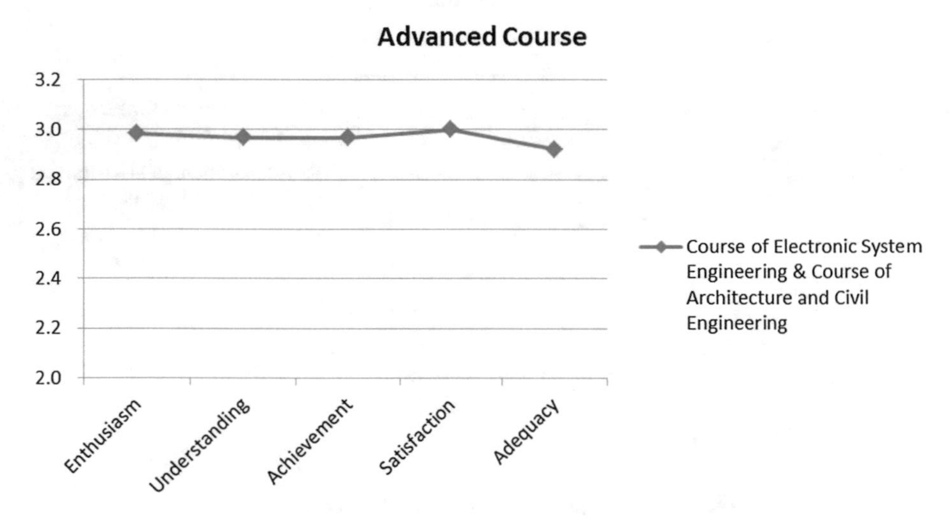

Fig. 5. The aggregate data of the advanced course students

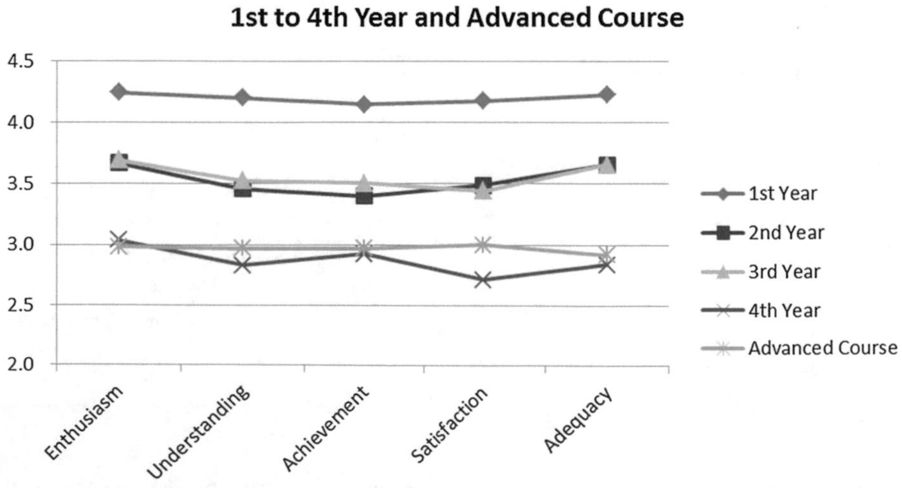

Fig. 6. The overall aggregate data

3 Feedback from Teachers on Smart Education

In our college, as part of FD, all the faculty members are required to submit a smart education practical report, and all the reports are shared among them. Writing a practical report on his or her own actual smart education is intended to provide respective teachers with a chance to review classes from his or her point of view. Also, they receive comments on their classes from other teachers through a report submitted after class observation of smart education. Teachers who inspect a class at work compile a report, where good points as well as points to be improved are described. After that, the comments are fed back to the teachers who conducted classes. This is performed as a "review" from other teachers' point of view.

As a "review" of smart education classes from his or her point of view, respective teachers are supposed to compile a report on the activities, writing about 2 pages in A4-size format for one class practiced. The format includes items which require self-analysis based on objective indexes. Specifically, they are the five items shown below: Creating a place for learning, Human interaction, Structuring, Consensus Formation and Information Sharing. Respective teachers are supposed to select one from among three levels and input it in the format for each of the five items. The aggregate analysis is used in FD sessions when some good examples of class are picked up as a theme there. Also, it is shared among the teachers for upward spirals.

- Creating a place for learning: Did your instruction succeed in creating a place for learning for students?
- Human interaction: Did your instruction succeed in developing human interaction between teachers and students as well as among students?
- Structuring: Was the structuring of your smart education classes dealt with in an appropriate manner?
- Consensus formation: Did your instruction succeed in forming a consensus on how to conduct your smart education classes between you and your students?
- Information sharing: Are you fully sharing information on digital materials with students?

The following is the overall aggregate data of "the checklist of facilitation skills drawn up by the teachers in charge of smart education". (Answers, a total of 73 subjects) (Fig. 7).

In general, the self-assessment values by teachers as to the facilitation skills are not high. "Neutral" answer is predominant. Most teachers are not satisfied with his or her facilitation skills necessary for smart education classes.

Figures 8 and 9 show the relatively high percentage of "Satisfied" in "I. Creating a place for learning" and "II. Human interaction". It seems that lively exchange of opinions was created among students and that a good place of learning was formed, compared with the conventional style of classes, or the style of classes conscious of smart education. On the other hand, there is a tendency for the percentage of "Somewhat dissatisfied" to increase as their year in college progresses. It seems to be important to share know-how for making a good place of learning, in case a teacher fails to activate discussions among students in groups.

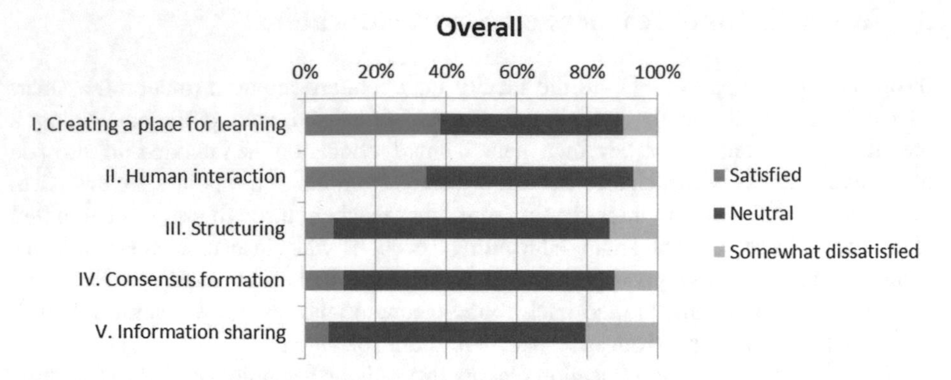

Fig. 7. The overall aggregate data

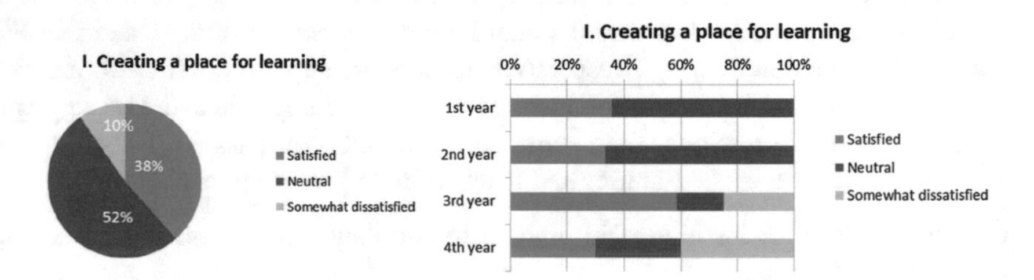

Fig. 8. The aggregate data of "Creating a place for learning"

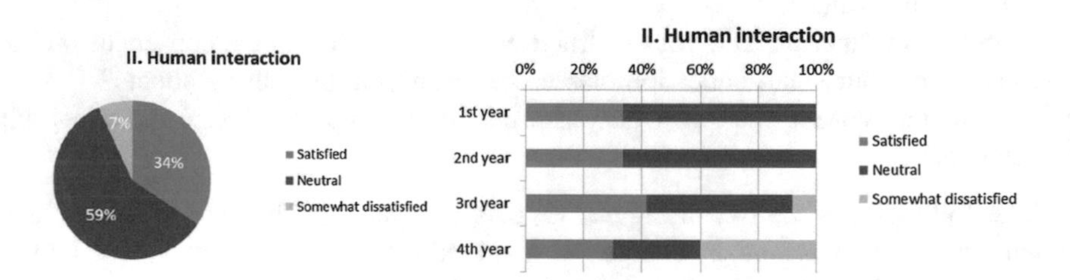

Fig. 9. The aggregate data of "Human interaction"

Figures 10, 11 and 12 show that the percentage of "Somewhat dissatisfied" is relatively higher than that of "Satisfied" in "III. Structuring", "IV. Consensus Formation" and "V. Information Sharing". There is a tendency for the percentage of "Somewhat dissatisfied" to increase as their year in college progresses. Therefore, it is hoped that we will practice smart education class where a teacher copes with differences in students' level of understanding created by the increase in the difficulty level of their learning content.

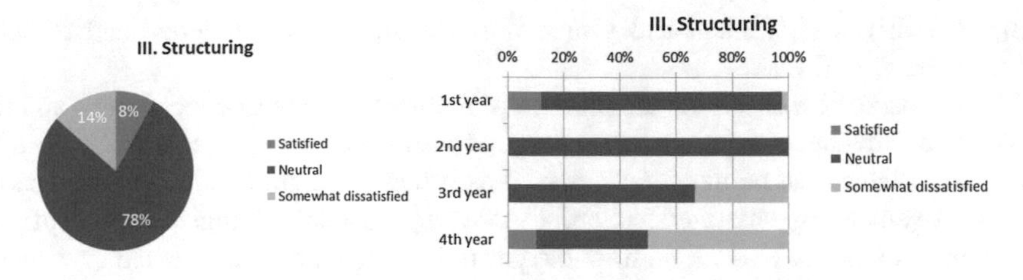

Fig. 10. The aggregate data of "Structuring"

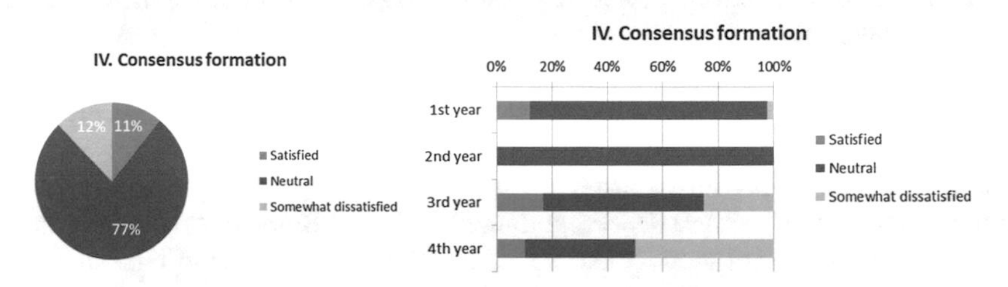

Fig. 11. The aggregate data of "Consensus formation"

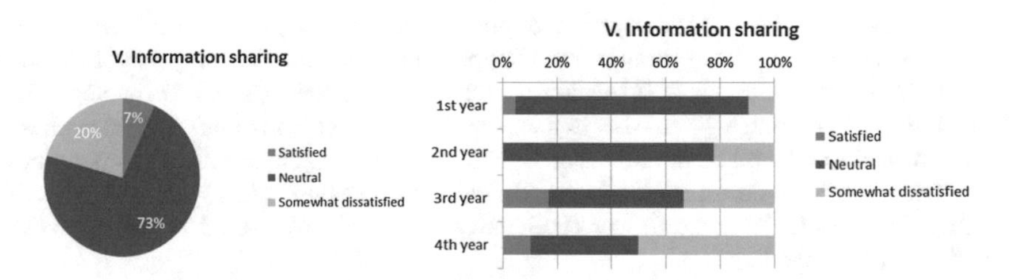

Fig. 12. The aggregate data of "Information sharing"

4 Conclusions and Future Steps

Universities and colleges, as the core of scholarly activities, are intended to give students a wide knowledge, make a deep study of specialized academic matters, and have students acquire intellectual, moral and applicative skills. Also, the qualifications and abilities necessary for social and vocational independence, as well as knowledge and skills related to respective specialized fields, lie originally in achievement acquired through education and life at universities/colleges (knowledge, skills, attitude, intentionality, etc.).

Actually, universities and colleges are providing education according to respective characteristics and academic disciplines in formal curriculum, and support for students' vocational consciousness formation through extracurricular student support. These activities are intended not only to commence employment on graduation but also to develop skills to find a problem and a way of solving it for oneself. In view of this, our two collegewide projects, promotion of smart education associated with active learning using ICT-driven equipment and visualizing education from students' and teachers' point of view, will make sense.

In this paper, we described the practice of feedback on smart education as an effort to visualize education from teachers' point of view. Regarding the practice as a "re-view" from their own point of view as well as others', we are conducting it as part of the improvement of smart education. Classifying "Keep" (things to be kept) and "Problem" (things to be dealt with) in the periodical "upward spirals cycle", respective teachers are considering "Try" (measures for improvement) and putting them into execution. In addition, we maintain the "upward spirals environment" by sharing the points of a "review", "effects of a review" and "measures for improvement" among all the faculty members.

Though smart education itself can be regarded as a future global standard, the same can be applied to the efforts to introduce, advance and improve it. Since the university/college system itself is international, we need to pay attention to various international trends. As education and research activities are developing across national borders, such as the "ERASMUS Programme" and the "Bologna Process" in Europe, we need to be more and more conscious of the superiority of smart education. From now on, teachers are expected to promote improvement of the smart education class with recognition of its background.

Acknowledgements. For the purpose of promoting researches and practices of smart education, we were financially supported through the "Support Program for Contemporary Educational Needs (GP)" by the Ministry of Education, Culture, Sports, Science and Technology, Japan (MEXT) from 2004 through 2006. Also, as a result of the appreciation of ongoing researches and practices related to GP and its development to date, we are being financially supported through the "Acceleration Program for Rebuilding of University Education" (AP) by MEXT from 2014 through 2019 in order to make further accelerative advances. We would like to express our gratitude for the support.

References

1. Uskov, V.L., Bakken, J.P., Pandey, A., Singh, U., Yalamanchili, M., Penumatsa, A.: Smart university taxonomy: features, components, systems. In: Uskov, V.L., Howlett, R.J., Jain, L. C. (eds.) Smart Education and e-Learning 2016, vol. 59, pp. 3–14. Springer, Cham (2016)
2. Hwang, G.J.: Definition, framework and research issues of smart learning environments - a context-aware ubiquitous learning perspective. Smart Learn. Environ. **1**, 4 (2014). A Springer Open Journal. Springer
3. Coccoli, M., Guercio, A., Maresca, P., Stanganelli, L.: Smarter universities: a vision for the fast changing digital era. J. Vis. Lang. Comput. **25**, 1003–1011 (2014). Elsevier
4. Bergmann, J., Sams, A.: Flip your classroom: reach every student in every class every day. In: International Society for Technology in Education. ISBN 1564843157 (2012)
5. Bonwell, C.C., Eison, J.A.: Active Learning: Creating Excitement in the Classroom. School of Education and Human Development, George Washington University, Washington, DC (1991)
6. Renkl, A., Atkinson, R.K., Maier, U.H., Staley, R.: From example study to problem solving: Smooth transitions help learning. J. Exp. Educ. **70**(4), 293–315 (2002)

7. Brant, G., Hooper, E., Sugrue, B.: Which comes first: The simulation or the lecture? J. Educ. Comput. Res. **7**(4), 469–481 (1991)
8. Kapur, M., Bielaczyc, K.: Designing for productive failure. J. Learn. Sci. **21**(1), 45–83 (2012)
9. Westermann, K., Rummel, N.: Delaying instruction: evidence from a study in a university relearning setting. Instr. Sci. **40**(4), 673–689 (2012)
10. Hake, R.R.: Interactive-engagement versus traditional methods: a six-thousand-student survey of mechanics test data for introductory physics courses. Am. J. Phys. **66**, 64 (1998)
11. Hoellwarth, C., Moelter, M.J.: The implications of a robust curriculum in introductory mechanics. Am. J. Phys. **79**, 540 (2011)
12. Khan Academy: "Khan Academy" (2006). https://www.khanacademy.org/
13. Lage, M., Platt, G., Treglia, M.: Inverting the classroom: a gateway to creating an inclusive learning environment. J. Econ. Educ. **31**(1), 30–43 (2000)
14. Ogawa, N., Shimizu, A.: Innovative approaches toward smart education at National Institute of Technology, Gifu College. In: Uskov, V.L., Howlett, R.J., Jain, L.C. (eds.) Smart Education and e-Learning 2016, vol. 59, pp. 29–38. Springer, Cham (2016)
15. Ogawa, N., Shimizu, A.: Visualization of extra curriculum education for promoting active learning. In: IEEE International Conference on Teaching, Assessment, and Learning for Engineering, Conference Proceedings, pp. 101–108, ISBN 978-1-5090-5597-5 (2016)
16. Ogawa, N., Shimizu, A.: Collegewide promotion of e-Learning/active learning and faculty development. In: Proceedings of the International Conference e-Learning 2016 Organised by IADIS International Association for Development of the Information Society, pp. 179–184, ISBN 978-989-8533-51-7 (2016)

Analysis of Students' Behaviors in Programming Exercises Using Deep Learning

Toshiyasu Kato[1](✉), Yasushi Kambayashi[1], Yuki Terawaki[2], and Yasushi Kodama[3]

[1] Nippon Institute of Technology, Saitama, Japan
{katoto,yasushi}@nit.ac.jp
[2] Hollywood Graduate School of Beauty Business, Tokyo, Japan
terawaki@hollywood.ac.jp
[3] Hosei University, Tokyo, Japan
yass@hosei.ac.jp

Abstract. Programming exercises are time-consuming activities for many students. Therefore, many classes provide meticulous supports for students, in the form of teaching assistants (TAs). However, individual students' programming behaviors are quite different from each other, even when they are solving the same problem. It can be hard for TAs to understand the unique features of each student's programming behavior. Using data mining (specifically autoencoding of deep learning), we have analyzed students' programming behaviors in order to determine their varied features. The purpose of this study is to present such behavioral features for TAs, to improve the effectiveness of the assistance they can provide.

Keywords: Programming exercises · Teaching assistants · Learning analytics · Deep learning

1 Introduction

During programming exercises at higher educational institutions, students individually tackle assigned problems. Students progress very differently, depending on their individual skill levels [1]. Certain students can easily solve problems, while there are many students who spend too much time on single problems. For this reason, most institutions provide teaching assistant (TA) supports in programming classes. TAs are expected not only to answer student questions, but also to assist the students who need specialized help [2]. TAs therefore need to understand who need assistance, what kind of assistance they need, and in what situations they need assistance [3].

Students' progress likely differs based on individual skill levels [1]. Some of them, who experience programming difficulties, do not know and thus do not follow certain programming modes. Students who are not making progress are often in this situation because they lack good programming modes. Such students occasionally come to a standstill, and may feel unmotivated [4]. It is thus important to be able to determine which students are likely to have such programming difficulties at an early stage [2].

V.L. Uskov et al. (eds.), *Smart Education and e-Learning 2017*, Smart Innovation, Systems and Technologies 75, DOI 10.1007/978-3-319-59451-4_4

This report presents characteristics of different types of student programming behaviors in order to help TAs understand students' contexts. In particular, we focus on the characteristic programming behaviors of students who take too long duration time for problem solving in programming exercises. This report will help TAs to understand students' programming behaviors without having prior experience instructing them. The authors have used their own programming exercise support system to obtain the programming behaviors discussed herein [5].

2 Related Work

Research on behaviors involved in learning programming has covered topics including debugging training, and programming tutorials. Ryan studied how to improve students' debugging skills [6], and used debugging and development logs to construct a model by which students can improve their skills. The present study also analyzes debugging logs, along with other programming records.

Alex proposed the employment of tutors who can support students' programming develop systematic [7]. Such tutors examine whether students are on the right track, and give students advice when their programs are incorrect. For example, tutors may give students hints on how to refactor their programs. The present study aims to support TAs in giving students problem-solving techniques, and thereby enables students to solve their own problems.

In some past studies aimed at learning how to support students taking their learning history, teaching materials were analyzed using statistical methods and Support Vector Machines (SVM). For example, Yamada used logistic regression to detect pages that were difficult for students to understand [8]. In this study, they created models from six explanatory variables: latent time, gaze point distance moved, gazing point movement speed, mouse distance moved, mouse speed, the amount of mouse wheel rotation, and correct discrimination. This evaluation experiment was able to achieve a discrimination rate of 81%. This study used machine learning and time series data to improve programming behaviors such as compile and execution times.

Nakamura used SVM to identify the learners' subjective perceptions of the difficulty level of teaching materials [9]. In the study, they have estimated subjective difficulty levels from six explanatory variables including face inclination angle, gaze point position, and the intervals of mouse operation. This evaluation experiment achieved an average discrimination rate of 85%.

3 Analysis of Programming Behaviors

In this study, we prepared data on the programming behaviors of students ($n = 80$) in a 2015 class analyzed in the previous paper [10]. We have analyzed seven explanatory variables: the number of compilations, the number of trial executions, the number of errors, the number of error repetitions, the average intervals of compilations, and the average intervals of executions. We have measured all of these analysis targets by using the programming exercise support system we have implemented [5].

We used Ward's method of hierarchical clustering to process the data [11, 12]. Clustering is a process of grouping objects into classes of similar objects [13]. It is an unsupervised classification and we have partitioned the patterns (observations, data items, or feature vectors) into groups or subsets (clusters) based on their locality and connectivity within an n-dimensional space. Figure 1 shows the results of cluster analysis of the seven explanatory variables. Table 1 shows the length of problem-solving time for each cluster in Fig. 1.

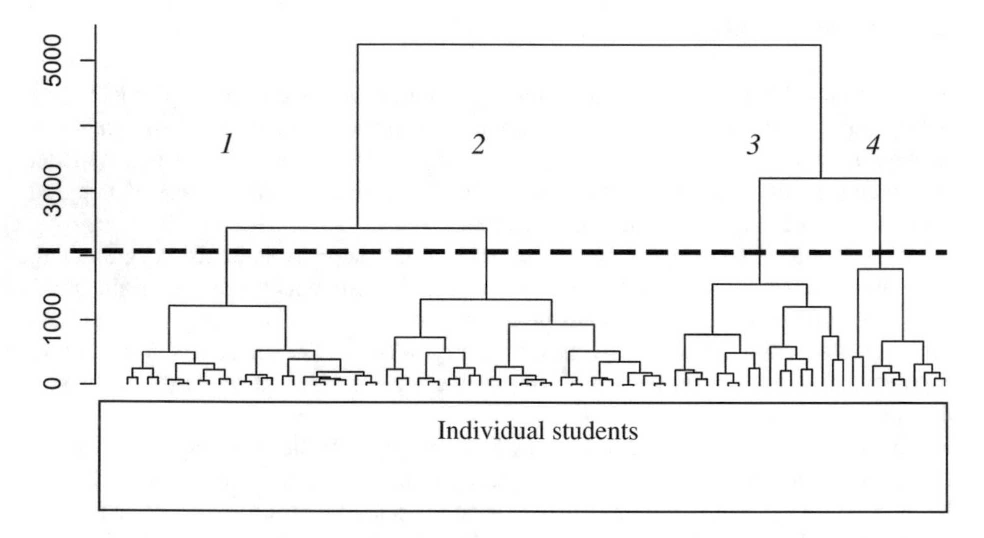

Fig. 1. Cluster analysis using Ward's method.

We have observed the following clusters:

Cluster 1: Problem-solving time is short. Students understand each error they encounter while programming.

Cluster 2: Problem-solving time is short. Students are programming without understanding the details of their errors. Note that clusters 1 and 2 (Table 1) have students with longer than average problem-solving times.

Cluster 3: Problem-solving time is long. Students compile by correcting the program without fully understanding their errors.

Cluster 4: Problem-solving time is long. Students repeat compiling without understanding anything.

Students in each cluster should perhaps receive different types of guidance. Instructors and TAs must therefore discover the characteristics of the students in their classes, paying special attention to those who have long problem-solving time. Additionally, cluster analysis cannot estimate the problem-solving time for new exercises. Therefore, TAs cannot know which students will need specific types of guidance at the beginning of exercises, or before class.

Table 1. Problem-solving time by cluster.

Cluster	Average solving time [s]	Longer than average [%]
1	528	20.0
2	812	17.5
3	1405	35.0
4	1506	60.0

Therefore, we then conducted machine-learning analyses, focusing on the following points:

- Length estimation of problem-solving time.

1. Analysis using data from 2014 and 2015
2. Analysis using time series data for 2012

- Characteristics of programming behaviors.

3. Extracted from 2015 data

4 Analysis by Machine Learning

4.1 Estimation of Problem-Solving Time

We have employed estimation time for problem solving from the programming behaviors of the 2014 class as training data from which to predict the behavior of the 2015 class on this metric. Table 2 shows the programming behavior data for 2014 and 2015 for each explanatory variable. These data are average values (excluding the number of students). Students received two assignments each year, and the contents of the assignment are the same.

We used logistic regression to determine whether students' problem-solving time was long or short. Logistic regression is a multivariate analysis method that was developed in the United States in 1948, to analyze the effects of multiple risk factors on disease [14]. The regression equation for the logistic regression is shown in the Eq. (1), where p is the probability estimate of the objective variable. In this study, we solved for the probability of a long problem-solving time. β_0 is a constant, β_1, β_2, β_k, are regression coefficients, and χ_1, χ_2, χ_k are explanatory variables.

$$p = \frac{1}{1 + e^{-(\beta_0 + \beta_1\chi_1 + \beta_2\chi_2 + \cdots + \beta_k\chi_k)}} \tag{1}$$

We first applied the logistic regression to the training data, to find the constant and regression coefficients. We have used Tensor Flow, a learning tool used to determine deep learning [15]. In the learning error function, we used cross entropy for the classification (identification) problem. We have carried out the optimization using the steepest descent method [16]. Since the objective variable of a logistic regression must

Table 2. Programming behaviors in 2014 and 2015.

	2014	2015
Number of students	85	80
Problem-solving time [s]	815	936
Average compile interval [s]	295	352
Average execution interval [s]	513	584
Compile times	10	8
Execution times	5	5
Number of errors until problem resolution	7	3
Number of errors of the same type that are most frequent	4	3
Number of students with long problem-solving time	32	35

be represented by a 0/1 binary, we set the teacher data to 0 when it was longer than the average problem resolution time. Teacher data was set to 1 if it was short.

We then applied the obtained coefficients to the 2015 data. We observed the discrimination rate 82.5%, a learning rate 0.015, and the number of learning iterations was 70,000. Figure 2 shows the relationship between the number of students used for the training data and the discrimination rate.

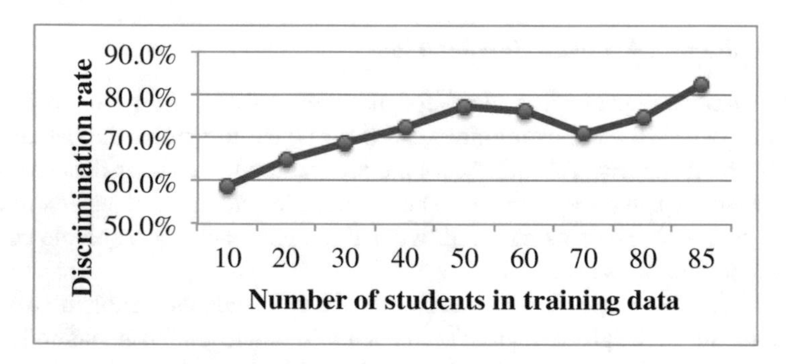

Fig. 2. The number of students in the training data and discrimination rate.

4.2 Analysis of Problem-Solving Time Using Time Series Data

We then estimated students' problem-solving time from the 2012 time series data (explanatory variable 91 = Effort time [1/min]) using the methods outlined in Sect. 4.1. The data are presented in Table 3.

Using thirty-five students as training data and six students as experimental data, the identification rate was 83.3%, the learning rate was 0.02, and the number of learning instances was 112,300. Figure 3 shows the relationship between the number of students used for as training data and the discrimination rate in the 2012 dataset.

Table 3. Time series data of compile times for 2012 data.

Number of students	41
Effort time [min]	91
Problem resolution time [min]	53
Average compilation count	23
Number of students with long problem-solving time	25

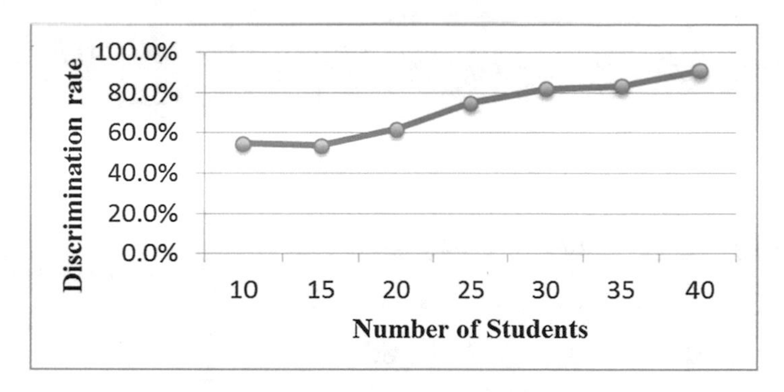

Fig. 3. The relationship between the number of students used as training data and the discrimination rate in the 2012 dataset.

4.3 Extraction of the Characteristics of Programming Behaviors

We applied the autoencoder to the 2015 programming behavior data (feature variable 7) in order to extract the characteristics of students' programming behaviors. The autoencoder is an unsupervised machine learning method used for dimension reduction. Figure 4 shows the normalized distribution (maximum value 1, minimum value 0). Students with long execution intervals and students with long resolution time emerged in Fig. 4.

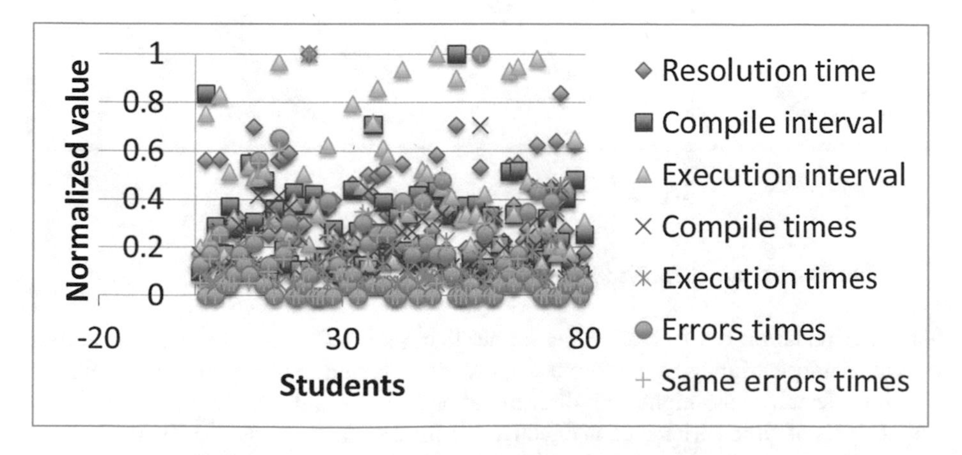

Fig. 4. Normalized distribution of programming behaviors.

Figure 5 shows the results of the autoencoder. In Fig. 5, all features are averaged and the layers are separated. Furthermore, when we set the middle layer to 3, we extracted 29 students with all the programming behaviors (Fig. 6). Both the matching with the cluster analysis result by the preliminary analysis contained 100% of the students in Cluster 3.

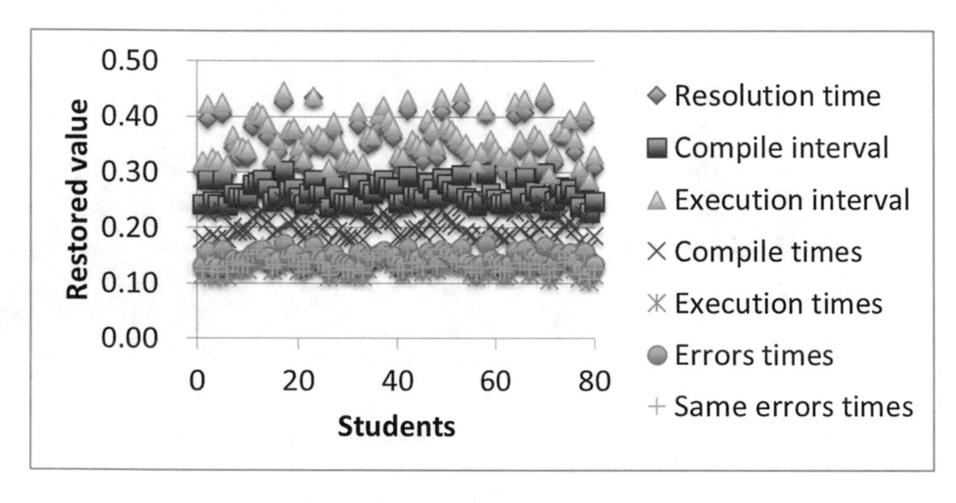

Fig. 5. Distribution after Autoencoder.

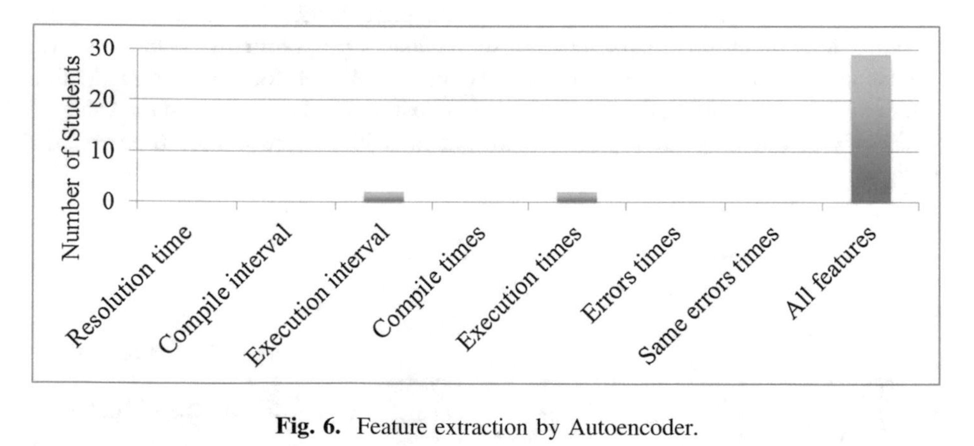

Fig. 6. Feature extraction by Autoencoder.

5 Discussion of Analysis Results

The discrimination rate in Sect. 4.2 is higher than that in Sect. 4.1 because the number of explanatory variables in the former largely exceeds that in the latter; it is known that the more features, the higher the discrimination rate. Therefore, we can consider that the analysis of time series data is effective for finding features. However, the learning takes longer time. It is a trade-off. In addition, the number of learning instances in Sect. 4.1 was large, whereas we observed many complicated patterns in Sect. 4.2. It therefore seems that the analysis using the intermediate layer in Sect. 4.3 was appropriate.

We can observe a pattern in the time series data in Sect. 4.2. We employed cluster analysis to try to decipher the pattern. Figure 7 shows the results of cluster analysis of the number of compilations, and Table 4 shows the cluster breakdown. We could classify the clusters into two groups. There was a significant difference (10%) in the proportion of students with long problem-solving time. In the case of an identification problem, if Cluster 1 contains a student with a long task resolution time, the identification rate is 43.9% (18/41). From this, we can say that the cluster analysis for identifying the length of problem-solving times is not very accurate.

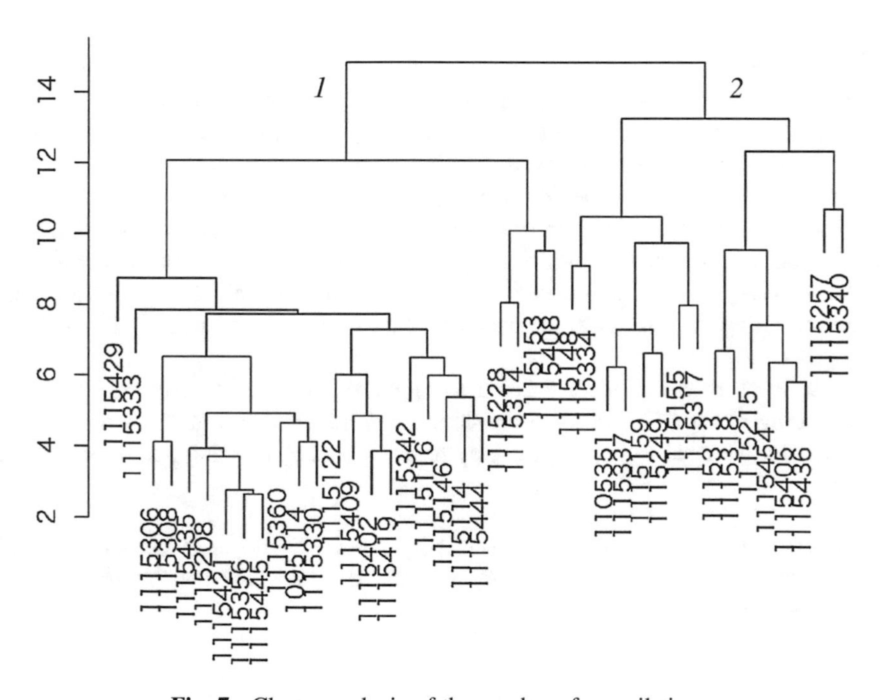

Fig. 7. Cluster analysis of the number of compilations.

In Sect. 4.3, we examined the programming behaviors of the students from the previous year. We found that we had improved the identification rate by increasing the volume of training data analyzed. We also found that error convergence required a minimum learning period. From this, we can conclude that we can improve the discrimination rate by using the intermediate layer.

Extraction of the characteristics of programming behaviors showed that students in one particular cluster were performing trial and error and making little progress. This shows that we can find students who are willing to learn and potentially good students

Table 4. Cluster breakdown from Fig. 7.

Cluster	Number of students	Number of students with long problem-solving time	Proportion
1	25	18	72.0%
2	16	7	43.8%

if they can receive appropriate guidance. We have successfully extracted the behaviors of students who needed help from TAs. With these results, TAs can grasp students who need guidance without checking the number of compiling times and the number of errors that individual students made.

6 Conclusion

In this paper, we reported the results of data mining of students' programming behaviors. We have carried out these analyses to support TAs in assisting students with programming exercises. We applied machine learning to construct a model to represent the programming behaviors of a class. We then applied that model to the programming behavior from students in the following year. We achieved discrimination rates of 82.5% for students with long problem-solving time. Additionally, as analyzing time series data, we achieved a discrimination rate of 83.3% for students with long problem-solving time. Furthermore, we applied the autoencoder to programming behaviors. Using the above means, we extracted the students who need help and want to progress.

As a future research, we are going to apply this method to contexts other than programming classes. We need to consider the kind of data that we should collect. Additionally, we need to explore the way to use the deep learning and to analyze the results.

Acknowledgments. This work was supported by Japan Society for Promotion of Science (JSPS), with the basic research program (C) (No. 15K01094 and 26240008), Grant-in-Aid for Scientific Research.

References

1. Horiguchi, S., Igaki, H., Inoue, A., et al.: Progress management metrics for programming education of HTML-based learning material. J. Inf. Process. Soc. Jpn **53**(1), 61–71 (2012). (in Japanese)
2. Sagisaka, T., Watanabe, S.: Investigations of beginners in programming course based on learning strategies and gradual level test, and development of support-rules. J. Jpn Soc. Inf. Syst. Educ. **26**(1), 5–15 (2009). (in Japanese)
3. Yasuda, K., Inoue, A., Ichimura, S.: Programming education system that can share problem-solving processes between students and teaching assistants. J. Inf. Process. Soc. Jpn **53**(1), 81–89 (2012). (in Japanese)

4. Igaki, H., Saito, S., Inoue, A., et al.: Programming process visualization for supporting students in programming exercise. J. Inf. Process. Soc. Jpn **54**, 1 (2013). (in Japanese)
5. Kato, T., Ishikawa, T.: Design and evaluation of support functions of course management systems for assessing learning conditions in programming practicums. In: ICALT 2012, pp. 205–207 (2012)
6. Ryan, C., Michael, C.L.: Debugging: from novice to expert. ACM SIGCSE Bull. **36**(1), 17–21 (2004)
7. Alex, G., Johan, J., Bastiaan, H.: An interactive functional programming tutor. In: Proceedings of the 17th ITiCSE 2012, pp. 250–255. ACM (2012)
8. Yamada, T., Nakamichi, N., Matsui, T.: Detection of low usability web pages with logistic regression. In: GN Workshop 2010, pp. 57–62 (2010). (in Japanese)
9. Nakamura, K., Kakusho, K., Murakami, M., Minoh, M.: Estimating learners' subjective impressions of the difficulty of course materials by observing their behaviors in e-Learning. In: MIRU 2007, vol. 6 (2007). (in Japanese)
10. Kato, T., Kambayashi, Y., Kodama, Y.: Data mining of students' behaviors in programming exercises. Smart Educ. e-Learning **2016**(59), 121–133 (2016)
11. Kamishima, T.: A survey of recent clustering methods for data mining (part 1): try clustering! J. Jpn. Soc. Artif. Intell. **18**(1), 59–65 (2003). (in Japanese)
12. Michael, R.A.: Cluster Analysis for Applications. Academic Press (1973)
13. Jain, A.K., Murty, M.N., Flynn, P.J.: Data clustering: a review. ACM Comput. Surv. **31**(3), 264–323 (1999)
14. Uchida, O.: Logistic Regression Analysis by SPSS, Ohmsa (2011). (in Japanese)
15. Tensor Flow, Google. https://www.tensorflow.org
16. Ozawa, Y., Kohno, Y., Takahashi, T.: Generalization method with satisficing on ANN. In: The 29th Annual Conference of the Japanese Society for Artificial Intelligence, vol. 29, pp. 1–4 (2015). (in Japanese)

A Serious Game to Promote Environmental Attitude

Veronica Rossano[1(✉)], Teresa Roselli[1], and Gabriella Calvano[2]

[1] Department of Computer Science, University of Bari via Orabona, 4, 70125 Bari, Italy
{veronica.rossano, teresa.roselli}@uniba.it
[2] Department of Biology, University of Bari via Orabona, 4, 70125 Bari, Italy
gabriella.calvano@uniba.it

Abstract. Environmental attitudes are essential for the today's world in which shrinking natural resources and pollution are one of the biggest problems of the society. The research challenges go in two directions: on the one hand, the environmental researchers try to reduce the impact of human actions on the environment and, on the other, human researchers try to educate people to adopt ecological behaviour. This is the main issue faced in this research to sensitize young people to the environmental health using multimedia technologies. The paper presents a serious game addressed to primary school pupils aimed at transferring knowledge about marine litter and four species of Mediterranean Sea that are estimated to be at risk of extinction. The aim is to foster knowledge about the marine life and to explain people the problems that marine litter can cause. The results of a pilot study that investigates the usability of the game is presented.

Keywords: Serious game · Game-based learning · Environmental education

1 Introduction

How many people know how much an airplane pollutes, or how long a plastic bottle remains in the sea before degrading? Why should we make the differentiated collection of waste? To save the planet, is it helpful to use a natural gas for our cars?

The environmental challenge, linked to the preservation of our planet and its resources, today is no longer avoidable. Through Agenda 2030 [24] all people of the world (even the poorest) are called to make choices radically different from those made in the past. These choices should be distant from the traditional production model, directed towards a new economic model that respects the environment, oriented to a society that does not produce waste but creates wealth and well-being trough the re-use and the regeneration of resources.

To make this happen, it needs a profound cultural change in people, in institutions, in the business world, associations. This change also calls into question the schools.

Environmental education wants to create learning experiences that, through thoughts, emotions and actions, lead students (pupils, teachers, parents) to discover, to rediscover or, more simply, to the affirm values and ethical behaviour respectful of

V.L. Uskov et al. (eds.), *Smart Education and e-Learning 2017*, Smart Innovation, Systems and Technologies 75, DOI 10.1007/978-3-319-59451-4_5

environment. Today it is especially important during training and growth, since childhood.

The themes of exhaustion of natural resources waste and excessive waste production should be addressed already at the stage where the children learn to relate to the world around them. Only in this way, it will be possible to raise responsible attitudes towards the environment.

To achieve the most important objective of an educational process, the integral development of the person, it is essential to develop since the early childhood an authentic man-environment relationship. In order to fit in each person an authentic global consciousness is fundamental the training from the earliest years of life oriented to promote sound principles and values, such as respect, care and attention to the environment around us.

Environmental education helps to promote the necessary skills to challenge existing models and trigger virtuous processes of overall change of behaviours and lifestyles.

In European Union, environmental education has become an integral part of the curricular activities of primary and secondary school: several Member States have introduced it in their schools. The study of the environment is essential to prepare students to build a green future and to live in a sustainable society.

Internationally, the UNECE strategy (United Nations Economic Commission for Europe) for Education for Sustainable Development [20] has defined environmental education as a requirement for sustainable development, as a tool for good governance and for decision-making.

One of the fundamental aspects of environmental education is to be an education for the future, change-oriented education, which is able to promote pathways oriented to the quality of life, to human relationships, to relationships between humans and the planet. Condition for this change is a different way of thinking, a different culture, which, as demonstrated during this work, can also be favoured using serious games.

The paper is organized as follows: the next section describes the game-based learning and the serious games; Sect. 3 describes some related works, Sect. 4 presents the solution implemented; Sect. 5 illustrates a pilot study aimed at evaluating its usability. Finally, some conclusions and future works are described.

2 Game-Based Learning and Serious Game

The game-based learning has been largely recognized as one of the best approaches for learning for both children and adults [1–6, 14, 15]. Usually, the game dimension makes the learning easier, student-centred, fun and engaging and thus more powerful. Several works have investigated on the effectiveness of the application of this approach on specific contexts in which the engagement and motivation is essential to promote knowledge and skill acquisition [7, 11, 15]. As stated in [10], games support such pedagogical principles making the learning experience successful. Each level of the game is personalised on the basis of the player's abilities, feedbacks are immediate, the player has an active role in the game, and the game engages players and improves motivation in learning. Until some time ago, the game dimension was devoted mainly to young students, because it was thought that it was pleasing and effective only for this kind of people and the game was not suitable for serious contexts, but it has been proved that also adults can be attracted to the games [4–6, 13]. Thus, the serious games are defined by Chen & Michael [3] as *Games that do not have entertainment, enjoyment or fun as their primary purpose*. In the last decade, the serious games have been largely implied in professional training and in informal learning contexts to promote knowledge/skills acquisition [17–19], engagement and motivation [12]. In this view, the research herein described adopt the serious game to develop environmental attitude in primary school students.

3 Serious Game for Environmental Education: Related Works

Video games have become part of our everyday life and our habits. Through technical and persuasive contents, they can transform personal and social behaviours of players.

This is also true in the field of environmental sustainability. Malone & Lepper [7], in fact, emphasize how the scenarios in video games can provide players analogies with some elements of the real world. This allows players to observe cause-effect processes tested in a real prospect.

The idea that games can be a tool to improve the people interest in environmental issues, led to the creation or modification of some existing games like Minecraft (https://minecraft.net/it-it/). It is a game in which the player has to construct buildings or structures, often setting fire to other elements in the scene. Minecraft has been modified by adding the visible effects caused by the carbon dioxide produced by burning. The player can help the environment by planting trees. The serious games dealing with water pollution and marine ecology are not very widespread, and few are the ones with educational purposes and able to promote environmentally-friendly behaviour. Usually, the few available are implemented or distributed by non-profit organizations that deal with children's education or environmental protection. To design and implement the serious game for our research, we started with the analysis of some serious games for children related to the topic of environmental education, and the pollution of the sea and the Marine Litter.

Clean Up the River (http://cleanup.noco2.com.au/) has been chosen for the interest that it has for familiar environment to children-players. In fact, our game is set in the Mediterranean Sea, which is the most known sea among students, and the characters are local marine species. The side effect will be the knowledge acquisition of the local environment.

WaterLife (http://games.noaa.gov/oscar/) was a model for the issue of marine waste. Our game invites to reflect on the consequences of waste in the sea and how these wastes end up in the products we eat.

Wild Kratts Habitats: Ocean (http://pbskids.org/wildkratts/habitats/ocean/) was considered because it provides some information of marine species. The goal of the game is the acquisition of knowledge about the marine species that inhabit the sea.

4 SeAdventure

The serious game named SeAdventure, is a platform game in which the player using an avatar swims through waste lost in the sea (Fig. 1). The intended users are primary school pupils aged between 8 to 10 years old, and the game was designed to be used in classroom as supporting tools for environmental education. The serious game set is the Mediterranean Sea and the characters are four species estimated to be at risk of extinction and the most interesting species for the intended users. The four species are: red tuna, great white shark, turtle (Caretta caretta) and the hippocampus (Fig. 2). Language, graphic, and interaction were accurately defined to meet users' needs and characteristics. The mission of the player is to help the character to swim and reach the final point avoiding junk and eating fish. When the character eats fish, the player collects healthy points, otherwise if the character eats junk unhealthy points are collected and the character swims more slowly for a few seconds. Both the points are used to build the final rank of the game (Fig. 3). To motivate the player to pay attention to the junk, a life is lost if too marine litter is eaten. Moreover, during the play, some knowledge pills about habitat, diet, lifestyle and dangers of the character are given. To improve the final rank, the player could answer to some quizzes (Fig. 4). This allows to motivate the user in reading the text.

Fig. 1. The Red Tuna swimming in the sea

Fig. 2. The characters in the game (Red Tuna, Great white shark, Turtle Caretta caretta and the hippocampus)

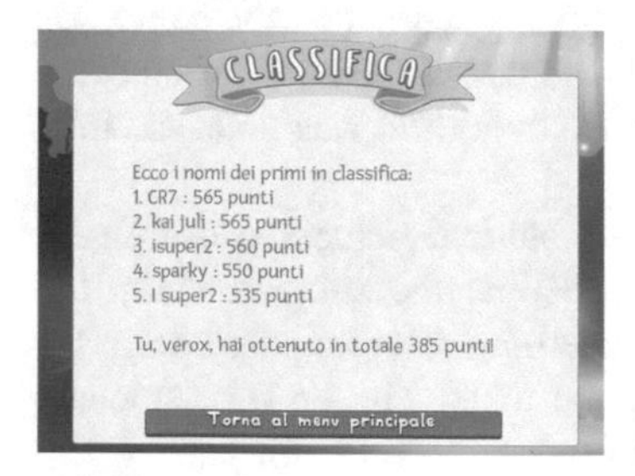

Fig. 3. Final rank of the game

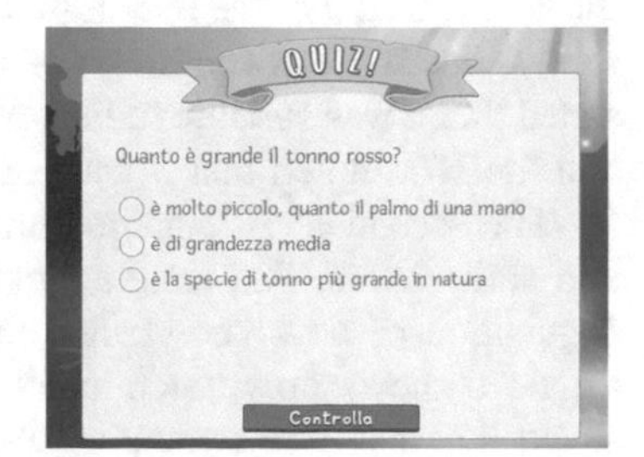

Fig. 4. An item of a final quiz about Red Tuna

Since to develop a serious game to foster environmental attitude is a complex process that involves different kind of professional figures, a software prototyping approach to develop the game has been used. It is a software development model that enables to understand customer requirements at an early stage of development. It helps get valuable feedback from the users and helps software designers and developers to understand about what exactly is expected from the product under development. Thus the first version of the serious game was submitted to a pilot study to evaluate its usability.

5 Usability Test: Procedure and Results

As said before, a first release of the serious game was submitted to a sample of primary school children aged between 8 and 10 years old to evaluate its usability. The pilot study involved 46 pupils attending the two classrooms of the 4th grade of an Italian primary school. The main goal of the study was to evaluate the usability and the usefulness of multimedia technologies and Game-based learning in fostering ecological skills. All the pupils (46) were new to ecological and environmental issues. The pilot study was conducted in the classroom equipped with notebook and a smart board. The marine litter problem was introduced by the teachers using a discussion and an explainer video which lasts 2 min. The students were very interested in discussing this topic, since all of them have found waste on the sand and in the sea during their holidays. The discussion was useful also to point out their feeling about this important issue. All pupils were very impressed by the un-ecological behaviour of people. This first step lasted 20 min.

Then the pupils were invited to use the serious game in pairs at least for four times to explore all the characters. This step lasted about 1 h.

Then two questionnaires aiming at evaluating the usability of the SeAdventure and its usefulness in acquiring knowledge were proposed. The questionnaires were very short and this step lasted about 10 min.

The first evaluable result was that users used the game without any kind of inconvenience, this to prove that both graphic and text were designed to meet typical children requirements. The result was confirmed also by the analysis of data collected using the questionnaire. Most students (86.9%) stated that the game was easy to use, the 84.7% said that the button functions were clear, the 95.7% stated that the game was funny. For what concerning the length of texts in the game: the 39.1% of students stated that it was suitable to their reading skills, the 43.5% thought it was quite suitable, and the 13% stated that it was too much.

The second questionnaire aimed at investigating the student perceived usefulness and satisfaction. The results are satisfying since the 83% of the subjects claim the acquisition of new knowledge, the 17% said that something was acquired (Fig. 5). Figure 6 shows how much attention students paid to the texts. The 88% of students stated that they will remember the information read in the game (Fig. 7).

The main result of this pilot study was the engagement of users in doing the activities. Teachers said that they rarely saw all students so interested in something and

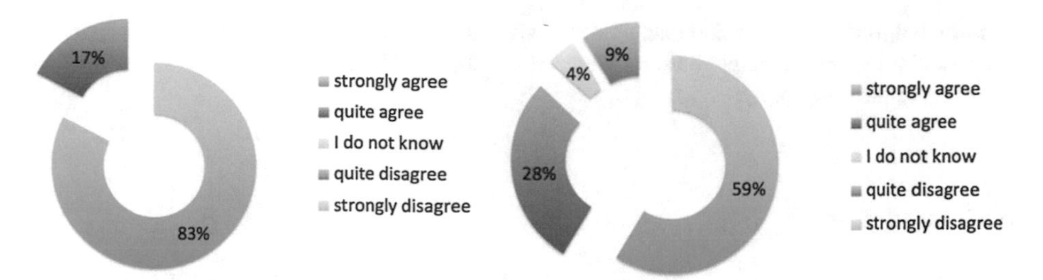

Fig. 5. I have learned using both the video and the game

Fig. 6. I have paid attention to the texts

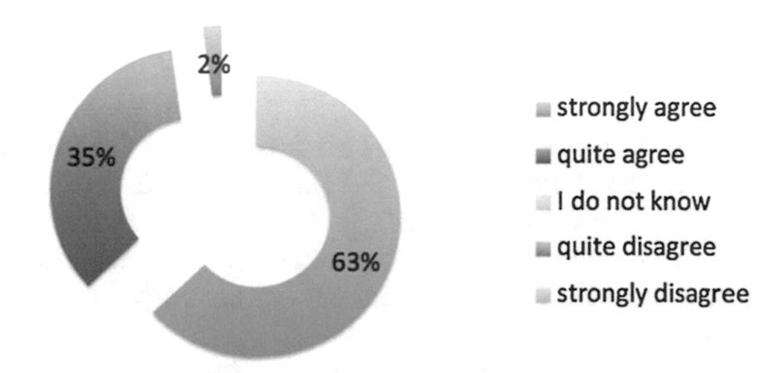

Fig. 7. I will remember the information read in the game

noticed the attention paid by students in reading the texts. Some of them, the most competitive, took notes to answer the final quiz and earned a good ranking.

6 Conclusions and Future Works

To promote environmental attitude has become one of the most important issue in education of next generations. It is important that young pupils become aware of the danger that un-ecological behaviour can cause to the whole ecosystem, including humans. In this context, it is essential not only to supply information but also change the human behaviours. The game-based learning and video game have successfully implied in the past such useful tool in both formal and informal learning contexts. In this view, the research presents a serious game addressed to primary school pupils to foster ecological skills and environmental attitudes. The pilot study revealed that students and teachers appreciated the approach, and the preliminary data gave good results also in the students' perception of usefulness and satisfaction. This allow us to go ahead in this research and a larger study will be performed in order to measure the learning effectiveness. Moreover, according with the software prototyping approach other components of the serious game will be developed.

Acknowledgments. We would thank the student Giulia Alfarano who build the serious game in her bachelor thesis and the pupils and teachers of the 14° Circle Re David Primary school in Bari involved in the pilot study.

References

1. Bakhuys Roozeboom, M., Visschedijk, G., Oprins, E.: The effectiveness of three serious games measuring generic learning features. Br. J. Educ. Technol. **48**, 83–100 (2015)
2. Barab, S., Scott, B., Siyahhan, S., Goldstone, R., Ingram-Goble, A., Zuiker, S., et al.: Transformational play as a curricular scaffold: using videogames to support science education. J. Sci. Educ. Technol. **18**(4), 305–320 (2009). doi:10.1007/s10956-009-9171-5
3. Chen, S., Michael, D.: Serious Games: Games that Educate, Train and Inform. Thomson Course Technology, USA (2005)
4. Di Bitonto, P., Roselli, T., Rossano, V., Frezza, E., Piccinno, E.: An educational game to learn type 1 diabetes management. In: The 18th International Conference on Distributed Multimedia Systems, Miami Beach, USA, 9–11 August 2012, pp. 139–143. KSI Press, Skokie, Illinois (2012). ISBN 1-891706-32-2
5. Jun Kiat Ong, M.: Gamification and its effect on employee engagement and performance in a perceptual diagnosis task. Master dissertation, Master of Science in Applied Psychology, University of Canterbury
6. Kato, P.M., Cole, S.W., et al.: A video game improves behavioral outcomes in adolescents and young adults with cancer: a randomized trial. Pediatrics **122**(2), 305–317 (2008)
7. Lepper, M.R., Malone, T.W.: Intrinsic motivation and instructional effectiveness in computer-based education. Aptitude Learn. Instr. **3**, 255–286 (1987)
8. Malone, T., Lepper: Making learning fun: a taxonomy of intrinsic motivations for learning. In: Snow, R., Farr, M.J. (eds.) Aptitude, Learning, and Instruction. Conative and Affective Process Analyses, vol. 3. Hillsdale, NJ (1987)
9. Mohler, J.L.: Using interactive multimedia technologies to improve student understanding of spatially-dependent engineering concepts. In: Proceeding of the International Conference on Computer Geometry and Graphics (Graphicon 2001), pp. 292–300 (2001)
10. Oblinger, D.G.: The next generation of educational engagement. J. Interact. Media Educ. **2004**(8), 1–18 (2004)
11. Papastergiou, M.: Digital game-based learning in high school computer science education: impact on educational effectiveness and student motivation. Comput. Educ. **52**(1), 1–12 (2009)
12. Pesare, E., Roselli, T., Corriero, N., Rossano, V.: Game-based learning and gamification to promote engagement and motivation in medical learning contexts. Smart Learn. Environ. **3** (1), 5 (2016)
13. Piccinno, E., Vendemiale, M., Tummolo, A., Ortolani, F., Frezza, E., Torelli, C., Di Bitonto, P., Rossano, V., Roselli, T.: New technologies for promoting hypoglycaemia self-management in type 1 diabetic children. In: 9th Joint Meeting of Paediatric Endocrinology, Milan, 19–22 September 2013 (2013)
14. Prensky, M.: The Digital Game-Based Learning Revolution. Digital Game-Based Learning. McGraw-Hill, New York (2001)
15. Reschly, A.L., Christenson, S.L.: Jingle, Jangle, and Conceptual Haziness: Evolution and Future Directions of the Engagement Construct. Handbook of Research on Student Engagement, pp. 3–19 (2012)

16. Rieber, L.P.: Seriously considering play: designing interactive learning environments based on the blending of microworlds, simulations, and games. Educ. Technol. Res. Dev. **44**(2), 43–58 (1996)
17. Rosas, R., Nussbaum, M., Cumsille, P., Marianov, V., Correa, M., Flores, P., Rodriguez, P.: Beyond Nintendo: design and assessment of educational video games for first and second grade students. Comput. Educ. **40**(1), 71–94 (2003)
18. Spires, H.A., Rowe, J.P., Mott, B.W., Lester, J.C.: Problem solving and game-based learning: effect of middle grade students' hypothesis testing strategies on learning outcome. J. Educ. Comput. Res. **44**(4), 453–472 (2011)
19. Squire, K., Barnett, M., Grant, J.M., Higginbotham, T.: Electromagnetism supercharged!: learning physics with digital simulation games. Paper presented at the Proceedings of the 6th International conference on Learning Sciences, Santa Monica, California (2004)
20. UNECE, Unece Strategy for Education for Sustainable Development, Vilnius (2005). https://www.unece.org/fileadmin/DAM/env/documents/2005/cep/ac.13/cep.ac.13.2005.3.rev.1.e.pdf
21. UNESCO, Shaping the Future We Want. UN Decade of Education for Sustainable Development (2005–2014). Final report, Paris pp. 9; 28–31 (2014)
22. UNESCO, Education for sustainable development: an expert review of processes and learning, Parigi (2011). http://unesdoc.unesco.org/images/0019/001914/191442e.pdf
23. UNESCO, Tbilisi Declaration. http://www.eenorthcarolina.org/Documents/tbilisi_declaration.pdf
24. UNITED NATIONS, Transforming our world: the 2030 Agenda for Sustainable Development (2015). https://sustainabledevelopment.un.org/content/documents/21252030%20Agenda%20for%20Sustainable%20Development%20web.pdf
25. World Commission on Environment and Development, Our Common Future. http://www.un-documents.net/our-common-future.pdf

Making MOOCs More Effective and Adaptive on Basis of SAT and Game Mechanics

Lubov S. Lisitsyna and Evgenii A. Efimchik(✉)

ITMO University, Kronvrkskiy pr. 49, Saint Petersburg 197101, Russia
lisizina@mail.ifmo.ru, efimchick@cde.ifmo.ru

Abstract. The paper describes an approach to improve the efficiency and adaptability of the MOOCs, using RLCP-compatible virtual laboratories for implementation of the interactive practical exercises. The proposed approach is based on the organization of systematic situational awareness training (SAT) of students both before the start of the course and in the process of studying the MOOC. Software is proposed for SAT which provides not only the growth of cognitive functions (attention and thinking speed), but also affects the overall performance on e-learning. For systematic SAT during the MOOC a solution is provided to develop RLCP-compatible virtual laboratories based on a standardized interfacial forms and special demonstration mode. The use of this approach significantly reduces execution time of practical exercises in assessment trials as showed by the results of experimental studies on the MOOC of ITMO University on a National Platform of Open Education of the Russian Federation https://openedu.ru. To increase motivation to achieve better learning outcomes in the MOOC the technique of the game mechanics of perfectionism is suggested. The technique is based on complexity management RLCP-compatible virtual laboratories, allowing to provide new optional tasks of high complexity to the students who already passed mandatory assessment.

Keywords: MOOC · Interactive practical exercise · RLCP-compatible virtual laboratories · Virtual stand · Situation awareness training · Game mechanics of perfectionism

1 Introduction

One of the main problems of the practical use of Massive Open Online Courses (MOOCs) is their low efficiency, which is valued primarily as a ratio of the number of students who successfully completed the course to the total number of students who registered to the course. The project "National Platform of Open Education of the Russian Federation" (https://openedu.ru) continues to grow: at the start of February 2017 there were 140 courses from eight leading universities of Russia, including 22 course from the ITMO University. A distinguishing feature of ITMO University courses is the usage of RLCP-compatible virtual laboratories for the implementation of interactive practical exercises, which are perhaps the most important elements of the MOOCs responsible for forming, monitoring and evaluation of the expected learning outcomes. Analysis of the practical application of several courses of the ITMO

V.L. Uskov et al. (eds.), *Smart Education and e-Learning 2017*, Smart Innovation, Systems and Technologies 75, DOI 10.1007/978-3-319-59451-4_6

University ITMO [1–3] has revealed several reasons related to the problem of their effectiveness.

Firstly, students showed a potential unpreparedness to e-learning. During studying process students have to work with a variety of rapidly changing user interfaces, which should be closely and quickly monitored for changes of tasks, scales, tips, etc. Therefore, before the start of learning at any MOOC the students should gain situational awareness about working with electronic forms and high level of the basic cognitive functions such as attention, thinking speed, memory volume, and so on [4–7].

Second, there were significant difficulties for students in interactive practical exercises. Each practical exercise in the MOOC is a software module that automatically generates individual task of predetermined difficulty for each student and evaluates student's solution comparing it to the reference solution step by step in order to define the final rating. The development of such software modules is always a complicated technical challenge that must be solved considering the subject area of the MOOC. It is therefore necessary to apply the approaches aimed to the development of a unified technology solutions, providing systematic situational awareness training when performing various practical course exercises [2].

Third, there was a reduction of interest and activity of students during the session of the course. This is due not only to the increasing complexity of the learning process (training "from simple to complex"), but also to the lack of mechanisms that encourage students to take the course to the end. Traditionally, it is common to use ratings that reflect the progress of the student compared to other students to stimulate interest in the MOOCs. However, in real life [8–10], this approach is far from effective. To further stimulate the interest of students to achieve the better learning outcomes game mechanics based on perfectionism can be used, the implementation of which will require to generate tasks of extra complexity for practical exercises (the task has already been completed and credited, but a student can continue training with more complex tasks). This circumstance makes new demands on the development of software modules for implementing adaptive practical exercises and evaluation of the solutions in the MOOC.

This article is devoted to a discussion of approaches to the development of the MOOCs, aimed at increasing their efficiency and adaptability on the basis of situational awareness training and game mechanics of perfectionism.

2 Situational Awareness Training for the Formation of Readiness for e-Learning

Special software tools have been developed at the ITMO University for situation awareness trainings (SAT) (access to the demo of the software in English is available on http://de.ifmo.ru/psycholog/), which were deployed on the learning management system of University ITMO AcademicNT and applied in practice for the SAT of students of first and second years. These software tools have been developed on the basis of well-known collection of psychological tests for continuous operation (Continuous Performance Test, CPT) of Psychology Experiment Building Language (PEBL) Test Battery, Version 0.14, that are available at http://pebl.sf.net. Experimental studies have confirmed the fact that as a result of training the students not only shows the growth of the main indicators of cognitive functions (primarily attention focusing and thinking speed), but also improving e-learning performance (assessment results at the online exams for a number of disciplines have improved by 14%) [11, 12]. These software tools can be used in e-learning systems, all educational institutions, as well as special SAT modules on platforms of open education and online courses.

3 Situational Awareness Training and Practical Exercises in MOOSs Based on RLCP-Compatible Virtual Laboratories

To organize situational awareness training in practical exercises based on RLCP-compatible virtual laboratories, a unified style of the following standard interface forms was designed and implemented [2].

The standard interfacial form of the page of the individual task for the exercise determines the position of the following basic elements: a common task description, description of individual options of the task, control elements and virtual stand. Example of such a page is shown in Fig. 1.

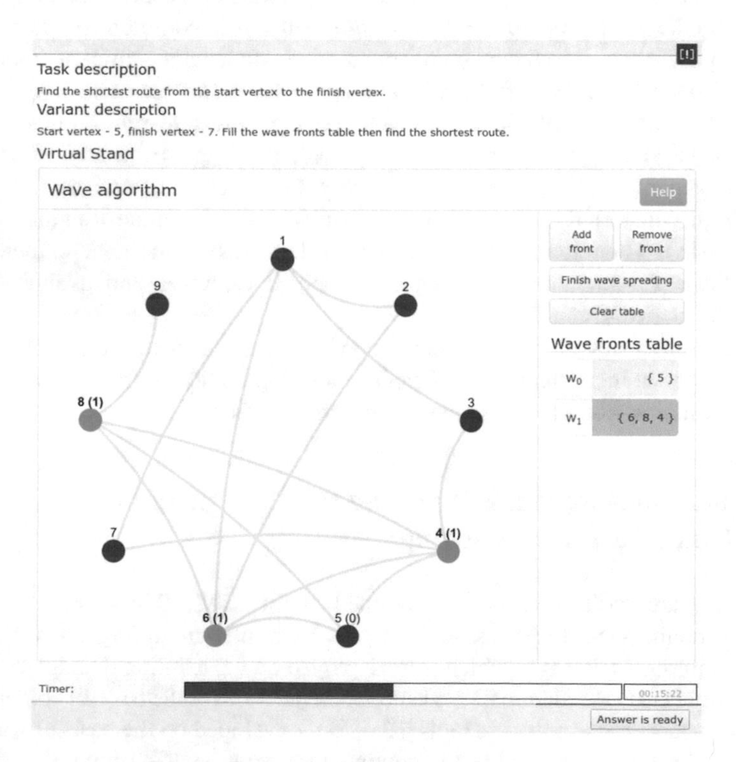

Fig. 1 Example of the individual task page

The standard form of the virtual stand with a fixed number of steps of the solution determines the layout of the following blocks: interactive picture element (visual image of a given graph, circuitry, etc.), the elements for entering intermediate solutions (tables, matrices, etc.) and area to enter a final result. In addition, each virtual stand is equipped with a tab "Help" with brief instructions on how to use it. Example of a virtual stand with a fixed number of solution steps is shown in Fig. 2.

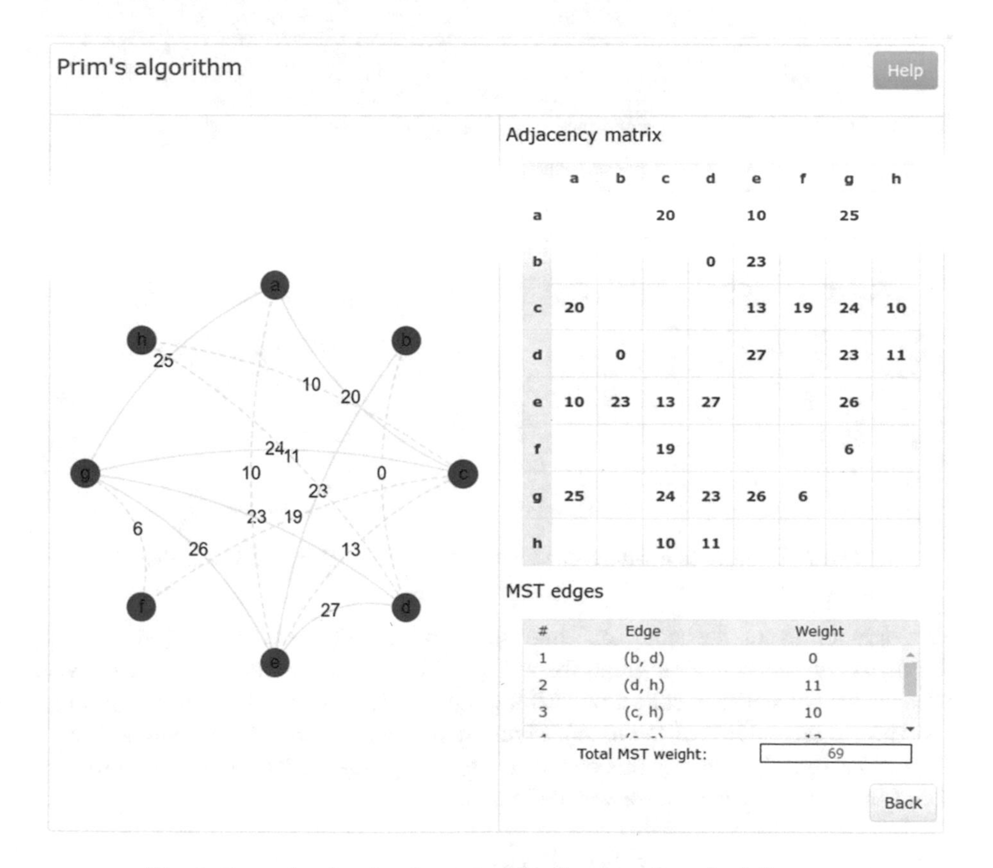

	a	b	c	d	e	f	g	h
a			20		10		25	
b				0	23			
c	20				13	19	24	10
d		0			27		23	11
e	10	23	13	27			26	
f			19				6	
g	25		24	23	26	6		
h			10	11				

#	Edge	Weight
1	(b, d)	0
2	(d, h)	11
3	(c, h)	10

Fig. 2 Example of a virtual stand with a fixed number of solution steps

The standard form of the virtual stand with a variable number of steps of solutions is a result of the further development of the previous form. The number of steps in such exercises is determined by the task variant and sometimes student's decisions. To control the element of the intermediate results it provides additional control elements to allow a student to add, remove and navigate through optional steps visual units. Example of a virtual stand with a variable number of decision steps shown in Fig. 3.

Research of the results of the practical application of these forms to implement practical exercises in the MOOC [2] confirmed presence of situational awareness training effect: taking next practical exercises of the course students knew locations of the main elements of the virtual stand and were aware of how to use them correctly.

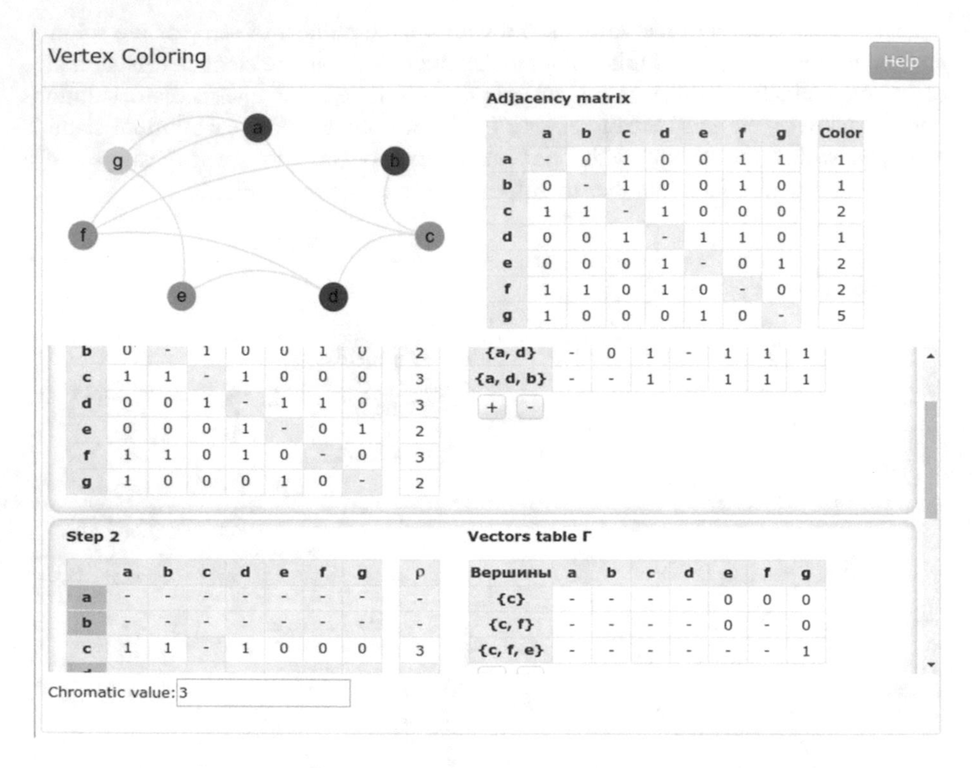

Fig. 3 Example of a virtual stand with a variable number of solution steps

Here are results of the study on time spent on practical exercises in the course "Methods and algorithms of graph theory" [1]. 1656 protocols for 152 students were selected for the statistics collection. All attempts were split into two equal parts (the first and the second half of the trials). Figure 4 is a plot of reducing execution time of 11 practical exercises used in this course. Here, the such reduction of execution time of each i-th exercise of the course was defined as:

$$p_i = \frac{\sum_{j=1}^{N} \frac{f_{ij} - s_{ij}}{a_{ij}}}{N} \times 100\%,$$

where j – the serial number of student who performed the i-th practical exercise, N – the total number of students who performed the i-th practical exercise, a_{ij} – average execution time for i-th exercise and j-th student, f_{ij} – average execution time for i-th exercise and j-th student at the first half of trials, s_{ij} – average execution time for i-th exercise and j-th student at the second half of trials.

The plot at Fig. 4 shows that the reduction of execution time when re-doing the exercises decreases sharply at the beginning of the course by SAT, and then set in a certain range. Some jump at exercise 5 and 8 are explained by the fact that these exercises had interfacial forms and control elements, which have not been used in the

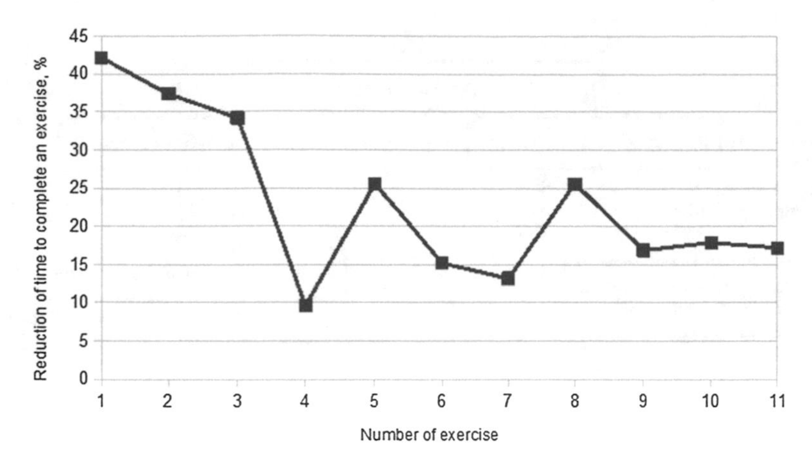

Fig. 4 Dynamics of the reduction of execution time (%) of practical exercises

previous exercises (required additional time for training with new forms). While there has been a positive impact, it was decided to continue work on providing situational awareness of students with exercises that use new forms. To achieve that, the corresponding virtual laboratories were provided with special demonstration mode to familiarize students with the work on the implementation of such exercises. This mode is suitable for initial contacts and can serve as a supplement to help working with a virtual stand. A preliminary study on such demonstration mode showed reduction of the execution time of practical exercises in the course [1] for another 10% as well as significant reduction of average number of attempts needed to successfully solve the task. Although, more research is still needed: only couple dozens of students participated in that study.

4 Managing the Complexity of Tasks in RLCP-Compatible Virtual Laboratory to Implement the Game Mechanics of Perfectionism

Existing interactive practical exercises of the course are implemented by RLCP-compatible virtual laboratories, which enable ability to adjust the complexity of the task. [13] Technology of RLCP-compatible virtual laboratories allows parameterized algorithms for composing practical exercises variants and formally define the complexity of algorithmic exercise variants. This means that it is possible to manage the complexity of the tasks given to the student based on information about the results that he already achieved.

Then, to get familiar with the virtual laboratory student can take tasks of lower complexity than those used in the assessment. And for the most motivated students the optional tasks of high complexity can be arranged, which allow them to better consolidate the skill of application of methods of graph theory to solve practical problems.

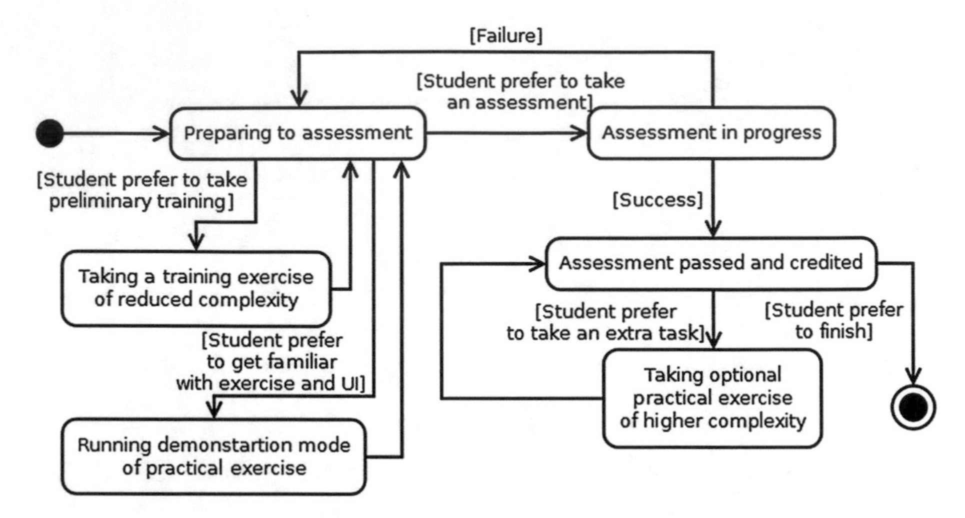

Fig. 5 Assessment process with optional preliminary trainings and available extra tasks

Thus, instead of a single control task the student may receive a set of similar tasks of varying complexity, and begin with the simple tasks, after which follow more complex, including any of the optional (Fig. 5). This assessment process allows to build a model of the student and appreciate his skill to solve practical exercises.

For example, in the exercise devoted to search of independent sets in graphs with the help of Magu-Weismann method complexity of the exercise depends on the number of independent sets in the generated graph and the number of edges in the graph. So, in typical tasks applied for assessment graphs with 8 vertices, 5–6 independent sets and 13–14 edges are generated. However, the graph generation algorithm is parametrized, so we can generate less complex tasks with graphs on 6 vertices with 3–4 independent sets and 8–10 edges for training or demonstration mode and extra complex tasks supplied with graphs of 10 vertices, 8–10 independent sets, 20 and more edges for more difficult optional exercises.

Motivation to carry out additional tasks should be maintained, while not included in the main assessment scale that can repel less motivated students. Game mechanics of perfectionism that provide information about individual achievements as a special part of student profile can be used to register solving of optional tasks. Achievements can be registered as set of badges, gradual rating of practical exercises passage (e.g. three-star system as in some popular mobile games) or filling a special additional "expert" scale of the course.

5 Conclusions

The analysis of the causes of the poor performance in MOOCs showed the lack of the readiness of students to e-learning: interactive practical exercises appeared to be a significant challenge for students, engagement level also reduced along the course.

To address the identified problems the following approaches are proposed at the ITMO University.

To prepare students for e-learning it is necessary to provide training of situational awareness at the beginning of the course, which requires special software (to get access to the demo of such software, developed at the University ITMO in English, please, follow http://de.ifmo.ru/psycholog/). Experimental studies have confirmed the fact that as a result of training the students showed the growth of the main indicators of cognitive functions (primarily attentiveness and speed of thinking) as well as improving performance in e-learning (for several disciplines grades for online exams have improved by 14%).

To eliminate difficulties in the implementation of interactive practical exercises it is better to use similar interfacial forms to design all exercises in the course. The proposed approach is used in design of 11 interactive practical exercises based on the RLCP-compatible virtual laboratories for the MOOC "Methods and algorithms of graph theory" on the National Platform of Open Education of the Russian Federation. Experimental study has confirmed the fact that this approach leads to reduction of the time required to solve practical exercises in the assessment attempts. This effect was achieved by effect of situational awareness training. It is shown that the preliminary training with virtual laboratory in demonstration mode increases the SAT effect.

To increase motivation of students to achieve better learning outcomes a game mechanic of the perfectionism is offered. It allows student to solve more complex optional tasks with the help of the same virtual laboratory after he succeeded in mandatory assessment. Experimental study on usage of such mechanics of perfectionism in the MOOC "Methods and algorithms of graph theory". is planned

Acknowledgment. This paper is supported by Government of Russian Federation (grant 074-U01).

References

1. Lisitsyna, L.S., Efimchik, E.A.: Designing and application of MOOC "Methods and algorithms of graph theory" on national platform of open education of Russian Federation. Smart Educ. e-Learning **59**, 145–154 (2016)
2. Lisitsyna, L.S., Efimchik, E.A.: An approach to development of practical exercises of MOOCs based on standard design forms and technologies. Lecture Notes of the Institute for Computer Sciences, Social Informatics and Telecommunications Engineering, vol. 180, pp. 28–35 (2017)
3. Lisitsyna, L., Lyamin, A.: Approach to development of effective e-learning courses. Front. Artif. Intell. Appl. **262**, 732–738 (2014)
4. Lisitsyna, L.S., Lyamin, A.V., Bystritsky, A.S., Martynikhin, I.A.: Problem support cognitive Functions in the e-Learning. Sci. Tech. J. Inf. Technol. Mech. Opt. **94**, 177–184 (2014). (in Russian)
5. Melby-Lervag, M., Hulme, C.: Is working memory training effective? A meta-analytic review. Dev. Psychol. **49**, 270–291 (2013)

6. Jaeggi, S.M., Buschkuehl, M., Shah, P., Jonides, J.: The role of individual differences in cognitive training and transfer. Mem. Cogn. **42**, 464–480 (2013)
7. Owen, A.M., Hampshire, A., Grahn, J.A., Stenton, R., Dajani, S., Burns, A.S., Ballard, C.G.: Putting brain training to the test. Nature **465**, 775–778 (2010)
8. Lisitsyna, L.S., Pershin, A.A., Kazakov, M.A.: Game mechanics used for achieving better results of massive online courses. Smart Innov. Syst. Technol. **41**, 183–193 (2015)
9. Connolly, T.M., Boyle, E.A., MacArthur, E., Hainey, T., Boyle, J.M.: A systematic literature review of empirical evidence on computer games and serious games. Comput. Educ. **59**, 661–686 (2012)
10. Peretz, C., Korczyn, A.D., Shatil, E., Aharonson, V., Birnboim, S., Giladi, N.: Computer-based, personalized cognitive training versus classical computer games: a randomized double-blind prospective trial of cognitive stimulation. Neuroepidemiology **36**, 91–99 (2011)
11. Lisitsyna, L.S., Lyamin, A.V., Martynikhin, I.A., Cherepovskaya, E.N.: Situation awareness training in e-learning. Smart Educ. Smart e-Learning **41**, 273–285 (2015)
12. Lisitsyna, L.S., Lyamin, A.V., Martynikhin, I.A., Cherepovskaya, E.N.: Cognitive trainings can improve intercommunication with e-Learning System. In: 6th IEEE International Conference Series on Cognitive Infocommunications, pp. 39–44 (2015)
13. Efimchik, E.A., Chezhin, M.S., Lyamin, A.V., Rusak, A.V.: Using automaton model to determine the complexity of algorithmic problems for virtual laboratories. In: Proceeding of 9th International Conference on Application of Information and Communication Technologies, pp. 541–545 (2015)

Smart e-Learning

The Development of the Critical Thinking as Strategy for Transforming a Traditional University into a Smart University

Adriana Burlea Schiopoiu[1](✉) and Dumitru Dan Burdescu[2]

[1] Department of Management, Marketing and Business Administration, University of Craiova, Craiova, Romania
adriana_burlea@yahoo.com
[2] Department of Computers and Information Technology, University of Craiova, Craiova, Romania
dburdescu@yahoo.com

Abstract. The research objectives of the article consist in the identification of the strategic elements which contribute to the transformation of TrU into SmU and of the characteristics of e-learning Tesys platform which improve the performances of the students and professors. In our research we have started from the premises that smart technology is a strategic tool for the development of critical thinking and the Smart curricula and the Smart pedagogy have to be created in order to develop the critical thinking for both students and professors. Our findings revealed that INCESA HUB has the possibility to optimize learning process and to transform the University of Craiova into a modern and SmU. The e-learning Tesys platform plays a very important role, because, through its flexibility, it is able to fulfil the Smart mission of SmU. Finally, we arrived at the conclusion that the solution resides in the involvement of all stakeholders at intern, local, national and international level in order to acquire and share the knowledge.

Keywords: E-learning tesys platform · INCESA HUB · Critical thinking · Strategy

1 Introduction

In a digital era, enabled by Smart technology, the Smart University (SmU) perform the same activities as traditional university (TrU), and adopts smart solutions as a result of a smart decision-making process in order to increase the satisfaction and to enhance the motivation of both students and professors [1].

A Smart University (SmU) is a traditional university (TrU) that had gradually implemented an interconnected system with a central control of the technological resources.

Starting with this definition, we have developed a strategy for transforming a Romanian traditional university (e.g. University of Craiova) in a Smart one.

V.L. Uskov et al. (eds.), *Smart Education and e-Learning 2017*, Smart Innovation, Systems and Technologies 75, DOI 10.1007/978-3-319-59451-4_7

It is difficult to transform a TrU into a SmU when the financial and human resources are limited. For this reason, first of all we have evaluated the existing resources. Secondly, we have limited our strategy to the social sciences domain.

The SmU, as a system, is very complex and it is difficult to find the starting point in a transformation strategy from TrU.

It is easier to develop a SmU than to transform a TrU into a SmU.

The objectives of our research are the following:

- identification of the strategic elements which contribute to the transformation of TrU into SmU, and
- identification of the characteristics of e-learning Tesys platform which improve the performances of the students and professors.

In our research we have started from the premises that smart technology is a strategic tool for the development of critical thinking.

The research questions are the following:

1. How does the smart technology develop the critical thinking of the students?
2. Why do the students prefer to use smart technology in classrooms?
3. How do the teachers from the others domains (economy, social sciences, law) adapt their learning style to the requirements of smart technology?

An excellent strategy combined with the competences of professors and students is important for the setup of the vision and mission of the SmU.

First of all, our analysis was oriented to smart curricula (e.g. adaptive programs of study and courses) and to smart pedagogy (e.g. learning by doing, collaborative learning and adaptive teaching).

Secondly, we have oriented our attention to software systems and hardware equipment.

2 The Impact of the Smart Technology on the Critical Thinking of the Students

Binkley and his colleagues [2] consider critical thinking as a combination of knowledge, skills, and attitudes/values/ethics (KSAVE) and emphasise the idea introduced by Rhodes [3] who associated the critical thinking to “A habit of mind”.

Critical thinking is a cognitive process which is built based on the interaction between information, ideas, facts and experiences.

The cognitive rigor is very important in the relationship between professor-student and e-learning Tesys platform is a tool used for problem-solving by evidence. In e-learning framework, the collaboration is not enough to develop the critical thinking process, because the partnership professor-students must operate following the same methods. Moreover, it has to prove solid evidence from shared information databases with the purpose of distinguishing beliefs from real knowledge and to draw a demarcation border between facts and judgments. Sometimes it is important to use genuine critical thinking together with Smart technology in order to overcoming the barriers of a peer-pressured and a peer-conscious environment.

The student, in the role of the critical thinker, uses the e-learning Tesys platform to define the problem, to examine the evidence and alternatively, to compare the results and to make the final decision [4]. At the same time, the student has to be open and receptive to different, perhaps, divergent opinions (e.g. a new point of view of the professors or a solid argument of another student), which may involve arguments and a new perspective. Therefore, the e-learning Tesys platform develops the critical thinking process of the students because it requires and engages them in an active participation to the learning development. A weak point of the e-learning Tesys platform consists in the fact that professors are looking for the right answers that come from e-courses content, and the critical thinking is sometimes difficult to put into value because it faces the student resistance. The critical thinking of the students must to be developed step by step, in close connection to the business environment, because companies do not always agree with the students who take initiative and work in their manner and not in a traditional one. For this reason, Smart technology helps students and professors to move in a direction that can be beneficial for both sides.

The role of a SmU is to overcome the resistance and any conflicts which may appear based on the dissimilarities between the academic and business environment.

The students use the smart technology to gain knowledge and they consider the professor as an actor who helps them to understand the problem and to solve it.

The final decision is made by the students and, in this context, depending on the complexity of the problem; the Smart technology helps them to change their attitude disposition.

The Tesys platform is a tool that integrates Smart technology and becomes a support for both professors and students who use e-learning techniques. It allows the students to have a real-time interaction with their professors, but also between them.

The e-learning Tesys application uses MVC architecture and the Controller is represented by Action, Bean, Helper, Java classes, MainServlet and Manager. The View is represented by Web-macro templates and DBMS represent the Model [5].

The INCESA HUB, due to its organizational structure, provides the instruments that help the professors and students to develop skills in the critical thinking process and to elaborate conclusions based on logical judgment.

3 Smart Technology: A Dilemma for Smart Universities

The SmU has to cope with a dilemma emerged from the relationship established between professors and students in e-learning system.

In e-learning process, students are using the Smart technology easily and regard the Tesys platform as a tool which contributes to their personal development and improves their learning experience [6].

The professors want to interact with the students in a classic way (e.g. face-to-face learning and/or blended learning) and consider the Tesys platform as being an ordinary tool for disseminating the information to students.

Previous research on e-learning Tesys platform has revealed that it is perceived as smart learning tool and a key social driver for the students and professors. University of

Craiova considers providing Smart services to students and professors to be a good indicator of an efficient strategy.

The e-learning Tesys platform was designed to offer a personalized feedback to its users and to help them to increase their performances by providing real-time solutions, explanations and recommendations on how to expand an idea [7].

The e-learning Tesys platform will be used as smart technology for the development of the ability of the students to distinguish belief from knowledge and to understand to what extent to which this belief has influenced their judgment.

The motivation of the students is related to innovation and creativity in the learning process. The motivation of the professors is based on the development of the critical thinking of the students.

The students want to have access to information anytime, anywhere and with low financial resources.

The professors want to focus on the communicational process and to interact with students directly, in real-time and without facing technological barriers.

The students consider smart technology as a universal panacea which creates many opportunities for their development.

The professors want to use smart technology to disseminate the information, but they are reluctant to use it to assess their students. The professor wants to be sure that the students do not cheat and to ensure a continuous monitoring of the learning environment, mechanism and processes.

4 The Strategy for Developing Smart University

The financial crisis has put its mark on the financial resources of public Romanian universities. Therefore, the financial resources for investments in Smart technology were reduced and oriented to the human factor (e.g. salaries).

The other challenge for the Romanian universities is the contradictory phenomenon concerning the number of the students. If, at global level the population is steadily increasing, at national level, the population, and consequently the number of the students, is continuously decreasing.

The faculties from social sciences domain are not interested in investing in Smart technology, but the students push them to keep up with the new teaching-learning-assessment methods.

How it is possible for a university, to become smarter in the absence of financial resources which would allow the adoption of smart technology?

The solution resides in the involvement of all stakeholders at intern, local, national and international level in order to acquire and share the knowledge.

If you have technology you need people to manage it, and if you have people who can use technology, you need a Smart technology [8].

Taking into account the particularities of the economic and social Romanian framework, the strategic decisions require realistic projections and forecasting derived not only from statistical data, but also from university's own experience.

Smart decisions involve important resources which impact the university structure and are often made in response to requirements of the academic and business environments [9].

In public academic environment, it is very difficult to implement the strategy, because of limited possibilities to plan the financial resources. For this reason, the planning horizon will cover short-term operational period and it will be oriented toward the university competencies, the evaluation of the important shifts in smart technology and toward a real evaluation of the opportunities and risks (e.g. competition for students and for financial founds from national and international projects; changes in the needs of professors and students). Finally, the strategic Smart objectives shall define the university's strategic needs and to take into account risk factors framed by the national and international legislation.

5 The INCESA HUB a Relay for Smart University of Craiova

The strategy is to build around The Research HUB for Applied Sciences (INCESA HUB) as a center that puts together both resources (human and technology) in order to capitalize the human expertise and the technology (Fig. 1).

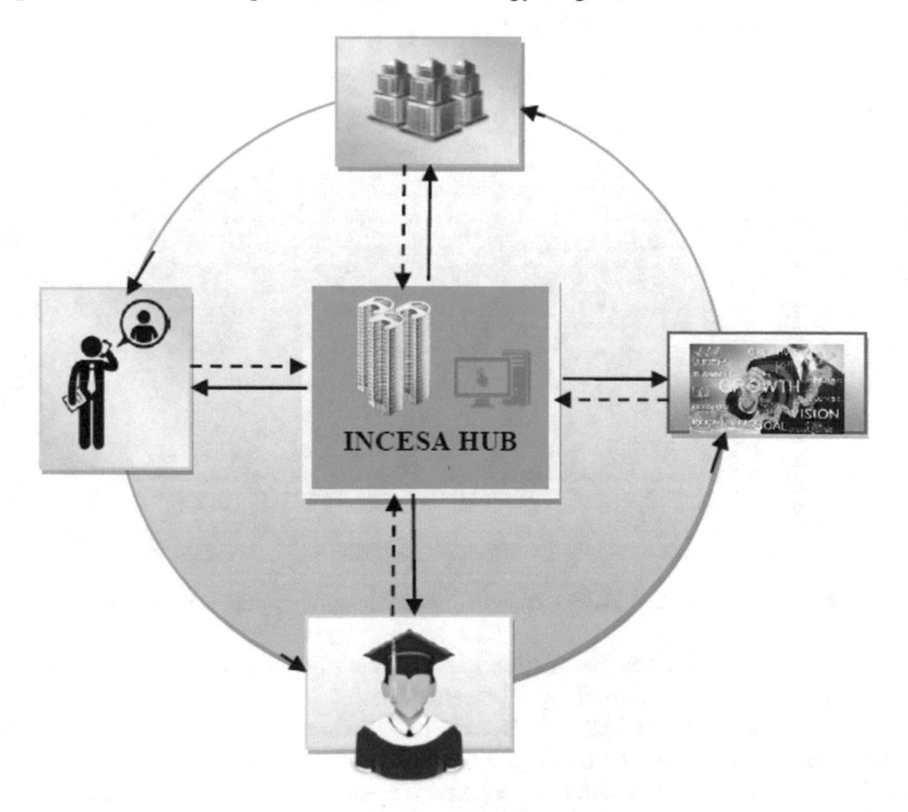

Fig. 1. The INCESA HUB a relay for Craiova Smart University (Source: https://www.dreamstime.com), (Source: http://ww25.iconfider.com)

The Centre for Technology Transfer (CTT) will provide viable solutions to problems arisen from the usage of Smart technology by professors, students and business environment.

Smart Strategy Implementation is a complex process and every step is a challenge for both INCESA HUB and the University of Craiova.

The action plan is build around the critical risks and it translated into operational activities taking into account the limited financial resources.

A successful point is represented by key performance indicators (e.g. number of online courses, number of equipments and software used in online courses for an individualized approach to Smart curricula through Smart technology).

The INCESA HUB has the possibility to optimize learning process and to transform the University of Craiova into a modern and SmU.

The mission of the INCESA HUB is to develop the University of Craiova as SmU. It bring together national and international scientists for joint research activities and it is oriented towards providing professional training and development services in the form of internship – placement – volunteering as schemes for Master's and doctoral students.

The outcomes of the INCESA HUB research obtained in 13 smart labs and 4 RDI centres, all equipped with state-of-the-art technology, as well as a Centre for Technology Transfer – CTT INCESA, are shared with the scope of benefiting both academic and business environment [10] (Fig. 2).

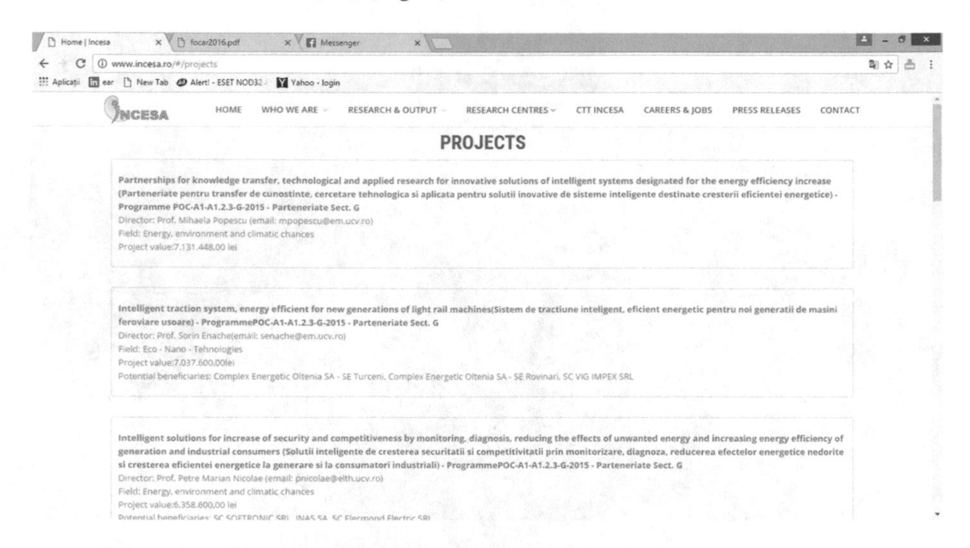

Fig. 2. The INCESA HUB - Projects (Source: http://www.incesa.ro/#/projects)

The INCESA HUB has the mission to solve this dilemma by putting together the three pillars: students – technology – professor.

As a central driver, INCESA HUB takes initiative on what aspects to explore and develop in the academic and business environment. Thus, the flexibility and interoperability between academic and business environment become strategic challenges for SmU, because the culture of cooperation based on autonomy is important to succeed [11] (Fig. 3).

Fig. 3. The INCESA HUB driving the agenda forward (Source: http://www.incesa.ro/#/gallery)

The University of Craiova, as a SmU, needs to be prepared to face changes and challenges it cannot completely predict and to formulate a case scenario planning together with a set of measures (e.g. strategic radar of the assumptions is an important Smart tool that rapidly develops the new Smart capabilities for challenges and puts into value the relationship between planning versus implementing and executing the Smart strategy).

The formulation and implementation of the INCESA HUB strategy is a top-down process that includes risk optimisation against future competitors. An important resource for both SmU and INCESA HUB is the talented strategic thinker – either professor or student – who gather around him/her intelligent people.

Benefiting of the help of the INCESA HUB strategy, the SmU will be developed step by step, progressively, but, with no doubt, taking into account the purpose of being (the mission of the SmU) and long-term goal (the goals and the vision of the SmU).

It is important that SmU through INCESA HUB to be dynamic enough to adapt when conditions change in the academic and business environment. To achieve this, agility and a cross functional approach that leads to actions, evaluation and continuous improvement are required.

The SmU strategy must focus on the student needs and to develop and use practice performance metrics and benchmarks to ensure that the resources are effectively used and will, ultimately, lead to success.

The role of INCESA HUB is to improve the confidence of people in the ability of SmU to value critical thinkers, visionary persons or their ideas. Therefore, SmU develops an active participation of both professors and students in the learning process.

6 Conclusions

We have arrived at the conclusion that Smart curricula and the Smart pedagogy have to be created in order to develop the critical thinking for both students and professors. The e-learning Tesys platform plays a very important role in the development of SmU, because, through its flexibility, it is able to fulfil the smart mission.

The on-line courses that integrate the critical thinking follow the new technology development, and the professors who are creating the courses integrate the cultural aspects, making the difference between the need of the students and the need of the community. The students have the possibility to define the problem, to examine the evidence, and, finally to take into account alternative interpretations when they analyze the assumptions and underline the evidence.

The team management of the University of Craiova accepts the idea that INCESA HUB requires autonomy in the administration of its resources in order to succeed. Therefore, dual empowerment, with clear attributes, is a solution for University of Craiova to become a SmU. The strengths of this structure resides on various building block such as administrative, technical and human capabilities (e.g. it has university support, the smart technology and smart pedagogy can be integrated into the faculties curricula, the direct access and real-time connectivity between all actors involved).

The INCESA HUB is a good strategic advisor because it has the possibility to bring together talented people (professors and students) and the Smart technology.

References

1. Uskov, V.L., Howlet, R. Jain, L. (eds.): Smart education and smart e-Learning. In: Proceedings of the 2nd International Conference on Smart Education and e-Learning SEEL-2016, 17–19 June 2015, Sorrento, Italy. Springer, Berlin (2015)
2. Binkley, M., Erstad, O., Herman, J., Raizen, S., Ripley, M., Rumble, M.: Defining 21st century skills. In: Griffin, P., McGaw, B., Care, E. (eds.), Assessment and Teaching of 21st Century Skills, pp. 17–66. Springer, New York (2012)
3. Rhodes, T.L. (ed.): Assessing outcomes and improving achievement: tips and tools for using rubrics. Association of American Colleges and Universities, Washington, DC (2010)
4. Burlea Schiopoiu, A., Burdescu, D.D.: An Integrative Approach of e-Learning: From Consumer to Prosumer, in Smart Innovation, Systems and Technology, pp. 269–280. Springer (2016)
5. Burdescu, D.D.: TESYS: An e-Learning System. In: Proceedings of the 9th International Scientific Conference Quality and Efficiency in e-Learning (2013)
6. Burlea Schiopoiu, A., Badica, A., Radu, C.: The evolution of e-Learning platform TESYS user preferences during the training processes, ECEL 2011, Brighton, UK, pp. 754–761, 11–12 November 2011
7. Burlea Schiopoiu, A.: The complexity of an e-Learning system: a paradigm for the human factor. In: Proceedings of the Inter-Networked World: ISD Theory, Practice and Education, vol. 2, pp. 267–278. Springer (2008)
8. Cervera, M.G., Johnson, L.: Education and technology: new learning environments from a transformative perspective. Rusc-Univ. Knowl. Soc. J. **12**(2), 1–13 (2015)
9. Coccoli, M., Guercio, A., Maresca, P., Stanganelli, L.: Smart universities: a vision for the fast changing digital era. J. Vis. Lang. Comput. **25**(6), 1003–1011 (2014)
10. INCESA – projects. http://www.incesa.ro/#/projects. Accessed Jan 2017
11. INCESA – The agenda. http://www.incesa.ro/#/gallery. Accessed Jan 2017

RLCP-Compatible Virtual Laboratories with 3D-Models and Demonstration Mode: Development and Application in e-Learning

Lubov S. Lisitsyna, Evgenii A. Efimchik(✉), and Svetlana A. Izgareva

ITMO University, Kronverkskiy pr. 49, Saint Petersburg 197101, Russia
lisizina@mail.ifmo.ru, efimchik@cde.ifmo.ru

Abstract. The article describes the features of the application of 3D-models in the design and development of RLCP-compatible virtual laboratories for interactive practical exercises in MOOC giving an example of laboratory devoted to the gradient descent method (backtracking line search) to define a multivariable function. Also features of application of the virtual laboratory in visualizer mode to preview and preliminary theoretical training of students are described. The article also shows the results of a study on pilot application of the developed virtual laboratory in practical training of the discipline "Methods of optimization". It is shown that the use of 3D-models and the implementation of the visualizer mode helps to achieve better learning outcomes (a 10% correct answers increase, significant reduction of the execution time and number of the attempts at the assessment).

Keywords: MOOC · RLCP-compatible laboratories · Virtual labs · e-Learning · 3D-models · Visualizers

1 Introduction

Technology of RLCP-compatible virtual laboratories [1] has become the base for the implementation of interactive practical exercises in Massive Open Online Courses (MOOCs) offered by the ITMO University at the National Platform of Open Education of the Russian Federation (https://openedu.ru). About half of these courses uses this technology based on the RLCP (Remote Laboratory Control Protocol), that allows to use virtual laboratories in fully automated mode which is very important for MOOCs. So, virtual laboratories that implements that protocol are called RLCP-compatible. However, an analysis of the practical application of this technology has revealed the following problems. Virtual stands of such laboratories only use interactive 2-D graphics drawings [2, 3], which limits the use of this technology in the subject areas, in which three-dimensional graphics is more appropriate to use. In addition, the question of application of RLCP-compatible virtual laboratories in the preliminary training mode remains. Today practical exercises that are implemented in MOOCs of the ITMO University ITMO as RLCP-compatible virtual laboratories are used only in assessment mode.

V.L. Uskov et al. (eds.), *Smart Education and e-Learning 2017*, Smart Innovation, Systems and Technologies 75, DOI 10.1007/978-3-319-59451-4_8

This article is devoted to the research on expanding the scope of technology RLCP-compatible virtual laboratories by designing of interactive practical exercises in MOOCs using three-dimensional graphics, as well as the possibility of their usage not only for the assessment of students, but also for theoretical preliminary training. The research results are based on the experience of developing RLCP-compatible virtual laboratory for the study of the well-known gradient descent (backtracking line search) method to determine a multivariable function [4, 5] and automatic evaluation of its application by a student to solve this problem.

2 The Technological Solution to Include 3D-Models in RLCP-Compatible Virtual Laboratory

The composition of any RLCP-compatible laboratory includes virtual stand and RLCP-server [1]. A virtual stand must be integrated into modern platforms of online courses that are focused mainly on the web, so it is created on the basis of HTML5. Virtual stand provides user interface of a RLCP-compatible virtual laboratory. Typically, it has two main functions: to provide the student with personal task in a convenient form (in this case a three-dimensional object) and allow to manipulate this object to solve the problem by means of appropriate control elements. These features meet the standard of RLCP-compatible virtual laboratories assessment mode.

Such object can be prepared using a three-dimensional graphics technology [6, 7], for example, Adobe Flash or WebGL. However, to display three-dimensional graphics in HTML5 there is no special tag such as for other media: image, video or audio. For a long time, the only solution was to use embeddable objects based on Flash, Java Applet, Silverlight technologies in HTML-page, but today modern standards and technologies propose to use WebGL for these purposes, a technology which has significant advantages, because it is an open, free and secure platform. There are several software libraries on WebGL basis. In the development of the described virtual laboratory three.js is used, which greatly reduces the efforts to write the code for creating of 3D-models.

Figure 1 shows an example of a virtual stand with a three-dimensional model. In carrying out the task the student has to repeat the action of a given algorithm, so the virtual stand maintains a set of necessary commands. In addition, the student can interact with 3D-model, choosing a comfortable viewing angle, and the result of the actions of the algorithm is displayed on the model. For example, if you save intermediate results by entering the required values, and then click "Save Step", a three-dimensional model of the trajectory of the descent will be shown from the previous point to the new specified point (Fig. 2).

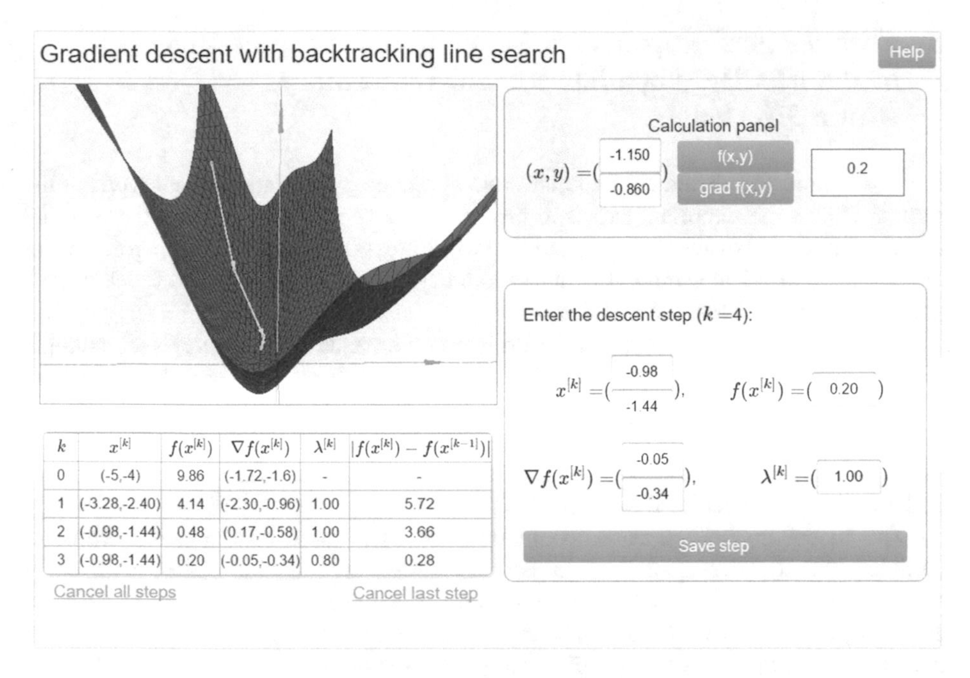

Fig. 1. Example of virtual stand using 3D graphics

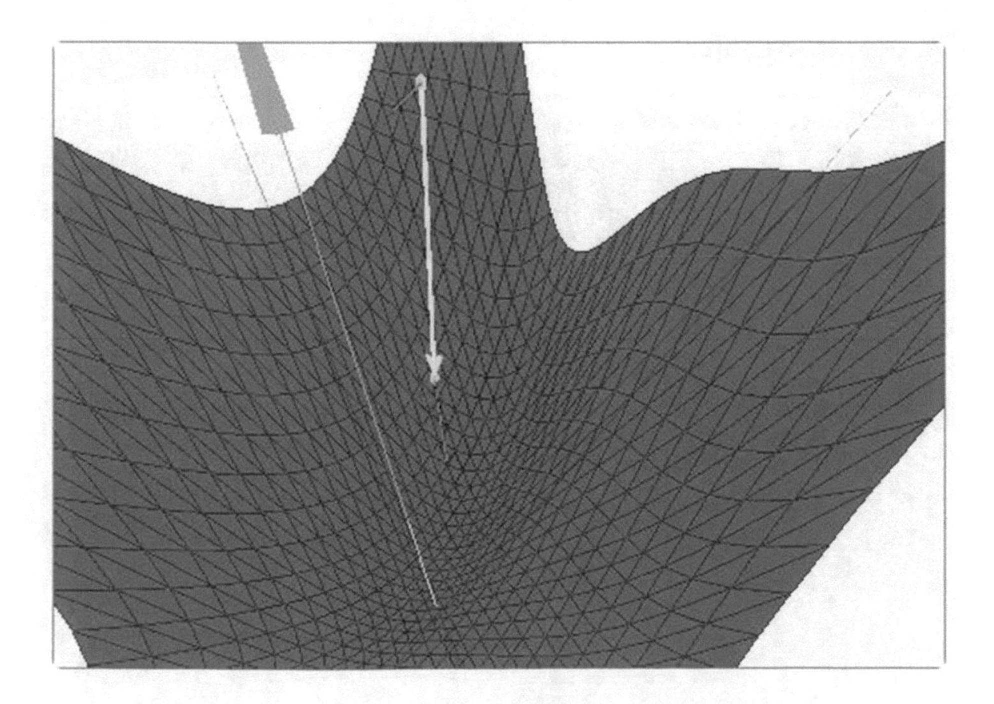

Fig. 2. Example of visualization of trajectory of descent to minimum point

3 Technological Implementation of Demonstration Mode of the RLCP-Compatible Virtual Laboratory with a 3D-Model

In the e-learning an extremely important role is played by the independent work of the student [8], so the important task is to provide a convenient way to learn the material and prepare for assessment. So, in addition to the text description of the algorithm for the solving a problem using the virtual laboratory, visualization is useful to let student get familiar with virtual lab interface.

Virtual laboratory in demo mode is an interactive visualizer that shows an example of task solving with the specified algorithm. Using three-dimensional graphics in real-time allows the student to interact with the model and study the objects according to his own pace, as well as in the assessment mode – to choose a comfortable viewing angle, which may produce further the interest in learning and ensure the correct understanding of theoretical material.

The model in the demonstration mode (Fig. 3) is more informative, because it includes grid signatures, gradient projection on the function plot, etc. In addition, it allows to add the necessary background information, such as the concept of gradient and antigradient of the function. When you click on an object model (point, descent trajectory, antigradient) corresponding information about the object shows up, as well as the necessary background information, in this case of course of mathematical analysis.

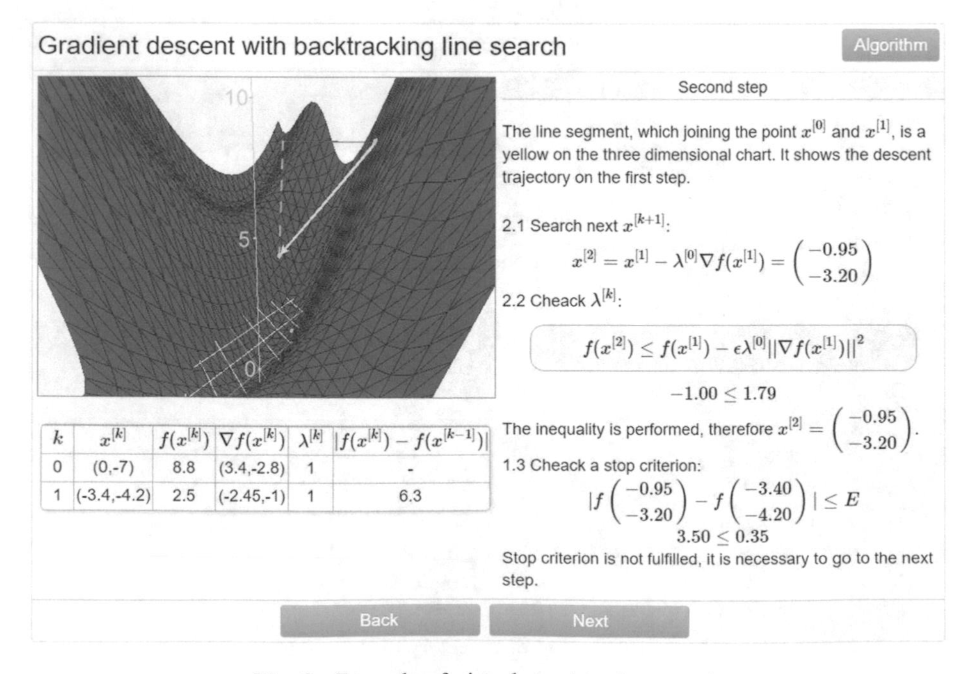

Fig. 3. Example of virtual stand in demo mode

4 Application Results

As part of the assessment events on the discipline “Methods of optimization” students of the 3rd year of the Department of Computer Educational Technologies after listening to a lecture entitled “Minimization of multivariable functions” took practical exercises on the basis of the virtual laboratory to find the minimum of a multivariable function by means of the gradient descent with backtracking line search. The aim of the study was the use of a virtual laboratory in two modes (demonstration and assessment modes), which expands the scope of application of the RLCP-compatible virtual labs.

During the experiment, the students were divided into two groups - control and active. Both groups were equal in size and performance. The approximate equality of achievement by groups determined by the average score of the groups (Table 1) in the disciplines of “Mathematics”, “Extras of mathematics” and “Discrete mathematics”. This choice of subjects is due to the following considerations. In the study of the first two courses students have acquired knowledge necessary for the study of the method (the concept of a function of several variables, graph of a function of two variables, partial derivatives, gradient, line level and others.). For the learning discrete mathematics students had used MOOC “Methods and algorithms of graph theory” and had experience with RLCP-compatible virtual laboratories [2].

Table 1. Average score of students

#	Discipline	Group #1	Group #2
1	Mathematics	76,38	73,39
2	Extras of mathematics	80,29	85,43
3	Discrete mathematics	84,33	87,06
Total		80,33	81,96

To generate and evaluate learning outcomes students in control group (#1 group) used a standard approach, implemented in other MOOC with virtual laboratories: first, they were invited to examine the text of the algorithm and see the step by step example of solutions (represented in the form of text with graphic pictures, diagrams, tables) then, having examined the proposed materials, they began the practical exercise with virtual laboratory in the assessment mode. The active group (#2) was provided with an example in the virtual laboratory in the demonstration mode instead of text example. For both groups time to study was not limited, and 60 min was given for assessment (Table 2).

Both groups spend about half an hour on learning algorithm and studying of the example of solution. However, students of the #2 group were spending 7.2 min less for assessment than #1 group (execution time decreased by about 28%). At the same time, the average number of wasted attempts for group #2 was 1.29 and the number to get the maximum from the first attempt - 71% (against 2.00 attempts and 29% of students in the #1 group).

Table 2. Assessment results

#	Result	Group #1	Group #2
1	Avg preparation time (min)	32,43	31,14
2	**Avg execution time per trial (min)**	**25,8**	**18,6**
3	**Avg trials**	**2,00**	**1,29**
4	**Success at first trial**	**29%**	**71%**
5	Avg final result	67%	76%

According to the assessment results it is clear that the group of students that used a virtual lab in a demonstration mode demonstrated a better performance. At the same time both groups had spent similar amount of time to prepare for the assessment. Assuming preliminary equality of the groups it may indicate better understanding of the studied material, which is achieved by using an interactive three-dimensional graphics in demonstration mode.

5 Conclusion

The proposed technical solutions allow to extend the scope of application of RLCP-compatible laboratories technology to design interactive practical exercises for MOOCs, where it is necessary to use 3D graphics. Experimental study of the developed RLCP-compliant laboratory for the discipline of "Methods of Optimization" on gradient descent method with backtracking line search proved efficiency of these solutions: not only the problem solving time was reduced in the active group, but also a number of attempts to achieve success that points to the efficiency of RLCP-compatible laboratories in demonstration mode to study theoretical material of practical exercises.

References

1. Lyamin, A.V., Efimchik, E.A.: RLCP-compatible virtual laboratories. In: International Conference on E-Learning and E-Technologies in Education, pp. 59–64 (2012)
2. Lisitsyna, L.S., Efimchik, E.A.: Designing and application of MOOC "Methods and algorithms of graph theory" on national platform of open education of Russian federation. Smart Educ. e-Learn. **59**, 145–154 (2016)
3. Lisitsyna, L.S., Efimchik, E.A.: An approach to development of practical exercises of MOOCs based on standard design forms and technologies. Lecture Notes of the Institute for Computer Sciences, Social Informatics and Telecommunications Engineering, vol. 180, pp. 28–35 (2017)
4. Boyd, S., Vandenberghe, L.: Convex Optimization, Chap. 9. Cambridge University Press, Cambridge (2004)
5. Jouili, K., Benhadj Braiek, N.: A gradient descent control for output tracking of a class of non-minimum phase nonlinear systems original research article. J. Appl. Res. Technol. **14**(6), 383–395 (2016)

6. Karsakov, A., Bilyatdinova, A., Bezgodov, A.: Improving visualization courses in Russian higher education in computational science and high performance computing. Procedia Comput. Sci. **66**, 730–736 (2015)
7. Pandey, A.V., Manivannan, A., Nov, O., Satterthwaite, M., Bertini, E.: The persuasive power of data visualization. IEEE Trans. Vis. Comput. Graph. **20**, 2211–2220 (2014)
8. Lisitsyna, L., Lyamin, A.: Approach to Development of Effective E-Learning Courses. Frontiers in Artificial Intelligence and Applications, vol. 262, pp. 732–738 (2014)

Remote Laboratory Environments for Smart E-Learning

Hugh Considine[1], Andrew Nafalski[2], and Zorica Nedic[1(✉)]

[1] School of Engineering, University of South Australia, Adelaide, Australia
{Hugh.Considine,Zorica.Nedic}@unisa.edu.au
[2] School of Education, University of South Australia, Adelaide, Australia
Andrew.Nafalski@unisa.edu.au

Abstract. The paper reviews existing representative remote laboratory environments in the context of e-learning. The focus is then on the NetLab, a remote laboratory developed at the University of South Australia (UniSA) and in operation since 2002. Details of the NetLab configuration and operation are outlined with recommendations of its use for effective e-learning in engineering education. Further developments of an Intelligent Tutoring System to improve NetLab as a smart learning environment are highlighted.

Keywords: Smart e-learning · Remote laboratories · Intelligent tutoring systems

1 Introduction

E-learning environments have long been a reality in most educational institutions in the world, including learning management systems (LMS) such as Moodle and Blackboard, and typically more locally developed student management systems (SMS), increasingly integrated with each other.

In engineering and science education, remote laboratories have been implemented for over two decades, either on a free access basis or on a pay-as-you go basis, allowing online access to either unique or ubiquitous real equipment and measuring instruments in experiments, closely resembling real student laboratory experience.

In this paper, we review the most representative examples of remote experiment environments for e-learning in variety of areas of engineering education.

The remote laboratory NetLab developed at the University of South Australia (UniSA) is then described in more detail, as an example of a successful, open-access laboratory, flexible and realistic e-learning environment.

2 Remote Laboratories

Engineering and science laboratories can be generally categorised into three classes: real (physical), virtual and remote, although some overlaps and intertwining exist. Real laboratories allow students to participate in scheduled classes under a human supervision. Repetition of experiments is rarely permitted but a human supervisor gives the

V.L. Uskov et al. (eds.), *Smart Education and e-Learning 2017*, Smart Innovation, Systems and Technologies 75, DOI 10.1007/978-3-319-59451-4_9

students timely feedback and help. Sometimes in this context students use the opportunity to conduct their experiments by their supervisors.

In virtual (simulated) environments students can experiment as many times they wish to obtain meaningful results, help is usually limited as is the 'realism' of the experimental environment, such as idealistic 'non-resistive/non-impedance' connection points.

In remote laboratories, students use real experimental equipment, with its non-ideal, real world engineering realities. The human help in the experiments is usually limited, leading to so called 'student centered learning' where the learning process and its outcomes are in the hands of the learners. It may inspire an intelligent tutoring system to be proposed later in the paper.

Many remote laboratories have already been created, allowing different experiments to be conducted in e-learning environments. Some representative examples of them are provided in Table 1 below.

Table 1. Some examples of remote laboratory environments

Laboratory	Experiment area	Summary
Telerobot (University of Western Australia) Established 1994 [1]	Robotics	An industrial robot arm was connected to a server PC, with a web interface to control it and a camera to feed back live images. This was the first remote laboratory in Australia, and one of the first in the world
iLab (MIT, USA) Established 1998 [2]	Semiconductor devices	An expensive semiconductor analyser in a research laboratory was not fully utilised. A web interface to the semiconductor analyser was created, allowing students to carry out practical work in an area that they had previously only been able to do theoretical learning in. Within 10 years of the creation of the laboratory, over 5400 students had conducted experiments with the remote laboratory
NetLab (UniSA, Australia) Established 2002 [3]	Electrical circuit theory	Provides students with access to a signal generator, passive components and basic instrumentation. The user interface is designed to closely resemble real equipment, making the transition to real equipment easier. NetLab allows students to conduct a wide range of experiments involving passive devices and can connect them together in any logical way. The system has recorded over 8000 user sessions between 2009 and 2016

(*continued*)

Table 1. (*continued*)

Laboratory	Experiment area	Summary
VISIR (BTH, Sweden) Established 1999 [4]	Electronics	As with NetLab, the VISIR electronics lab provides students with basic electronic components and test equipment. The user interface for the test equipment is very close to that used in real equipment. The system is batched in operation, allowing it to support a large number of users but taking away some control over the experiments
MEFLab (UniSA, Australia) Established 2008 [5]	Microelectronics	This system provided access to equipment that must normally be in a clean room. Students were able to measure the voltage-current curves of discrete components on the silicon wafer. A key feature of this system is its ability to precisely position measurement probes on the microelectronic circuit with an accuracy of 10 μm
Remote FPGA lab (or RFL) (National University of Ireland, Galway) Established 2011 [6]	Digital electronics	A Field Programmable Gate Array (FPGA) is a programmable component that can be used to create any digital circuit. Students can reconfigure the FPGA in any way they wish, limited only by the hardware that they have access to. A live video of the hardware is streamed to the users. To assess the impact of the RFL, the authors analysed student usage information and final student results
DIESEL (University of Ulster, Derry, Northern Ireland, UK) Established 2002 [7]	Digital electronics	The DIESEL laboratory provides a wide range of digital electronics experiments in microcontroller and FPGA programming. Novel features include an augmented reality based user interface. The system also includes a basic student assistance tool, which can assist with simpler issues but not with major misconceptions
RoboLab (University of Alicante, Spain) Established 1999 [8]	Robotics	Students learn how to manipulate a robot arm, and can interact with either a real robotic arm or a 3D simulated version of that arm. The laboratory

(*continued*)

Table 1. (*continued*)

Laboratory	Experiment area	Summary
		setup includes two different models of arms, which can be operated through the same user interface. A key feature here is the combination of the remote laboratory with a virtual laboratory – this allows students to test their understanding of the theory before they move on to experiments with real equipment
Remote laboratory for advanced motion control Experiments (UniSA, Australia) Established 2014 [9]	Advanced motion control	Allows students to implement different control algorithms in a reconfigurable hardware and software controller. Students could then test the algorithms and see how more classical existing algorithms compare with newer techniques, such as H–infinity
Distributed online remote laboratory (University of Southern Queensland, Australia Established 2014 [10]	Robotics	A remote laboratory is the peer to peer system where students create experimental setups of their own that other students can interact with. This system will automatically switch between batched and interactive modes based on the quality of the connection between the user and the experiment

3 Remote Laboratory NetLab

NetLab is a remote laboratory developed at the University of South Australia that has been used in teaching several courses including Electrical Circuit Theory, Introduction to Electrical Engineering, Electricity and Electronics, Signals and Systems, Monographic Lecture on Remote Laboratories and IT Tools in Research, since 2002, with tens of thousands of participants from various campuses of UniSA in South Australia, Singapore, Sri Lanka, as well as in Sweden and Poland. It has gone through many redevelopments and improvements based on students' evaluations in many years of its implementation by generations of Australian and foreign students from Bachelor, through Masters to PhD levels.

NetLab has been designed to resemble the students' experience in a real laboratory as closely as possible; this includes a realistic Graphical User Interface, a Circuit Builder for intuitively configuring a circuit connection and a provision for student collaboration in teams.

A remote laboratory environment of NetLab allows any participants to log on and conduct the experiments remotely on real equipment. Netlab is also a collaborative learning environment that enables concurrent users to jointly control laboratory equipment. There are very few collaborative remote facilities set-ups in the world. This is unfortunate as such an environment would allow the students to network and collaborate. NetLab creates an exciting world without borders for all willing or encouraged to become part of the framework, to become engaged with students from different locations, cultures and work habits. These generic skills are becoming increasingly important for professional engineers to become effective international team members.

The NetLab remote laboratory is considered an ideal base for testing the new remote laboratory intelligent tutoring system (ITS) proposed in this paper. NetLab has a similar architecture to other remote laboratories, meaning that tools, such as ITS, which work in NetLab should be applicable in other laboratories.

3.1 Physical Location

Physically, the laboratory is located in the Sir Charles Todd Building at the Mawson Lakes Campus of the University of South Australia (UniSA). NetLab can be accessed through the Internet at http://netlab.unisa.edu.au/ or http://netlab2.unisa.edu.au/.

3.2 Access

NetLab is an open access system, i.e. anybody with Internet availability, can use it after creating his/her own account by registering and booking time slot(s) using the booking system. This allows the UniSA onshore and offshore students and any users from any location in the world to conduct experiments in NetLab. There is a limit of 3 h per week to be used by any student, alone or in a team of up to three collaborators. Technically, the number of collaborating students is not limited, but experiences have shown that allowing more than three participants (except in research trials) makes collaboration very difficult.

3.3 Architecture

NetLab has its own dedicated server which is connected on the one side to the Internet allowing users to access the remote laboratory. On the other end, the server communicates with a number of programmable laboratory instruments via the IEEE 488.2 standard interface, also known as the General Purpose Interface Bus (GPIB). These instruments include a digital oscilloscope, a function generator and a digital multimeter. All the instruments and components are connected to a 16 × 16 Agilent programmable matrix relay switch which provides the user with an option to wire and configure various electrical circuits from the library of available components and instruments.

In Fig. 1, on the left, the server's screen is shown, followed by the instruments (including the oscilloscope which shows the identical display as the NetLab interface screen), the switching matrix, and variable components shown. All the variable components are controllable via decade interfaces (Fig. 2), very similar to real laboratory components, in this case changeable by the mouse movements.

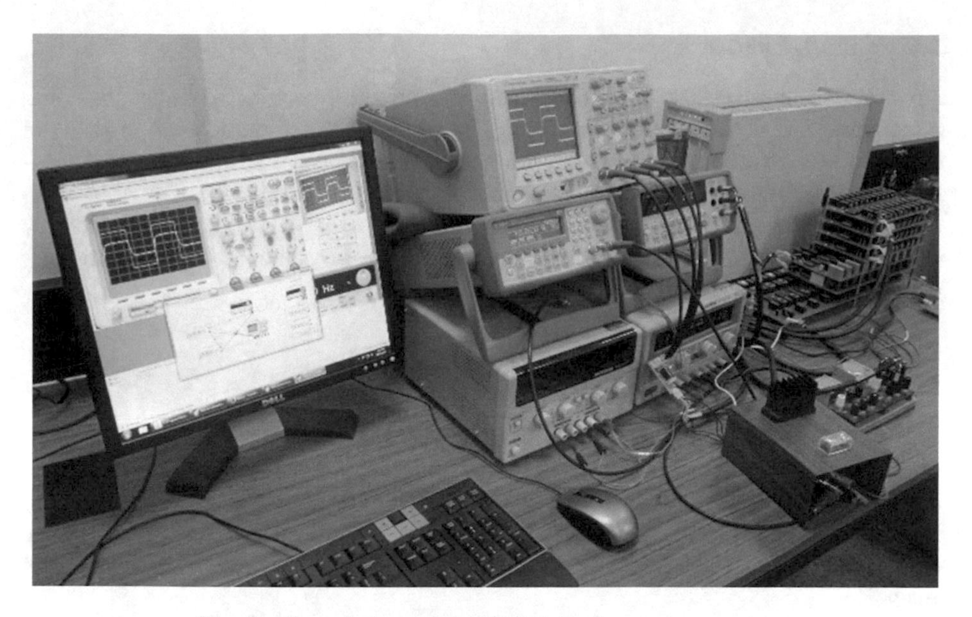

Fig. 1. General view of the NetLab hardware arrangement

Fig. 2. Interface of variable resistor

3.4 Software

The software is substantial, and consists of three major components:

1. The Web Interface, which provides access to the booking and user account management tools through a HTML-based interface.
2. The NetLab Server, which runs on the server and provides remote access to the hardware. The server ensures that only valid users with current bookings are able to access the hardware. The server is implemented in Java, with the use of some native code for low level communication. The server communicates with the web interface through a shared database.
3. The NetLab Client, which runs on the users' computers. This provides the graphical user interface with which users can control experiments.

3.5 Circuit Builder

An intuitive interface has been developed to configure electrical circuits, including instruments and variable components (Fig. 3), resembling real-world configuration, in this case of a series RC circuit, with a function generator and a digital storage oscilloscope. Required components and instruments are dragged from the library of components on the right to the configuring area on the left, then appropriately connected by clicking on terminals to be connected. Physical wiring is established by the switching matrix by clicking on the button 'Configure'.

3.6 An Exemplary Experiment

One of the existing NetLab experiments covers transients in a RC circuit (configuration shown in Fig. 3). The function generator was set to a rectangular voltage of 1 V-peak-to-peak. That voltage together with the charging and discharging voltages of the capacitor through the resistor are shown on the NetLab oscilloscope GUI (Fig. 4). Oscilloscope signals can be saved to the users' computer for further processing and analysis.

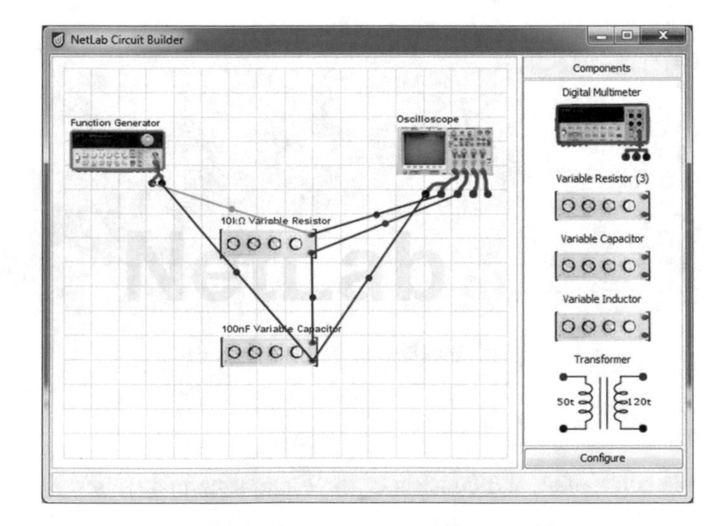

Fig. 3. NetLab circuit builder

For comparison, the camera (with its own web server), that includes pan, tilt and zoom functions and preprogrammed positions pointing to most common objects in the physical laboratory, shows voltage signals on the real oscilloscope (right-upper corner). The video feed from the camera is not a part of any experiment and can be switched off to save on the bandwidth. However, it is quite important as it provides users with the telepresence in the laboratory. It persuades some skeptical students that the experiments are on real equipment and are not simulations.

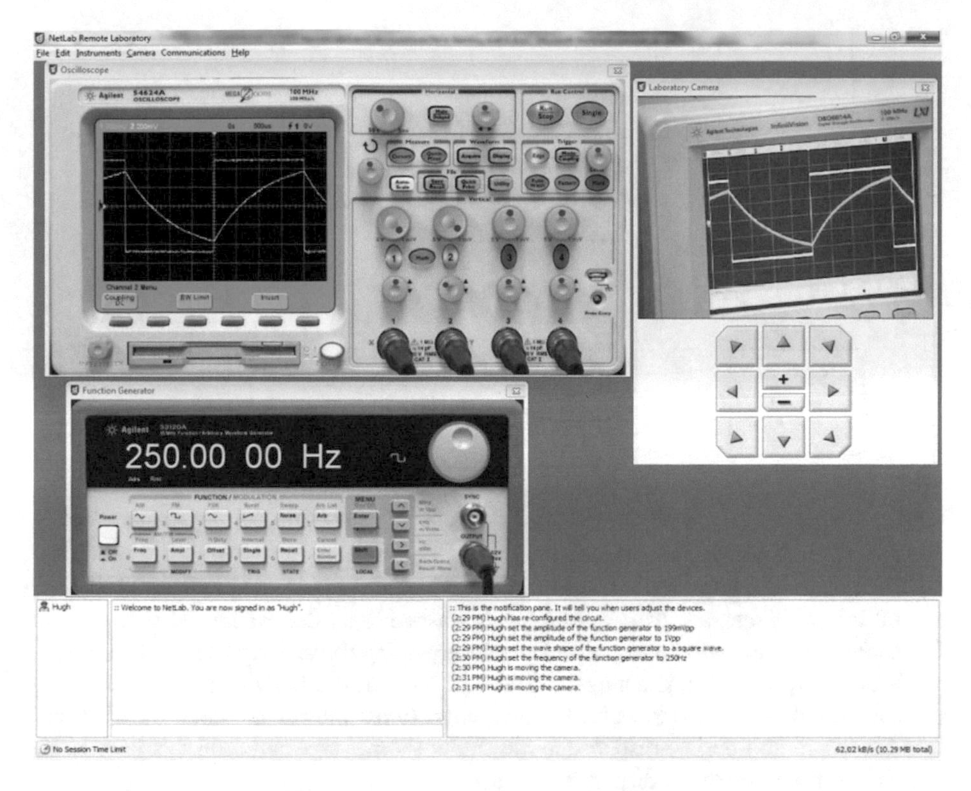

Fig. 4. GUI of RC transients experiment

The left-lower window in the NetLab GUI (Fig. 4) is used for text communication between users (video and graphical communications are also available) and shows who is logged in. The window to the right of it, shows all actions of users, changing circuit configuration and its parameters. These actions are recorded and are part of the learning analytics data to be used with the intelligent tutoring system.

4 Intelligent Tutoring System

An Intelligent Tutoring System is an intelligent software agent capable of tutoring a student. There are many differences between existing tutoring systems, but all have key features in common. All of them use some knowledge base relevant to their teaching area, and can present this knowledge to a student when necessary. The system typically has a model of what the student knows and should know, and uses this to increase the student's knowledge.

A tutoring system is currently under development for use with NetLab. The tutoring system will be integrated with the collaborative chat function in NetLab, and will be designed to replicate a human tutor as closely as possible. Communication between the intelligent tutoring system and the student will be through text and natural language processing. The overall system architecture is shown in Fig. 5.

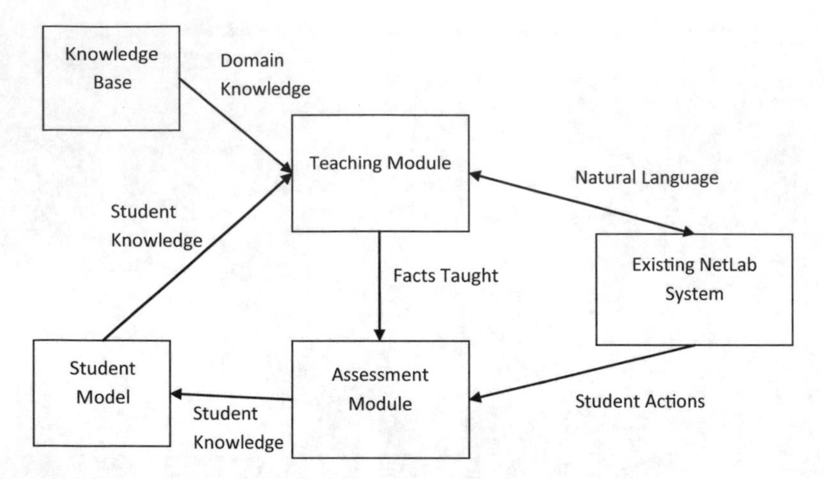

Fig. 5. High level tutoring system architecture

The core of the proposed intelligent tutoring system is the teaching module. When a natural language message from the student is passed in by the existing NetLab system, the teaching module will be responsible for interpreting the message. Once the message has been interpreted, the teaching module will look at the knowledge base (what the student should know) and the student knowledge (what the student does know). It will then decide an appropriate response to the student's message and send that back to the existing NetLab system to display to the user.

The assessment module is the other key part of the system. This part will be responsible for applying learning analytics to student actions in the NetLab system, and translating them into the student's knowledge. It will then be responsible for updating the student model to show what the student's current state of knowledge is.

The knowledge base will store all of the necessary information that the tutoring system requires for teaching. The exact structure of the knowledge base will depend on the teaching method employed.

5 Conclusions

In this paper we have presented 10 remote laboratory environments for use in e-learning strategies for various fields of engineering. Our NetLab system shares many key design features with other laboratories, but with a high level of reconfigurability and flexibility that aims to provide a similar learning experience to face-to-face and e-learners. With the addition of an intelligent tutoring system using learning analytics, we aim to develop NetLab into a smart e-learning environment that will further reduce the gap between face-to-face and e-learning.

Acknowledgement. We would like to acknowledge the Commonwealth Government of Australia funding, received under the Research Training Program (RTP), supporting the PhD project.

References

1. Taylor, K., Trevelyan, J.: Australia's telerobot on the web. In: Proceedings of the International Symposium on Industrial Robots. International Federation of Robotics & Robotic Industries, pp. 39–44 (1995)
2. Harward, V.J., del Alamo, J.A., et al.: The iLab shared architecture: a web services infrastructure to build communities of internet accessible laboratories. Proc. IEEE **96**(6), 931–950 (2008)
3. Teng, M., Considine, H., Nedic, Z., Nafalski, A.: Current and future developments in the remote laboratory NetLab. Int. J. Online Eng. **12**(8), 4–12 (2016)
4. Gustavsson, I., Zackrisson, J., et al.: The VISIR project–an open source software initiative for distributed online laboratories. In: REV 2007, pp. 1–6, Porto, Portugal (2007)
5. Mohtar, A.A.: Remote laboratory for testing microelectronic circuits on silicon wafers. Ph.D. thesis, School of Electrical and Information Engineering, University of South Australia (2009)
6. Morgan, F., Cawley, S., Newell, D.: Remote FPGA lab for enhancing learning of digital systems. ACM Trans. Reconfigurable Technol. Syst. **5**(3), 1–13 (2012)
7. Callaghan, M.J., Harkin, J., McGinnity, T.M., Maguire, L.P.: Intelligent user support in autonomous remote experimentation environments. IEEE Trans. Industr. Electron. **55**(6), 2355–2367 (2008)
8. Torres, F., Candelas, F.A., et al.: Experiences with virtual environment and remote laboratory for teaching and learning robotics at the University of Alicante. Int. J. Eng. Educ. **22**(4), 766–776 (2006)
9. Gadzhanov, S.D., Nafalski, A., Nedic, Z.: Remote laboratory for advanced motion control experiments. Int. J. Online Eng. **10**(5), 43–51 (2014)
10. Maiti, A.: Enabling peer-to-peer remote experimentation in distributed online remote laboratories. Ph.D. thesis, School of Mechanical and Electrical Engineering, University of Southern Queensland (2016)

An Automata Model for an Adaptive Course Development

Andrey V. Lyamin and Elena N. Cherepovskaya(✉)

ITMO University, Saint Petersburg, Russia
lyamin@mail.ifmo.ru, cherepovskaya@cde.ifmo.ru

Abstract. Adaptive course development is a complicated task. Application of the traditional approaches leads to the increase of laboriousness during the development and evaluation of the course. This paper presents an approach based on automata model that helps to develop adaptive courses and provide a high quality assessment of the related knowledge and skills. The presented model uses smart analysis in order to build a proper course scheme.

Keywords: e-Learning · Adaptive learning · Course development · Automata model

1 Introduction

Education is one of the most important activities in person's life as people learn new facts every day that is known as a lifelong education. Nowadays many specific tools that help to create perfect conditions for receiving necessary knowledge came into being. These tools include specific applications on smartphones and computers that allow accessing different information resources. Moreover, many universities and schools apply smart learning management systems (LMS) [1, 2] that provide an opportunity to organize learning process and develop and store various types of assessment tools, such as online tests, virtual labs [3, 4], etc. as well as demonstration videos, synopses, etc. A specification that is commonly applied for constructing an e-learning system is called Sharable Content Object Reference Model, or SCORM, [5, 6]. This set of standards states the requirements for an informational system and suggests a way to organize big educational data. During the last few years, people obtained an opportunity to access Massive Open Online Courses (MOOCs) [7] on different platforms that can be passed by a student even if he is living in a far-located place that requires much time to get to a city to pass the test in person. Proctoring systems [8] are used in such cases in order to provide a complete person identification procedure and confirm person's skills with a verified certificate. As MOOCs maintain necessary capabilities of verifying a level of student's skills, some universities are including these courses in education process.

In order to make the way of mastering learning material more convenient to a student, an adaptive learning trajectory for building personalized learning can be constructed. A learning trajectory is a set of courses that fulfills the requirements of program's discipline, i.e. a set of prerequisites and a set of learning outcomes of the

V.L. Uskov et al. (eds.), *Smart Education and e-Learning 2017*, Smart Innovation, Systems and Technologies 75, DOI 10.1007/978-3-319-59451-4_10

discipline. Each course contains various educational units, such as scripts, tests, virtual labs that are usually organized and presented to a student in a specific order. All people differ in a way they comprehend the learning material and speed of its comprehension. Hence, it is advisable to adapt learning materials to the current level of student's skills.

Many universities apply SCORM model in their learning management systems providing learning resources to students according to the selected part in the table of contents. This model specifies basic rules of constructing a course. However, a SCORM model assumes a straight relation between a content item and a resource associated with an item that does not allow tuning complexity of an item to a particular student or requires duplicating items in that case. Moreover, a resource has to include only one asset or a sharable content object. Another specific thing is that because of that an activity tree of a SCORM model is strict and strongly related to the content organization. Hence, a SCORM model applies usual programming paradigms, i.e. imperative, object-oriented, etc., that makes a process of building adaptive learning trajectories and learning modules of the system more complicated. In order to simplify the process of course programming, an automata model can be applied as it is more transparent that helps course programmers to determine and locate failures control the course easier.

In this paper, an automata model for an adaptive course development is presented. It assumes the predefined sequence of passing course's elements based on a person's state in the course. The second section presents the automata model itself with different levels of course elements it is applied to. The third section is related to the description of model's application in the real system based on the provided example. The last section contains concluding statements.

2 Method

2.1 Description of the Learning Components

An automata model is an abstract mathematical model that includes a set of states, a process can possess, a set of transitions between states, input and output automata's alphabets and functions for sets of transitions and outputs. One of the most well-known automata is Turing machine [9]. A process that can be described with an automata model is mathematically comprehensive that helps to develop its programming representation in a simplified way. A SCORM model of representing learning materials is strongly related to the content organization that leads to the complexity of implementing this model.

ITMO University had developed an information model called AcademicRM that extends SCORM specification with providing a more convenient approach of controlling learning outcomes using the automata models presented further. Learning management system AcademicNT [3, 4, 8, 10], where AcademicRM is applied, is used in several universities in Russia. AcademicNT has its own developed learning and methodology complex, which contains learning materials that are necessary to organize learning process for a discipline. This complex is based on the educational program of the related discipline that may consist of many electronic courses. A course itself includes lots of scripts that can be presented as electronic tests, virtual labs, synopses, practical works, or information resources. The scripts are used to describe interaction between an LMS and a student. The lowest level of representation are frames (or pages for synopses and information resources) contained within a script. The structure of the complex is presented in Fig. 1.

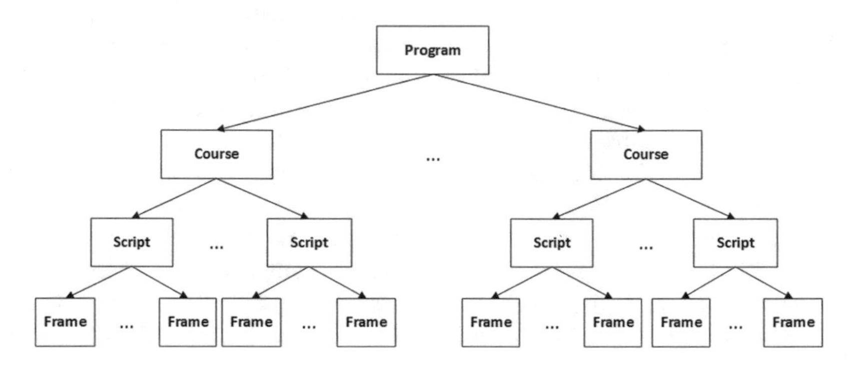

Fig. 1. The structure of the learning and methodology complex

The main components of the electronic course are metadata, which describes a name of the course, author, organization that developed the course, etc., a table of contents, a tree of course's variables and a list of possible states. The table of contents is a hierarchical structure, where each module can be presented as a container for other modules or a link to a particular script, and is used for orienting a student along the course. Course's variables are applied for computing students' rating for a course and in case of checking the conditions of changing expertise states. A set of course's variables is presented as a syntactic tree, where the variable's value can be defined either as a number or as a string. The states are defined according to the learning outcomes of a course. These states form a list of scripts that are available for a student as well as rules of changing values of course variables.

2.2 Automata Models for Different Courses

The state of the course consists of the following components:

- Events (initialization of the module, completing the script);
- Conditions (a number of script's launches, amount of spent time, points achieved during all the assignments, values of course's variables);
- Actions (a change of the value of a course variable, switching to a new state, launch of the script).

During the learning process and dealing with the course, a student develops certain competences, and based on the achieved results, a switch between states is performed. An automata model describes the way of programming these switches, having only one initial state and one ending state that indicates mastering of the related competence.

A switch of the states is available during either the module initialization or completing the script. The result of the automata model evaluation presents the updated value of student's rating.

A graphical representation of the algorithm of passing an electronic course with two assignments that should be completed in a defined order is depicted in Fig. 2. The states and assignments together with their identifiers are presented as rectangles and circles respectively. A colored circle is used, when an assignment is allowed to be opened by a student. Switches between the states are presented as arrows showing the identifier of an assignment that caused corresponding switch. The upper indexes of assignments' identifiers states for the following:

- "*" is used to describe a switch caused by entering an assignment by a student;
- "+" is used for defining a switch caused by completing assignment successfully;
- "-" is used for a switch caused by unsuccessful completion of assignment.

In Fig. 2 x_0 is student's initial state in the course, when the first assignment is available to be passed. x_1 is a state that student is switched to, when he is passing a particular assignment and opening any of the other course elements is prohibited. When student fails to complete an assignment successfully, he is switched back to the state x_0, otherwise, a student is transferred to the state x_3, when an access to the following course elements is provided. x_4 is a state that student is switched to, when he is passing the second assignment. When a student fails to complete an assignment successfully, he is switched back to the state x_3, otherwise, a student is transferred to the final state x_6 that indicates that all the necessary assignments were successfully passed. The presented course has two states x_2 and x_5 that are used to provide additional attempts for

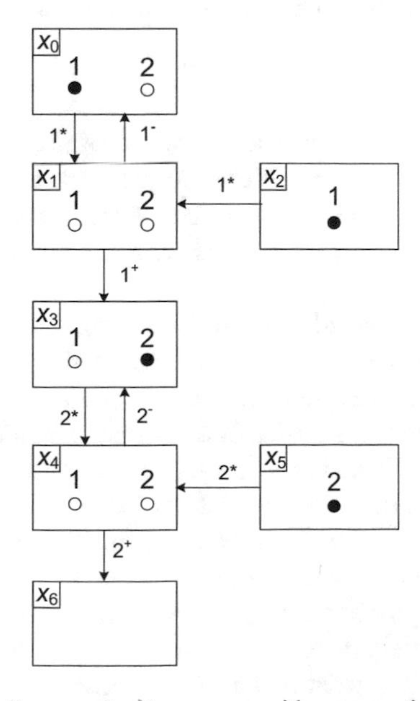

Fig. 2. State diagram for the course with consecutive assignments

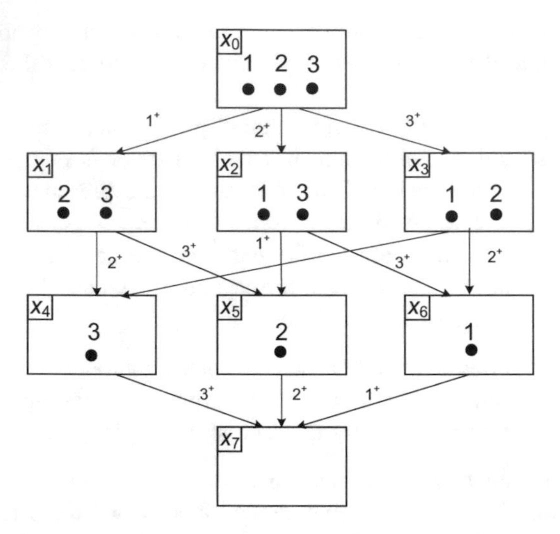

Fig. 3. State diagram for the course with an arbitrary way of passing assignments

passing the first and the second course assignments respectively. Transferring to these states are performed according to teacher's actions.

Figure 3 contains graphical representation of the algorithm of passing an electronic course with three assignments. The presented course allows passing the assignments in any order, a student wants. Its automata model does not consider states for switching student's state or providing additional assignment attempts, and, therefore, it includes only eight states that fully describe the learning process. The formula that is used to calculate a number of model's states N in this case is stated as:

$$N = 2^n, \tag{1}$$

where n is a number of assignments. The fast speed of increasing a number of model's states increases code complexity that is used to program a model. In order to simplify an automata model, it should be expanded by defining special conditions that check values of the course variable, which is used to store a number of successfully completed assignments.

An automata model that considers conditions is presented in Fig. 4. x_0 is student's initial state in the course, when all assignments are available for a student. x_1 is a state that student is switched to, when he is passing particular assignment and opening any of the other course elements is prohibited. When a student fails to complete an assignment successfully, he is switched back to the state x_0, otherwise, a condition of passing two antecedent assignments successfully is checked. The condition is stated in Fig. 4 as $T = 2$, where T is a number of assignments. If all the assignments were successfully passed, T is incremented, and a student is switched to the final state x_3, when all assignments are prohibited to pass. Another case is, when a student did not successfully complete an assignment and, hence, T is less than two. In this case, he is switched back to the state x_0, when he has a possibility to pass any other assignment,

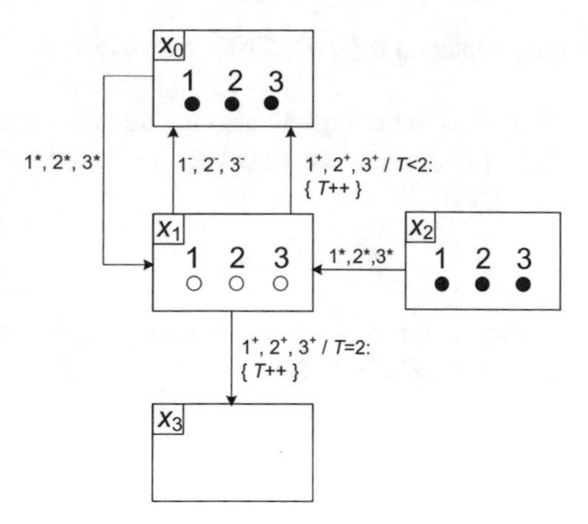

Fig. 4. State diagram for the course with an arbitrary way of passing assignments considering conditions

T is incremented, when an assignment was finished successfully. This automata model considers the state x_2 of giving an additional attempt for an assignment according to teacher's actions, when he can provide a possibility to pass an assignment even without any restrictions, e.g. a number of attempts. Hence, comparing two models presented in Figs. 3 and 4, it can be concluded that the second example maintain smaller amount of model's states. AcademicNT LMS, where an automata model is successfully applied, contains electronic courses that combines elements with a consecutive and an arbitrary ways of passing assignments.

2.3 Student's Rating

Student's rating is computed based on the weights for frames in a script that represent questions of the electronic tests, virtual labs, etc. An answer can be fully correct, when all the options are fulfilled, partially correct, indicating that only a limited number of options were completed, or incorrect. The final number of points for a script is calculated by the following formula:

$$s = \frac{\sum_{i=1}^{n} r_i w_i}{\sum_{i=1}^{n} w_i} f(t) \tag{2}$$

where $w_i \in [0, \propto)$ is frame's weight, r_i states for the rate of the passing the frame correctly and is calculated as $r_i = \frac{N^+}{N_\Sigma^+}$, where N^+ is a number of the fulfilled or chosen correct options and N_Σ^+ is a whole number of correct options. $f(t) = \sum_{j=1}^{m} h(t - t_j)k_j$ is a function that lies within an interval $0 \leq f(t) \leq 100$, $t \in [0, \infty)$, when $h(t)$ is a Heaviside step function.

The list of things that affect the rating can also include the amount of time spent on the assignments, as well as number of attempts used to complete the task. Hence, the formula is stated as following:

$$R = k \cdot S \cdot g(l), \tag{3}$$

where k is a proportionality factor, S is a rating for a script, $0 \leq g(l) \leq 1$ is a function with an argument of a number of spent attempts l.

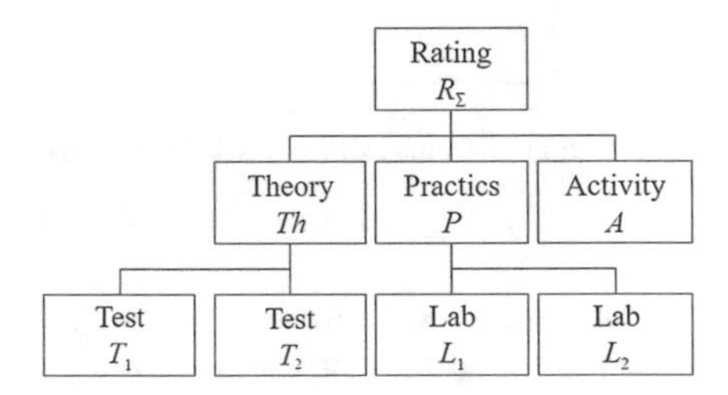

Fig. 5. Syntactic tree of course variables

Figure 5 presents an example of a syntactic tree of the variables for a particular course. The overall rating for this course according to the variables is computed by the following formula:

$$R_\Sigma = A + R_{T_1} + R_{T_2} + R_{L_1} + R_{L_2} \tag{4}$$

3 Evaluation

The presented automata model is successfully applied in AcademicNT LMS. An evaluation of the model is presented as a Stateflow diagram constructed in MATLAB's Simulink module. The Stateflow environment provides an opportunity to create examples of finite automata models in a comprehensive way. Hence, this module was chosen in order to show the logic of an electronic course similar to the described above. A model for the course that includes three assignments is presented in Fig. 6. The keys control the evaluation of the Stateflow diagram, input values include the results for three assignments (*Result_test1*, *Result_test2* and *Result_test3*), amount of time (*Time*) for each of the assignments and number of attempts (*Tries*). The main output produced by a model is rating (*Rating*) together with system's messages demonstrating the availability of the course element (*Message*). If a message shows "0", it indicates that an element is prohibited to be accessed, "1" states that element is allowed to be dealt with.

The Stateflow diagram of the described course is presented in Fig. 7. The states here are used to define stages of passing the assignments. Transitions between states are

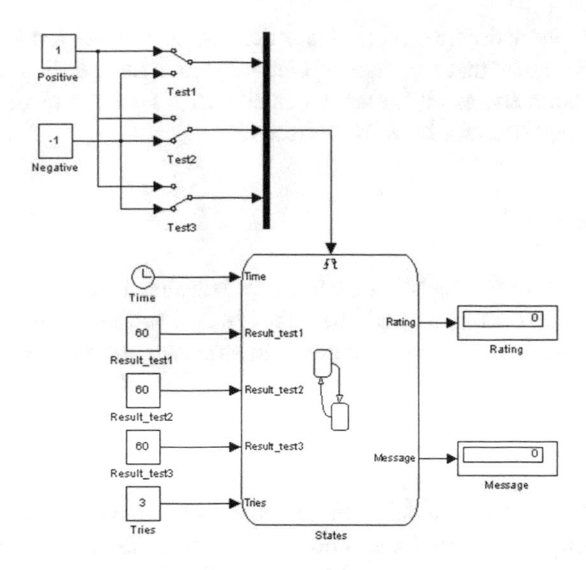

Fig. 6. Model of the course

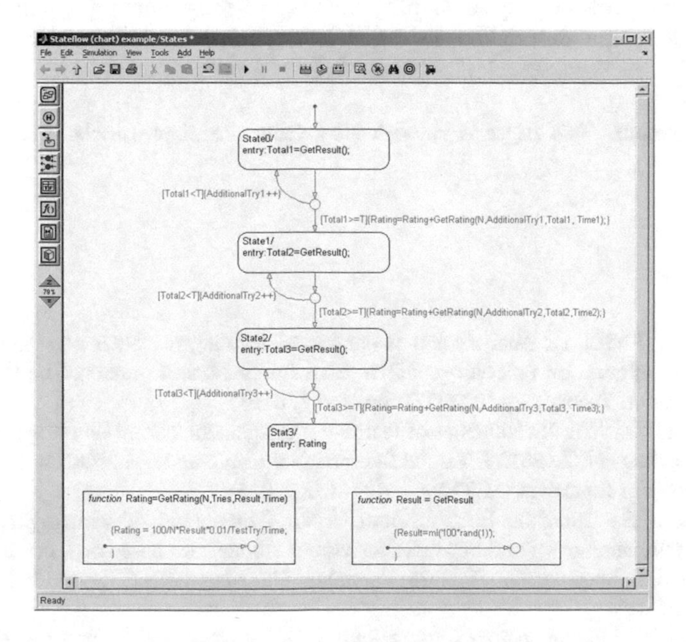

Fig. 7. State flow diagram for the course

displayed as arrows and define the conditions of transferring from one state to the other. These conditions comprise a number of attempts spent on passing an assignment, time, and a condition to check a number of points being appropriate to consider an assignment completed successfully. Moreover, functions that are used to compute rating within a state and between states are defined. Automata model and the related Stateflow diagram show that a course became easy to control during its evaluation for the course programmers, as all the complicated actions related to the course maintenance are simplified and can be easily fulfilled.

4 Conclusion

As people understand the meaning of lifelong learning in a deep way, it becomes necessary to develop courses that will be effective and will be effortless to maintain by the course programmers. This paper presents an automata model that helps to construct courses and is easy to be programmed in order to be included in learning management systems. The described approach defines a model that assumes different states of the course and its variables. The model reveals the most important conditions affecting a level of student's competences and rating. An assessment of the states' number is based on the rules of accessing modules of an electronic course. The presented model assumes the minimal number of states that simplifies creation and maintenance of the course.

The provided automata model can become the basis for the development of new algorithms and tools helping to simplify the way of programming a course and modifying its structural description in order to decrease laboriousness of developing courses.

Acknowledgement. This paper is supported by Government of Russian Federation (grant 074-U01).

References

1. Saroha, K., Mehta, P.: Analysis and evaluation of learning management system using data mining techniques. In: Proceedings of the 2016 International Conference on Recent Trends in Information Technology (ICRTIT), pp. 1–5 (2016)
2. Venugopal, G., Jain, R.: Influence of learning management system on student engagement. In: Proceedings of the 2015 IEEE 3rd International Conference on MOOCs, Innovation and Technology in Education (MITE), pp. 427–432 (2015)
3. Chezhin, M.S., Efimchik, E.A., Lyamin, A.V.: Automation of variant preparation and solving estimation of algorithmic tasks for virtual laboratories based on automata model. In: Vincenti, G., Bucciero, A., Vaz de Carvalho, C. (eds.) LNICST, vol. 160, pp. 35–43. Springer (2016)
4. Efimchik, E., Lyamin, A.: RLCP-compatible virtual laboratories. In: Proceedings of the International Conference on E-Learning and E-Technologies in Education (ICEEE 2012), pp. 59–64 (2012)
5. Advanced Distributed Learning (ADL), Sharable Content Object Reference Model (SCORM®) 2004, 4th Edition Content Aggregation Model (CAM) Version 1.1 (2009)
6. Advanced Distributed Learning (ADL), Sharable Content Object Reference Model (SCORM®) 2004 4th Edition Sequencing and Navigation (SN) Version 1.1 (2009)

7. Sammour, G., Al-Zoubi, A., Gladun, A., Khala, K., Schreurs, J.: MOOCs in universities: intelligent model for delivering online learning content. In: Proceedings of the 2015 IEEE Seventh International Conference on Intelligent Computing and Information Systems (ICICIS), pp. 167–171 (2015)
8. Belashenkova, N.N., Cherepovskaya, E.N., Lyamin, A.V., Skshidlevsky, A.A.: Protection methods of assessment procedures used in e-Learning. In: Proceedings of the 13th International Conference on Emerging e-Learning Technologies and Applications (ICETA), pp. 27–32 (2015)
9. Turing, A.M.: On computable numbers, with an application to the entscheidungs problem. Proc. Lond. Math. Soc. **2**(42), 230–265 (1937)
10. Uskov, V., Lyamin, A., Lisitsyna, L., Sekar, B.: Smart e-Learning as a Student-Centered Biotechnical System. In: Vincenti, G., Bucciero, A., Vaz de Carvalho, C. (eds.) LNICST, vol. 138, pp. 167–175. Springer (2014)

Algorithm of Contextual Information Formation for Smart Learning Object

Asta Slotkienė(✉)

Department of Information Technology, Siauliai University, Šiauliai, Lithuania
asta.slotkiene@su.lt

Abstract. This paper presents a contextual information formation algorithm for smart learning object, which enables to comprise learning content, process of active learning activities and the essential aspects of the problematic learning situations. It presents how to get the specification of a smart learning object by using context of learning content and learning activities. This paper also analyzes the characteristic of proposed algorithm by comparing two smart learning objects.

Keywords: Smart learning object · Active learning · Contextual information · Contextual modeling · Learning situation

1 Introduction

The educational content and its representation are ensured not only by the available hardware and software but also by the creation of purposefully selected e-learning situations and their application, enhancing an active participation of the learner. Active learning in e–learning environments is implemented by applying a variety of active learning methods: virtual experiments, assessment tests, simulation–based learning, practical tasks, interactive presentations, discussions, etc. Researchers M.D. Merril, O. Sheel and J. Magenheim [7, 8] who analyzed active learning implementation methods in learning objects noticed that they do not provide solutions for describing active learning situations that would involve the majority of features of the learning process, e.g. difficulty, flexibility, goals, learner's activity level and the scope of knowledge and concepts required for task solving. Jovanovic et al. [5] suggests solving the issues with LO by presenting the context as a unique set of interrelated data that describes the learning situations. On these grounds, the article suggests an algorithm of contextual information selection and formation, before implementing the learning situation in a smart learning object (SLO). Also in the article, we explain how we use contextual information of learning content in order to get a smart learning object specification. We also present experimental results of the algorithm application - comparison characteristics of two smart learning objects.

V.L. Uskov et al. (eds.), *Smart Education and e-Learning 2017*, Smart Innovation, Systems and Technologies 75, DOI 10.1007/978-3-319-59451-4_11

2 Smart Learning Objects for Active E-Learning and Their Characteristics

According to J. Bang and C. Dalsgaard [1], "learning objects become useful in the learning process only when the learner applies them usefully. The essential thing is the creation of respective learning activity". LO is normally understood as a set of smallest undividable objects of learning content designed for reaching a specific learning goal. Nowadays, this concept neglects the evolution of technology. New technology, education and context awareness brings forward a new conception of LO: active LO, generative LO, smart LO, intelligent LO. We observe the rise of smart LO, that is incorporated within the learning context together with the specification to support flexibility, reusability, and interoperability [10]. However, such adaptation of LO perception together with the concept of a smart learning object contradicts the fundamentals of active learning: active learning is based on the learner's activity when the learning content is perceived by interactive meaningful actions which convey and develop knowledge [9]. This system includes interactive actions such as the change of object position, selection of input values to stimulate events, structure formation of the given objects, etc. During its application, a smart learning object performs the following functions: stimulates the application of the experience; conveys the learning material through problematic situations; encourages the performance of interactive actions and assessment of reasonability of problematic situation solution. According to Jonassen and Štuikys, SLO is formed by taking into consideration the following statements [4, 10]

- SLO enables implementation of generative aspects though the explicit representation of learning variability
- SLO is not clear granulated information, i.e. information units that comprise it are interrelated and interlinked semantically.
- SLO is not merely the object of practice or evaluation. The ability to apply knowledge cannot be elaborately compound of an exact sequence of actions; it must reflect substantial quality changes in thinking.
- SLO as a unit is not a reusable object in different e-learning environments; it rather provides a possibility of reusability on the level of action or in the case of the change in learning circumstances.

3 Formation Algorithm of Contextual Information

The importance of context has been underscored by numerous researchers. Decontextualized courses lead to inert knowledge, i.e. the knowledge that can be reproduced, but not applied to real-world problem solving while the context information can provide learning authenticity and immediacy of purpose [3, 11]. There are two general approaches for contextual information used in the learning content. The first approach attempts to use characteristics of learning content to the requirements and needs of individual learners. It not only describes the learning content based on learner's primary bits of knowledge, but it also considers only static elements of the context, which do not allow a deeper adaptation of learning processes. The second approach focuses on delivering the most appropriate effective sequence of learning content (learning situation and relationships between them). It enables a wide contextualization of LO on the learning situations and learner actions with them.

The proposed contextual information formation algorithm is comprised of two main stages: analysis of contextual data and description of the contextual information of one learning topic and formation of learning situations.

The main and the most important stage is the analysis of contextual data of learning content for SLO. It creates the smallest part of learning content, that we identify as a contextual element. For this purpose, scholars [2, 10] propose SLO information models to be used for linking learning components comprising of SLO with semantic relations thus creating applicable scenarios of e–learning and for providing learning objects according to the sequence determined beforehand or according to the sequence selected by the system of individual abilities. However, their essential noticeable limitation is that they do not describe the semantic structure of the SLO reusability part. This affects the flexibility of SLO usage and its reusability by author's modification of the smart learning object, e.g. by adjusting it to other learning goals [6, 9]

Since during the application of a smart learning object the learning material is conveyed to the learner by making different problematic situations in the learning task, in order to encourage the learner to make different interactive actions, we suggest using contextual modeling and its concepts for SLO analysis [9, 10]. The main concepts of contextual modeling applied in the method are the entity, contextual element describing the entity, action and a rule linking of the aforesaid elements together; the designation of the mentioned rule is to describe the problematic situation and learner's actions stimulated by it. The essential stage in the creation of problematic situations from the learning material is filling in the rule and formation of its interdependence correlations; therefore we will explain its structure in more detail and present it in Backus-Naur form (BNF).

```
<rule>::=if<logical-phenomenon>then<activity>
    [else<activity>]
<logical-phenomenon>::=<entity>"."<contextual-element>
(<ratio-operation-symbol>)<contextual-element -value>
<ratio-operation-symbol>::=
<equal>|<notequal>|<more>|<less>|
<more-or-qual>|<less-or-equal>
<activity>::=[<action>{,<action>}]|<new-rule>|<partial-
goal>
```

Since active learning is based on the learner's activity which is stimulated by a variety of circumstances indirectly conveyed in the problem-based learning situation and actions they evoke, we present a rule filling algorithm in Fig. 1 [10]. One rule is designed for the description of one problem-based situation. It is possible that several situations are provided for the solution of SLO task; in such case the rules are interrelated.

By using a created rule filling algorithm, an expert on the subject must perform the following three major steps [10]

1. Entering the logical phenomenon. The expert uses it for specifying the problem-based situation which will be evaluated by the learner in the course of the learning process by applying laws of the already known learning material or earlier experience in the subject field.
2. Selection of activity. According to the verification of the result of logical phenomenon (i.e. the value of a logical phenomenon is the truth) interactive actions to be performed by the learner are selected. Activity for the situation where the value of a logical phenomenon is not the truth is selected in an anologous way. According to the actions performed in the task provided by the learner, the expert of the subject may assess whether the problem-based situation was identified and solution methods were chosen correctly
3. Identification of correlations of problem-based situations. An expert on the subject, after forming the rule, selects one of the following three cases:

   – The problem-based learning situation is completed and the task is completed, i.e. all problems intended for the learner's evaluation have already been described. After completing it the learner proceeds to another SLO task.
   – A problem-based learning situation needs another expanding situation, i.e. for the task to be completely solved, additional evaluation and solutions of the circumstances are required.

A problem-based learning situation is completed, but for the task to be solved, evaluation of other learning situations is required. These problem-based situations make the whole of one task; however, the solution of one situation does not affect the solution of another situation.

After the expert of the subject fills in the rules with problem-based situations expressing the learning material, the move is made to another stage of SLO formation algorithm, i.e. to the formation of SLO specification. In this stage, SLO specification is obtained, which considers the following characteristics of the learning process:

The difficulty of the learning material conveyed through SLO, i.e. the number of problem-based learning situations necessary to be solved by the learner in the task during learning. The scope of the learning content which comprises the SLO. Subject to the knowledge and skills of the learners, the same SLO shall possess the possibility to cover the different amount of the learning topic content. Goals of SLO application. Upon the change of learning achievements imposed for the learning topic, SLO often needs to be changed in order to achieve the necessary intended goals.

A problem-based learning situation is completed, but for the task to be solved, evaluation of other learning situations is required. These problem-based situations make the whole of one task; however the solution of one situation does not affect the solution of another situation.

After the expert of the subject fills in the rules with problem-based situations expressing the learning material, the move is made to another stage of SLO formation

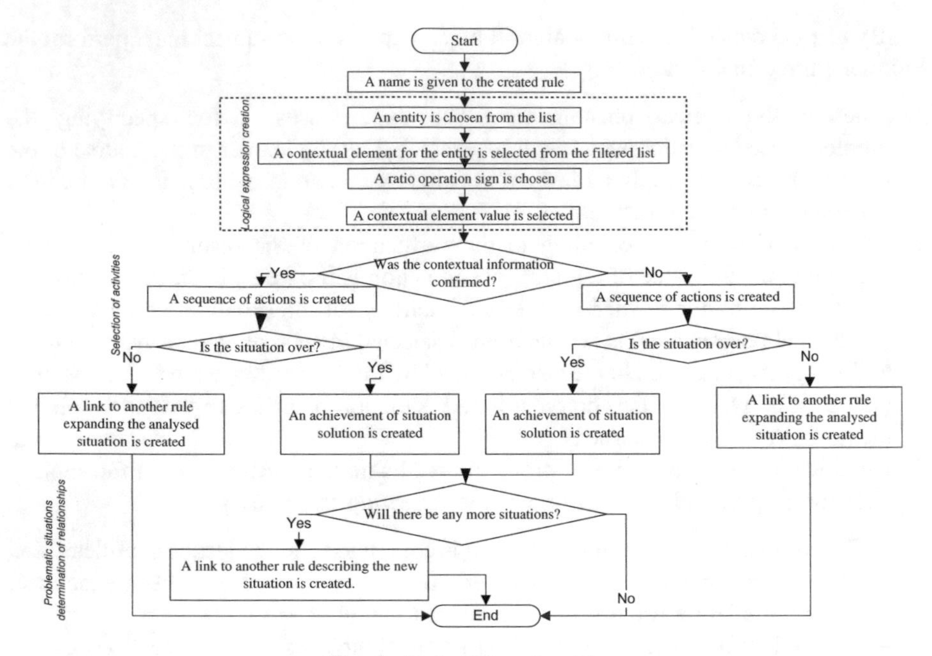

Fig. 1. Rule filling algorithm

algorithm, i.e. to the formation of SLO specification. On this stage, SLO specification is obtained, which considers the following characteristics of the learning process:

Difficulty of the learning material conveyed through SLO, i.e. the number of problem-based learning situations necessary to be solved by the learner in the task during learning. Scope of the learning content which comprises the SLO. Subject to the knowledge and skills of the learners, the same SLO shall possess the possibility to cover the different amount of the learning topic content. Goals of SLO application. Upon the change of learning achievements imposed for the learning topic, SLO often needs to be changed in order to achieve the necessary intended goals.

Specification of smart learning object consists (see Fig. 2) of the description of problem-based learning situations of every task and the list of active actions. The learning situation in the SLO is described by contextual information, i.e. the units of specific entities, contextual elements and their values.

An expert on the subject creates the complete SLO specification and may create several incomplete SLO specifications in case of the need to cover only part of one learning topic content according to the learning competence or demands of the learners' group. When having an SLO specification, an author of SLO (teaching expert, lecturer) according to it implements the smart learning object in the chosen software and applies the developed SLO in a variety of contexts of the application by providing possibilities of flexible learning.

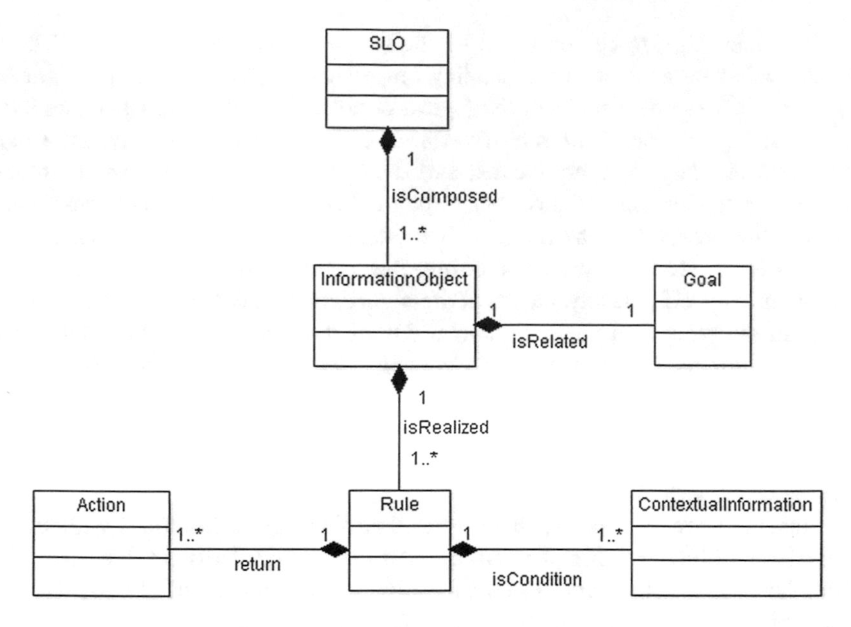

Fig. 2. Structure of SLO specification

4 Experimental of SLO Formation Using Contextual Information

For the formation of the SLO specification, we have implemented the software prototype of the proposed algorithm, upon the application whereof the expert of the subject obtains SLO specification of the learning topic in XML format. As it was already mentioned earlier, the essential stage is the designing of problem-based learning situations, that are implemented in the algorithm by filling-in the created rule. Here with we will provide the example of the rule of the proposed algorithm by showing "Algorithm of removing the element from the B–tree" of the learning subject.

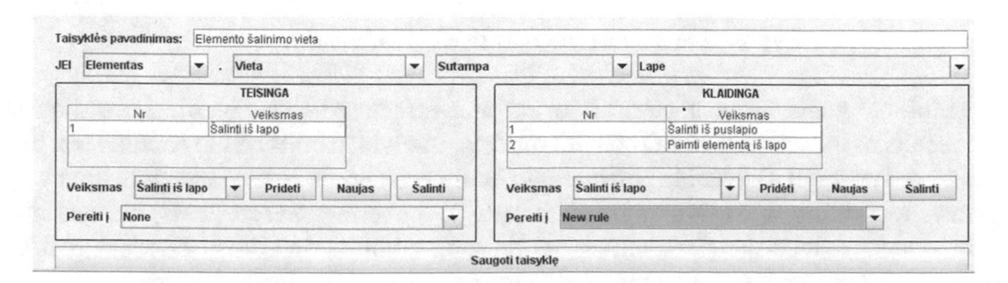

Fig. 3. Windows of filling in the algorithm software prototype rule

As it is seen in Fig. 3, the rule is started by creating the contextual information (this is: entity is *Element* and contextual element is *place,* contextual information is *Element.place* = *Leaf*) and by selecting active actions (*remove form leaf*) to each logical result (true and false) to be performed by the learner when solving the task. The rule may end or initiate another rule depending on the action (see Fig. 3 in the right New rule), as in the example provided, the logical condition is false when two actions are executed and go to the *Filter* rule. By forming the rules in such way, information objects of a learning topic are created and they are comprised of several situations where each situation has its own cause (contextual information) and consequences (action). After creating all the rules, the expert obtains 9 information objects for the mentioned topic which are presented during the learning process.

In order to specify the advantages of the algorithm application, we have conducted a study. In the process of the study two SLOs of the algorithm course with learning content (algorithms for *Algorithm of removing the element from the B–tree*) were created. Their major difference was that the first SLO was created without formation algorithm, the other was created by using the creation stages of the developed SLO's contextual information generation algorithm and its learning components were formed according to the specifications of SLO.

In the study, we hypothesize that a smart learning object developed without these proposed algorithms is characterized by stronger dissimilarity of the interrelated learning content, which is subject to the intuitive implementation of SLO by the expert of the subject.

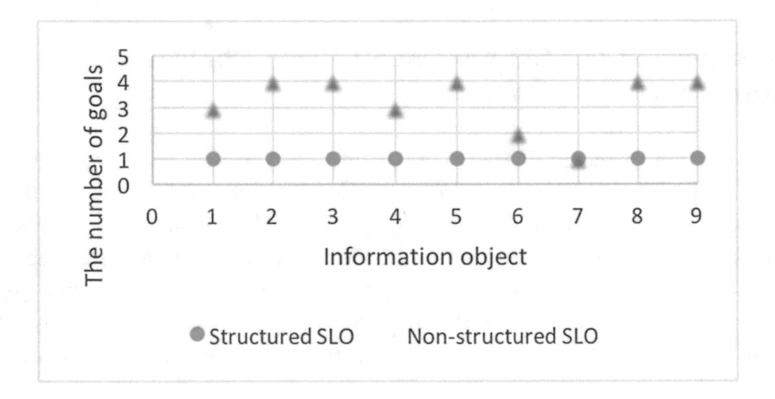

Fig. 4. The number of goals in SLO information model

Multiple goals are characteristic to the non–structured (see Fig. 4). According to the results of the analysis of LO content models, undivided units of LO should seek one goal, i.e. that an SLO learning component should seek no more than one goal, that way greater reusability of SLO may be achieved. By covering several goals, the learning components reduce the possibilities of their reusability, since they become part of several learning topics. An aggregated learning object is designed for this.

During the study, a similarity matrix of active actions (e.g. entering/selection of sum storage sentence, entering/selection of logical phenomenon, employment of data entry block etc.) existing in both above-mentioned tasks of smart learning objects related to the learning content have been developed. In order to study this relationship, we have applied the method of correlation analysis, i.e. the object dissimilarity and similarity score. It enables one to quantifiably define the dependence or independence degree of SLO tasks.

Suppose learning tasks defining X_I may carry p characteristics which describe the validity or invalidity of the reusability of active actions. When an active action related to the learning content is applied, then $x_i^{(s)} = 1$; when it is not applied, $x_i^{(s)} = 0$. Then the object X_i is described by a vector:

$$X_i = (x_i^{(1)}, x_i^{(2)}, \ldots, x_i^{(p)}), \text{here}\, x_i^{(s)} \in \{0, 1\} \tag{1}$$

For the creation of similarity scores of two learning SLO tasks, frequencies of $(x_i^{(s)}, x_j^{(s)})$ values specifying the similarity or dissimilarity of a feature are used. Since the aim of the learning component assessment is to show their dissimilarity within the limits of SLO, the absence of the relevant feature, i.e. the frequencies of the following pairs is (0; 0), (0; 1), (1; 0). We suggest evaluating this by applying the Lance-Williams dissimilarity score.

$$r(X_i, X_j) = \frac{v_{ij}^{(01)} + v_{ij}^{(10)}}{(2v_{ij}^{00} + v_{ij}^{01} + v_{ij}^{10})} \tag{2}$$

The obtained results of dissimilarity score of both SLO tasks have shown that creation of SLO with the use of SLO specification where its semantic structure is designed enables us to achieve greater inter-similarity of the realized learning components with respect to the active actions.

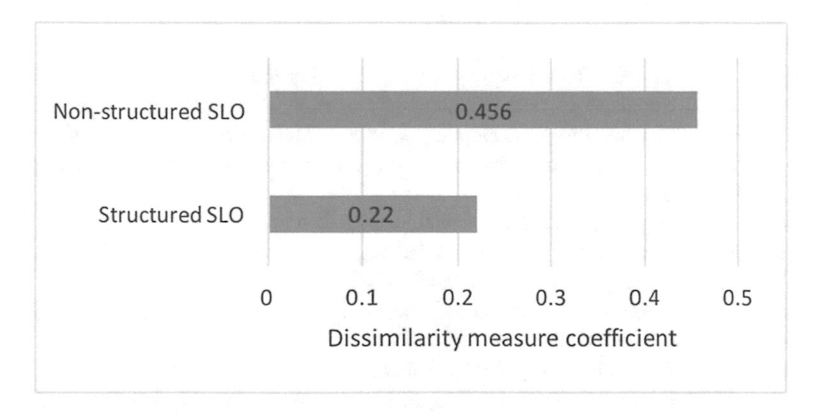

Fig. 5. Comparison of average values of SLO dissimilarity criteria

The diagram (see Fig. 5) shows that the average value of dissimilarity criterion of interactive actions related to the learning content of the designed SLO tasks is no less than that of non-structured SLO (it is traditional SLO, which developed without designing). This demonstrates that it would be complicated to apply the latter SLO for flexible learning, i.e. to adjust the SLO to the changed characteristics of the learning process, since interrelation of its tasks in the learning content is weaker.

Summarizing the results of the study we may state that the study hypothesis was confirmed. If an expert of the subject develops an SLO without a prior analysis and description of its content, tasks of such implemented SLO are two times weaker in regard to the learning content than those of the designed SLO.

5 Conclusions

A thorough analysis of the subject field prior to the implementation of the object or software under development is an important stage of the engineering process determining the results of successive stages. The conducted study proves that this is true in regard to SLOs as well. When an expert of the subject, first of all, fills information to an SLO, i.e. after a thorough analysis of the learning topic content prepares an SLO specification, he then creates SLO tasks which are more similar to each other in the learning content. This initial process of SLO development enables us to increase the reusability of SLOs, since during the preparation of specification an expert of the subject takes into consideration different values of the learning process characteristics (difficulty, the scope of learning material, learning goals).

Thus the time spent on the implementation of single learning tasks (SLO developed without formation of contextual information), characterized by large variability of the learning content, may be used for SLO formating and for the developing of flexible SLO from the obtained specification.

References

1. Bang, J., Dalsgaard, C.: Rethinking e–learning. Shifting the focus to learning activities, In: O Murchú, D., Sorensen, E. (eds.) Enhancing Learning Through Technology. Information Science Publishing, pp. 184–202. Hershey, London, Melbourne, Singapore (2006)
2. Bouzeghoub, A., et al.: A knowledge–based approach to describe and adapt learning objects. Int. J. E-Learn. **5**(1), 95–102 (2006)
3. Burbaitė, R., Bespalova, K., Damaševičius, R., Štuikys, V.: Context-aware generative learning objects for teaching computer science. Int. J. Eng. Educ. **30**(4), 929–936 (2014). TEMPUS Publications, Dublin. ISSN 0949-149X
4. Jonassen, D., Churchill, D.: Is there a learning orientation in learning objects? Int. J. E-Learn. **3**(2), 32–41 (2003)
5. Jovanovich, J.: Generating context-related feedback for teachers. Int. J. Technol. Enhanced Learn. (IJTEL) **1**(1/2), 47–69 (2008). Lin, F. (Oscar): Scenario analysis using Petri nets. In: Wang, P.P., Shih, T.K. (eds.) Intelligent Virtual World: Technologies and Applications in Distributed Virtual Environments, pp. 213–224. World Scientific Publishing Co. Inc. (2003)
6. Ignatova, V., Dagienė, V., Kubilinskienė, S.: ICT based learning personalization affordance in the context of implementation of constructionist learning activities. Inform. Educ. Int. J. **14**(1), 51–65 (2015)
7. Magenheim, J., Scheel, O.: Integrating learning objects into an open learning environment: evaluation of learning processes in an informatics learning lab. In: Proceedings of the 13th International World Wide Web Conference on Alternate Track Papers & Posters, pp. 450–451 (2004)

8. Merrill, M.D.: Knowledge Objects and Mental–Models. The Instructional Use of Learning Objects: Online Version, pp. 261–280 (2002). http://id2.usu.edu/Papers/KOMM.PDF
9. Slotkiene, A., Baniulis, K., Paulikas, G.: Desining reusable active learning object by using its information model. Jaunųjų mokslininkų darbai. **3**(24), 107–113 (2009). ISSN 1648-8776
10. Slotkiene, A.: Design Methods of Active Learning Object and Research on this Method. Dissertation. Technologija, KTU (2009)
11. Stuikys, V.: Smart Learning Objects for Smart Education in Computer Science: Theory, Methodology and Robot-Based Implementation. Springer, Heidelberg (2015)

Information Technologies in Musical and Art Education of Children

Nataliya G. Tagiltseva[1(✉)], Svetlana A. Konovalova[1], Nataliya I. Kashina[1], Elvira M. Valeeva[2], Oksana A. Ovsyannikova[3], and Sergey I. Mokrousov[3]

[1] Ural State Pedagogical University, Yekaterinburg, Russia
{musis52nt, konovsvetlana, koranata}@mail.ru
[2] Southern Ural State University, Chelyabinsk, Russia
valelya@mail.ru
[3] Tyumen State University, Tyumen, Russia
sergeiovsiannikov@yandex.ru, yory67@mail.ru

Abstract. The article discusses the ways of Smart-e-Learning usage in the system of education of children in institutions of musical and art education. In relying on the ideas of contemporary Russian [5, 8, 9] authors determined how to incorporate into the learning process of children's musical and art schools such modern forms of learning such as e-tutorial, presentation, individual sites, multimedia programs for music, music recording, the tempo changes, etc. Based on the test scores of students in the subject of "Musical Literature", "Pop vocal" survey of teachers of children's musical school, as well as the results of the competitive activities of students proved that these areas are effective in teaching children of teenage age.

Keywords: Musical education · Art education · Smart-e-Learning · Information technology · Children's musical schools · Art schools

1 Introduction

At the present stage of development of education in Russia and abroad, there has been a widespread use of new opportunities associated with the development of science and technology, current approaches to learning content, innovative technological developments.

At present time in Russia a major reform of art education is carried out, during which problems are solved by improving the quality and accessibility for every child, update its content in accordance with the interests of children, families and the needs of society, which is impossible without modern information technologies.

They are tools that contribute to the popularization of the system of art and musical education of children and their involvement in the process of creating of art products and the perception of art, creative self-realization in different types of musical activities. In this context, issues of development of methods of introduction of modern Smart-e-Learning in the content of the institutions of art and musical education are extremely important.

V.L. Uskov et al. (eds.), *Smart Education and e-Learning 2017*, Smart Innovation, Systems and Technologies 75, DOI 10.1007/978-3-319-59451-4_12

2 Theoretical Foundation

Information technologies today, by definition of many authors [1, 4, 6] are the set of techniques and methods for producing, processing and reporting, aimed at changing the status information, properties, form, content, which are carried out in the interests of users. Modern technologies are used today in all spheres of public life, including pedagogy.

An analysis of numerous works on the introduction of information technologies in teaching science, namely in this industry it is didactic, and especially art and musical didactics, leads to the conclusion of the great pedagogical potential of information technologies. It is defined by: a complex impact on the perception of the user, including his emotional sphere (through the synthesis of visual, auditory and motor images in a single object of communication); creating the illusion of three-dimensional material objects, an illusion of movement; Creating a view by using visual objects, the existence of which is impossible in objective reality, the last renovation of the world, designing the future of the world (virtual simulation) [5]; create a sense of presence of the user at the presentation of artistic events (exhibitions, concerts, performances), in different buildings of the past - the Gothic cathedrals, Orthodox churches, palaces Petrine era, etc., in order to simulate that environment, for which the music was originally created to intensify associative thinking of the child. The result of this is the development of a child's ability to a holistic and artistic perception that determines the level of his musical and art culture.

Smart-e-Learning, IT applied with success in the musical education of children. Analysis of studies of various authors [5, 8, 10, 11] leads to the conclusion that these technologies are used to: organize the educational process; create training manuals; learn a new musical material; monitor students' knowledge of music by incorporating computer-based testing; obtain information from the Internet, and further work with this information; create of a site of children's musical school, which allows you to link together all the subjects of the educational process: students, parents, teachers and inform them about the extraordinary artistic events of the school [12]. It is also a great potential for IT to be used for training of pop singers. So, you can use: programs designed for listening to music, music recordings; applications for smartphones (virtual vocal processor), giving the opportunity to rehearse and prepare for performance, while allowing one to store and multi-channel recording; virtual piano keyboard to perform songs in which the singer begins the execution before he or she will hear the soundtrack accompaniment in a situation of absence in the immediate vicinity of the instrument; Programs that change the tempo without affecting on other characteristics – tone quality, tonality, etc., programs that reproduce the metronome function.

3 Experiments

This article describes those IT directions in the implementation of musical education of students who were tested in the children's musical school with a separate group of students on the subject of "Musical Literature". The same article revealed the possibility of using multi-media programs in the learning process by pop singers of musical and computer technologies.

Consideration of the first direction is associated with a specific discipline, which is available in the curriculum of musical school, including IT, contributing to the formation of knowledge and skills of students, expand their musical experience. It's - "Music Literature", which is taught in the children's musical school from the fourth to the graduating class, in the course of development which students get acquainted with the best samples of world music, the main trends in the development of musical culture of different eras, biographies and works of a number of domestic and foreign composers.

The primary mean of training on the given discipline is sounding music that can be performed by the teacher, as well as presented in audio and video recordings, when a piece of music performed by great musicians, conductors, vocalists.

But IT can be claimed on the discipline not only to listen to music, and for subsequent analysis, but for the realization of the work the student with musical scores studied music since the simultaneous perception of the musical text and immediately the sound of a musical work gives you the opportunity to study music to the subject (the recipient) holistic coverage of musical form and artistic content. Such a simultaneous hearing and "monitor" child musical text "in cycles," perhaps because of IT, when sound is complemented by musical notation appearing on the computer screen at the same time with the music of a particular product.

On the subject of "Music Literature", students of art schools and musical schools get acquainted not only with musical masterpieces, but also with other forms of art, such as painting, choreography, film, theater, etc. Simultaneous exposure to the arts contributes to the listener's emotional response to bright artwork. Such polyart approach [7, 8] is impossible without modern IT, when a student in the children's musical school while listening to music can find himself or herself in the halls of art galleries, exhibitions, concert hall, hall of the Opera House, etc., which not only promotes comprehension the emotional tone of the work, but also thinking about the content of the music listener, conflicts musical form, the means of artistic expression and, most importantly, a sense and understanding of the spiritual component of a piece of music and music in general.

The teacher at the subject of "Music Literature" can arrange a viewing of recordings of TV channel "Culture" or the perception of online music contest, which will allow students to experience the atmosphere of a real musical action. IT enable the student to hear the music in a concert performance of many orchestras, conductors, and individual artists, representing the creativity the best concert halls and opera houses of the world, such as the Golden Hall of the Vienna Philharmonic Society, Grand Opera, Covent Garden, and others. Such a view helps to create a positive emotional atmosphere in the classroom in the children's musical school of arts or children's musical school.

Using modern smartphones, children can find answers to questions about the mysteries of musical works of the composer, the main or side topics on symphonies, about the best options of choral arrangements, etc.

The perception of the behavior of the audience in the concert hall will form the students' skills to conduct in cultural activities, a positive attitude to the work of musicians. The use of role-playing games in the "concert hall", "reporting on the musical event", "familiar with the work of his friends of the composer," allow using of IT components to form a creative approach to music in children, the organization of musical events.

Students of musical schools in the perception of the different interpretations of musical works represented by different musicians can find the most appropriate, in their opinion, the composer's intention. The study of different kinds of music interpretations allow each child to identify those special moments of specific performance, which the child can show the audience at academic concert exam or offset.

Effective instructional techniques, implementation of which is possible only with the use of IT, is to compare the performance of vocal numbers in Russian and in the original language. Using examples from the archives of the channel "Culture" will allow the teacher to find a varied operatic and chamber vocal material for comparing performances of singers in different languages. In the development of this method it is possible to offer students free discussion on the theme "Music and Lyrics: main and side", "Do I need to transfer Opera foreign author of vocal works," etc.

Another direction of Smart-e-Learning, information technologies not only on the subject of "Musical Literature", but also in other disciplines in the children's musical school is to teach students independent search for the information they need. For this purpose, the teacher can introduce students to Internet sites containing information in the field of culture and art, as well as recorded music to mp3; video of opera and ballet performances, symphony and chamber concerts, concerts of folk music ensembles; reproductions of paintings, conformable to listening music (for example, for a cycle of piano pieces of Mussorgsky's "Pictures at an Exhibition" used drawings VA Hartmann, under the impression that the composer and created his cycle); information about the life and work of composers, Musicians, artists, and others. Here are some of them. They are: Site "Belcanto" (Electronic resource - http://www.belcanto.ru/), which presents the following sections: news, work, theaters and halls, festivals, celebrities, bands, encyclopedia. In the "News" section you can see the events taking place in the field of ballet, opera, classical music and culture. "Works", composed as a musical encyclopedia, contains information on the operas, ballets, symphonies, chamber music, choral and vocal works. In the "Theatre and Hall" students can get information about the best musical theaters and concert halls of the world. "Festivals" is a list of international music festivals (the history of their origin, the direction of live performances, repertoire, and outstanding performers). "People" gives information about the life and work of outstanding performers, conductors, theatrical figures of the past and the present. Thus, this site is an electronic music encyclopedia that allows students to expand their horizons, to deepen the personal musical experience.

Another site where students can get all the information necessary in the study of the subject of "Musical Literature" is the "Classic-online" (Electronic resource - http://classic-online.ru). It presents biographical information about composers and their creative way. One of the advantages of this site is the fact that it is organically combined auditory and visual information. Home page of this site has two sections: "Composers" and "Artist". Alphabet navigation allows you to find the right information about musical works; they listen to different interpretations, download media. This site also presents non-fiction and feature films and books about composers and performers.

Website “Intermezzo” (Electronic resource - http://www.aveclassics.net/board) allows you to get acquainted with quite a large number of educational, methodical, scientific and popular literature on the art of music - books, notes, piano. For user convenience this literature can not only be read and viewed online but also downloaded to the media.

Quite often, students who are interested in certain pieces of music, look for their notes. In this case, they can take advantage of a musical material, which is offered on the websites: http://www.notarhiv.ru/vokal.html; http://www.musicalarhive.ru/; http://xn–80aerctagto8a3d.xn–p1ai/ and http://notes.tarakanov.net/. Listen to and download music from sites you can: http://mp3ostrov.com/; http://gidmusic.net/classical and http://classic.chubrik.ru/ [6].

Smart-e-Learning today is a necessary component of the pedagogical activity of children’s musical schools’ teachers. [8, 10] For example, teachers create electronic textbook, which includes audio and video composers (performers, conductors) and their creativity, musical scores of music, audio and video recordings of music or opera and ballet performances, or fragments thereof, a test material (quizzes, tests on the passed themes), music videos, or fragments thereof [9–11].

E. Prisyazhnaya one of the authors of this article specifically for the effective teaching of the textbook M.I. Shornikova “Music literature”. The development of Western music “2-year training” has created an electronictutorial on the Microsoft Power Point program [8]. When creating electronic aids were used specific cognitive and expressive possibilities of modern computer technologies: interactivity, hypertextuality, virtuality and multimedia.

An example of such manual is a fragment devoted to the work of I.S Bach. The first slide contains an original menu of the manual, which includes: a biography by I.S Bach, the art video on the life of the composer, audio works (Sonata C-dur for Violin, Prelude and Fugue in C-dur Prelude and Fugue in c-moll), information on the creative heritage of I.S Bach, music and texts of test material. In this tutorial, electronic composer’s biography is presented in four sections: I.S Bach - childhood, the Weimar period, Ketensky period and Leipzig period. The teacher and students using this textbook can use hyperlinks to move to any of the sections.

When using an electronic textbook in the process of listening to a piece of music a student can open the musical text and follow the development of the main themes that, as it is already indicated, significantly activates the auditory and visual perception and allows better understanding of the musical form, musical highlight features I.S. Bach’s Language. Inclusion into the electronic manual of artistic movie about the life and work of the IS Bach allows students to present it as a living contemporary, and not representative of a past age, to lure students into the life and culture of the era when the composer lived.

In the section “Tests” there is a test and a crossword puzzle. The test includes 10 questions: to check the fact-based material - important dates in the life of the composer; names of towns where he was born, lived and worked; tests on the knowledge of the key concepts associated with the work of I.S Bach - Polyphonic, suite, etc. The role of a crossword puzzle is the signification of the studied material’s evaluation. The answers to the crossword questions proposed by the students - “The name of the German bipartite dance, which was a mandatory part of any suite for harpsichord,” “What do you call the old church tunes that I.S Bach handles, creating on the basis of their beautiful melodies small organ pieces?" "As you know, four compulsory dance included in the old suite. What do you call a fast dance of English sailors, which usually ends with this suite," etc. Allow to demonstrate their musical background, including events in the life and work of the composer.

Another form of IT is a computer presentation, which is a way of studying the material at the musical literature lessons. Computer presentation allows to combine academic work with visual, textual and auditory information. In addition, this presentation activates mental abilities of children, as well as the process of storing the educational material, to formulate ideas of the main lesson, its subject, content and concept of music.

IT used not only in the education of children on the subject of "Music Literature", which aims introduction to academic art, but also to such discipline as "Variety solo singing." In particular, the following multimedia programs:

1. Programs created for listening to music. For example, such as Winamp (program for music and video playback), AIMP (a program for listening to music, developed by Russian programmers, and supports all popular audio formats, has the ability to convert to any formats of different audio files), iTunes (the program to play music for computers and smartphones based on iOs).
2. Musical editor: Sibelius (one of the most user-friendly editors that allows you to create and edit notes, store them in a convenient format - both graphical and audio), Finale (allows you to work with large scores), MuseScore (allows you to record music material using virtual keys, has a Russian version); ScoreCleaner Notes (can be installed on your smartphone, you can record a melody note text without using the virtual keyboard, and directly from the voice).
3. The application for VocaLive smartphones that allows the vocalist to rehearse and prepare for performance, allowing you to store single or multi-channel recording. When connecting a microphone and headphones artist can record and listen to the sound of his or her own voice, download the melody and put on the voice, shaped by special effects. Any procedure processing vocals and imposing voice on the audio series available to users of this powerful vocal processor. Mark such effects for vocals as: vocal effects that are available in real time, when there is a simultaneous voice recording with instant processing and applying tone compensation, harmonizer polyphonic accompaniment modifier usage, voice doubler, and de-esser to remove non-natural noises - wheeze and hiss when recording voice; sound effects to be applied to already saved voice recording, including parametric EQ, reverb, delay the effect of the sound track, a compressor, an imitation of the choral sound, the effect of "floating" sound and avtokvaker. Also, using the application can be downloaded VocaLive ready soundtracks from the library smartphone or PC for voice recording with the finished soundtrack, singing to the "minus" if you disable the voice recording mode when playing music in karaoke, set the metronome for singing warm-up etc.
4. Virtual piano keyboard used during rehearsals for inning a tone, play songs and tunes a cappella songs in which the singer begins execution before heard "minus one" in the situation of absence in the immediate vicinity of the instrument; Perfect

Piano (software for mobile phones based on iOs, meets the requirements above), the program "Small piano" (for mobile phones based on Android).

5. Programs that change the pace, or rather to recast the track into a different tempo without affecting the timbre, tonality, etc. Cubase SX meet the above characteristics, SoundShifter changes pace in real time and allows you to listen to the preliminary results.
6. Programs which function as Metronome (Tempo) that can set the tempo, if you set the right number of beats or choose Italian term (from Lento to the Allegro). The program stores the necessary pace for certain songs, you can select from existing 35, and to find the right rhythm pattern, such as a quarter, triplets or sixteenth. You can also specify a certain rhythmic pattern of the metronome sound. Also it gives an option to select voice work, even vocal.

Thus, these multimedia programs allow building an effective learning process from the beginning of work on vocal pieces to get the full artistic product - the concert performance. They contribute to greater development of their creative self-realization that is the main landmark of the national education system's modernization, as defined in the "National Doctrine of Education in the Russian Federation", "Concept of the Federal target program of education development for 2016–2020 years" [2, 3].

Such presented forms as: electronic textbook, computer presentation materials taken from various sites, and multimedia programs presented above were tested in the educational process of the children's musical school № 2. Named after MI Glinka, children's musical school at the secondary school № 32, the children's school of arts. Named after N.A. Rimsky-Korsakov, Yekaterinburg.

As a result of usage of IT on the subject of "Musical Literature", a large number of positive marks at the exam on the subject than the students' groups where IT was used minimally (only for creating presentations by children). The percentage of positive marks in the group where the IT was used by 88%, whereas in the group where IT was occasionally used by 57%. The use of these forms of IT positive impact on motivation and learning music, to maintain its interest in perception and execution, to intensify the attention of children to the art of music in general.

The introduction of these IT in the music school at the secondary school № 32 also affected the learning outcomes of students. Percentage of more successfully passed exams in the experimental group was 76%, and in control group was 62%.

At the exam in the children's school of arts. Named after N.A Rimsky-Korsakov students of the experimental group received excellent marks for 23% more than in the control group. In addition, the competitive activities of the experimental group of students were on 18% more efficient. They got diplomas at such competitions as "Hope Coast" (2014–2016), All-Russian Olympiad on vocals (2014–2016), and others.

4 Conclusions

The results of the implementation directed on usage of Smart-e-Learning in training musical-theoretical disciplines - "Musical Literature", executive discipline "Pop vocal" showed: contribute to the systematization of musical-theoretical knowledge of students;

they effect positively on retention of the theoretical concepts and the formation of practical skills and performance analysis of musical works; expand the limits of the fixed lesson time and entrance to the concert halls and opera houses of the world Philharmonic; stimulate research and creative activity of students and teachers within the various musical and theoretical subjects in the children's musical school; contribute to a more effective alignment of the rehearsal process, creative self-realization of students in performing activities.

The proposed areas of work can be used in musical education in secondary schools, having thc ability to create a variety of vocal groups and studios, clubs, fans of classical music, music lounges and children's Philharmonic.

References

1. Griban, O.N.: Information technology in the learning process. http://www.griban.ru/blog/14-informacionnye-tehnologii-v-processeobuchenija.html. (in Russian)
2. Kashina, N.I.: The problem of development of creative self-realization by students of college of culture and arts in the process of musical composition activities. In: Kashin, N.I., Pavlov, D.N. (eds.) Innovative Projects and Programs in Education, no. 3, pp. 11–15 (2016). (in Russian)
3. Konovalova, S.A.: Musical-creative development of children, based on a common approach in primary and secondary education. In: Teaching Experience: the Theory, Methodology, Practice, vol. 2, no. 3(4), pp. 14–19 (2015). (in Russian)
4. Bukharkina, M.U., Moiseeva, M., Petrov, A.E, Polat, E.S.: New pedagogical and information technologies in the education system: textbook for students of pedagogical universities and in-service training of teaching staff. In: Polat, E.S. (ed.), 272 p. Academy, Mumbai (2000). (in Russian)
5. Polozov, S.P.: Educational Computer Technologies and Musical Education, 208 p. Publishing House of Saratov. University Press, Saratov (2002). (in Russian)
6. Ragulin, P.G.: Information Technologies. Electronic Textbook, 208 p. TIDOT Dalnevost. University Press, Vladivostok (2004). (in Russian)
7. Tagiltseva, N.G.: Polyart approach in the development of educational strategies. In: Teacher Education in Russia, no. 12. pp. 91–95 (2015). (in Russian)
8. Tagiltseva, N.G., Prisyazhnaya, E.A.: Information technologies in professional work of the teacher of an additional education. In: MUNICIPAL Education: Innovation and Experiment, no. 6. pp. 12–16 (2016). (in Russian)
9. Taraeva, G.R.: Computer and innovation in musical pedagogy. In: Book 1: Strategies and Techniques, 128 p. Publishing House "Classic – XXI" (2007)
10. Taraeva, G.R.: Computer and innovation in musical pedagogy. In: Book 2: Presentation Technology, 120 p. Publishing House "Classic – XXI" (2007)
11. Taraeva, G.R.: Computer and innovation in musical pedagogy. In: Book 3: Interactive Testing, 128 p. Publishing House "Classic – XXI" (2007)
12. Uskov, I.V.: Information and communication technologies as a means of motivation of educational activity of students: Abstract. Dis. on degree seeking of step. cand. ped. Sciences (13.00.01), 24 p. Uskov Ivan Vasilievich, Ryazan (2006)

A SPOC Produced by Sophomores for Their Junior Counterparts

François Kany(✉) and Bruno Louédoc

ISEN-Ouest (Institut Supérieur d'Electonique et du Numérique),
20 Rue Cuirassé Bretagne, CS 42807, 29228 Brest Cedex 2, Brest, France
kanyfrancois@hotmail.com

Abstract. Fifteen sophomore students are creating a SPOC on probabilities for their freshmen/women mates. This SPOC is based on 80 exercises to solve first numerically then analytically. For each exercise, the sophomore group have prepared a notebook, a video tutorial to explain the problem and another video clip to comment on the solution.

Keywords: Peer production · Creation of a SPOC · Teaching peer to peer

1 Introduction

As underlined by V. Uskov [1], modern and smart technologies enable unprecedented opportunities for academic organizations to create higher standards in teaching and learning strategies. We present hereunder the approach which led us to have a SPOC produced by sophomores about probabilities calculus. Our motivations to achieve such a project and the choice of a particular methodology are explained as follows.

2 Taking up the Challenge

In May 2016, Mr. Jean-Michel Rolland, head of ISEN-NÎmes, gathered teachers of the Yncréa Federation[1] and let them know that the group had been selected by the French National Research Agency when they launched the call for projects IDEFI-N[2]. Those IDEFI-N [2] aim at backing ambitious projects which might appear to be the most strategic ones to initiate a new dynamic, pushing forward digital training offer for French Higher Education.

The project the Federation hold out is entitled FR2I[3]. It aims at initiating trans-disciplinary pedagogical innovation: on-line courses, tutorials and distant

[1] Yncréa (formerly HEI-ISA-ISEN Federation) is the largest association of private engineering schools in France. This structure is made up of 5,500 students.

[2] *Initiatives d'Excellence en Formations Innovantes Numériques* = Premium Initiatives In Developing Digital Break-trough Tutorials.

[3] *Formation en Réseau d'Ingénieurs Internationaux* = Network Training for International Engineers.

V.L. Uskov et al. (eds.), *Smart Education and e-Learning 2017*, Smart Innovation, Systems and Technologies 75, DOI 10.1007/978-3-319-59451-4_13

evaluation, virtual learning-devoted spaces, etc - including notably the production of 18 "Massive Open On-line Course" (MOOC) or "Small Private On-line Course" (SPOC) over a 4-year span. The specifications for these MOOC or SPOC require them to meet the Bachelor level (L1 to L3); that they will be multilingual and that each of them will be worth 2 credits in the "European Credit Transfer System" (ECTS) which equals to a student's full-time working week.

Even though the production of these 18 MOOC/SPOC is dispatched over the six schooling sites of the Federation, the load of work is huge. To produce a MOOC, you need in between 600 to 1000 h of work [3]; Rémi Bachelet acknowledges that 10 h of preparation are needed to have a one-hour video [4]. As far as costs are concerned, they may rise up dramatically, especially if you use dedicated external societies.

3 A cMOOC to Produce an xMOOC

There are two types of MOOC: the xMOOC and the cMOOC.

An xMOOC recaps, digitally and on wider scale, the grounds of a tutorial class. The learners attend a course (viewing of video sequences, reading of documents, etc.) and, then, they have to produce individually activities which can be checked automatically (validation of an MCQ, execution of a computing code, etc.) or checked by hand (grading by peers, etc.).

On the contrary, in a cMOOC, activities are not carried out individually but in a collaborative way.

Inside cMOOC, you may see, on the one hand, connectivist classes, without deadlines, which are similar to a debate or an exchange of ideas and, on the other hand, classes based on team projects (in that case, MOOC with precise specifications set by teachers with a deadline are to be distinguished from MOOC where theme focus and final production are freely chosen by learners).

To rise to the challenge set by Jean-Michel Rolland, we chose to achieve a prescriptive cSPOC in order to produce an xSPOC matching the FR2I project specifications.

4 Sophomores Providing Tuition for Junior Counterparts

The sophomores who took computer science as an option were given the opportunity to participate in a cSPOC in order to create an xSPOC for the whole junior form.

The point is to provide the sophomores with the necessary pedagogical support they need to produce the xSPOC contents and to ask them to make, in a collaborative way, video and computing supports.

Three relevant issues are thus met.

Firstly, the teacher is able to focus on the production of a good quality learning material as he is free from the most time-consuming tasks: conceiving the xSPOC scenario; making explanatory videos; integrating digital support; animating the xSPOC, etc.

Secondly, as sophomores have to explain the exercises in short videos and be active on the forum answering questions, they need to get higher training. Besides, setting up a digital workspace and taking in numerical supports allow them to practice their computing knowledge.

Thirdly, the junior class will appreciate to get further explanations from other students along with the learning material provided by their teacher. However didactic a teacher may be, his/her message is sometimes misunderstood by his/her students. When a course appears as a MOOC or a SPOC, there is a higher risk of misunderstanding, because the teacher cannot have the direct feedback from his/her class. In that context, the explanatory videos made by the sophomores is a real added-value. Those students are in the best position to explain the difficult points they had to face and solve by doing an exercise.

5 Profile of the Sophomores Attending Computer Science Option Class

At ISEN graduate-school, L1 and L2 students (the sophomores and their juniors) have to choose freely an option among the four available: economics, biology, energy and computer science. This option, which comes on top of the CPGE[4] regular time-table, allows them to get a hint of the major options they have to choose when entering the course of engineering studies (M1 and M2). These class hours are taught out of the National Education frame; teachers are no longer bound to follow a syllabus, they can teach and grade their students as they wish. The point is to discover a subject which is not at all taught in CPGE but useful in engineering studies (economics or biology) or, on the contrary, to explore deeper one part of the CPGE syllabus (energy or computer science).

This option equals a two-hour implement per week; generally, students will keep the same option for their L1 and L2 years. On average, the groups of students would amount to fourteen. The year before, the fifteen ones studying the L2 computer science option had studied a course about OCaml (a functional language [5]) except for two new students who joined ISEN graduate-school after attending one L1 year in another Engineering school; they are acquainted only with the language shared in the common syllabus: Python.

The group of students are not necessarily the best in their form (some of them having finished the L1 year ranking in the last third of their form, two of them repeating their L2 school year).

6 A Carrot (Better Than a Stick)

In order that students choose their options according to their own main focuses (and not according to the mark they may hope to get), teachers decided that

[4] *Classe Préparatoire aux Grandes Écoles* = a two-year-intensive program in math and physics after the “Baccalauréat” (A Level).

the medium rate for the four options would be lined out on the same mark (in order to avoid any overbidding in the marking process).

Though the elaboration of the xSPOC is shared out amongst 15 students, the load of work asked from the students with the computer science option is more important than the one asked from students with other options (all the more that they may not even claim a boost on their pass mark). It was then important to offer a "carrot" in order to rally the students to the achievement of the project.

From the beginning, we told the students that their works would form the matter of a communication during an international conference on pedagogics to be held in Grenoble (France). If that publication was selected by the reading committee, the students themselves would go and talk about their works. (We see with deep regret that very few students take part in these kind of conferences especially when the "big talk" nowadays is to consider the student as "the core of the school-system"). Even if all the students do not see at once the point to have one's name in a publication, every one of them is motivated by the opportunity to make a trip.

7 Choosing the Theme of the xSPOC: Probabilities

The choice of a theme for the xSPOC was made by using a part of the mathematics syllabus of the CPGE first year: probabilities.

The point is as follows: most of the time, you do not need any mathematical prerequisites to understand the terms of the problems. For instance: *A jar contains six black marbles, six red marbles, ten green marbles and twelve blue marbles. If six marbles are drawn from the jar, what is the probability that one gets, at least, one marble of each color?*

In spite of the simple terms of the problems, one may be facing logical difficulties as solving them. Analysis mistakes are very frequent. If the conditional probabilities are not properly taken into account (when propositions are not independent), it will lead to many paradoxes. Nevertheless, the mathematical tools needed do not demand excessive technical skills: addition or multiplication of basic probabilities (in worst cases, the calculus of an integral or of an eigenvector of the transition matrix of a Markov chain). The calculus are seldom more complicated; the student does not need to master a wide range of technical calculi when stepping into probability problems.

Rather than the Kolmogorov's axiomatic approach, we prefer to interpret probabilities as frequentism and to define them as the limit of the statistical frequency when repeating a random experience set a very great number of times in the same set conditions. This approach allows, on the one hand, to build a bridge with statistics and, on the other hand, to perform numerical simulations.

Thus, students are able to valid automatically, on a server, numerical experiments to compute probabilities. In that context, computing as a tool reveals to bring a real added value: a simple Monte-Carlo simulation will allow to check the numerical value of a probability rapidly before proceeding to a rigorous demonstration.

8 Organization

8.1 Preparing

The time span from June to September 2016 was used to prepare the pedagogical material that would be later given to the students. This material consist of a probability course set as methodological sheets (Fig. 1) and a list of 84 exercises covering different themes. Some of these themes directly follow the usual chapters of a probability course: enumeration, random numbers, Monte-Carlo methods, statistics laws, geometrical problems, integral calculus, random walk, ...; other themes are more specific to computer science: Las Vegas method, Atlantic City method; and, finally, others are on purpose extracted from the mathematical course to show how probabilities apply to physics: statistical physics, quantum physics and information theory.

8.2 Setting up the Collaborative Plate-Form for the cSPOC

Early September 2016, the project was brought to the attention of the sophomores (L2 students) who were given the tools that would enable them to work on a collaborative way. The numerical platform of the cSPOC is a JupyterHub [6] which allows to execute code on-line as on the France-IOI server [7] or on the OCaml MOOC on FUN[5] [8]. It was set up with the help of open-source software - Ansible and Vagrant - and it was hosted on a school server. It foreshadows the final version of the future site that will be available for L1 students when working on the xSPOC.

The L2 students dispatched the work amongst themselves by registering in a Google Doc displaying all the exercises. This document allowed everyone to get a real-time visibility on the global on-going of the project.

8.3 Producing Deliverables

Here is what students are asked to do for each of the 84 exercises:

- to prepare a notebook (an interactive document in which computing codes can be executed) [see Fig. 2] from the terms of a problem and its correction in Python.
- to create unit tests so that the notebook can be automatically graded by using nbgrader extension [9].
- to translate the code into OCaml. This step may seem useless: most of the L1 students who will take part in the xSPOC will indeed code in Python. Nevertheless the step done while translating the code is meaningful for L2 students, allowing them, on the one hand, to practice OCaml (the initial purpose of their computer science option) and, on the other hand, to get a better understanding of the difficult items of the code (copying the solution in Python is meaningless on a pedagogical point of view).

[5] *France Université Numérique* = France Numerical University.

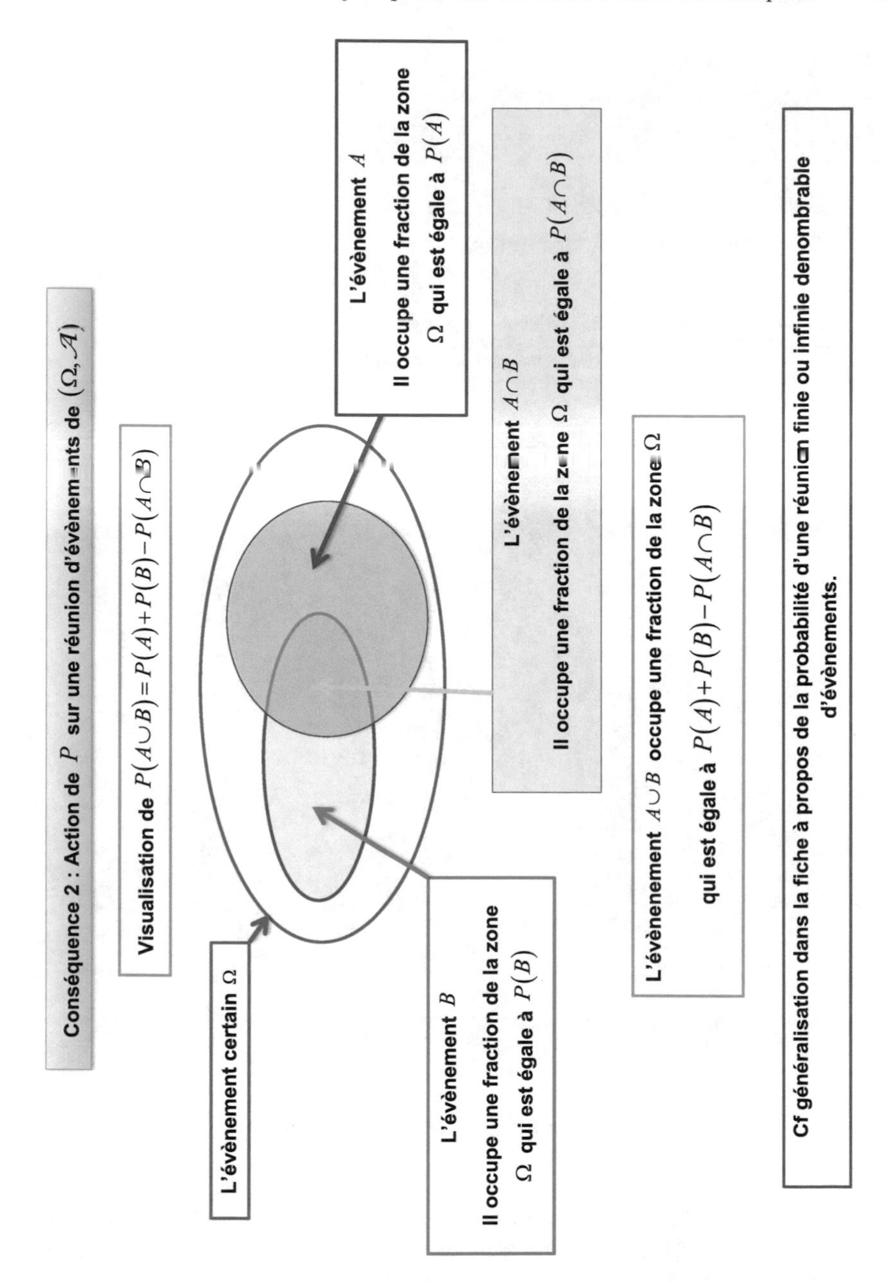

Fig. 1. Example of a methodological sheet

- to produce a video tutorial (lasting 2 or 3 min) to help the future students taking part in the xSPOC by explaining the problem and tricky issues (that is why the previous step is necessary).
- to produce a short video explaining the correction.

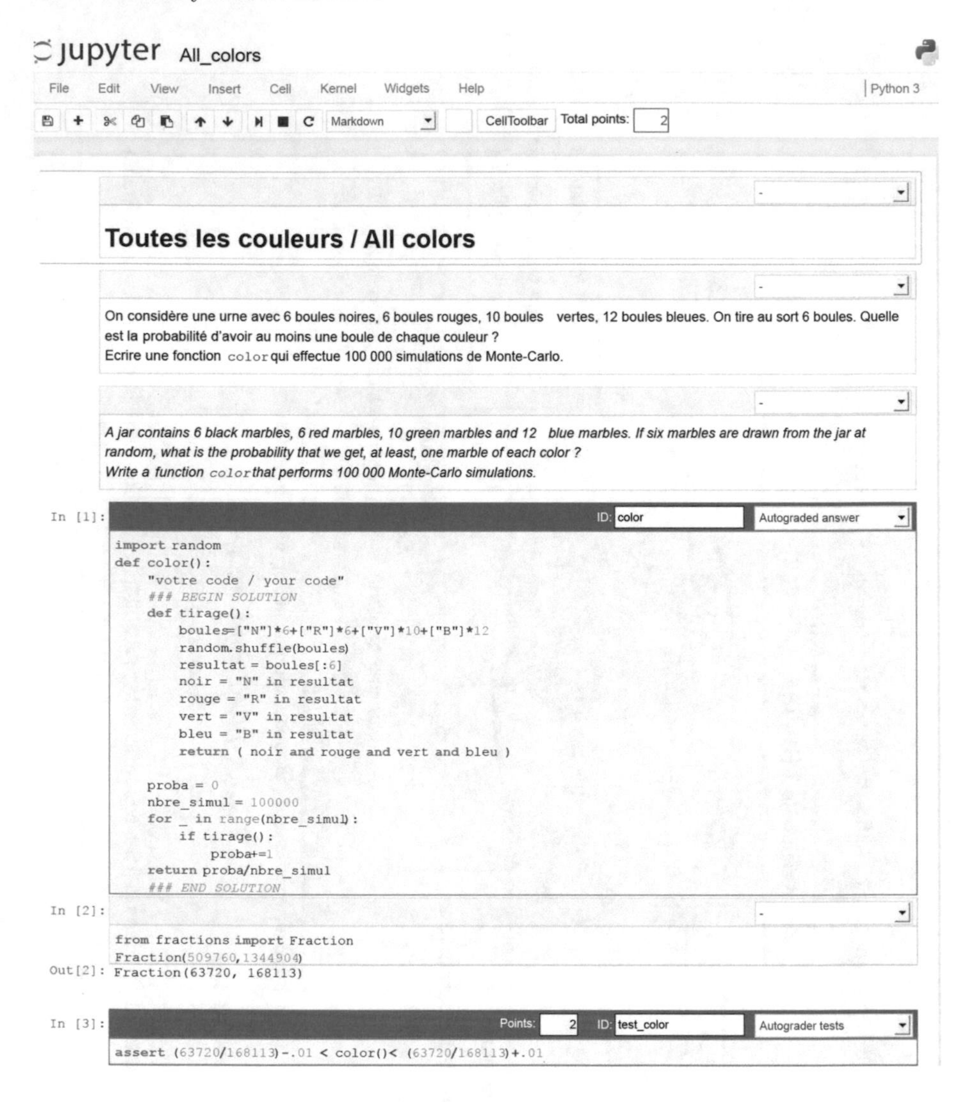

Fig. 2. Example of instructor's notebook

- to write the verbatim of those two videos so that they may be subtitled and then translated (the verbatim will be read over by the teacher in order to check any mistake).
- to put their production on-line on the server.

9 Results and Prospects

The students committed to the project, being studious and productive. A glossary has been created to centralize the notebooks conversion methods so that all the interactive documents have been generated quicker and quicker. The students spontaneously made use of the school resources to create quality video media by borrowing audio equipment from the students' movie club, using the language laboratory as a recording studio, etc.

To this day (two months and a half after starting) nearly the whole of the notebooks have been produced and a dozen of exercises completed with the videos. By the end of May (a dead-line for options), we hope to be over with the whole of videos and tests. For the beginning of the 2017 academic year, the xSPOC will be played for the first time and the students (then in L3) will have to manage the forum, answering the participants' questions.

10 Conclusion

The production of an xSPOC by students, on such a scale and at that level, is an unprecedented pedagogical experiment. The results are conclusive as far as L2 (sophomores) students are concerned: the students developed their computing knowledge; they will be able to re-invest their competence during the second-year probability course. The verbatim reports of the videos did not entail major corrections: students fully understood the exercises and were able to explain them with their own words.

We are looking forward to seeing how this xSPOC will be taken in next year by L1 students.

Acknowledgments. We would like to thank Denis Pinsard, computer science teacher at the Chateaubriand High School of Rennes who helped us to set up the JupyterHub server.

We are grateful to Serge Bays, who teaches mathematics at "Les Eucalytpus" High School of Nice, for fruitful discussions about prediction interval and confidence interval.

Annex: Automatic Grading of a Monte-Carlo Simulation

In an assignment, the validation of a computation is usually done by performing a comparison between the results of a student's code and the results of the instructor's code with well-chosen input parameters.

Most of the time, one has only to check the strict equality of both results. The situation is a little more tricky in the case of the evaluation of a Monte-Carlo simulation.

By definition, a random experiment (with a random seed) is not reproducible. One cannot validate a computation by checking a simple equality. On the contrary, one must perform a double comparison in order to check if the result falls in a given interval (carefully chosen by the instructor).

For instance, let us suppose that the instructor wants his/her students to solve a problem on probabilities whose theoretical solution is p_{th}. He/She asks his/her students to make N trials, to compute the n favorable cases and to return $m_{exp} = \frac{n}{N}$. (m_{exp} is an approximation of p_{th} if N is large enough).

Let us suppose that 100 students post correct codes, we want the server to validate, *on average*, 95 of these codes. (The instructor may test the unvalidated codes a second time to detect the false-negative ones; the risk of two successive false-negatives is then 0.25%).

The instructor has to build a prediction interval (not a confidence interval) to estimate, with a 95% probability, that the future students' values will fall within the limits.

m_{exp} will fall in $\left[p_{th} - 1.96.\sqrt{\frac{p_{th}.(1-p_{th})}{N}}, p_{th} + 1.96.\sqrt{\frac{p_{th}.(1-p_{th})}{N}}\right]$ with a 95% probability if $N.p_{th} \geq 5$ and $N.(1 - p_{th}) \geq 5$ (if $N \geq 30$; which is always the cases in a Monte-Carlo simulation).

As $p_{th}.(1 - p_{th}) \leq 0.25$, this prediction interval can be simplified to $\left[p_{th} - \frac{1}{\sqrt{N}}, p_{th} + \frac{1}{\sqrt{N}}\right]$ (slightly larger).

When a student makes N trials, he/she realizes a *sample*. (If 100 students post correct codes, this gives 100 *samples*). Each *sample* has 95% chance to fall in the prediction interval. On average, for 100 correct codes, the server will validate at least 95 codes (but it could occasionally validate only 94 or 93 codes).

More precisely, if the probability to validate a correct code is 95%, then the percentage of correct codes validated by the server will fall in the prediction interval $\left[0.95 - 1.96.\sqrt{\frac{0.95 \times 0.05}{100}}, 0.95 + 1.96.\sqrt{\frac{0.95 \times 0.05}{100}}\right]$ (which is approximately $[0.907, 0.993]$) with a 95% probability. Thus, the server will validate, in 95% cases, between 93 and 97 codes.

References

1. Uskov, V.L., Bakken, J.P., Pandey, A., Singh, U., Yalamanchili, M., Penumatsa, A.: Smart university taxonomy: features, components, systems. In: Uskov, V.L., Howlett, R.J., Jain, L.C. (eds.) Smart Education and e-Learning. Springer (2016)
2. Programme <<*Investissement d'avenir*>> (2015). http://www.agence-nationale-recherche.fr/fileadmin/aap/2014/selection/IDEFI-N-selection-2014.pdf
3. Cisel, M.: Monter un MOOC: combien ça coûte? = Create a MOOC: how much does it cost? (2013). http://blog.educpros.fr/matthieu-cisel/2013/11/12/monter-un-mooc-combien-ca-coute/
4. Bachelet, R.: *MOOC: les profs face aux nouveaux cours en ligne.* = MOOC: teachers react to new on-line courses (2013). http://www.franceculture.fr/emissions/pixel-13-14/mooc-les-profs-face-aux-nouveaux-cours-en-ligne
5. OCaml is a programming language developed for more than 20 years at Inria by a group of leading researchers. It emphasis on speed, expressiveness and safety
6. Pinsard, D.: Jupyterhub server (Jupyter: Python3 + OCaml + SQL + Processing.py) (2016). https://framagit.org/dpinsard/jhubserver/
7. The International Olympiad in Informatics (IOI) is an annual competitive programming competition for secondary school students. http://www.france-ioi.org/
8. OCaml, MOOC de l'Université Paris Diderot, France Université Numérique (2016). https://www.fun-mooc.fr/courses/parisdiderot/56002S02/session02/about
9. Nbgrader is a jupyter extension that enables an automatic grading of a notebook. http://github.com/jupyter/nbgrader.git

An Outcome-Based Framework for Developing Learning Trajectories

Andrey V. Lyamin(✉), Elena N. Cherepovskaya, and Mikhail S. Chezhin

ITMO University, Saint Petersburg, Russia
lyamin@mail.ifmo.ru, {cherepovskaya,msch}@cde.ifmo.ru

Abstract. Many universities apply e-learning and learning management systems to control the learning process. Such systems require educational programs to be uploaded inside the system as specific structures. Nowadays almost all of the educational programs are based on a competency model with specified learning outcomes, which can also be determined for the courses that are parts of these programs. This paper provides a discussion of the three most widespread learning technology standards for describing, referencing, and sharing competency and learning outcome definitions in the context of online and distributed learning. It also describes a mathematical model that helps to develop a correct educational program based on the newly specified requirements, and provides information about the results of an application of the model in the learning management system of ITMO University.

Keywords: Model of educational program · Learning trajectories · e-Learning

1 Introduction

Modern educational approaches comprise different types of learning that could be realized in educational institutions, e.g. classroom learning process, e-Learning, Massive Open Online Courses (MOOCs) [1] with verified certificates and gamification tools [2, 3] and so on. e-Learning is essentially growing nowadays since many institutions figured out the benefits it brings to the learning process. e-Learning is aimed to provide different education levels for all interested students, even if they live in distant places and cannot attend institutions themselves. As well, e-learning simplifies the way of implementing and approving new education methods as they can be stored in institutions' smart learning management systems (LMS) [3, 4]. Many unique training instruments appeared in the last few years, such as online computer simulators for different courses of the educational programs, virtual labs [5, 6] used for the learning outcome assessment, etc. Ones of the most popular of the currently discussed learning technologies are personalized and adaptive learning [7].

Adaptive and personalized learning [8] states an idea of providing personalized educational programs and smart curriculum to the students, as well as an application of adaptive assessment tools. Adaptive learning also helps a person to fully comprehend learning material in accordance with his current functional and emotional state [9, 10]. These learning types help to develop personalized learning trajectories [11], though

V.L. Uskov et al. (eds.), *Smart Education and e-Learning 2017*, Smart Innovation, Systems and Technologies 75, DOI 10.1007/978-3-319-59451-4_14

maintaining almost all the learning outcomes stated for the courses and educational programs themselves.

Modern educational programs realize the competences' approach coupled with focusing on the prospective learning outcomes. A competency of a graduating student means his capability to apply all the obtained knowledge, skills and personal qualities in the context of professional activities, dealing with different social and professional tasks. The graduating competences of an educational program, which are developed by the universities' departments together with the employers, form a competency model of the graduating student. A set of graduating competences together with the planned outcomes form an overall learning outcome of the course and a basis for structuring a competency-oriented content. Analyzing competences in the course helps to determine the aims of the course and its syllabus. A set of competences define topics of course sections and, therefore, its structure, as well as the cause-and-effect relations between competences' outcomes state the sequence of studying these sections. The next stage of detailing course content depends on the types of a teaching load and planned outcomes and defines quantitative characteristics of various teaching units, e.g. lectures, labs, practical classes, essays, etc. The next step for a program developer is to specify procedures of the current, mid-term and final assessment.

It should also be stated that educational programs might consider MOOCs as courses of the program, if it is approved by the university. MOOCs can be used in the main learning process in different ways: (a) it can be embedded as a mandatory program's unit or fully replace the program module; (b) as an alternative way of mastering an educational program, which can be chosen by students themselves; (c) as a free minor module, which can be chosen by students simultaneously with the other program modules. Nowadays many universities use MOOCs inside educational program as a specific course providing certain competencies and learning outcomes.

A complex structure should be realized for each of the educational programs. Hence, a model can be defined in order to provide a mathematically formalized description of the program. In our paper we present a mathematical representation of the results of mastering an educational program, which is used in AcademicNT LMS of ITMO University. The second section provides a comparison of the three standards developed to define a competency structure. The third section describes a structure of the AcademicNT LMS, as well as components of the structure of educational programs. The fourth section is used to define basic principles, newly defined requirements of an educational program, and presents the mathematical model itself. The fifth section presents experimental results of an application of the described model and examples.

2 Comparing Different Educational Standards

In the world of modern education, there are many standards that help to specify the structure of an educational program. These standards have many things in common, however, distinctions also exist.

2.1 Integrating Learning Outcomes and Competences (InLOC)

The first standard to be observed is an open standard called Integrating Learning Outcomes and Competences, or InLOC [12]. InLOC is a European project that describes the ways to control and exchange learning outcomes and competences. InLOC specification strictly determines a necessity to use a unique identifier of a learning outcome and competency (LOC), as one of the main advantages of InLOC standard is a possibility to use the same name for different LOCs. In order to extend description of the learning outcomes and competences the next information, as well as the creator's, version's and creation and editing dates' metadata, is added according to InLOC:

- Title and description (a record can be presented in different languages);
- Levels and credits (are used to describe the learning progress and reveal differences between proficiency levels);
- Topic (a category, which LOC refers to).

A majority of LOCs do not exist separately, but they are combined in one structure, e.g. as a tree. InLOC specification described associations and rules of grouping for them. The information model of this specification consists of three main classes, which extend the LOC class:

- LOCdefinition. Main properties of the class: primaryStructure (an instance of LOCstructure for the current LOC);
- LOCstructure. Main properties of the class: comprisesAssociation (an instance of LOCassociation for the current structure);
- LOCassociation. Main properties of the class: type (type of relationship), hasSubject (a subject of relations or combined attributes), hasScheme (a scheme of associations), hasObject (an object of relations or combined attributes), number (a positive number to describe levels and credits).

All of the described three classes extend the abstract class LOC with the following properties: id, language, abbr (abbreviation), created, issued, title, description, rights, validityStart/validityEnd (the interval of LOC's validity), etc.

2.2 ISO/IEC 20006

ISO/IEC 20006 standard [13, 14] is appropriate for those, who develop and use learning and human resource management systems. It is divided into two parts:

- ISO/IEC 20006-1: Information technology for learning, education and training – Information model for competency. Part 1: Competency general framework and information model [13];
- ISO/IEC 20006-2: Information technology for learning, education and training – Information model for competency. Part 2: Proficiency level information model. [14];

The standard itself describes the following definitions and aspects:

- The basics for working with competencies with the use of IT in learning education and training;

- A model of controlling and exchanging information about competencies and the related objects;
- An information model used to describe competencies and the related objects, including principles of their composition;
- Examples that may help in the development of the whole structure and information model for competency.

As the same competencies can be used in different information systems, it is necessary to specify their structure and forms of their transferring between different systems. Hence, HR-XML, RDCEO, RDC standards that allow controlling and transferring the identifying data for competencies. However, in order to provide full comprehension of the competency model more information about it should be provided. Therefore, there are two types of the information about competencies, i.e. a semantic information model and proficiency level, described in two parts of the standard correspondingly.

A semantic information model consists of two blocks of data as well as three main parameters, i.e. identifier, description and title:

- Competency meaning information
  - Essential unit
  - Scenario unit
- Competency situation information
  - placement
  - relatedCompetency
  - level
  - criteria

The second part of the standard describes information model used to express semantics of the competency' formation levels' as well as examples that help to create this information model. Hence, three data structures had been specified: (a) proficiency composition model; (b) proficiency information model; (c) level information model. The described ISO/IEC 20006 standard provides specific information about the competencies and the resulting outcomes. It considers not the exact learning outcomes of a course/program, but a proficiency level of a person.

2.3 IEEE 1484.20.1

The group of standards IEEE 1484 describes the rules and specifications of e-Learning technologies and education aspects. As we are interested in competencies' description, let us focus on 1484.20.1-2007 – IEEE Standard for Learning Technology – Data Model for Reusable Competency Definitions [15].

IEEE 1484.20.1 standard defines a data model for describing and transferring definitions of the competencies mainly in the field of distributed e-Learning. It specifies a way of presenting key characteristics of competencies without relation to the context they are applied in. This standard helps to ensure the interoperability of e-Learning systems that deal with competencies by providing links to the necessary definitions and descriptions. IEEE 1484.20.1 allows encoding and transferring information about competencies. A competency definition according to this standard contains the following properties:

- identifier
- title
- description
- definition:
  - model_source
  - statement:
    - statement_id
    - statement_name
    - statement_text
    - statement_token (vocabulary_type with 'source' and 'value' properties)
- metadata:
  - rcd_schema (A schema that defines and describes rcd)
  - rcd_schema_version
  - additional_metadata

IEEE 1484.20.1 standard provides a data model for the competency definition, which is able to be reused in different systems, as well as it can be uniquely identified in all systems it is applied in.

2.4 Comparison Conclusions

We considered three of the most widespread educational standards. Each of them has its own benefits and limitations. Table 1 provides a short summary for all the three standards.

Table 1. A comparison of the competency definition standards

Standard	Benefits	Limitations
Integrating Learning Outcomes and Competences (InLOC)	The definitions are simple and provide all the necessary information about competencies and learning outcomes and corresponding relations	Data corruption may occur
ISO/IEC 20006	It provides very exact information about competencies, describing different structure levels and aspects of a competency and proficiency level	It does not consider separate learning outcomes, but focuses on the proficiency levels. Data corruption may occur
IEEE 1484.20.1	Describes a competency definition that can be simply reused in different systems	The standard does not provide information about what category does a competency refer to, e.g. a skill, knowledge or ability. Data corruption may occur

According to all the presented information about different standards, we can conclude that all of these specifications are used to describe competencies in different degrees of detail. As we can see, all of the presented standards include definitions of learning outcomes and the relations between them. Hence, we can highlight the following most important aspects:

- An information model, which is used to describe the learning outcomes and competences, is a hierarchical model that can be presented as a set of trees, where the vertices are learning outcomes and edges are the relations between them.
- A learning outcome and a competence are characterized by an identifier, a title, a description, an action (usually presented as a verb), an object that is affected by action, credits, mastering level, subject area and prerequisites.

However, the provided description does not specify rules of setting relations between learning outcomes, so that program developers can possibly set a redundant relation. It may prevent determining the sequence of the learning outcomes correctly, e.g. a cycle can be set that will cause data corruption and make the model invalid. Hence, it is important to control the integrity of the formed information model.

3 The Structure of an Educational Program

This section is related to the description of an educational program's structure applied in AcademicNT LMS [3, 5, 6, 9, 10]. AcademicNT LMS was established in 1999 in ITMO University, and since then it had been significantly developing and is now used in ITMO University and some other universities in Russia. The educational program structure supported by AcademicNT is presented in Fig. 1.

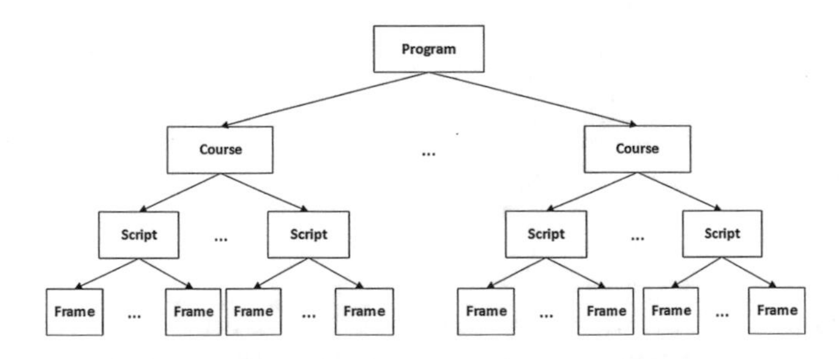

Fig. 1. The structure of an educational program

The main unit of the educational program is a program, which may consist of many courses used to produce their own learning outcomes and fulfil program's learning outcomes. A course is a specific unit, which is divided in parts in order to present learning material in a more comprehensive way. These parts of the courses are presented as scripts. Student's state in the course can change according to his progress and, hence, a state defines the list of scripts that can be currently presented to a student. One of the possible list of states can be defined as "Learning", "Assessment", "Review", and "Evaluation".

A script is a unit, which defines a description of an interaction between a student and LMS. It contains the sequence of learning and assessment materials that should be presented to students, time limits, and criteria for the grading. Moreover, a script defines whether to display a number of the correct answers to students, system reactions to user's answers, an opportunity to miss the question and return to it later. A script is a container for frames. A frame can represent a particular page of the electronic synopsis, a question in the online test or a container for an embedded virtual lab. Each frame can have its own settings, which can be different from the script it is included in, and, in that case, frame settings will have higher priority than the script's ones.

AcademicNT allows using personalized assessment of knowledge and skills with tasks of different complexity levels. In order to apply this, a definition of a task's weight is introduced. Task's weight is presented as a number in range (1, 9), when the basic task's result in range (0, 1). The resulting points for an assessment frame are calculated as multiplication of the weight to the basic result.

4 A Model of the Educational Program

This section presents a mathematical formalization of the educational program's structure. The model is based on the main principles of an educational program, namely:

- Learning outcomes and competences of the program and courses;
- Course's prerequisites and planned outcomes;
- Learning modules of the course and student's experience.

Let us define a set V of the learning outcomes and competences (LOC), which can be represented as $V = \{LOC_i : i \in I_{LOC} \subset \mathbb{N}\}, |I_{LOC}| = q$. After that on the set V we can define a part-whole relation W:

$$W = \{(LOC_i, LOC_j) : LOC_i, LOC_j \in V, W, (i, j)\} \tag{1}$$

As it is important to control the integrity of the formed information model, so that relations between competencies could be uniquely identified, rules should be defined in order to prevent data corruption. A relation W should possess the following properties:

- Irreflexivity (there is no relation from LOC_i to itself):

$$\forall i \neg W(i, i) \tag{2}$$

- Asymmetry (there are no pair relations):

$$\forall i \forall j \, W(i, j) \rightarrow \neg W(j, i) \tag{3}$$

- Intransitivity (when the third learning outcome depends on the second that can be accessed only after the first, there should be no relation between the third and the first LOC):

$$\forall i \forall j \forall k\, W(i,j) \wedge W(j,k) \rightarrow \neg W(i,k) \tag{4}$$

- Acyclicity (when the transitive closure W^+ is irreflexive)

It is assumed that each jth *LOC* ($\forall i \neg W(i,j)$) is univariate and corresponds with a specific latent characteristic of the student. In that case, the results of every two tests will be independent random variables:

$$P(X_i < x', X_j < x'') = P(X_i < x')P(X_j < x''), \tag{5}$$

where X_i, X_j are the results of the ith and j^{th} tests.

Each of the educational programs is directed to the formation of the necessary learning outcomes based on the time limit $\tau_0 > 0$ and, hence, the prerequisites for a program $V^b \subset V$ as well as planned outcomes $V^f \subset V$ can be defined. Both prerequisites and planned outcomes of the program should be unique, and, therefore, expediency of educational program can be defined as following:

$$\begin{gathered}(\forall LOC_i \in V^f)(\forall LOC_j \in V^b) \\ ((LOC_i, LOC_j) \notin W^+ \vee i \neq j)\end{gathered} \tag{6}$$

Each educational program can be associated with a tuple $\langle V^b, V^f, \tau_0 \rangle$. The next stage is to present mathematical formalization for each course of the educational program. A course consists of the set of learning, informational and assessment modules and can be defined as $LC_r \in C = \{LC_i : i \in I_{LC} \subset \mathbb{N}\}$ associated with a tuple $LC_r = \langle V_r^b, V_r^f, \tau_r \rangle$. Both prerequisites and planned learning outcomes of the course should be unique, and, therefore, the reasonability of the course can be defined as:

$$\begin{gathered}(\forall LOC_i \in V_r^f)(\forall LOC_j \in V_r^b) \\ ((LOC_i, LOC_j) \notin W^+ \vee i \neq j)\end{gathered} \tag{7}$$

A set of courses $\{LC_{j_1}, LC_{j_2}, \ldots, LC_{j_p}\}, j_k \in I_{LC}, k = 1, 2, \ldots, p$ related to the educational program should meet the following conditions:

- The effectiveness of the set of courses:

$$\tau_0 \geq \sum_{s=1}^{p} \tau_{js} \tag{8}$$

- The productivity of the set of courses:

$$(\forall LOC_i \in V^f)\left(\exists LOC_j \in \bigcup_{s=1}^{p} V_{js}^f\right)(i = j) \tag{9}$$

- The validity of courses' prerequisites:

$$\left(\forall LOC_i \in \bigcup_{s=1}^{p} V_{js}^{b}\right)\left(\exists LOC_j \in V^b \cup \left(\bigcup_{s=1}^{p} V_{js}^{f}\right)\right)(i=j) \tag{10}$$

- The relevance of courses' learning outcomes:

$$\left(\forall LOC_i \in \bigcup_{s=1}^{p} V_{js}^{f}\right)\left(\exists LOC_j \in V^f \cup \left(\bigcup_{s=1}^{p} V_{js}^{b}\right)\right)(i=j) \tag{11}$$

- The uniqueness of learning outcomes of the set of courses:

$$\bigcap_{s=1}^{p} V_{is}^{f} = \varnothing \tag{12}$$

- The acyclic property of the set of courses (as irreflexivity of all possible relations composition):

$$RF_{jk} = V_{jk}^{b} \times V_{jk}^{f}, j_k \in I_{LC}, k = 1, 2, \ldots, p \tag{13}$$

A course can be divided into the learning modules, and each of the modules includes a collection of learning and assessment materials, assigned to the assessment of learning outcomes. Hence, a module $LU_r \in M = \{LU_i : i \in I_{LU} \subset \mathbb{N}\}$ is associated with a tuple $LU_r = \langle V_r^b, l_r, \tau_r \rangle$, where V_r^b is a set of requirements for student's previous experience (prerequisites of the module), l_r is an identifier of the learning outcomes $LOC_{l_r} \in V, \tau_r > 0$ is a general laboriousness of the learning module. Prerequisites of the module should be unique, and, therefore, reasonability of the module can be defined as:

$$\left(\forall LOC_j \in V_r^b\right)\left(\left(LOC_{l_r}, LOC_j\right) \notin W^+ \vee l_r \neq j\right) \tag{14}$$

Student's competency L is characterized by a set $LA = \{\theta_i : \theta_i \in R, i \in I_{LOC} \subset \mathbb{N}\}$, where $\theta_i \in L$ is a latent variable that corresponds to the proficiency level of learning outcome LOC_i. The main aim of the educational program's realization $\langle V^b, V^f, \tau_0 \rangle$ is to construct learning process, so that during the time $t \leq \tau_0$ the requirements of $\theta_i(t) \geq \theta_i^*, LOC_i \in V^f$ will be met, with the fact that at the beginning of training the conditions of $\theta_i(0) \geq \theta_i^*, LOC_i \in V^b$ were met. θ_i^* is a threshold competency level for learning outcome $LOC_i \in V$. Hence, we should define a set of threshold values $LA^* = \left\{\theta_i^* : \theta_i^* \in R, i \in I_{LOC} \subset \mathbb{N}\right\}$ in addition to the set V. Aims of the courses, modules, etc., can be defined in the same way.

If all the defined requirements for the program and courses had been met, then the developed program will be correct and no data corruption will occur. Hence, the described model helps to create a fully correct educational program.

Based on the defined structure of an educational program we can conclude that a set of courses depends on the set of prerequisites or, in other words, learning outcomes obtained by a student before he started the program. Hence, a learning trajectory can be formed based on the related prerequisites and outcomes of the educational program. To provide an automatic formation of the program that can consider different types of courses including MOOCs that can be used in programs, it is necessary to develop an algorithm that analyzes all the prerequisites and outcomes simultaneously, and returns a list of all possible learning trajectories of the educational program, which meet the stated requirements. This algorithm consist of the following stages:

- The courses that ensure all the prerequisites are chosen from the whole set of courses C in a database and form a special set of courses (set A_n: $A_n = \{LC_{j_1}, LC_{j_2}, \ldots, LC_{j_q}\}, LC_{j_k} : j_{j_k} \in I_{LC}, LC_{j_k} \in C$ associated with a tuple $LC_{j_k} = \left\langle V^b_{j_k}, V^f_{j_k}, \tau_{j_k} \right\rangle$, where $V^b_{j_k}$ is a set of prerequisites for the course LC_{j_k}, and $V^f_{j_k}$ is a set of related learning outcomes). Each of the presented courses in the set can be considered as a starting course for the learning trajectory.
- The second step is to form a set of courses ensuring the stated learning outcomes of the program (set B_m: $B_m = \{LC_{j_1}, LC_{j_2}, \ldots, LC_{j_r}\}$, $LC_{j_k} : j_k \in I_{LC}, LC_{j_k} \in C$, each course of the set B is associated with a tuple $LC_{j_k} = \left\langle V^b_{j_k}, V^f_{j_k}, \tau_{j_k} \right\rangle$).
- For each pair A_n and B_m, a set of courses LT_{nm} (a specific learning trajectory) that helps to fulfil the learning trajectory is formed by the following rule: $LT_{nm} = \{LC_{j_1}, LC_{j_2}, \ldots, LC_{j_p}\}, LC_{j_k} : j_k \in I_{LC}$, where $V^b_{j_s} \in \left(V^f_{j_{s-1}} \cup V^f_{j_{s-2}} \cup \ldots \right)$. The whole set LT_{nm} consists only of the unique courses, and corresponds with the requirements and properties of the educational program. The prerequisites of the following course in a trajectory belongs to a set of learning outcomes of the previous course or a union of the sets of learning outcomes related to a number of previous courses.

When the prerequisites and learning outcomes of the educational program are stated, a set of learning trajectories LT_{nm} is formed using the described algorithm. Hence, after the algorithm had analyzed a set of all courses according to the requirements of the educational program, a specified learning trajectory LT can be chosen from a set of learning trajectories LT_{nm}.

5 Modeling of the Educational Program

The presented educational program's structure is successfully applied in AcademicNT LMS. During more than 15 years of system's application, many educational programs had been developed and realized. Number of almost all types of program's structures stored in LMS is presented in Table 2.

Let us consider an example of a learning trajectory built for a particular educational program. These trajectories are based on the experience a student had obtained before starting the current educational program. An example of the formed trajectory using a described mathematical model is presented in Fig. 2.

According to the learning trajectory presented in Fig. 2, the educational program could be started, when student's experience satisfies the two obligatory prerequisites

Table 2. Number of educational program elements

Elements	Number of instances
Courses	1587
Information resources	3904
Virtual labs	348
Tests	4131
Synopses	1648
Practical works	998

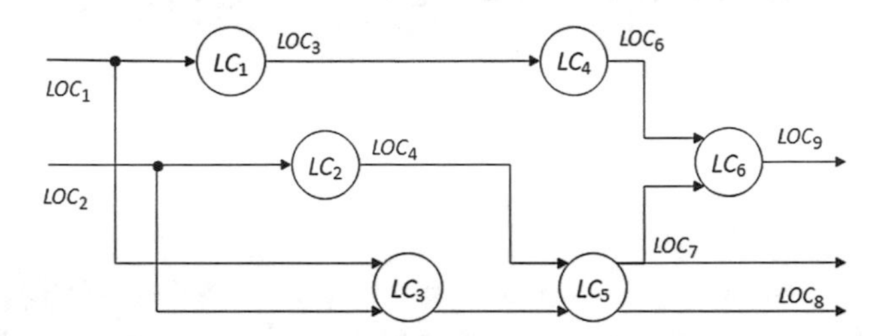

Fig. 2. The first learning trajectory with specified learning outcomes

$\{LOC_1, LOC_2\}$ that consist of six courses $\{LC_1, LC_2, \ldots, LC_6\}$ with total number of the resulting learning outcomes equal three $\{LOC_7, LOC_8, LOC_9\}$. Some of the courses can be approved MOOCs as stated in the introduction of the paper. It can be seen, that learning outcomes of previous courses are used as prerequisites for the next ones. Hence, based on the presented learning trajectory the curriculum for the specified educational program presented in Fig. 3 can be formed.

Described learning trajectory states that learning outcomes of the first course have influence on the permission to pass the fourth course, and fifth course can be started only after successful completion of the second and the third one. Moreover, the sixth course uses learning outcomes of the fourth and the fifth ones as its prerequisites. The formed curriculum for the program assumes the parallel passage of the 1st, 2nd and 3rd courses, as well as simultaneous passage of the 4th and 5th courses when the first three

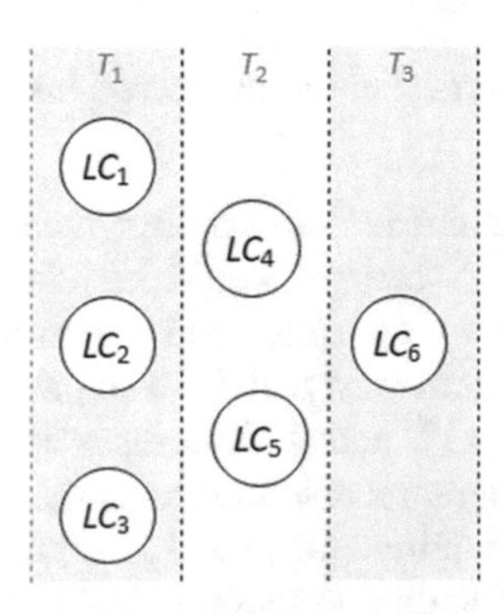

Fig. 3. The curriculum for the first learning trajectory

ones had been successfully completed. Outcomes of the program are formed after all the courses were finished.

As well as the learning trajectory presented in Fig. 2, another learning trajectory for the assumed educational program presented in Fig. 4 can be defined.

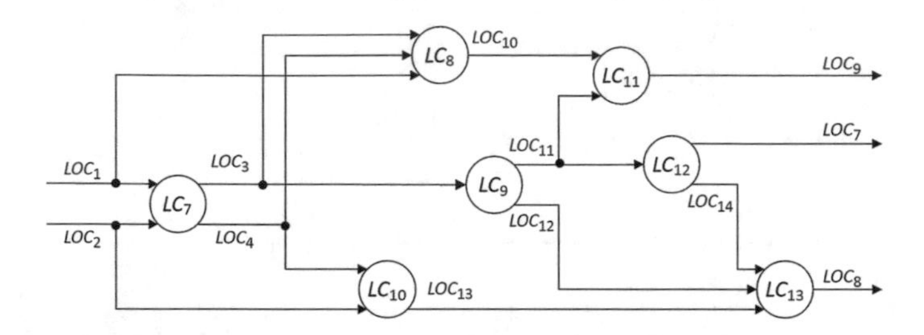

Fig. 4. The second learning trajectory with specified learning outcomes

The formed trajectory is based on the same prerequisites $\{LOC_1, LOC_2\}$ and provides the same three learning outcomes $\{LOC_7, LOC_8, LOC_9\}$. However, a set of courses that are used to achieve the stated outcomes differs between two trajectories. The second learning trajectory presented in Fig. 4 consists of 7 courses $\{LC_7, LC_8, \ldots, LC_{13}\}$ that result in different learning outcomes each, but the final outcomes of the formed trajectory correspond with program's outcomes. Hence, a different curriculum can be formed (Fig. 5).

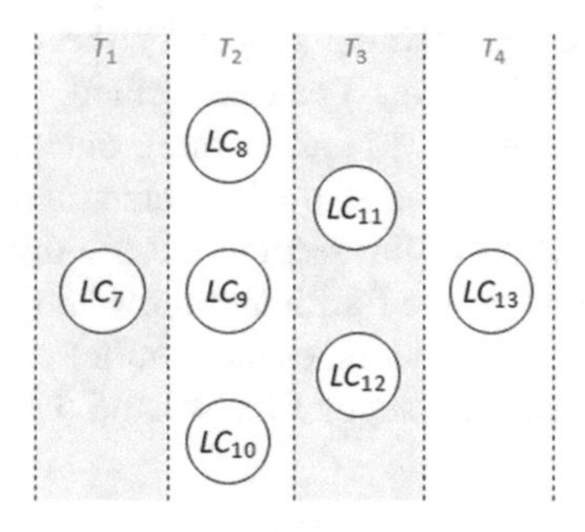

Fig. 5. The curriculum for the second learning trajectory

According to the presented curriculum scheme, the sequence of the courses can be described as following. The 7th course should be taken as the starting course as it provides learning outcomes that are further used as prerequisites in the 8th, 9th and 10th courses that are launched simultaneously. After a student had taken all these courses, a simultaneous passage of the 11th and 12th courses is provided. The 11th and 12th courses provide two of the program's resulting outcomes LOC_7 and LOC_9, however, in order to fulfill the program's requirements, the 13th course should be included after the 12th one for obtaining the last learning outcome LOC_8. Having passed all the courses, the whole list of the educational program's requirements are completed.

6 Conclusion

e-Learning has become one of the most perspective directions in modern education. It provides many opportunities for storing and managing learning materials and assessment tools. Learning management systems are widely used nowadays in order to simplify interaction with stored information and provide access to all available materials for students. This paper is related to a description of the educational program's structure, which is applied in AcademicNT learning management system. Moreover, it defines a unique set of the requirements for the educational program that helps a program to be formed correctly without any data corruption. A mathematical model, which helps to describe all the related rules, is presented. The provided description of an application of the presented model in AcademicNT shows all the practical advantages of the model. Nowadays many MOOCs have come into being, and a possibility of their application in educational programs at universities had been stated. The presented model aimed to consider all types of courses, as it is based on the analysis of the prerequisites and learning outcomes of a course. Hence, the use of the mathematical model helps to develop an educational program with all types of different courses, if they are not prohibited by an educational institution.

Acknowledgement. This paper is supported by Government of Russian Federation (grant 074-U01).

References

1. Sammour, G., Al-Zoubi, A., Gladun, A., Khala, K., Schreurs, J.: MOOCs in universities: Intelligent model for delivering online learning content. In: Proceedings of the 2015 IEEE Seventh International Conference on Intelligent Computing and Information Systems (ICICIS), pp. 167–171 (2015)
2. Freire, M., Martínez-Ortiz, I., Moreno-Ger, P., Fernández-Manjón, B.: Requirements for educational games in MOOCs. In: Proceedings of the IEEE Global Engineering Education Conference (EDUCON), pp. 993–997 (2015)
3. Kopylov, D., Fedoreeva, M., Lyamin, A.: Game-based approach for retaining student's knowledge and learning outcomes. In: Proceedings of the 2015 IEEE 9th International Conference on Application of Information and Communication Technologies (AICT), pp. 609–613 (2015)
4. Venugopal, G., Jain, R.: Influence of learning management system on student engagement. In: Proceedings of the 2015 IEEE 3rd International Conference on MOOCs, Innovation and Technology in Education (MITE), pp. 427–432 (2015)
5. Chezhin, M.S., Efimchik, E.A., Lyamin, A.V.: Automation of variant preparation and solving estimation of algorithmic tasks for virtual laboratories based on automata model. In: Vincenti, G., Bucciero, A., Vaz de Carvalho, C. (eds.) E-Learning, E-Education, and Online Training. LNICST, vol. 160, pp. 35–43. Springer (2016)
6. Efimchik, E., Lyamin, A.: RLCP-compatible virtual laboratories. In: Proceedings of the International Conference on E-Learning and E-Technologies in Education (ICEEE 2012), pp. 59–64 (2012)

7. Johnson, L., Adams Becker, S., Cummins, M., Estrada, V., Freeman, A., Hall, C.: NMC Horizon Report: 2016 Higher Education Edition. The New Media Consortium, Austin (2016)
8. Toth, P.: New possibilities for adaptive online learning in engineering education. In: Proceedings of the 2015 International Conference on Industrial Engineering and Operations Management (IEOM), pp. 1–5 (2015)
9. Uskov, V., Lyamin, A., Lisitsyna, L., Sekar, B.: Smart e-Learning as a student-centered biotechnical system. In: Vincenti, G., Bucciero, A., Vaz de Carvalho, C. (eds.) E-Learning, E-Education, and Online Training. LNICST, vol. 138, pp. 167–175. Springer (2014)
10. Lisitsyna, L., Lyamin, A., Skshidlevsky, A.: Estimation of student functional state in learning management system by heart rate variability method. Front. Artif. Intell. Appl. **262**, 726–731 (2014)
11. Laksitowening, K.A., Hasibuan, Z.A.: Personalized e-learning architecture in standard-based education. In: Proceedings of the 2015 International Conference on Science in Information Technology (ICSITech), pp. 110–114 (2015)
12. InLOC Standard: Integrating Learning Outcomes and Competences (2013)
13. Information technology for learning, education and training: Information model for competency, Part 1: Competency general framework and information model (ISO/IEC 20006-1) (2014)
14. Information technology for learning, education and training: Information model for competency, Part 2: Proficiency level information model (ISO/IEC 20006-2) (2014)
15. IEEE Standard for Learning Technology: Data Model for Reusable Competency Definitions (1484.20.1-2007) (2008)

A Simple MVC-Framework for Local Management of Online Course Material

Frode Eika Sandnes[1,2](✉) and Evelyn Eika[1]

[1] Faculty of Technology, Art and Design,
Oslo and Akershus University College of Applied Sciences, Oslo, Norway
{Frode-Eika.Sandnes,Evelyn.Eika}@hioa.no
[2] Westerdals Oslo School of Art, Communication and Technology,
Oslo, Norway

Abstract. Managing online materials for large classes can be time-consuming and error prone. In particular, it can be challenging to manage long lists of students, lecture progress, and auditorium schedules as these often change on a daily basis. We therefore introduce a simple Model-View-Controller (MVC) framework implemented in Excel that can help teachers handle daily tasks more efficiently. Examples include how to generate lecture plans, student presentation schedules, and peer-review plans for students. The authors have successfully used the system for more than five years in several courses. The framework simplifies the task of reusing material from one teaching semester to another. Teachers only need to focus on the content and not the visual appearance.

Keywords: Model-View-Controller · Course management · Students · Higher education · Content management

1 Introduction

When teaching a course at university level, the teacher faces several challenges. First, courses often comprise many students. Some students enroll late, some students drop off in the middle of the course, and others become ill or need extensions or certain treatment for various reasons. Consequently, the list of students changes frequently.

Second, classrooms and lecture theatres are often a scarce resource. Limited resources are even more a challenge when different classes follow different teaching paradigms. For example, semester programs require the same rooms on a weekly basis throughout the semester, while intensive courses require rooms in concentrated intervals at certain time of the semester. Therefore, students and teachers must relate to classes scheduled at different weekdays, different times, and in different locations across different weeks.

Third, it is common practice to follow a teaching plan. Experienced teachers will deviate from the plan depending on the interaction and events in lectures and classes.

If a class becomes engaged in a fruitful discussion, the teacher can postpone material to later lectures, while if there is little discussion, the teacher can cover more material than planned.

We assume that a learning management system (LMS) is the main channel for formal information and communication between the teacher and the students [1–4].

V.L. Uskov et al. (eds.), *Smart Education and e-Learning 2017*, Smart Innovation, Systems and Technologies 75, DOI 10.1007/978-3-319-59451-4_15

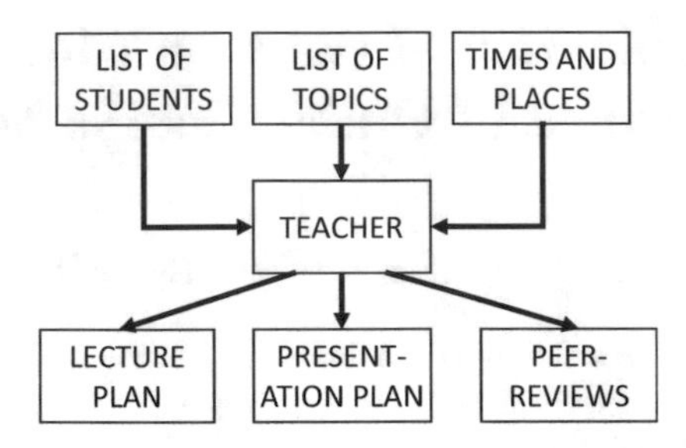

Fig. 1. The dynamic nature of micro managing a course

Regular changes to the teaching plan, lists of students, and room-schedule take up much of teachers' time as they need to update the online material published on an institutional LMS. Often, these systems do not provide the functionality needed by teachers and students. For instance, if there are changes to the progress of lectures, the published teaching plan needs updating. A teaching plan usually comprises topics and dates. These are important as students use these to know which lectures to attend.

Moreover, some courses include compulsory student presentations, where each student, or a group of students, must present an assigned presentation at a designated time-slot. Several time-slots, perhaps across multiple weeks, are needed if there are many students. The teacher needs to organize and publish the presentation schedules and update these due to changes in the list of students or the assigned rooms.

A course may also utilize peer-review where students give feedback on each other's work. Such peer-review activities usually require careful planning and the task can be quite complicated and error-prone when done manually.

Figure 1 shows how the teacher needs to act on changes in the list of students, in-class progress, and allocated auditoriums. This example includes lecture plans, student presentation plans, and peer-reviews; naturally other documents could also be included.

The examples above require focus and attention as mistakes can easily occur and may lead to misunderstandings. This work proposes a framework to assist teachers with managing updated material for publishing to students. The framework presented is created using the Microsoft Excel spreadsheet software as Microsoft Office is widely available. Moreover, one may substitute Microsoft Excel with any other spreadsheet application such as OpenOffice Calc as the same principles apply.

2 Spreadsheet MVC Framework

The following sections present the MVC framework.

2.1 The MVC Pattern

The framework presented herein is inspired by the MVC pattern introduced by Reenskaug [5]. This pattern is widely used in all areas of user interfaces from native applications [6] to web-applications accessible via browsers and smartphones [7] and

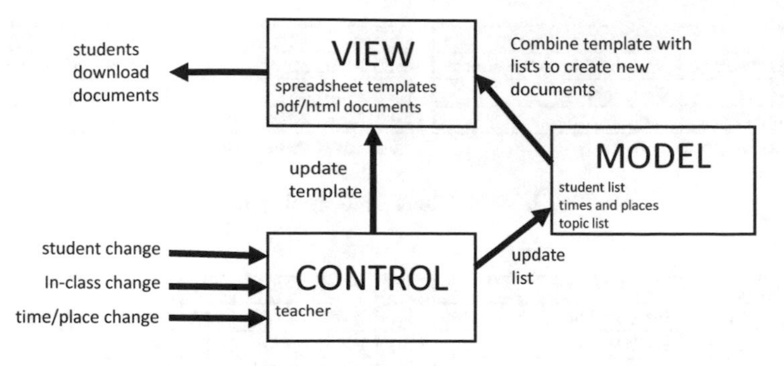

Fig. 2. MVC adopted for semi-automatic content micro management

even LMSs [8]. The main advantage of the MVC pattern is to separate the visual appearance from the application logic, which again is separated from the data. It is then easier to make improvements in the various parts independently. As an example, a recent trend is that online resources should be universally accessible and satisfy the WCAG accessibility guidelines [9, 10]. Also, it is possible to experiment with the visual presentation of the published materials that enhance students' motivation and interest [11, 12]. The MVC pattern is the de facto standard for interactive systems.

Figure 2 illustrates how course contents are changed dynamically, inspired by the MVC pattern. Unlike the traditional MVC pattern, the framework is semi-automatic as it relies on manual intervention from the teacher. For instance, changes to the list of students usually come in various forms such as emails from students or via oral messages in, or outside, class. Teachers follow the teaching progress and room allocations closely during the semester. Commonly teachers have to use a special room-booking system to find room changes.

Next, the list of students, the lecture topics, and the times and places represent the model. The views are static pdf or html-documents. Excel templates are used to generate these documents whenever the model is changed. Although document generation is automatic, the teacher must initiate the generation process. The following sections describe how to (a) use the spreadsheet to maintain lists and templates and (b) generate the static documents.

2.2 Mapping Topics to Times

Students often consult lecture plans to find out what contents teachers will cover in upcoming lectures. In our framework, a cell represents each lecture session with a textual descriptions. A column of lecture cells represents series of lecture sessions. The dates for the respective lectures are put into the cells of an adjoining column. Excel facilitates simple arithmetic on dates such as adding 7 to say that a cell is one week later, or adding 1 if it is the next day.

The teacher only has to worry about the first date and then let Excel handle the day of the week, the month, number of days per month, etc. Figure 3 illustrates how to schedule three weekly lectures.

	A	B
1	21.08.2017	Introduction to the course
2	=A1+7	Basic theory
3	=A2+7	More theory...

Cells as they are input

	A	B
1	Mon. 21. Aug, 2017	Introduction to the course
2	Mon. 28. Aug, 2017	Basic theory
3	Mon. 4. Sep, 2017	More theory...

Cells as they are displayed

Fig. 3. Scheduling three weekly lectures

	A	B
1	21.08.2017	Introduction to the course
2	=A1+2	Basic theory
3	=A1+7	More theory...
4	=A2+7	Some practice

Cells as they are input

	A	B
1	Mon. 21. Aug, 2017	Introduction to the course
2	Wed. 23. Aug, 2017	Basic theory
3	Mon. 28. Aug, 2017	More theory...
4	Wed. 30. Aug, 2017	Some practice

Cells as they are displayed

Fig. 4. Scheduling several lectures during a week with across-week references

	A	B
1	21.08.2017	Introduction to the course
2	=A1+2	Basic theory
3	=A1+7	More theory...
4	=A3+2	Some practice

Cells as they are input

	A	B
1	Mon. 21. Aug, 2017	Introduction to the course
2	Wed. 23. Aug, 2017	Basic theory
3	Mon. 28. Aug, 2017	More theory...
4	Wed. 30. Aug, 2017	Some practice

Cells as they are displayed

Fig. 5. Scheduling several lectures during a week with within-week references

The first lecture is scheduled at a specific date, in this instance, Monday August 21, 2017 (cell *A1*). The second lecture is August 28, 2017 (*A1* + 1 in cell *A2*). The third lecture is scheduled for Monday, September 4, 2017.

Figure 4 shows how to schedule courses on the Monday and Wednesday, recurring subsequent weeks. Here, both cells *A3* and *A4* refer back seven days. Figure 5 shows an alternative where cell *A4* refers back two days within the same week instead. Note that the cell references are updated correspondingly when using copy and paste in the spreadsheet.

It is easy to schedule regular courses. For irregularly scheduled courses, each cell is given a specific date according to the room-booking schedule. If there are changes to either the times or the lecture progress, it is trivial to update the lists by moving cells around or modifying cells according to needs. When reusing the material for subsequent semesters, it may be sufficient simply to alter the start date to update the entire teaching plan with correct days of week and dates. The next section illustrates how to generate the documents.

2.3 Generating HTML-Documents

The lists, including topic-time assignments and students, are stored in separate workbook sheets as lectures, students, etc. Each html document is constructed in a separate sheet. Once the sheet is constructed, all its contents are copied into a blank html-file with the extension html. The teacher can inspect the html-files using a web browser.

	A	B	C	D	E
1	=header				
2	=beginTable				
3	=beginLine	=lecture!a1	=separator	=lecture!b1	=endLine
4	=beginLine	=lecture!a2	=separator	=lecture!b2	=endLine
5	=beginLine	=lecture!a3	=separator	=lecture!b3	=endLine
6	=endTable				
7	=footer				

Fig. 6. Building html-markup

	A	B
1	=header	<html><body>
2	=beginTable	<table>
3	=beginLine	[illegible]
4	=separator	=</tr></td><tr><td>
5	=endLine	=</tr></td>
6	=endTable	</table>
7	=footer	</body></html>

Fig. 7. HTML-definitions assigned to variables

The html document in the html sheet is built using a mixture of html markup and content from the lists. Figure 6 shows how to build an html document from the lecture example in Fig. 3. Cells in column A are references to spreadsheet variables. The teacher can define commonly-used markup in just one place, namely, a separate style sheet. Figure 7 shows an example of a sheet with html-definitions. Changes to these definitions affect all the documents. Note that this example is stripped for certain markup-details and styling to simplify the presentation herein.

The labels in column *A* are provided for easy reference. To create an actual variable in excel, the cell needs to be given a name. Figure 8 illustrates how to give cell *B4* the label *dateStart* (see the cell label field below the highlighted file menu). The content from the lecture lists is referred to in column *B* and *D*, respectively. Figure 9 shows how the resulting document appears in the html sheet. These cells appear as ordinary html-markup when pasted into a text document.

2.4 Generating PDF-Documents

We use Google Chrome to convert html-documents to pdf as Google chrome contains a convenient function for generating pdf-documents. Alternatively, one may use any other html-to-pdf converter.

2.5 Conditional Styling

Documents that are more complicated may require conditional styling of elements. For example, conditional styling is used if one wants certain elements to be marked as a deadline and no-lectures using consistent color and formatting.

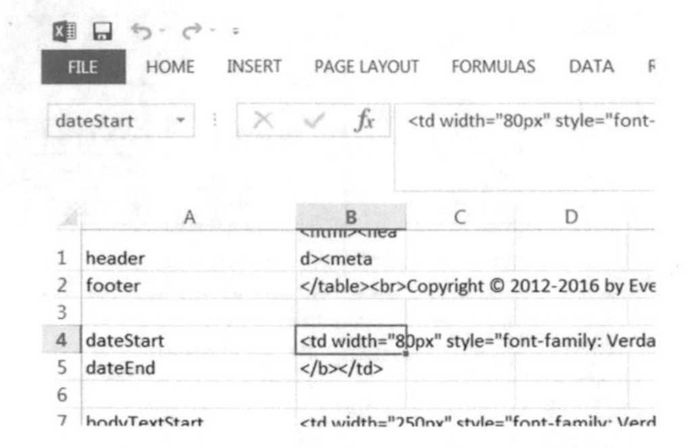

Fig. 8. Assigning variable names to cells in Excel

	A	B	C	D	E
1	=<html><body>				
2	=<table>				
3	=<tr><td>	Mon. 21. Aug, 2017	=</tr></td><tr><td>	Introduction to the course	=</tr></td>
4	=<tr><td>	Mon. 28. Aug, 2017	=</tr></td><tr><td>	Basic theory	=</tr></td>
5	=<tr><td>	Mon. 4. Sep, 2017	=</tr></td><tr><td>	More theory..	=</tr></td>
6	=</table>				
7	=</body></html>				

Fig. 9. The appearance of the spreadsheet codes (from Fig. 6)

A	B	C
21.08.2017		Introduction to the course
23.08.2017		Basic theory
28.08.2017	Holiday	Public holiday
30.08.2017	deadline	Some practice

Fig. 10. Controlling conditional markup

Figure 10 shows an example of conditional markup of a lecture plan where column *A* contains the dates and column *C* the topics, while column *B* indicates the conditional markup. The corresponding html sheet may contain the following:

```
=IF(schedule!B4="holiday";holidayStart;IF(schedule!B4="
deadline";deadlineStart;normalStart))
```

This definition ensures the conditional formatting by checking if the corresponding cell (*B4*) in the schedule sheet contains the tag “holiday” or “deadline”, in which case the *holidayStart* markup or *deadlineStart* markup is included.

Figure 11 illustrates how to generate a lecture schedule with the framework. This example uses purple colored texts for deadlines and grey colored texts for holidays. This example also contains decorated formatting of dates to assist students’ rapid comprehension. In addition, the framework automatically inserts month headings. The following definition achieves this.

```
=IF(tmp!F4<>tmp!F3;CONCATENATE(monthSeparatorBegin;UPPE
R(tmp!F4);monthSeparatorEnd);"")
```

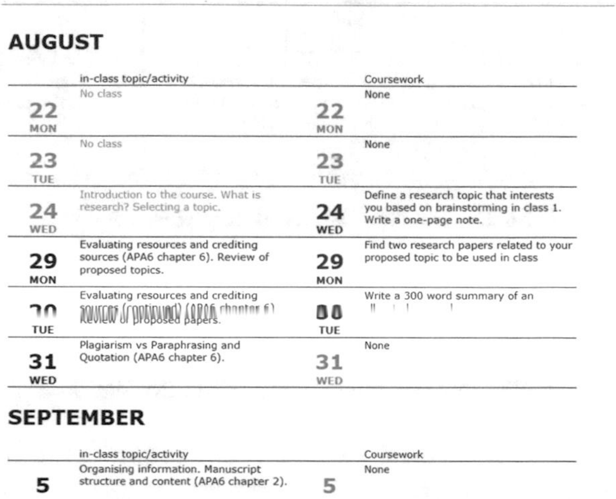

AUGUST

	in-class topic/activity		Coursework
22 MON	No class	22 MON	None
23 TUE	No class	23 TUE	None
24 WED	Introduction to the course. What is research? Selecting a topic.	24 WED	Define a research topic that interests you based on brainstorming in class 1. Write a one-page note.
29 MON	Evaluating resources and crediting sources (APA6 chapter 6). Review of proposed topics.	29 MON	Find two research papers related to your proposed topic to be used in class
30 TUE	Evaluating resources and crediting sources (continued) (APA6 chapter 6) Review of proposed papers.	30 TUE	Write a 300 word summary of an
31 WED	Plagiarism vs Paraphrasing and Quotation (APA6 chapter 6).	31 WED	None

SEPTEMBER

	in-class topic/activity		Coursework
5	Organising information. Manuscript structure and content (APA6 chapter 2).	5	None

Fig. 11. Rendering of a lecture schedule with conditional formatting

Group presentations

September 28, 2016	Erna Solberg and Jens Stoltenberg
September 29, 2016	Torbjørn jagland and Siv Jensen Kontra Bass and Electric Gitar
October 12, 2016	Ben Affleck and Emma Thomson Jonas Gahr Støre and Trine Skei Grande
October 13, 2016	Donald Duck and Mickey Mouse Erna Solberg and Jens Stoltenberg
October 19, 2016	Erna Solberg and Jens Stoltenberg Torbjørn jagland and Siv Jensen

Individual presentations

October 20, 2016	Bjarne Banan Erna Solberg Trond Giske
October 26, 2016	Kåre Willock Jens Stoltenberg Clemic Thommessen
October 27, 2016	Marit Nybakk Kennet Svendsen Svein Roald Hansen

Fig. 12. A fictitious presentation schedule with both individual and group presentations

The definition refers to a tmp sheet with the month parts isolated from the dates. It is then easy to check if months of two neighboring cells are equal or different. The monthSeperatorBegin and monthSeparatorEnd variables contain the styling, and the Excel UPPER function is used to convert the month name to uppercase for esthetical purposes.

	A	B	C	D	E	F	G	H
1	1							
2	=IF(LEN(C1)>0;A1+C1;A1)	=IF(LEN(C1)>0;schedule!A1;"")	=IF(ISNUMBER(SEARCH("group present";schedule!C1));IF(LEN(schedule!C1)<20;2*pairs;pairs);"")	=E5&F5&G5&H5	=IF($C2>E$1;INDIRECT("tmppair!a"&($A2+E$1))&delimiter;"")	=IF($C2>F$1;INDIRECT("tmppair!a"&($A2+F$1))&delimiter;"")		
3	=IF(LEN(C2)>0;A2+C2;A2)	=IF(LEN(C1)>0;schedule!A1;"")	=IF(ISNUMBER(SEARCH("group present";schedule!C2));IF(LEN(schedule!C2)<20;2*pairs;pairs);"")	=E5&F5&G5&H5	=IF($C3>E$1;INDIRECT("tmppair!a"&($A4+E$1))&delimiter;"")	=IF($C3>F$1;INDIRECT("tmppair!a"&($A3+F$1))&delimiter;"")		

Fig. 13. Structure for scheduling presentations

2.6 Scheduling Presentations

Figure 12 shows a document containing the presentation schedule for students enrolled onto a course. We rendered this document using both information in the lecture list and the student list.

In short, the method first identifies timeslots allocated for presentations. Presentation slots are numbered consecutively. These numbers are used to pull names from the student list using the INDIRECT excel function as the INDIRECT function parameterizes cell references. The process is only performed if the lecture cell is not empty.

This means that the presentation sheet contains more rows than what are actually used. Such empty rows pose no problem as html-browsers do not display blank space. An instance of the word "presentations" in the topic description of the schedule-sheet signals a presentation. If the "presentations" keyword is the only word, the entire session is allocated for (more) presentations; otherwise, half of the timeslot is allocated for the lecture and the rest is allocated for (fewer) presentations.

The first column of a student-sheet lists the students with one student per row. Similarly, a list of pairs can be set up where students are combined randomly or systematically.

Next, the main engine for generating the presentation schedule is a temporary *tmppres*-sheet. In this sheet, one column (in this case column *C*) is connected to the schedule-sheet. If the schedule sheet contains the keywords "group-presentation" or "individual presentation", the corresponding cell is given a number indicating the number of students to present for the given date. This means that lecture slots without presentations will remain blank, or zero. For all cells in column-*C* which have a value greater than zero, the corresponding date in schedule is copied into the neighboring cell in column *B*. Column *A* is used to indicate the start index of the student to present from the list. It is updated by using the current start index with the number of students in cell *C* of the previous row added to it.

The current student is copied into a list in column *D*. This list is built by copying the names from the student list in the students-sheet to the remaining column of the tmppres-sheet. The start index in column A for a given row indicates the first student. The structure of the tmppres-sheet is provided in Fig. 13.

Finally, the presentation-output-sheet copies the cells in column *B* from the *tmppres*-sheet, namely, the date for the presentation, or nothing if there is no presentation.

	A	B	C	D	E	F
1	=RAND()	=RAND()	=RANK(A1;A:A)	=RANK(B1;B:B)	=INDIRECT(CONCATENATE("stduent!a";A1))	=INDIRECT(CONCATENATE("stduent!a";B1))
2	=RAND()	=RAND()	=RANK(A2;A:A)	=RANK(B2;B:B)	=INDIRECT(CONCATENATE("stduent!a";A2))	=INDIRECT(CONCATENATE("stduent!a";B2))
3	=RAND()	=RAND()	=RANK(A3;A:A)	=RANK(B3;B:B)	=INDIRECT(CONCATENATE("stduent!a";A3))	=INDIRECT(CONCATENATE("stduent!a";B3))

Fig. 14. Excel commands to generate a peer-review list

	A	B	C	D	E	F
1	0,453	0,563	2	3	John	Beth
2	0,785	0,888	1	2	Lisa	John
3	0,123	0,931	3	1	Beth	Lisa

Fig. 15. The appearance of the peer-review Excel commands (depending on random seed)

Html styling is added if the cell in column *B* for the given row is non-empty. For each row, the student is copied from cell *D* if the date field is non-empty.

Although the procedure may appear complicated, it only needs to be set up once. Given such a structure, the teacher can simply edit the progress plan, parameters controlling the number of students per lecture, and student lists. The framework performs the fitting of students into the lecture sessions. Certain fine-tuning may be needed as student numbers vary from year to year, while the amount of lectures usually remains fixed. However, with the framework, the teacher can easily adjust the parameters until the students fill up the slots.

2.7 Organizing Peer-Review

The last example illustrates how to generate peer review lists using Excel. For each student, the peer review list shows the other students who are to check the work. Clearly, students should not review their own work and no student should review the same work twice. Two reviewers are common, and the example presented herein assumes two reviewers.

Row *A* in the student-sheet contains the names of the students. For each student, a random number is generated for each reviewer. Given two reviewers, two random numbers are generated. To make the routine more generic, a random number is only generated if the corresponding cell in the student list is not empty. Next, the Excel RANK function is used to generate a rank of the random numbers in a new row for each row of the random numbers. This rank gives the index to the reviewing student in the student list. The student is thus accessed using the INDIRECT function based on the rank. The structure of the peer review sheet is shown in Fig. 14. Figure 15 displays how this may appear (depending on the random number seed).

Note that only the reviewers' names are included here for simplicity. The reviewees' names are copied directly from the student list.

For each review, three simple checks are performed to ensure that the list is correct. First, we check if the two reviewers are identical. Second, we verify if the first reviewer is identical to the reviewee. Lastly, we inspect if the second reviewer is identical to reviewee. The results of all these checks are summed, where true is 1 and false is 0. If this sum is larger than zero, the sheet is refreshed giving it a new random seed. The process is repeated until there are no errors. It is usually sufficient to refresh the sheet around 3–5 times before the result is free of errors.

3 Conclusions

This paper presented a simple framework for managing course information. A spreadsheet was used to implement the framework, as spreadsheet software is commonly available. Although some aspects of the implementation are intricate, it only needs to be implemented once. The teacher can configure the framework to suit a specific teaching scenario without having to know programming of html. The framework ensures that information is only stored in one place. Changes, such as updated student lists and lecture schedule, propagate automatically to all affected documents. Furthermore, the framework separates the visual layout from the content, giving teachers freedom to tailor the visual appearance of teaching material without affecting the content and vice versa. The framework helps teachers reduce time and errors.

References

1. Wang, Q., Woo, H.L., Quek, C.L., Yang, Y., Liu, M.: Using the Facebook group as a learning management system: an exploratory study. Br. J. Educ. Technol. **43**, 428–438 (2012)
2. McGill, T.J., Klobas, J.E.: A task–technology fit view of learning management system impact. Comput. Educ. **52**, 496–508 (2009)
3. Beatty, B., Ulasewicz, C.: Faculty perspectives on moving from Blackboard to the Moodle learning management system. TechTrends **50**, 36–45 (2006)
4. Weaver, D., Spratt, C., Nair, C.S.: Academic and student use of a learning management system: implications for quality. Australas. J. Educ. Technol. **24**, 30–41 (2008)
5. Reenskaug, T.: Thing-model-view-editor—An example from a planning system. Technical note, Xerox Parc (1979)
6. Krasner, G.E., Pope, S.T.: A description of the model-view-controller user interface paradigm in the smalltalk-80 system. J. Object Oriented Program. **1**(3), 26–49 (1988)
7. Leff, A., Rayfield, J.T.: Web-application development using the model/view/controller design pattern. In: Fifth IEEE International Proceedings on Enterprise Distributed Object Computing Conference, EDOC 2001, pp. 118–127. IEEE (2001)
8. Yu, Z.Q., Ran, S.Y., Li, S.: Design and implementation of teaching material management system based on MVC pattern. Comput. Technol. Dev. **1**, 58 (2006)
9. Eika, E., Sandnes, F.E.: Authoring WCAG2. 0-compliant texts for the web through text readability visualization. In: Antona, M., Stephanidis, C. (eds.) Proceedings of HCI International 2016, Universal Access in Human-Computer Interaction. Methods, Techniques, and Best Practices. LNCS vol. 9737, pp. 49–58. Springer (2016)

10. Eika, E., Sandnes, F.E.: Assessing the reading level of web texts for WCAG2. 0 compliance —Can it be done automatically? In: Di Bucchianico, G., Kercher, P. (eds.) Advances in Design for Inclusion, pp. 361–371. Springer (2016)
11. Jian, H.L., Sandnes, F.E., Law, K.M., Huang, Y.P., Huang, Y.M.: The role of electronic pocket dictionaries as an English learning tool among Chinese students. J. Comput. Assist. Learn. **25**, 503–514 (2009)
12. Jian, H.L., Sandnes, F.E., Huang, Y.P., Cai, L., Law, K.M.: On students' strategy-preferences for managing difficult course work. IEEE Trans. Educ. **51**, 157–165 (2008)

Multi-agent Smart-System of Distance Learning for People with Vision Disabilities

Galina Samigulina[1(✉)], Adlet Nyussupov[2], and Assem Shayakhmetova[3]

[1] Doctor of Technical Sciences, Chief of Laboratory of the Institute of Information and Computational Technologies, Almaty, Kazakhstan
galinasamigulina@mail.ru

[2] Engineer of Laboratory of the Institute of Information and Computational Technologies, Almaty, Kazakhstan
moniumverse@outlook.com

[3] Kazakh National Research Technical University Named After K. I. Satpayev, Almaty, Kazakhstan
asemshayakhmetova@mail.ru

Abstract. This article focuses on the design of smart-system of distance learning (DL) for people with impaired vision (PIV) based on multi-agent approach. There are considered the most common multi-agent platforms and based on them practical applications. Particular attention is paid to Java Agent Development Framework platform (JADE), there were analyzed the advantages and disadvantages. There were considered main problems of DL for people with impaired vision to provide high-quality engineering education in the shared laboratories (SL) with modern equipment. There was offered an integrated solution to this problem by the use of cognitive, ontological, statistical, intellectual and multi-agent approaches. Were developed the structure and the algorithm of functioning of the smart-system of DL for PIV using different agents.

Keywords: Distance learning · People with impaired vision · Multi-agent smart-system · JADE · Integrated approach

1 Introduction

Nowadays, distance learning (DL) is one of the most important form of education, which during the process of formation united in himself all the best from the teaching and learning experience [1]. Thanks to the flexibility and universal approach, DL becomes competitive with conventional systems of learning, offering new methods of modernization of educational activities based on the use of the latest technology, the latest achievements of computer technology and modern telecommunication devices [2, 3]. The introduction of modern approaches and technologies improve the quality of education and increase its effectiveness. The most promising is the DL based on intellectual systems and smart-technologies that have broad capabilities to ensure the highest level of learning [4]. Modern smart-systems of DL offer affordable environment

V.L. Uskov et al. (eds.), *Smart Education and e-Learning 2017*, Smart Innovation, Systems and Technologies 75, DOI 10.1007/978-3-319-59451-4_16

of knowledge supply and are characterized by fast processing of multi-dimensional data, the prediction of the learning results, by the work based on fuzzy logic, usability, etc. [5, 6].

Actual is the use of smart-systems of DL for people with various disabilities, especially for people with pathologies of vision [7]. This category of people is in need of specialized systems, because PIV face the problem of learning in conventional educational institutions and traditional systems of DL due to peculiarities of perception. An important task is the organization of a special environment of DL available to people with visual impairments, as this type of learning involves long-term work with the computer, which greatly increases the load on the visual apparatus and reduces the level of perception of the educational information [8]. There is a need for the introduction of innovative integrated approaches that could improve the quality of the systems of DL for PIV. The most convenient way to implement these tasks is on the basis of multi-agent systems (MAS). In multi-agent systems, the solution is obtained automatically as a result of the interaction of many separate targeted software modules - the so-called agents. [9–11]. Each agent can carry out certain tasks and pursues defined goals. Agents can work together or alone and make their own decisions on the basis of modern approaches [12]. In DL these agents can perform a variety of functions based on different intellectual approaches.

Different agent platforms (AP) were used to build multi-agent environment of DL. Similar AP [13] are the software solution with graphical interface or connected libraries for such programming languages as: Java, C#, Rubby, Python, etc. There are many AP that have a number of advantages and disadvantages. They are divided into open AP and commercial. Eve platform is multi-functional and based on web technology non-commercial platform [14]. It operates in an open and dynamic environment, where agents can interact at any location: in cloud storage, smartphones, desktops, browsers, robots, device, home automation, etc. The article [15] describes the semi-closed AP AnyLogic used to build applications in the field of general use, particularly: logistics, marketing, transport modeling, server optimizing, etc. In work [16] there is considered the activity modeling of individuals and groups behavior on the basis of the commercial AP Brahms. Provided communication and patterns over a long period of time give the ability to build a reliable synchronization model. The most convenient platform for the solution of complex problems of DL for PIV is multi-agent platform JADE (Java Agent Development Framework).

2 Statement of the Problem

Statement of the problem is formulated as follows: it is necessary to implement multi-agent smart-system based on the JADE (Java Agent Development Framework) platform for distance learning for people with impaired vision, in order to obtain high-quality engineering education in a shared laboratory (SL).

We introduce the following definition: as a shared laboratory, there are considered any labs with complete infrastructure (equipment, hardware and software) for distance learning and for quality of engineering education of people with impaired vision.

3 Multi-agent Platform Java Agent Development Framework

Widely used JADE platform is an open, flexible and effective tool for developing applications based on agent-oriented technology [17]. It includes a runtime environment of the agents, a set of graphical tools to monitor the activity of agents and the library, which contains the working classes. Software environment JADE integrates with various applications on Java language. The construction of JADE agents can vary in different ranges - from simple, only reacting, to complex - mental. JADE platform consists of a main container and a number of agent containers [18], each of them may be shared between different hosts (Fig. 1).

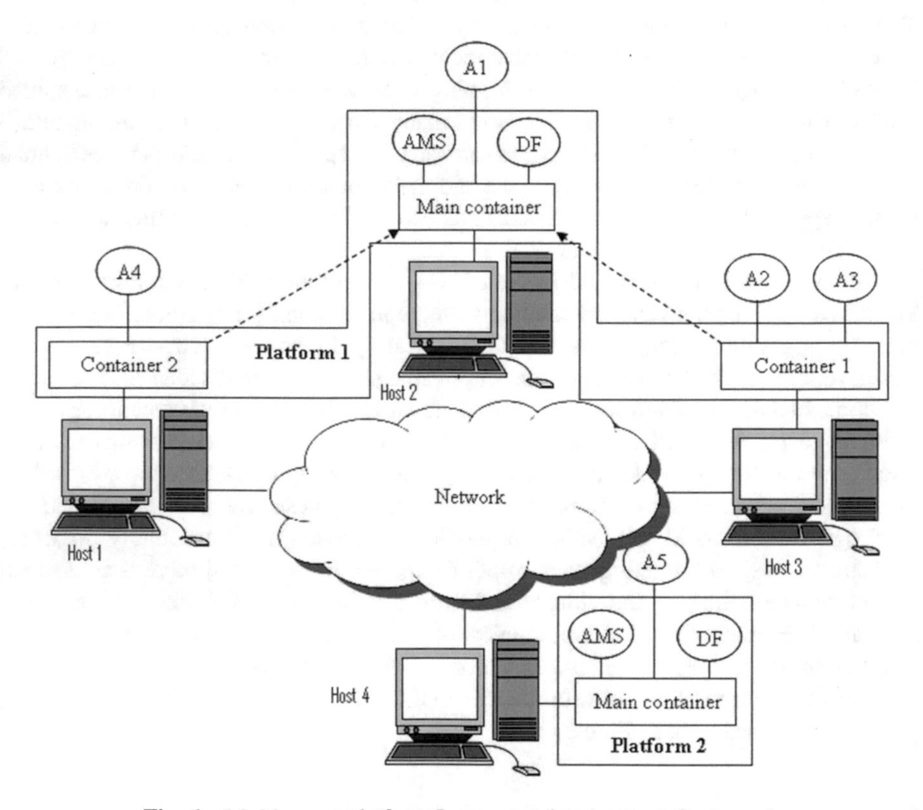

Fig. 1. Multi-agent platform Java agent development framework

Working with multi-agent platform JADE there are used standard agents Dummy Agent and Sniffer Agent, which are system controllers responsible for the work of the main container. For the transmission of information between the agents there is used a ready-agent tool of Dummy Agent, the parameters of which are set in the additional user interface. Another ready tool of Sniffer Agent allows you to view the transmitted data between agents and record them in a Log Manager.

The main container includes an Agent Management System (AMS) and Directory Facilitator (DF) of the agent. AMS system monitors the platform, manages the lifecycle of all agents within it, and provides white pages service (register of currently available agents, including their identifiers Agent Identifiers, which are used for communication). DF directory provides additional yellow pages service (agent service registry, registered in the AMS). Agents can communicate in the platform, irrespective of whether they are - in the same container (e.g., A2 and A3) or in different platforms (e.g., A1 and A5).

Communication is based on the paradigm of asynchronous messaging. The message format is based on the language of ACL (Agent Communication Language), which is composed of such fields as sender, recipient, communication and content. Communication Field informs about the intention to transfer the information to another recipient and sends a message through the sender. The Content Field contains a type of processed information. The JADE message elements can be of a primitive type (logical expression, figure, line) or of an aggregate type (user-defined structure, consisting of primitive or set of elements). Aggregated elements are usually presented as Java, XML and bytecode classes.

To program the configuration of created agents there is used compiled environment, which includes classes and libraries of JADE platform. Writing and editing of environment code is performed on Java language in an integrated environment of IDE Eclipse development. This environment provides with many different possibilities that can be found in commercial IDE: syntax highlighting in the editor, compile code, source-level debugger, the classes browser, file manager and project manager, interfaces for standard source control systems. It also contains a number of unique features: code refactoring, code automatic updates and collection, a task list, support for unit testing capabilities, as well as integration with application assembly tool.

There was developed a large number of MAS based on the JADE platform. The work [19] describes a MAS that is designed to ensure the safety of mobile devices and their protection against malicious programs. Particular emphasis is placed on the monitoring of SMS (Short Message Services) services. Agents act as behavioral analysis by identifying the correlation between the suspicious SMS-messages and their own profiles. The work [20] presents the agent-oriented software module for the development and coordination of smart-objects that represent the digital objects of information system, complemented by computational and accumulative capabilities. Smart-objects are assigned with individual agents, which bind to each other, and can do the organized overall work. The article [21] shows a multi-agent software model, which is used to index the research and supply of reference educational materials stored on different servers. This model is developed with the help of agent JADE platform and metadata LOM (Learning Object Metadata) standard, which has a hierarchical structure and provides quick access to the set information. Multi-agent technology improves the accuracy and relevance of the search. In work [22] there is shown a system of information processes modeling using JADE. There is used a library that integrates in JADE and allows proper allocation of resources. The results of comparative test showed that in the practical implementation such agent-based modeling gives the best results.

Therefore, the use of JADE platform in DL for PIV has great prospects. There is actual the developing of multi-agent smart-system of DL that allows PIV to work with the latest equipment in the SL and create accessible learning environment. The main advantages of JADE are such factors as: stability, clear interface, extensive user base, easiness of creating agents. The disadvantages are resource-intensive of computing processes.

4 Creating a Specific Information Perception Environment in DL for PIV

Different degrees of visual impairment (the nature of disease) affect the physiological characteristics of PIV in DL. It is actual the study of psycho-physiological characteristics of the visual system of PIV and the development of an integrated approach to the organization to the DL process. Problems of visual impairment lead to a quality decrease in the perceived information by the students. People with impaired vision need to be adapted to the learning environment with the help of special software and the creation of new principles of operation of DL [23]. During the learning of PIV, it is especially important to pay attention to the development of their intellectual abilities in the engineering specialties learning on the modern complex equipment and the study of advanced information technologies. Students with impaired vision have well developed hearing, therefore so actively used this channel of information perception. Considering PIV with major defects of vision (hyperopia, myopia and astigmatism), the important point is different form of the educational material supply from the computer monitor screen, including various color schemes, location area of important information. All these aspects require the implementation of specific information systems of DL for PIV.

During the organization of accessible learning environments for PIV there are used different intellectual approaches. Actual the use of the cognitive approach for the identification of perception features and information perception by PIV. Self-regulating cognitive processes in DL for PIV have a significant impact on the quality of education [24]. There are different approaches in the use of many organizational learning planning procedures in conjunction with such cognitive strategies as: metacognitive monitoring, emotions and motivations regulating. Cognitive methods used in DL for PIV allow significantly to improve the perception of learning information and to increase the effectiveness of learning. There are different approaches in the use of organizational learning planning procedures in conjunction with cognitive strategies such as: metacognitive monitoring, regulation of emotions and motivations [25–27].

Smart-system implemented on the basis of the cognitive approach is aimed at creating an adaptive learning environment, taking into account the individual characteristics of students and visual defects. Cognitive approaches in DL of PIV can significantly improve the awareness of learning information and increase the effectiveness of learning. The combination of intelligent methods (artificial neural networks [25], evolutionary algorithms [26], artificial immune systems [27], neuro fuzzy logic [28], etc.) and statistical approaches extend the capabilities of the operation of the DL for PIV system on processing of multidimensional data in real time.

In work [31] for the multidimensional data processing the is used intelligent methods of fuzzy and neuro-fuzzy logic and factor analysis. Formation of a database of informative features is performed using the cognitive approach. In order to choose the informative features, characterizing each student, there is used the method of principal components. On the basis of fuzzy logic there is selected a learning model adapted to the model of a student. Prediction of learning outcomes is based on neuro-fuzzy logic.

During the creation DL for PIV based on intelligent methods there are widely used ontological models [29]. The work [32] describes the development of combined OWL (Web Ontology Language) model for implementing intelligent innovative technology and for construction of Smart-system of distance education of people with vision disabilities. The proposed combined OWL model includes an ontological model of a student, training and shared laboratory. Software implementation of these models in the ontology editor Protégé in presented in work [33].

The development of these systems has its own specific features. Ontology models allow creating effective intelligent information systems and implementing the interaction between complex structured and formalized data. The ontological approach makes it possible to allocate resources efficiently and to improve the performance of the information system DL for PIV. Application of modern editors such as Protégé and construction of OWL (Web Ontology Language) models greatly facilitates the solution of this problem.

5 The Architecture of the Multi-agent Smart-System of Distance Learning for People with Impaired Vision

In the design of multi-agent smart-system of DL for PIV there are formed following tasks, which are realized by the agents:

- the use of the cognitive approach for feature extraction characterizing PIV with various defects of vision (myopia, hyperopia, astigmatism);
- the construction of a special information perception environment by PIV (using scoring opportunities, change the background color and text style to improve the perception of educational information) and the development of personalized learning trajectory;
- the application of the ontological approach to systematize the subject area of DL for PIV and implementation of interactions between complex structured and formalized data;
- statistical processing of multi-dimensional data on the basis of factor analysis to extract informative features of PIV students;
- intelligent processing of multidimensional data on the results of the PIV survey to predict the level and learning outcomes;
- providing information assistance to PIV in learning (guidelines and tips);
- improving the information perception by PIV;
- the introduction of the game approach in order to increase the level of perception and understanding of the information;
- the introduction of collective learning function for PIV groups;
- the introduction of the cloud approach to improve the speed of information processing in DL for PIV;

– the development of the state educational regulation (implementation of regulations, administration and control of the learning process, certification).

In order to implement these tasks with the help of multi-agent JADE platform there are created the following agents: cognitive (CA), personal (PA), ontological (OA),

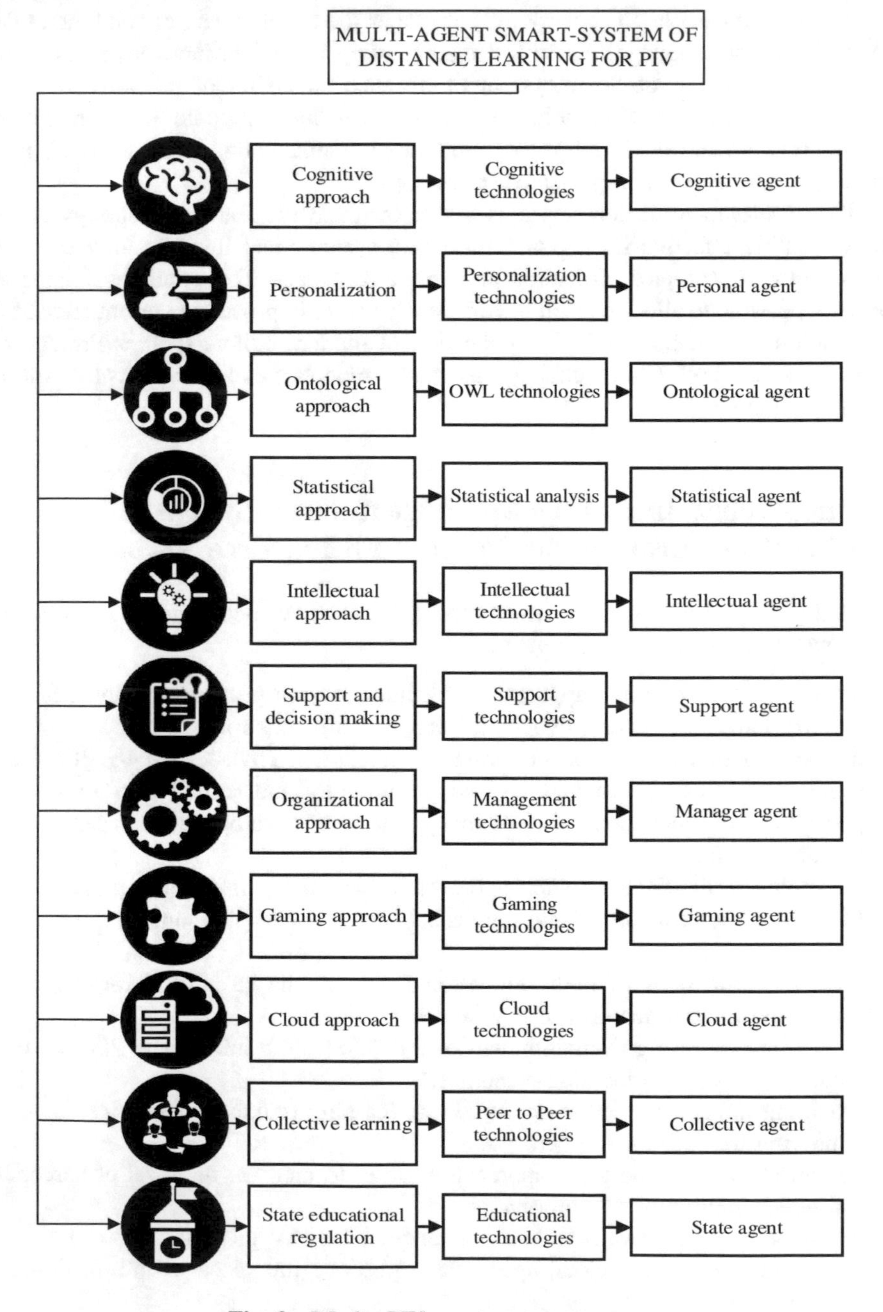

Fig. 2. DL for PIV smart-system structure

statistical (SA), intellectual (IA), support and decision-making (SuA), the manager (MA), gaming (GA), cloud (CA), collective (CoA) and the state agent (GoA). The structure of multi-agent smart-system of DL for PIV, which consists of created JADE agents, is shown in Fig. 2.

Table 1 shows the function of agents of smart-system of distance learning for people with impaired vision. Visual interface of the smart-system DL for PIV is displayed by Manager agent. During the registration PIV in the graphic system, ontological agent writes his personal information in a single structural OWL (Ontology Web Language) database. Later PIV has a test, after which with the help of an intellectual agent is determined his personalized learning trajectory. The construction such a database is made by the use of Protégé editor. The use of cognitive agent takes into account physiological factors and individual characteristics of vision perception of PIV. Manager agent defines a special environment of PIV perception, color mode, etc., and connects the editor of screen reader JAWS (Job Access With Speech) for scoring of educational materials. In the process of learning of PIV there can be used game and collective learning regimes. Gaming agent allows to master the chosen course in a game mode. Collective agent organizes to the student cooperative learning together with other students, compares the results with each other, thereby increasing the motivation to acquire knowledge. In order to control the input and output information the Statistical agent constantly monitors the activity of the students, registrate a system errors and analyzes the statistics of information system usage by the students at the time. If there will be any problem of user with the information system of DL, there is provided the support and decision-making agent, which can give tips and recommendations on choosing the necessary learning materials, various methods of laboratory and practical work on modern equipment.

Table 1. Functions of agents of smart-system of DL for PIV

Agent	Function
Cognitive	- Identifying the level of perception of PIV - Identification and implementation of cognitive techniques in DL
Personal	- The organization of individual learning trajectory of PIV based on vision defects - Selection of individual learning material for a particular student
Ontological	- Systematization of the subject area - Construction of OWL model of PIV - Construction of OWL learning model of PIV - Construction of OWL model of SL - Construction of a combined OWL model for the structuring of the system input and output data - Construction the links between different parts of the information system
Statistical	- Registration of statistical data about the users - Record event logs and system errors - Selection of informative features of PIV

(*continued*)

Table 1. (*continued*)

Agent	Function
	- Statistical analysis
Intellectual	- Processing of multidimensional data based on intelligent methods - Analysis and prediction of learning outcomes of PIV - Determination of the learning level of PIV
Support and decision-making	- Supply with the recommendation information by the students - Assistance to students for choosing teaching materials - Support the learning process through the tips and notifications - Support with the personal recommendations for learning regime
Manager	- Supply with the educational material, concretized on certain PIV taking into account features of the vision defects, the level of intelligence, level of basic knowledge in the chosen discipline - Change the background color - Change the text style - Speech Synthesis - Distribution of information on the screen - The choice of presenting information (in graphical, tabular, animation, video, etc.)
Gaming	- The introduction of gaming techniques in learning - Displaying game objects
Cloud	- Work with the use of cloud technologies - Processing of data in the cloud server
Collective	- Development of learning techniques for PIV by the vision defects in groups - Analysis of the impact of cooperative learning between several students and the development of joint forms of learning - The formation of a collective environment of learning materials support
State	- Communication of information system of DL for PIV with government agencies - The organization of the legal database (regulations, certification, etc.) - The organization of a common base of knowledge between education institutions

Cloud agent improves system capacity in SL with the use of high-performance remote servers. Cloud approach improves the speed of smart-system and ensures the stability of its work. The implementation of government agents is important for communication with government agencies and creating with them a common legal framework, as well as the ability to use common educational resources in the operation of smart-system of DL for PIV. At the end of each course, PIV has a test, according to the result of which learning outcomes are predicted by the intelligent agent and is performed operational adjustment of the process of knowledge acquiring.

6 Conclusion

The scientific novelty of the proposed smart-system is an integrated approach in the design of DL for PIV based on combined use of intellectual and statistical methods for the multi-dimensional data processing, and taking into account psycho-physiological peculiarities of perception of learning information by people with impaired vision, based on the cognitive and ontological approach, implemented on the basis of the MAS.

Therefore, the advantage of the developed smart- system of DL for PIV based on multifunctional agents are:

1. Identification of physiological, intellectual and psycho-physiological peculiarities of perception and awareness of the information by PIV (for people suffering from myopia, astigmatism and hyperopia) on the basis of the cognitive agent.
2. Construction of the different OWL models of PIV and systematization of subject area using ontology agent.
3. Processing of statistical data and the selection of informative features of PIV using statistical agent.
4. Intellectual data analysis and prediction of learning outcomes of PIV based on intellectual agent.
5. Formation of the special perception environment of PIV, registration of the supply and location of educational information on the screen, depending on the visual defects, as well as a choice of color schemes, best for people with various diseases using an agent manager.
6. The ability to improve certain parts of the smart-system of DL for PIV independently from the other, with the help of the modernization of the necessary agent.
7. Getting a quality engineering education of PIV on modern equipment with the required special software in the SL. As an example, there was offered the course "Design of Information Systems" for national scientific SL of information and space technologies in KazNTRU named after K.I. Satpayev.

For the developed software (www.dlspiv.kz) there was received the certificate of state registration of rights to the object of copyright MJ RK [30] "DLS_PIV (Distance learning system for people with impaired vision) for distance education of people with impaired vision" and the act of implementation in Almaty branch of the public association "Kazakh society of blind" (Almaty).

The work is executed under the grant of the Committee of Science of the Ministry of Education and Science of the Republic of Kazakhstan №GR0215RK01472 (2015-2017), on the theme "Development of information technology, algorithms, software and hardware for intelligent complex objects control systems with unknown parameters."

References

1. Zawacki-Richter, O.: Distance Education. Encyclopedia of Educational Philosophy and Theory, pp. 1–7. Springer, Singapore (2015)
2. HongYan, Z., JiKui, W.: Study on learning performance evaluation of distance continuing education. In: Advanced Technology in Teaching, vol. 163, pp. 255–260. Springer, Heidelberg (2013)
3. Gouvêa Rezende, S.R., Campos, V.G., dos Reis Pereira Fantone, P., Brasil, M.M.: Interaction in distance education: student, teaching material, information technology and communication. In: HCI International 2013 - Posters' Extended Abstracts, vol. 374, pp. 76–79. Springer, Heidelberg (2013)
4. Samigulina, G., Samigulina, Z.: Intelligent system of distance education of engineers, based on modern innovative technologies. In: Proceedings of the II International Conference on Higher Education Advances, HEAd 2016, pp. 229–236, Valencia, Spain. Elsevier (2016). J. Social and Behavioral Sciences. **228**
5. Tikhomirov, V., Dneprovskaya, N., Yankovskaya, E.: Three dimensions of smart education. In: Smart Education and Smart e-Learning, vol. 41, pp. 47–56. Springer, Cham (2015)
6. Štuikys, V.: Smart education in CS: a case study. In: Smart Learning Objects for Smart Education in Computer Science, pp. 287–310. Springer, Cham (2015)
7. Samigulina, G., Shayakhmetova, A.: The information system of distance learning for people with impaired vision on the basis of artificial intelligence approaches. In: Smart Education and Smart e-Learning, vol. 41, pp. 255–263. Springer, Cham (2015)
8. Jafri, R., Ali, S.A.: A multimodal tablet–based application for the visually impaired for detecting and recognizing objects in a home environment. In. Computers Helping People with Special Needs, vol. 8547, pp. 356–359. Springer, Cham (2014)
9. Gath, M., Herzog, O., Edelkamp, S.: Autonomous, adaptive, and self-organized multiagent systems for the optimization of decentralized industrial processes. In: Intelligent Agents in Data-Intensive Computing, vol. 14, pp. 71–98. Springer, Cham (2016)
10. Panagiotis, S., Ioannis, P., Christos, G., Achilles, K.: APLe: agents for personalized learning in distance learning. In: Computer Supported Education, vol. 583, pp. 37–56. Springer, Cham (2016)
11. Wei, W.: Research on fire distance education and training system based on multi-agent. In: Advances in Computer Science and Engineering, vol. 141, pp. 193–198. Springer, Heidelberg (2012)
12. Gluz, J.C., Vicari, R.M., Passerino, L.M.: An agent-based infrastructure for the support of learning objects life-cycle. In: Proceedings of the 11th International Conference, ITS 2012, Chania, Crete, Greece, June 14–18, vol. 7315, pp. 696–698. Springer, Heidelberg (2012)
13. Singh, D., Padgham, L., Logan, B.: Integrating BDI agents with agent-based simulation platforms. J. Auton. Agents Multi-Agent Syst. **30**(6), 1050–1071(2016). Springer, US
14. Stellingwerff, L., de Jong, J., Pazienza, G.E.: Practical applications of the web-based agent platform 'Eve'. In: Advances in Practical Applications of Heterogeneous Multi-Agent Systems. The PAAMS Collection, vol. 8473, pp. 268–278. Springer, Cham (2014)
15. Postema, B.F., Haverkort, B.R.: An anylogic simulation model for power and performance analysis of data centres. In: Computer Performance Engineering, vol. 9272, pp. 258–272. Springer, Cham (2015)
16. Sierhuis, M.: Multi-agent activity modeling with the brahms environment. In: Theory, Practice, and Applications of Rules on the Web, vol. 8035, pp. 34–35. Springer, Heidelberg (2013)

17. Bergenti, F., Iotti, E., Poggi, A.: Core features of an agent-oriented domain-specific language for JADE agents. In: Trends in Practical Applications of Scalable Multi-Agent Systems, the PAAMS Collection, vol. 473, pp. 213–224. Springer, Cham (2016)
18. JADE Architecture Overview. http://www.jade.tilab.com/
19. Alzahrani, A.J., Ghorbani, A.A.: A multi-agent system for smartphone intrusion detection framework. In: Proceedings of the 18th Asia Pacific Symposium on Intelligent and Evolutionary Systems, vol. 1, pp. 101–113. Springer, Cham (2015)
20. Fortino, G., Guerrieri, A., Lacopo, M., Lucia, M., Russo, W.: An agent-based middleware for cooperating smart objects. In: Highlights on Practical Applications of Agents and Multi-Agent Systems, vol. 365, pp. 387–398. Springer, Heidelberg (2013)
21. Rocha Campos, R.L., Comarella, R.L., Silveira R.A.: Multiagent based recommendation system model for indexing and retrieving learning objects. In: Highlights on Practical Applications of Agents and Multi-Agent Systems, vol. 365, pp. 328–339. Springer, Heidelberg (2013)
22. Cardoso, H.L.: SAJaS: enabling JADE-based simulations. In: Transactions on Computational Collective Intelligence XX, vol. 9420, pp. 158–178. Springer, Heidelberg (2015)
23. Nikashina, N.: Particular of development people with confined eyesight. Vector Sci. **2**, 223–226 (2012)
24. Azevedo, R., Harley, J., Trevors, G., Duffy, M., Feyzi-Behnagh, R., Bouchet, F., Landis, R.: Using trace data to examine the complex roles of cognitive, metacognitive, and emotional self-regulatory processes during learning with multi-agent systems. In: International Handbook of Metacognition and Learning Technologies, vol. 28, pp. 427–449. Springer, New York (2013)
25. Rana, H., Rajiv, Lal, M.: Role of artificial intelligence based technologies in e-learning. Int. J. Latest Trends Eng. Sci. Technol. **1**, 24–26 (2014)
26. Lam, R.L., Firestone, J.B.: The application of multiobjective evolutionary algorithms to an educational computational model of science information processing: a computational experiment in science education. Int. J. Sci. Math. Educ. **15**(3), 473–486 (2015). Springer, Netherlands
27. Samigulina, G.: Technology Immune – Network Modeling for Intellectual Control Systems and Forecasting of the Complex Objects. Monograph, 172 p. Science Book Publishing House, USA (2015)
28. Yildiz, O., Bal, A., Gulsecen, S.: Improved fuzzy modelling to predict the academic performance of distance education students. Int. Rev. Res. Open Distrib. Learn. **14**(5), 732–741 (2013)
29. Samigulina, G., Shayakhmetova, A.: Development of smart-system of distance learning of visually impaired people on the basis of the combined of OWL model. In: Smart Education and e-Learning, vol. 59, pp. 109–118. Springer, Cham (2016)
30. Samigulina, G., Shayakhmetova, A., Suleimen, O.: DLS_PIV (Distance learning system for people with impaired vision) for distance education of people with impaired vision. In: Certificate of State Registration of Rights to the Object of Copyright MJ RK, 0090, p. 25 (2016)
31. Samigulina, G.A., Shayakhmetova, A.S.: Smart-system of distance learning of visually impaired people based on approaches of artificial intelligence. J. Open Eng. **6**, 359–366 (2016)

32. Samigulina, G.A., Shayakhmetova, A.S.: Development of the smart - system of distance learning visually impaired people on the basis of the combined OWL model. In: Proceedings of the 3rd International KES Conference on Smart Education and E-learning (KES-SEEL 2016), Puerto de la Cruz, Tenerife, pp. 109–118. Springer, Spain (2016)
33. Samigulina, G.A., Shayakhmetova, A.S., Suleimen, O.: DLS_PIV (Distance learning sys-tem for people with impaired vision) for distance education of people with impaired vision. In: Certificate of State Registration of Rights on Object of Copyright MJ RK, Astana, 15 January 2016, № 0090, p. 25 (2016)

Use of Smart Technologies in the e-Learning Course Project Management

Martina Hedvicakova and Libuse Svobodova(✉)

Department of Economics, Faculty of Informatics and Management,
The University of Hradec Králové,
Rokitanského 62, 500 03 Hradec Králové, Czech Republic
{martina.hedvicakova, libuse.svobodova}@uhk.cz

Abstract. Modern technologies have entered into all area and levels of education. This paper aims to introduce the Project Management e-learning course which was primarily designed for public administration. The paper describes the current situation and the specifics of public administration and also focuses on the strengths and weaknesses of the application of project management in public administration. The practical demonstration of the course describes the smart technologies that are used in the course, SWOT analysis of the course and then, according to the questionnaire, the evaluation of the success of the course and its possible ways of improvements.

Keywords: LMS blackboard · Project management · Public administration · Effectiveness · Smart technologies

1 Introduction

Regional policy objectives are based on defining the primary regional problems and the concept of national economic policy. The objectives are usually expressed as a decrease in GDP level per capita, reduction of inter-regional disparities in average income, level of unemployment in each region. Partial objectives are to improve the skills of the regions with technical infrastructure, to support business activities in the regions and the like [1, 2].

In today's dynamic period full of changes, which also puts demands on a "just in time", project management is increasingly gaining importance. Project management is essentially a management of changes and it can be used in any organisation whether private or public as well as in any other field. Using project management can successfully help you to plan, manage and carry out projects.

The Project Management Institute (hereinafter referred to as PMI) [3] defines project management as "the application of knowledge, skills, tools and technologies to project activities so that they meet the requirements of the project".

Kerzner [4] presents project management as "a set of activities consisting of planning, organising, directing and checking resources of a company with a relatively short-term goal which was set for the implementation of specific goals and intentions."

Němec [5, p. 22] notes that it is "a certain philosophy of approach to project management with a clearly defined objective that must be achieved within a required

V.L. Uskov et al. (eds.), *Smart Education and e-Learning 2017*, Smart Innovation, Systems and Technologies 75, DOI 10.1007/978-3-319-59451-4_17

time, cost and quality while respecting the intended strategies and while taking advantage of specific project procedures, tools and techniques".

All of the above definitions show the importance of project management in achieving the objectives and plans of projects. With the use of project management, institutions can successfully implement their projects which save costs incurred and lead to meet deadlines. If public administration effectively uses the project management, it can lead to reducing regional disparities.

The Czech Republic Interior Ministry also realised the importance of project management for public administration and in 2011 launched a three-year project called "Preparation of project managers of state government and their certification by IPMA" which was funded by the ESF of the Operational Programme for Human Resources and Employment and the Czech Republic state budget. Currently, there are ongoing preparations for use of the project products in the state government in the coming years so that demonstrable gain of knowledge and skills for project management is an available tool of the personnel development policy.

The project's main objective was to strengthen the competencies of state government employees for project activities through educational activities. Knowledge and skills obtained have been validated via internationally recognised certification exam IPMA (International Project Management Association) with grade "D". The project was carried out in accordance with the needs of state government to fulfil the "Smart Administration Implementation Strategy for the period 2007-2015" where the emphasis is put on solving the tasks set by the projects. (The Czech Republic Interior Ministry, [6].)

Poulová and Šimonová [7] focused on individual learning styles and university students. Interesting results were also obtained by Moravec et al. [8] They compared two study groups. One without e-learning support and a second with the LMS Blackboard. Their research confirmed that providing an e-learning tool for students has a positive influence on their test results.

Smart learning Researchers have tried to define smart learning to reach a consensus on its definition. According to Noh, Joo, and Jung (2011) [16], smart learning is a human-centered and self-directed learning method which connects the smart information communication technology to the learning environment. Other researcher has claimed that smart learning is intelligent and adaptive learning that considers many learning types and abilities and enables learners to foster thinking, communication, and problem solving skills using various smart devices [17, 18].

2 Methods and Goals

Smart schools have been proposed as a solution to increase the capabilities of the new generation in the era of ICT. Recently, many smart schools have been established in Iran and other developing countries. [19]

The aim of the article is to introduce the course Project management that was taken 3 years ago by city managers of municipalities. The goal of the article is to present technologies that can use managers when they would like to study this course nowadays. Teachers from the Faculty of Informatics and Management at University of Hradec Králové are processing necessary changes in the eLearning course for use of

SMART technologies in the process of education. Changes will be not done only in the study materials but also in the next tools of LMS Blackboard and in the process of education.

The second goal of the article is to analyze the utilization of LMS Blackboard tools in the educational process in "Project Management for public administration" at the Faculty Informatics and Management at the University Hradec Králové. Project Management for public administration for the 2014/2015 academic year are presented.

The article is based on primary and secondary sources. The primary sources are represented by the statistics from the utilization of the LMS Blackboard and by the author's thoughts. Secondary sources include websites, technical literature, information gathered from professional journals, discussions and participation at professional seminars and conferences. It was then necessary to select, categorize and update available relevant information from the collected published material so that basic knowledge about the selected topic could be provided.

3 Current Situation and the Specifics in Public Administration

As indicated by OECD data, public administration in the Czech Republic isn't exercised in a very effective way. At the moment there is no comprehensive overview of what public administration actually carries out, there is no sophisticated concept of how and what public administration should actually provide. Individual agendas of the administration are often badly and spontaneously carried out; they are at the mercy of contradictory, ever-changing political priorities and in captivity of departmentalism. There are only slow and partial optimizations often without any inter-agency co-ordination or more sophisticated concepts.

For this reason, the "Effective Administration" project has been developed which is part of the pillar of the "Institutions" the International Competitiveness Strategy of the Czech Republic for the period 2012-2020 (SMK) [9].

3.1 Specifics of Public Administration

Public administration solves specific issues compared to the commercial entities in relation to process control. In the commercial sector, the organisation motivation is clear - to make a profit. This main objective is helped to be reached by efficiently configured management system. If the company is inflexible and doesn't respond to the wishes of customers', it will soon lose them due to strong competition. The customer wouldn't be satisfied and would go to a competitor. The situation is logically different in the public administration organisations. Public administration organisations are established by law and as such they are a monopoly. It tempts to certain "rigidity", i.e. reluctance to changes of the established procedures [10].

We can generally identify the following as the main problems and weaknesses in the public administration organisations in relation to process management [10]:

- absence of competition - absence of competition as a natural motivation for improvement
- low flexibility - due to the complexity of the legislation and the lack of a clear description of the process,
- high turnover and low motivation of file officers – caused by the remuneration system which leaves no room for the valuation of prospective employees and the absence of a clearly-set goals
- interaction of political influences - "four-year cycle" and instability in occupying senior management positions, dependence on the state budget and any budget constraints,
- customer - term of customer (citizen) is difficult to grasp, the citizen doesn't choose, the organisation chooses and must return even when he is dissatisfied with its work,
- strong functional approach to work - it is important to meet the expectations of the supervisor worker not the citizen (low pro-customer exposure of workers),
- strict adherence to organisational structures - processes are inherently horizontal and cross the borders of several organisational units, therefore, workers of a single organisation unit could be able to participate in multiple processes,
- low flexibility to carry out the agenda - due to long lead times needed for approval and low-profile border of some workers - often difficult substitutability,
- low flexibility of spending the entrusted funds - organisations have a given budget for the entire year, its dynamic increase or reduction is difficult and tied by a large number of legislative and procedural obstacles.

Conversely, the strengths of public administration for the implementation of process management include:

- stability and backgrounds for public institutions,
- stability settings of legislative procedures - any changes are long known in advance and it is possible to carry out redesign of the process in advance [10].

4 Project Management E-Learning Course

For the above reasons, the "Project Management For Public Administration" e-learning course was developed at the University of Hradec Kralove, which was created as part of the research Project Management for solutions for local and regional disparities, initiated by the Ministry for Regional Development in the Czech Republic. All course costs were paid to the participants from the research plan.

The course aims to provide basic knowledge regarding project management, its content, scope and application in practice. After successfully completing the course, the trainees will be well-versed in project management issue, from the project definition through to its final evaluation.

One of the main objectives is to enable students to connect and use the acquired knowledge and skills in practice (e.g. increasing the efficiency of the existing projects /

investments, grant programs, dotation/, searching internal savings, streamlining the implementation of investment projects, to well create schedules and timetables, extension of business opportunities in public administration, to effectively manage project risks, etc.).

The course consisted of 15 separate chapters devoted to the most important areas of project management and provides students a theoretical and practical perspective on the issue, focusing on public administration. The whole course was planned for three months, during which, three optional meetings will be held:

1. The first will be devoted to familiarisation with the form, content and distance learning course, logging into the system and becoming familiar with the WebCT environment and the newer Blackboard. There will also be a presentation of the course and defining the requirements for successful completion (discussion and a short final treatise on the use of the project management at the workplace of the learners).
2. The second meeting will take place in the middle of the course and will primarily be used to solve problems that arise during the study and practical examples of applications for the project management in the public administration.
3. The third and last meeting is held at the end of the course and serves to exchange the experiences between the participants of the course and handing to them the certificate.

All study materials are accessible online in a virtual environment and the PDF version for easier download of the files and the possibility to study independently from a computer connected to the network at any time. Each chapter takes about 30 min a week on average [11].

The Project Management course (hereinafter referred to as PM) is created in LMS (Learning Management System) Blackboard. The course builds on the previous six modules, produced under the project of the Czech Republic MRD WE-10-05 - “Improving the quality of management of municipalities up to five thousand inhabitants II” (Fig. 1).

Tools that LMS Blackboard [12] provides are:

- Syllabus
- Content, Information and Learning Modules
- Discussions
- Groups
- Announcements
- Assessments
- Assignments
- Calendar
- Chat
- Mail
- Media library
- Glossary
- Tools
- Web links
- My grades

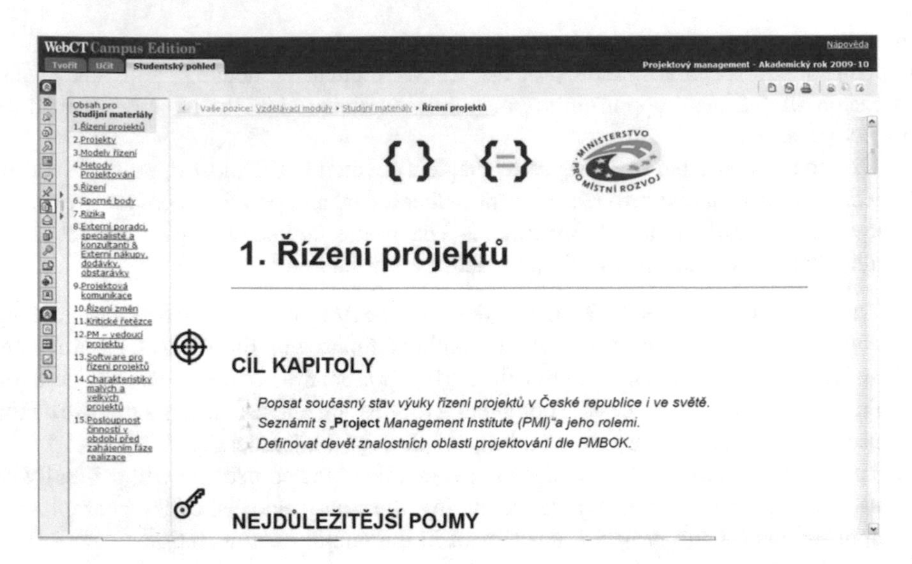

Fig. 1. Project management for public administration E-learning course (oliva.uhk.cz – course Project management in Czech language [12])

5 Smart Technology Enhanced Teaching and Learning

All materials are in PDF version of the course. In the future, we propose to combine PDF version with HTML. The second possibility is the use of the availability of 3G and 4G technologies coupled with tablet PCs and smart phones makes the communication and learning system easier. In engineering and project management education, these sophisticated technologies can bring immense advantages to the methods of delivering and acquiring knowledge [20].

- Live broadcast of a lecture to multi-class is realized by real time mixture.
- Students are highly in favor of annotation facility.
- High internet speed is required for live broadcast using real time mixture [20].

Given that Blackboard provides many tools to support online education within their LMS environment, statistics using individual instruments were monitored directly in this course. The survey results are distorted by the target group of students who unlike students at the University of Hradec Kralove, are unfamiliar with the LMS Blackboard environment and even after initial familiarisation with all features and tools, the representatives of public government were not willing to use all the opportunities of the course. One major factor was the temporal aspect.

The course pilot project was attended by 15 students. Over 4 months (3 months course + one month after the course had finished), after which the students had access to the course, they used the individual Blackboard tools as follows:

- 100% PDF download of materials
- 93% e-mail within the Blackboard
- 93% tasks

- 80% course assessment
- 80% calendar
- 60% contacts
- 40% discussion

Additional features such as instant messaging, blogs and diaries weren't used by any of the students. From the large labour deployment, it wasn't even possible to use teamwork.

Although in the final tutorial all the participants expressed those contacts with other students on the course was a strong part, only 60% of them used them within the LMS environment. Others preferred to take advantage of the possibility to exchange contacts personally at the tutorials [13, 14].

6 SWOT Analysis of the Course

After the first run of the course, a SWOT analysis of the course and subsequently a questionnaire survey were carried out regarding the student satisfaction with e-learning course (Table 1).

Table 1. Swot analysis of the course Project management

Strengths	Opportunities
LMS environment with many functions	Close co-operation with municipalities mayors
Feedback from the tutor and from virtual environment	Enrichment of knowledge and experience with competent persons
Image of an educational institution - university	A continuation of the e-learning course: Project Management II.
Experience with previous runs of courses	Lower fixed course costs
Students from the preceding six courses	Better Graphics of the course
Free course	The use of multimedia tools
Certificate of course completion	Links to the conference of the given focus
Primacy among universities	English version of the course
Exchange of information and knowledge from students from the area and the region as part of chatting and tutorials	Constantly changing legal environment - the need to update materials = the need to educate
Professionally qualified staff	
Weaknesses	Threats
Lack of practical experience of tutors with public administration	A large number of educational agencies on the market
A small number of course participants – it is not possible to exchange many experiences	Involving more universities in education in the given region
Insufficient support for chat and discussions	Charging rate after the project completion

(*continued*)

Table 1. (*continued*)

Weaknesses	Threats
Profiles of the participants were missing (important for all distance courses)	Loss of student's motivation due to the knowledge testing
Ignorance of LMS environment by the course participants	LMS competitive environment (e.g. MOODLE)
Small promotion of the course	References of the students
	Increased need to update educational materials for the public administration

Overall, the Project Management course for public administration could be evaluated as beneficial for public administration workers who don't have sufficient knowledge in this area.

We can evaluate especially the insufficient inter-connection and a focus on public administration which was the aim of the course, as the weak point. Some other students also complained about this fact. This fact may help other educational institution to focus on this target area and to create courses "tailored-made" for the public administration.

Although it is necessary to analyse the needs of public administration and constantly changing legal environment. If the courses are constantly updated, it will facilitate the work of public administration. Everything depends on practical knowledge of the given issue by the tutor and creative team [14].

7 Discussion

In the modern era of rapid changes of information and technology, the process of teaching and learning is changing. Using ICT in education has been proposed to be led to increase in education quality, expansion of learning chances and accessibility of education beyond the classroom [19].

During online learning, motivation to study is very important, as well as other factors. One of the major parts of the motivation is on the specific instructor that is teaching the students. A personal approach, helpful attitude, emphasis and enthusiasm about topics are most important. The most important factor for successful study is self-motivation. Well-prepared materials in the subject and a great instructor do not mean success.

Utilization of smart technologies and social networks in the process of the education is the next important issue in the nowadays turbulent time. The question should be: Why students do not use communication tools in the LMS instead of the other channels like Facebook, email, telephones, Skype, face-to-face etc.? [15]

The question for discussion is the selection of the project team within the public administration which is often bound by employment agreements and a project manager cannot pick their own people into the team or fire them if their performance of tasks is low. Most respondents of the course also complained about this fact.

8 Conclusion

Due the situation on the market in the connection with Education 4.0 which responds to the needs of Industry 4.0 it is possible to expect, that in future will be highlight collaborative, creative learning community in the process of the education. [21, 22] In the course Project management can participants create the content of the topics for example in the discussion where they can establish the topics that are important for them and discuss hot topics with other participants of the course.

Project Management can be applied wherever it is necessary to achieve the desired goal - to push through a change. Therefore, it can find use in all fields and also in commercial and private spheres.

For this reason, the e-learning course aimed at project management for the needs of public administration with regard to its specifics, originated at the University. The application of project management in the implementation of individual projects will increase their success and this can lead to a reduction in disparities between regions. Project Management is also supported by the selected Ministries.

After a successful first run of the e-learning course of Project Management for Public Administration, the students raised requirements for the continuation of the given course. It will be primarily regarding soft skills. These skills are currently needed for each employee in any professional direction, including the public administration. Students will also be able to try to work directly in the software designed for project management (e.g. Microsoft Office Project and others). Short video tutorials will also be developed to help students to understand the given issues.

These social and economic pressures have served to redefine the role of the university, the academic and the student in the learning transaction [21].

Acknowledgements. This study is supported by internal research project no. 2103 Investment under concept Industry 4.0 at Faculty of Informatics and Management, University of Hradec Kralove, Czech Republic. We would like to thank student Petra Henclova for cooperation in the processing of the article.

References

1. Macháček, J., Toth, P., Wokoun, R.: Regionální a municipální ekonomie. Oeconomica, Praha (2011). 199 p.
2. Homolka, J., Koulová, D.: Hodnocení hospodářského a sociálního rozvoje vybraných regionů. In: Proceedings of the International Scientific Conference Region in the Development of Society 2014, Mendelova univerzita v Brně, 2014, pp. 279–289 (2013)
3. Project Management Institute: A guide to the project management body of knowledge. PMBOK Guide, 4th edn. Newtown Square Publisher (2008)
4. Kerzner, H.: Project Management: A Systems Approach to Planning, Scheduling and Controlling. Wiley, New Jersey (2006)
5. Němec, V.: Projektový Management. Grada Publishing, Praha (2002). 184 p.
6. Novák, P.: Posílení projektového řízení ve státní správě. Ministerstvo vnitra České Republiky (2016). http://www.mvcr.cz/clanek/posileni-projektoveho-rizeni-ve-statni-sprave.aspx

7. Poulová, P., Šimonová, I.: Individual learning styles and university students. In: EDUCON 2012, IEEE Third Annual Global Engineering Education Conference, pp. 128–133. IEEE, US Piscataway (2013)
8. Moravec, T., Štěpánek, P., Valenta, P.: The influence of using e-learning tools on the results of students at the tests. Procedia – Soc. Behav. Sci. **176**, 81–86 (2015)
9. Ministry of Industry and Trade, Czech Republic, BusinessInfo.cz, Projekt 01, Efektivní veřejná správa (2016). http://www.businessinfo.cz/cs/clanky/smk-efektivni-verejna-sprava-7312.html
10. SOFO Group: Metodická příručka pro potřeby projektového řízení v organizacích ústřední státní správy, Centrální podpora projektového řízení projektů MVČR a jím řízených organizací (2012). http://www.smartadministration.cz/soubor/projekty-dokumenty-3-metodicka-prirucky-pro-potreby-projektoveho-rizeni-v-organizacich-ustredni-statni-spravy.aspx
11. Kala, T., Lacina, K., Mohelská, H., Bachmann, P.L., Hedvičáková, M., Dittrichová, J., Nekola, J.: Management pro řešení disparit mezi obcemi a regiony. Vyd. 1, 248 p. Professional publishing, Praha (2011)
12. Course Project Management. http://oliva.uhk.cz/
13. Hedvičáková, M.: Project management in public administration sector, innovations and advances in computer, information, systems sciences, and engineering. Lecture Notes in Electrical Engineering, vol. 152, pp. 741–749. Springer, New York (2013)
14. Hedvičáková, M., Svobodová, L.: Project management as a tool for solving regional disparities, pp. 254–260. Mendelova univerzita v Brně, Brno (2016)
15. Svobodová, L., Hedvičáková, M.: The Use of LMS Blackboard Tools by Students in subject Enterprise Accounting (in press)
16. Noh, K., Joo, S., Jung, J.T.: An exploratory study on concept and conditions for smart learning. J. Digital Policy **9**(2), 79–88 (2011)
17. Kwak, D.: Meaning and prospect for smart learning. In: Proceedings from the Seminar for Korea e-learning Industry Association, 13 December 2010 (2010)
18. Sung, M.: A study of adults' perception and needs for smart learning. Procedia – Soc. Behav. Sci. **191**, 115–120 (2015)
19. Taleb, Z., Hassanzadeh, F.: Toward smart school: a comparison between smart school and traditional school for mathematics learning. Procedia – Soc. Behav. Sci. **171**, 90–95 (2015)
20. Alelaiwi, A., Alghamdi, A., Shorfuzzaman, M., Rawashdeh, M., Hossain, M.S., Muhammad, G.: Enhanced engineering education using smart class environment. Comput. Hum. Behav. **51**(Part B), 852–856 (2015)
21. Fasso, W., Knight, B.A., Knight, C.: Development of individual agency within a collaborative, creative learning community. In: Encyclopedia of Information Science and Technology, 3rd edn., p. 10 (2015)
22. Fisk, P.: Education 4.0 … the future of learning will be dramatically different, in school and throughout life (2017). http://www.thegeniusworks.com/2017/01/future-education-young-everyone-taught-together/

Approaches to the Description of Model Massive Open Online Course Based on the Cloud Platform in the Educational Environment of the University

Veronika Zaporozhko, Denis Parfenov(✉), and Igor Parfenov

Orenburg State University, Orenburg, Russia
fdot@mail.osu.ru

Abstract. The use of cloud computing in open education resulted in the emergence of a new dynamic form of e-learning – massive open online courses. Massive open online courses are considered as large-scale web-based learning courses of a new generation that allow free admittance to the content of academic disciplines or subjects for a huge number of students from around the world at the same time. The article presents an analysis of their specific characteristics. To formalize the process of designing and creating online course the authors developed two models: four-level model describing the hierarchy of the basic levels of user access to the courses, and mathematical reflecting the interconnectedness of the course elements. The technique of course creating and placing on modern cloud platform is described in the abstract. The materials presented can be useful in the development of mass open online courses and their implementation in the educational process in the institutions of higher education.

Keywords: SMART environment · SMART e-learning · e-learning · Open education · Online course · Massive open online course · MOOC · Electronic learning content · Cloud computing · Cloud technologies · Cloud platform

1 Introduction

Nowadays the leading universities are realizing the consolidation of available educational platforms creating resources for layout of massive open online courses (MOOC). This work is conducting in order to popularize education and create an available SMART e-learning environment to get knowledge [1, 2], or to increase the competitiveness of universities and their integration into international educational space. The demand for online courses is increasing with every passing year. First of all, it is connected with their ability to provide a high level of education availability, virtual academic mobility and creating great opportunities for individual educational paths throughout all the life (the conception of life-long learning).

Massive open online courses represent large-scale web-based learning courses of the modern generation and allow mastering the content of educational disciplines or subjects without any charge for the huge number of students from all over the world at

V.L. Uskov et al. (eds.), *Smart Education and e-Learning 2017*, Smart Innovation, Systems and Technologies 75, DOI 10.1007/978-3-319-59451-4_18

the same time. «A course of study made available over the Internet without charge to a very large number of people. 'anyone who decides to take a MOOC simply logs on to the website and signs up'» [3].

There is no doubt that to control such large data flow a scalable solution is vital [4]. In the scope of the investigation, we developed the conception of the educational platform, which maintains MOOCs allocation based on the cloud calculation [5, 6]. We devised a four-level model of the basic levels of user access to the massive open online courses, and also described the processes of their development and allocation on the educational platform.

In the scope of the investigation, we analyzed the main distinctive features of massive open online courses in comparison with other solutions at the open education market [7–13].

First of all, for the open education organizing it is essential a significant quantity of participants studying simultaneously (up to thousand people) to be put down for the course.

The other feature of MOOC is its openness and availability for all the subjects of educational process whether it is a student, a whole university or a large corporation that promotes its flagship advancements for people through e-learning. During creation and usage of massive open online courses, it is necessary the following aim to be chased: access to the learning must be open for everybody as the reduction of barrier for getting education leads to development and consolidation of human resources. Studying in informal atmosphere lets being flexible in allocating of efforts for forming necessary competence. A free access to training for anyone who wants without formal requirement of basic level of education is not in the last position. Payment is performed only for voluntary certification.

The third feature of MOOC is permanent access to study material and educational process online with the usage of modern conveniences. The access to the course over the Internet provides for educating anytime, anyplace, when it is convenient and in a calm setting. Open logging for course enables any interested person to start learning instantly having received unique user accounts. An important feature is that course materials are placed in the educational platform and are available from any gadget whether it is a computer, a laptop, a tablet, or a smartphone. Distant access enables to create an individual schedule with taking into account the course realization form a student registered for (synchronous lesson – online/asynchronous lesson – video lecture recording). One more distinctive feature is cycle (periodicity) of a course, which allows to plan a convenient time for studying a material and doing tasks. Usually a course starts every 3–6 months. At the same time, periodicity enables to create definite studying period – the date of starting and finishing a course is fixed what disciplines students making virtual educational process look like a real one. Besides there are definite time scopes – lessons are divided into week cycles from 5–8 to 30 weeks depending on volume and complicatedness of an educational course.

The next feature of MOOC is qualitative electronic learning content, which is a structured subject content used in educational process. The next components can be a part of the structure of every course: short video lectures of from 5 till 20 min duration that can be carried out both online and offline, the elements of computer-assisted control of learning objectives usually by testing; practical exercises; questions for discussion; means of internetworking (educational forums, blogs, learning communities). Besides there is an opportunity to keep track of individual student's progress and network storage of portfolio of his made works. An electronic journal of achievements helps students to keep track of study progress, to see knowledge gaps.

High quality of created courses is achieved by attracting teachers of the leading universities and high-quality specialists tutor support in the open educational space. The leading universities teachers all over the world are involved into MOOC creation. By the competent construction and the content of the course adapted for the target group studying a material becomes more available to anyone who wants. Feedback with course authors is organized in the course scope to get consultations on the questions arose. Thus, students take part not only in studying materials but also in finishing and enhancing the course. The advantage of MOOC is the process of certification. The results of course completing are accepted by other universities. After successful finishing of educating unconfirmed (free) and confirmed (paid) electronic certificates are given.

Thus, the above specific features of massive open online courses enable us to make a conclusion that they represent didactic foundation of new e-learning form.

2 The Fourth Level Model of the Basic Levels of User Access to the Massive Open Online Courses

The hierarchical structure of access to MOOCs helps distribute the objects - the main components that make up a SMART e-learning environment of open education - through the levels, indicate relationships between objects and functions of these objects. In the scope of the investigation we created the four-level model of user access to MOOCs (Fig. 1). Let us describe every level of created model in detail.

The first level is formed MOOCs, which represent the structured electronic learning content and address obligatory internetworking between all the participants of educational process. Students use MOOCs for self-education in the scope of educational programs and for raising the level of the skills. Usually in the process of developing MOOC dozens or hundreds of people are involved.

The second level is represented by MOOC Developers – Universities are engaged directly in the design, creation and maintenance of MOOCs, in particular the preparation of e-learning content or an academic certification after successful finishing of education. If the educational institution developed MOOC places in their campus network, the access to them, as a rule, only a few thousand members, mostly students of the University. Nowadays there are a huge number of universities involved in MOOC creation, particularly, Stanford University, University of London/Goldsmiths, Princeton University, Harvard University, University of Pennsylvania, Massachusetts Institute of Technology (MIT), Johns Hopkins University, University of Michigan, Peking University, University of Oxford, Higher School of Economics (Russia).

On the third level, there are MOOC providers – companies which vendor electronic learning content. Massive open online courses delivery to the end users is based on different MOOC Provider Platforms (online platforms, cloud educational platforms), which provide with the technological component of e-learning and allowing to

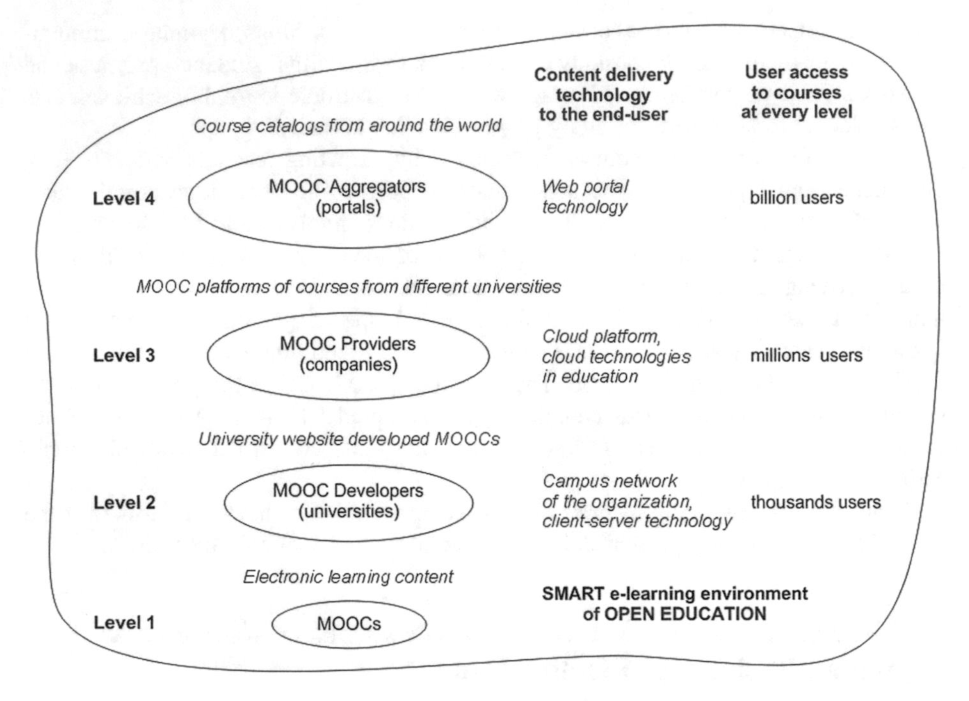

Fig. 1. The fourth level model of user access to MOOCs

significantly reduce the cost of creating, publishing and maintenance MOOC. In this case, access to MOOC gets millions of users from around the world. The most efficient MOOC platforms are Open EdX (edx.org), Coursera (coursera.org), Udacity (udacity.com), NovoEd (novoed.com), Udemy (udemy.com), Harvard Open Courses (extension.harvard.edu/academics/courses/course-catalog), MIT OpenCourseware (ocw.mit.edu), Khan Academy (khanacademy.org) [14–21].

The fourth level involves MOOC aggregators – specialized portals, which provide with the catalogue of wide variety of the best online courses from different providers, publish courses ratings and students' responses. They also provide for advanced course search by different parameters, and give the opportunity to create an individual schedule in a personal account of student, receive notifications as the planned date of course beginning approaches. Undoubtedly, in this case accesses to courses have billions of users around the world. Such online course aggregators are: Academic Earth (academicearth.org), Class Centra (class-central.com), CourseBuffet (coursebuffet.com), CourseTalk (coursetalk.org), Degreed (degreed.com), eclass (eclass.cc), EMMA – European Multiple MOOC Aggregator (platform.europeanmoocs.eu), LearningAdvisor (learningadvisor.com), MOOC List (mooc-list.com), Mooctivity (mooctivity.com) [22–31].

The categories of users are students, teachers, tutors, instructors, course managers, design team.

3 Structure Model of the Massive Open Online Course

To have clearer understanding of the main links between MOOC structural elements we developed the next mathematical model.

Let *MOOC* be the set of MOOCs put on the MOOC platform. Every MOOC consists of two disconnected sets.

$$MOOC = Required\, elements \cup Extra\, elements, \tag{1}$$

where *Required elements* – the set of required course elements, defining its minimal content, and *Extra elements* – the set of extra course elements, which enrich it, let students know about others possibilities (for example, Remote Proctor Solutions and Certificate program).

The set of required elements can be presented as

$$Required\, elements = (Welcome, Syllabus, Welcome\, Unit, Unit, Final\, Unit, Grades), \tag{2}$$

where *Welcome* – Welcoming video; *Syllabus* – Course Plan; *Welcome Unit* – Introduction; *Unit* – standard training Unit, $u_j \subset U, U_j = \overline{1 \ldots K}$; *Final Unit* – Final Course Reflection (for example, Video), Final assessment (Test/Project); *Grades* – User report, Accomplishments page, Educational achievements progress bar.

The set of extra course elements can be presented as

$$Extra\, elements = (Exam, Certificate), \tag{3}$$

where *Exam* – Final Exam, *Certificate* – Course Certificate.

The set of elements of *Welcome Unit* is

$$Welcome\, Unit = (Overview, Creators, Learning\, goals, Instruction), \tag{4}$$

where *Overview* – About this course; *Creators* – Information about course creators and instructors; *Learning goals* – Learning goals, Learning objectives, Roadmap; *Instruction* – References, Introduction Video.

Every standard training Unit also contains a definite set of elements. Let us describe it as

$$Unit = (Study\, guide, Type\, of\, resource, Type\, of\, activity, Feedback), \tag{5}$$

where *Study guide* – Guidelines for the student, Unit Objectivies.

Plenty elements of electronic learning resources *Type of resource* we describe as

$$Type\, of\, resource = (Video, Presentation, Reading, Weblinks, Multimedia), \tag{6}$$

where, *Video* – Video lectures; *Presentation* – Lecture Notes & Slides; *Reading* – Materials for independent learning (for example, tutorials/manuals, interactive textbooks, documents, lectures in pdf or html); *Weblinks* – Weblinks to additional resources (Digital Electronic Libraries, Educational Video channels, Articles, Maps, Photos, any other Social Media and Internet resources); *Multimedia* – multimedia components (3D animation, Animated video clips, Workbook audio, recording screencasts, virtual learning models, other).

The set of activity elements of the course *Type of activity* we describe as

$$Type\ of\ activity = (Questionnaire, Quiz, Assignment, Workshop, Glossary, Game), \tag{7}$$

where *Questionnaire* – Web questionnaire or Web poll; *Quiz* – Quick interactive quiz (test); *Assignment* – Self-report, Essay, Project, Mini action research, Exercise, Task, other; *Workshop* – Peer Assessment, Shared documents for collaborative work (project work) and peer tutoring; *Glossary* – for example, thesaurus, technical glossary, wordlist; *Game* – Educational game, including simulation video game, virtual worlds.

The set of feedback elements in the course *Feedback* we describe as

$$Feedback = (Forum, Blog, Social\ network, Video\ \&\ audio\ communication, other), \tag{8}$$

where *Forum* – Centralized discussion forum, Centralized consultation forum, Online learning community; *Blog* – Teaching or Learning Blog; *Social network* – distributed

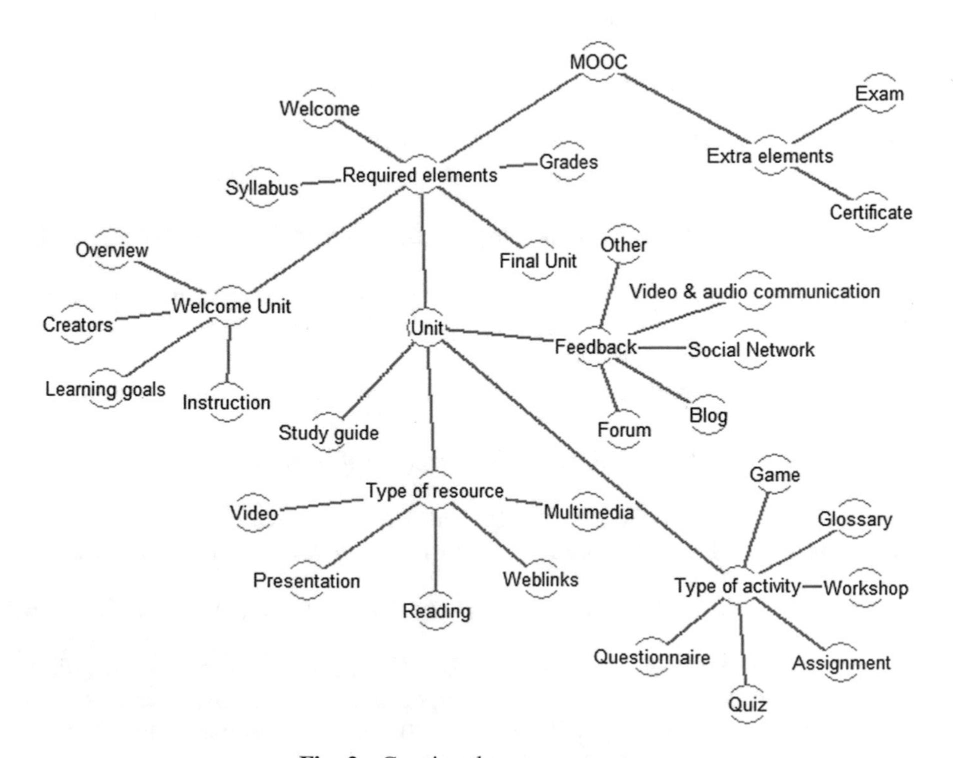

Fig. 2. Creating the course structure

open spaces (for example, Facebook, Twitter, Google+, YouTube, other); *Video & audio communication* – Webinar, Video and audio conference, Live webcasting, Online Meetings; *other* – Any other feedback (Virtual classroom, E-mail, Chat, Wiki, other shared writing/editing tools).

For graphical display of the structural MOOC model construct a connected graph that showing the elemental composition of the course and structural relationships between its elements (Fig. 2).

4 Massive Open Online Courses Development Process

Various aspects of MOOC development are presented in the following scientific publications [32–37]. Despite the depth of research and a variety articles in this field, the task of improving efficiency MOOC development process remains urgent and requires further decisions.

Having understood the structure of the course, we can discuss the development process of a standard MOOC. Based on the provisions of a modern model of instructional design ADDIE [38], International quality standard ISO/IEC 19796-1 Information technology — Learning, education and training — Quality management, assurance and metrics [39], we describe all the process in the next scheme (Fig. 3).

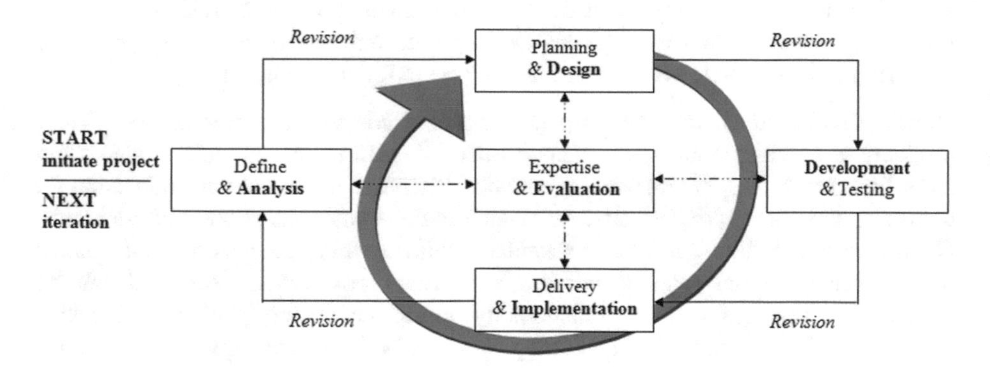

Fig. 3. The massive open online courses development process

Define and Analysis. Formulate the learning goals and learning objectives of the MOOC creation taking into account the characteristics of the target audience. Determination student learning outcomes. Formation and analysis of competencies, students' knowledge, skills and actions. Identifying requirements for the students' beginning level of training, etc. Clarification of the full workload of the MOOC and awarding of credit hours (units). Determination the recommended time for optimal study with this course. Analysis of human resources and other resources needed to develop the MOOC. Analysis functional and didactic capabilities of existing MOOC Platforms. The MOOC schedule production.

Planning and Design. Determination of the didactic concept of MOOC. What are the learning object? Description learning objectives (learning goals) in accordance with the revised Bloom's Taxonomy. What overall results or outcomes should be accomplished by students? Modeling the structure and content of the course in accordance with the proposed structural MOOC model (see Sect. 3), the creation of the course (content) tree. Planning the detailed MOOC program and description course structure where each element corresponds to a specific learning objective and contributes to the achievement of the overall course goal. Selecting teaching and learning strategies (distribution knowledge, interaction, collaboration), content delivery strategy to the end user (formats of educational and training resources and materials), strategy for the provision of electronic learning content and management of personal learning paths (static content, adaptive dynamic content), interaction strategy (student-content, student-student, student-tutor/lecturer), criteria based assessment strategy of learning outcomes (formative – assessment for learning, summative – assessment of learning), strategy for placing and maintaining the MOOC (on the server in the campus network, the cloud platform), the strategy for testing the quality of the MOOC (functional testing, non-functional testing, acceptance testing). Definition of teaching-learning methods (an expositive, active and interactive methods), forms of assessment of learning outcomes (self assessment, peer assessment, expert assessment). Clarification, how we will measure the effectiveness of e-learning programs? Define assessment and grading formula. Development plans for assessing students' achievement of learning outcomes. Is certification examination planned? Do you want to provide certification? Online proctoring solution. Preparing the MOOC content. Writing video lectures scenarios. Preparing guides to help setting and pursuing SMART learning goals.

Development and Course Testing. Development course interactive content. Video production. Creating multimedia components (3D animation, animated video clips, recording screencasts, virtual learning models, others). Development and allocation the course on the chosen MOOC platform, which is the technological basis of e-learning. Course design. Filling out all of its modules with teaching educational and training resources and materials, adding activities, assessment tools and services, feedback and interactions. Testing the course functionality on various devices/platforms. Checking the user interface. Testing the functioning correct of settings and appropriate operation modes of MOOC interactive quizzes and assignments.

Delivery and Implementation. The course approbation (a trial version of MOOC). The MOOC publication on campus network for the closed access to students. Learner engagement and management. Start course implementation. The MOOC placement in the cloud-based educational platform that provides ubiquitous and convenient network access for millions of Internet users and maximum computing performance when entering large data streams.

Expertise and Evaluation. The course evaluation by students, lectures, tutors, experts, developers. Determination of whether the achieved goals were achieved? Does the course meet the content requirements of the target audience? Qualitative and quantitative readjustment of the course structure and learning content. Evaluation of the quality of the course should take place at every phase of the development of MOOC. The results of which are being modernized (revision) rate based on identified short-comings and the need for updating the electronic learning content. Evaluation of the MOOC quality is carried out, for example, using criteria developed by specialists of the international project ECBCheck (www.ecb-check.net/downloads/).

All stages of the development of MOOC are linked, cyclically repeated to improve the quality and efficiency of online learning and to respond to new topical problems in this area.

The proposed technology has been tested by teaching staff of the Orenburg State University (Russian Federation) and has proven its effectiveness in the development of MOOCs and their practical use on the basis of online platform Stepic.org.

5 Conclusion

Thus, the development of cloud computing and the emergence of educational platforms have become key factors in massive open online course widespread. Cloud computing can significantly reduce the cost of creating, storing and maintenance MOOCs. MOOCs uniqueness is revealed through specifics detailed described above. It is quite difficult and time consuming to create a high-quality and effective online course. The article presents the four-level model of user access to MOOCs, which allows to design a favorable SMART e-learning environment for open education, and structural model presented in the form of a connected graph for a better understanding of the main elements links included in the MOOC. The presented model of MOOC development process corresponds to the requirements of International quality standard ISO/IEC 19796-1, but takes into account the specific features of MOOC and its use in the educational environment of the university.

References

1. Chang, M., Li, Y.: Smart learning environments, 219 p. Springer (2014)
2. Uskov, V.L., Bakken, J.P., Pandey, A., Singh, U., Yalamanchili, M., Penumatsa A.: Smart University taxonomy: features, components, systems. In: Smart Education and e-Learning, pp. 3–14 (2016)
3. English Oxford living Dictionaries. https://en.oxforddictionaries.com/definition/us/MOOC
4. Uvarov, Yu.: Why do we need MOOCs? Inform. Educ. **9**, 3–17 (2015). (in Russian)
5. Bolodurina, I., Parfenov, D., Shukhman, A.: Efficient access to multimedia resources in distributed systems of distance learning. In: Proceedings of IEEE Global Engineering Education Conference EDUCON, Berlin, Germany, 13–15 March, pp. 1228–1231 (2013)
6. Pireva, K., Kefalas, P., Dranidis, D., Hatziapostolou, T., Cowling, A.: Cloud e-Learning: a new challenge for multi-agent systems, agent and multi-agent systems. In: Technologies and Applications, pp. 277–287 (2014)

7. Konanchuk, D., Volkov, A.: Epoch "Greenfield" in education. In: Research SEDeC, SKOLKOVO Education Development Centre (SEDeC), 50 p. Moscow School of Management "Skolkovo" (2013). (in Russian)
8. Allen, E., Seaman, J.: Grade level: tracking online education in the United States, Babson Survey Research Group and Quahog Research Group (2015). http://www.onlinelearningsurvey.com/reports/gradelevel.pdf
9. Hollands, F.M., Tirthali, D.: MOOCs: expectations and reality: full report, Center for Benefit-Cost Studies of Education, Teachers College, Columbia University, USA, 211 p. (2014)
10. McAuley, A., Stewart, B., Cormier, D., Siemens, G.: The MOOC model for digital practice: social science and humanities research council's "Knowledge Synthesis Grants on the Digital Economy", University of Prince Edward Island, Canada, 63 p. (2010)
11. Mazurov, Y.: Massive open online courses in the context of modern educational processes within universities. Open Distance Educ. **1**(57), 20–26 (2015). (in Russian)
12. Gaebel, M.: MOOCs massive open online courses: European University association occasional papers (2013). http://www.eua.be/Libraries/publication/MOOCs_Update_January_2014.pdf?sfvrsn=2
13. Proceedings of the European Stakeholders Summit on experiences and best practices in and around MOOCs. In: Proceedings of Conference EMOOCS 2016, University of Graz, Austria, 557 p. (2016)
14. Open EdX. https://www.edx.org
15. Coursera. http://coursera.org
16. Udacity. https://www.udacity.com
17. NovoEd. http://novoed.com
18. Udemy. http://udemy.com
19. Harvard Open Courses. http://extension.harvard.edu/academics/courses/course-catalog
20. MIT OpenCourseware. https://ocw.mit.edu/index.htm
21. Khan Academy. https://www.khanacademy.org
22. Academic Earth. http://academicearth.org
23. Class Centra. https://www.class-central.com
24. CourseBuffet. https://www.coursebuffet.com
25. CourseTalk. https://www.coursetalk.com
26. Degreed. https://degreed.com
27. eclass. http://eclass.cc
28. EMMA – European Multiple MOOC Aggregator. https://platform.europeanmoocs.eu
29. LearningAdvisor. http://www.learningadvisor.com
30. MOOC List. https://www.mooc-list.com
31. Mooctivity. http://www.mooctivity.com
32. Lee, G., Keum, S., Kim, M., Choi, Y., Rha, I.: A study on the development of a MOOC design model. In: Educational Technology International, Seoul National University, Korea, vol. 17, no. 1, pp. 1–37 (2016)
33. Lisitsyna, L., Efimchik, E.: Design and application of MOOC "Methods and Algorithms of Graph Theory" on national platform of open education of Russian Federation. In: Proceedings of Smart Education and e-Learning, pp. 145–154 (2016)
34. Blackmon, S.J., Major, C.H.: MOOCs and higher education: implications for institutional research: new directions for institutional research, no. 167, 112 p. Wiley (2016)
35. Kim, P.: Massive Open Online Courses: The MOOC Revolution, 166 p. Routledge (2014)
36. E-learning methodologies: a guide a guide for designing and developing e-learning courses, food and agriculture organization of the United Nations, 141 p. (2011)

37. Sonwalkar, N.: The First Adaptive MOOC: A Case Study on Pedagogy Framework and Scalable Cloud Architecture Part I. Mary ann Liebert, Inc. (2013). http://online.liebertpub.com/doi/pdf/10.1089/mooc.2013.0007
38. Gustafson, K.L., Branch, R.M.: Survey of instructional development models. In: ERIC Clearinghouse on Information and Technology, Syracuse, NY, 92 p. (2002)
39. ISO/IEC 19796-1:2005 Information technology — Learning, education and training — Quality management, assurance and metrics. https://www.iso.org/standard/33934.html

Systems and Technology for Smart Education

Building Smart Learning Analytics System for Smart University

Vladimir L. Uskov[1(✉)], Jeffrey P. Bakken[2], Colleen Heinemann[1], Rama Rachakonda[1], Venkat Sumanth Guduru[1], Annie Benitha Thomas[1], and Durga Poojitha Bodduluri[1]

[1] Department of Computer Science and Information Systems, and InterLabs Research Institute, Bradley University, Peoria, IL, USA
uskov@fsmail.bradley.edu

[2] The Graduate School, Bradley University, Peoria, IL, USA
jbakken@fsmail.bradley.edu

Abstract. The performed analysis of innovative learning analytics systems clearly shows that in the near future those systems will be actively deployed by academic institutions. The on-going research project described here is focused on in-depth analysis of hierarchical levels of learning analytics and academic analytics, types of data to be collected, main features, and the conceptual design of smart learning analytics for smart university. Our vision is that modern analytics systems should strongly support smart university's "smartness" levels such as adaptivity, sensing, inferring, anticipation, self-learning, and self-organization. This paper presents the up-to-date research outcomes of a research project on the design and development of smart learning analytics systems for smart universities.

Keywords: Smart learning analytics · Smart university · Smart education

1 Introduction

Smart Education (SmE), Smart University (SmU), Smart Learning (SmL), and Smart Classroom (SmC) concepts are rapidly gaining popularity among the world's best universities because modern and sophisticated smart technologies, smart systems, smart devices, as well-data-driven and data-based strategies and solutions, create unique and unprecedented opportunities for academic and training organizations. Using those innovative concepts, universities and colleges may obtain higher standards and innovative approaches to (1) education, learning, and teaching strategies, (2) modeling of students/learners as objects and education/learning as processes, (3) unique and/or high technology-based services to local in-class and remote/online students, (4) set-ups of modern highly technological smart classrooms with easy Web-based audio/video interactions between local/remote students and faculty, and collaboration between in-class and remote students, (5) design and development of Web-based rich multimedia learning content with interactive presentations, video lectures, Web-based interactive quizzes and tests, instant knowledge assessment and automatic posting of attendance, class activities, and learning assessment outcomes on course web sites,

V.L. Uskov et al. (eds.), *Smart Education and e-Learning 2017*, Smart Innovation, Systems and Technologies 75, DOI 10.1007/978-3-319-59451-4_19

visualization of data in various forms including student, faculty and department dashboards, and many other advantageous features [1, 2].

Our vision is based on the idea that SmU, SmE, SmC, SmL – as smart systems – should implement and demonstrate significant maturity at various "smartness" levels or smart features, including (1) adaptation, (2) sensing (awareness), (3) inferring (logical reasoning), (4) self-learning, (5) anticipation, and (6) self-organization and re-structuring [3]. This is the reason that we consider emerging learning analytics (LA) as an integral part of SmE, SmL, and SmC concepts.

The Society for Learning Analytics Research defines LA as: "… the measurement, collection, analysis and reporting of data about learners and their contexts, for purposes of understanding and optimizing learning and the environments in which it occurs" [4].

"A recent report from the U.S. Department of Education makes the point that on the program and institutional level, learning analytics can play a role that is similar to that of already existing business intelligence departments and applications. Just as business intelligence may utilize demographic, behavioral and other information associated with a particular enterprise and its customers to inform decisions about marketing, service and strategy, learning analytics promises to do something similar in educational terms" [5].

1.1 Learning Analytics' Goals and Objectives: Literature Review

Various authors of available publications define the goals of LA in different ways. For example, in accordance with Siemens [6], "The broad goal of learning analytics is to apply the outcomes of analyzing data gathered by monitoring and measuring the learning process, as feedback to assist directing that same learning process. …Six objectives are distinguished: predicting learner performance and modeling learners, suggesting relevant learning resources, increasing reflection and awareness, enhancing social learning environments, detecting undesirable learner behaviors, and detecting affects of learners."

Suchithra et al. [7] believe that "The main purpose of Learning Analytics is to improve the performance of learners. Also, the environment of learning in which the learner undergoes is enhanced which will ultimately result in a quality education. Learning Analytics helps educator/teacher to understand the students. Learning capabilities can be improved for the learners. … The Learning Analytics aims at the curriculum design, predicting the students' performance, improving the teaching learning environment, decision support system for Higher Education Institutions, personalized approach to individual students, online and other learning modes including mobile, subject wise teaching and learning, subjects which has practical and evaluation process in the education system. … Learning Analytics is about the collection, analysis of data about the learners. It is an emerging field in research which uses data analysis on every tier of educational system".

Borkar and Rajeswari presented the following approach in [8]: "Learning analytics approaches in general offer different kinds of computational support for tracking learner behavior, managing educational data, visualizing patterns, and providing rapid feedback to both educators and learners".

Additionally, Tempelaar et al. [9] argue that "The prime data source for most learning analytic applications is data generated by learner activities, such as learner participation in continuous, formative assessments. That information is frequently supplemented by background data retrieved from learning management systems and other concern systems, as for example accounts of prior education".

1.2 Learning Analytics on University Level: Literature Review

In accordance with Suchithra et al. [10], "Institution can use academic analysis to know the success of the students. It can also be used to get the attention of public. Report of the analysis can be used for the publicity of the institution."

Mattingly et al. [11] argue that "Learning analytics in higher education is used to predict student success by examining how and what students learn and how success is supported by academic programs and institutions. … The focus is to explore the measurement, collection, analysis, and reporting of data as predictors of student success and drivers of departmental process and program curriculum".

Siemens et al. in [12] wrote: "It is envisaged that education systems that do make the transition towards data-informed planning, decision making, and teaching and learning will hold significant competitive and quality advantages over those that do not".

1.3 Learning Analytics on Course Level: Literature Review

Multiple publications are available regarding the approaches, concepts, and proposed framework for LA at the course level. For example, Dietz-Uhler and Hurn in [13] argue that "Goals that learning analytics address include predicting learner performance, suggesting to learners relevant learning resources, increased reflection and awareness on the part of the learner, detection of undesirable learning behaviors, and detecting affective states of the learner. … Data as login frequency, site engagement, student pace in the course, and assignment grades to predict course outcome. … Performance on course assignments and tests at various times in the course significantly predicted final grades."

In accordance with Dyckhoff et al. in [14], "Masses of data can be collected from different kinds of student actions, such as solving assignments, taking exams, online social interaction, participating in discussion forums, and extracurricular activities. This data can be used for Learning Analytics to extract valuable information, which might be helpful for teachers to reflect on their instructional design and management of their courses."

Ruiperez-Valiente et al. in [15] present the following vision of learning analytics: "The Khan Academy…platform provides an advanced learning analytics module with useful visualizations for teachers and students…the Khan Academy platform provides different learning analytic features by default."

1.4 Levels of Learning Analytics: Literature Review

In general, LA may have several levels or layers of hierarchy and/or maturity. For example, Siemens et al. in [16] introduced a general hierarchical framework for LA levels that include 3 levels for LAs: (1) LA on personal level (analytics on personal performance in relation to learning goals, learning resources, and study habits of other classmates), (2) LA on course level (social networks, conceptual development, discourse analysis, "intelligent curriculum"), and (3) LA on departmental level (predictive modeling, patterns of success/failure). Additionally, for academic analytics they introduced (4) LA on institutional level (learner profiles, performance of academics, knowledge flow, resource allocation), (5) LA on regional level (state/provincial): comparisons between systems, quality and standards, and (6) LA on national/international level.

On the other hand, Lynch et al. in [12] proposed an LA sophistication model that contains the following LA maturity levels: (1) awareness, (2) experimentation, (3) organization, students, faculty, (4) organizational transformation, and (5) sector transformation.

1.5 Smart Learning Analytics: Literature Review

The idea of smart learning analytics (SLA) is in an embryonic state at this moment; a thorough search on the Internet discovers a few relevant publications. For example, Giannakos et al. in [17] defined "Smart Learning Analytics as a subset of learning analytics that focuses on supporting the features and the processes of smart learning". The authors primarily concentrated on "recent foundations and developments [of Smart Learning Analytics] in the area of Video-Based Learning" [16]. On the other hand, Boulanger et al. introduced in [18] "… a framework called SCALE that tracks finer level learning experiences and translates them into opportunities for custom feedback. … Students have been provided with customized feedback to optimize their learning path in programming".

Multiple publications are available on various topics related to LA aspects. Unfortunately, the aforementioned and multiple additional analyzed publications do not provide detailed information about SLA from the smartness levels point of view, i.e. levels of (1) adaptivity, (2) sensing, (3) inferring, (4) anticipation, (5) self-learning, and (6) self-organization [1–3]. Additionally, the analyzed publications are focused on applications of LA to learning process of students and/or life-long learners; however, they do not emphasize the fact that LA should demonstrate "smartness" features (or be smart) and strongly support all designed smartness levels, including the "self-learning" level, i.e. an ability of a smart university "to learn" about itself and, therefore, be able "to self-optimize" its operation and main business functions.

2 Project Goal and Objectives

The overall goal of the on-going research, design, and development project at the InterLabs Research Institute at Bradley University (Peoria, IL, USA) is to use a systematic approach to identify, analyze, test, design, and eventually implement various components of SLA system for an entire SmU. In order to achieve this goal, the project team selected the following objectives:

- analysis of most recent innovative developments in LA and SLA areas;
- analysis of existing LA levels;
- analysis of available software systems that may support learning analytics, and potentially, SLA;
- identification of main features of SLA – types of data to be collected and processed, and main functionality of SLA system

A summary of up-to-date project findings and outcomes is presented below.

3 Smart Learning Analytics: Hierarchical Levels

Our vision of an SLA system is based on the concept that SLA should have a hierarchical layered structure and strongly support all major components of SmU, including

(1) SmU stakeholders, including students, faculty, professional staff, administrators, life-long learners, donors, alumni, etc.;
(2) SmU main smartness features, including adaptation, sensing, inferring, self-learning, anticipation, self-optimization or re-structuring (see Table 1 below);
(3) SmU curricula, i.e. a set of smart programs of study and smart courses at SmU – those that can, for example, change (or optimize) its structure or mode of learning content delivery in accordance with given or identified requirements (due to various types of students or learners);
(4) SmE and SmL at SmU – main processes and business functions at SmU;
(5) Smart Pedagogy (SmP), i.e. a set of modern pedagogical styles (strategies) to be used at SmU;
(6) smart learning environment at SmU, including smart classrooms, smart labs, smart departments and smart offices, etc.;
(7) smart software systems at SmU, i.e. a set of university-wide distinctive smart software systems at SmU – those that go well beyond those used at a traditional university;
(8) smart hardware at SmU, i.e. a set of university-wide smart hardware systems, devices, equipment and smart technologies used at SmU – those that go well beyond those used at a traditional university);
(9) smart technology, i.e. a set of university-wide smart technologies to facilitate main functions and features of SmU and smart campus, for example, Internet-of-Things, cloud computing, iSafety, ambient intelligence, etc.;
(10) SmU resources, including financial, technological, human, and other types of resources.

As a result, we proposed the following hierarchical levels of SLA system:

(1) **Personal level** is the lowest level of the SmU for various types of stakeholders – students, faculty, professional staff, administrators, long-life learners, etc.; for

Table 1. SmU smart features [1]

SmU features	Details
Adaptation	SmU ability to automatically modify its business functions, teaching/learning strategies, administrative, safety, physical, behavioral and other characteristics, etc. to better operate and perform its main business functions (teaching, learning, safety, management, maintenance, control, etc).
Sensing (awareness)	SmU ability to automatically use various sensors and identify, recognize, understand and/or become aware of various events, processes, objects, phenomenon, etc. that may have impact (positive or negative) on SmU's operation, infrastructure, or well-being of its components – students, faculty, staff, resources, properties, etc.
Inferring (logical reasoning)	SmU ability to automatically make logical conclusion(s) on the basis of raw data, processed information, observations, evidence, assumptions, rules, and logic reasoning
Self-learning	SmU ability to automatically obtain, acquire or formulate new or modify existing knowledge, experience, or behavior to improve its operation, business functions, performance, effectiveness, etc. (A note: Self-description, self-discovery and self-optimization features are a part of self-learning)
Anticipation	SmU ability to automatically think or reason to predict what is going to happen, how to address that event, or what to do next
Self-organization and configuration, re-structuring, and recovery	SmU ability automatically to change its internal structure (components), self-regenerate, and self-sustain in purposeful (non-random) manner under appropriate conditions but without an external agent/entity. (A note: Self-protection, self-matchmaking, and self-healing are a part of self-organization)

example, students can get data-driven dashboard on (a) current academic performance in a course, on programs of study, etc., and (b) progress of development of various skills, including analytical, technical, management, and communication skills;

(2) **Course level** for students, learners, faculty, department chair, etc.; for example, faculty can compare academic performance of a selected student with (a) all students in the same class during current semester, (b) with all students in this class during recent semesters – based on these data a faculty can predict student's learning ability, final grade, etc.; additionally, faculty can compare student academic performance in courses that use (a) various modes of learning content delivery (face-to-face, online, blended), (b) various learning strategies (learning-by-doing, games-based learning, flipped classroom, adaptive learning, etc.)

(3) **Concentration/minor program level** (i.e. a level of a group of specialized courses) for students, faculty, administrators; for example, a department chair

can assess academic performance of (a) majors from his/her department, and (b) students from other departments who take a selected concentration, certificate or minor program at his/her department;

(4) **Departmental/program of study/curriculum level** for students, faculty and administrators; for example, chair of department can perform predictive analysis for various students and faculty; identify patterns of success for various types of students and faculty; perform data-driven control of enrollment into departmental programs - certificate, concentration, minor and major programs;

(5) **University level** is the highest level of a SmU for administrators, alumni, donors, etc.; for example, provost's office can monitor (a) student academic performance in various courses and programs, (b) retention rate in a major, department, college, university, etc.

4 Smart Learning Analytics: Types of Data to Be Collected/Processed

An SLA system should collect numerous pieces of data in order to support the idea of the *Data* ↦ *Information* ↦ *Knowledge* ↦ *Smartness* continuum at a SmU. Several examples of types of data to be collected by SLA system are presented in Table 2.

Table 2. Types of data to be collected by SLA system at SmU (examples)

SLA level	Types of data to be collected	Types of data to be collected (cont.)
University level	• Campus-related data • College or school-related data • Department-related data • Programs of study-related data (major, minor, concentration, certificate, etc.) • Student enrollment-related data • Student data (name, gender, DOB, DOA, credits obtains, course taken, grades, attendance, etc.) • Total numbers of students (full-time, part-time, undergraduate, graduate, etc.)	• Retention rate (%) • Graduation on time rate % • Faculty-related data (name, start date, level of education, promotion/tenure dates and decisions, etc.) • Student demographics data • Professional staff data • Demographic characteristics etc.
Program of study (curriculum) level	• Numbers of various students in a program (majors, non-majors, etc.) • Student data • Courses in a program • Retention/successful completion rates	• Faculty data • Post-graduation data (in program's area) etc.

(*continued*)

Table 2. (*continued*)

SLA level	Types of data to be collected	Types of data to be collected (cont.)
Course	• Number of students who regularly engage in course • Student goals • Course learning goals, objectives and expected outcomes • Course pre-requites • Course software and hardware systems, smart classroom and technology needed • Number of lecture views, logins to course web site, course LMS, PowerPoint slides, tutorials, assignments, discussion forums, etc.	• Student individual grades/scores on various course assignments • Class average grades/scores on various course assignments • Mean and standard deviation of scores • Attendance • Student behavior in a class • Student feedback etc.

5 Smart Learning Analytics: User Requirements

The introduced concept of SLA hierarchical levels and types of to-be-collected data enabled us to create a comprehensive list of user requirements to SLA system functionality. Table 3 below contains examples of such requirements from students, faculty, and department chair for one hierarchical level – LA on course level.

Table 3. SLA course level: SmU user requirements (examples)

Users type	Desired functions in SLA system (examples)
Faculty	• Keep track of student attendance and activities in the class • Keep track on student grades/scores for various course assignments • Collect data on of all standard calculations for grades for each course assignment (mean, standard deviation, etc.) • Compare scores and grades, student attendance, etc., of current students vs students in previous semesters of the same class and/or the same class with different faculty • Predict student academic performance, specific score, final grade, etc. • Model student success in a course, and his/her analytical, technical, management and communications skills
Depart-ment chair	• Keep track of course instructor accomplishments versus other faculty at the department, or at the same course (but other sections) • Compare student academic performance in a given course vs. student performance in the same course but in other sections, in previous semester, etc. • Access data of course academic performance of any student

(*continued*)

Table 3. (*continued*)

Users type	Desired functions in SLA system (examples)
Students	• Keep track of grades on tests, quizzes, labs, etc. • Keep track of what assignments are due when, when exams are, etc. • Keep all information about one specific class in the same place • Keep track of individual grades compared to other students in the class (anonymous comparative analysis) • Calculate what grade must be gotten on different assignments/tests/labs in order to maintain or get a grade of A, B, C, etc. • Calculate GPA depending on what grade is gotten in the class • Input experience level in information covered in the class to output expected grade • Give projected grade in class based on the grades in the grade book at any given time • Request help from professor/TA/etc. if not doing well in a course

6 Analytics Systems Analyzed

Multiple software systems used in the LA area were reviewed or analyzed with the purpose of identifying those suitable for SLA; the obtained outcomes are presented in Table 4.

Table 4. Existing analytics systems reviewed or analyzed (non-comprehensive list)

#	Name of analyzed system	System's developer	Technical platform	Ref. #
Commercial systems				
1	Google analytics	Google	Windows/Mac	[19]
2	Clicky	Clicky	Windows/Mac/Linux	[19]
3	Panoramaed		Windows/Mac	[19]
4	Moz Keyword Explorer	MOZ	Windows/Mac	[19]
5	SolutionPath		Windows/Mac	[19]
6	Chartbeat	Chartbeat	Windows/Mac	[19]
7	EdSurge		Windows/Mac	[19]
8	Adobe analytics	Adobe	Windows/Mac	[19]
9	Church analytics		Windows/Mac	[19]
10	Woopra	iFusion Labs LLC	Windows/Mac	[19]
Open source (free) systems				
1	Piwik	Piwik Pro	Windows/Mac/Linux	[19]
2	Open web analytics	OWA	Windows/Mac/Linux	[19]

(*continued*)

Table 4. (*continued*)

#	Name of analyzed system	System's developer	Technical platform	Ref. #
3	Klass data	Klass	Windows/Mac	[19]
4	Cyfe	Cyfe	Windows/Mac	[19]
5	eAnalytics		Windows/Mac	[19]
6	Countly	Countly	Windows/Mac	[19]
7	Unicon		Windows/Mac	[19]
8	Open dashboard		Windows/Mac	[19]
9	OpenReports		Windows/Mac/Linux	[19]
10	Site Meter	Site Meter	Windows/Mac	[19]
11	Ipoll	Griffith University	Mac, Windows, Linux	[7]
12	Moodog	University of California	Mac, Windows, Linux	[7]
13	Equella	University of Wollongong	Mac, Windows, Linux	[7]
14	E2coach	University of Michigan	Mac, Windows, Linux	[7]
15	Signals	Purdue University	Mac, Windows, Linux	[7]
Systems developed by authors of analyzed papers				
1	SPAM system		Windows	[20]
2	STEP UP!		Mobile App	[21]
3	SRES		Windows/Mac	[22]
4	HOU2LEARN		Windows/Mac	[23]
5	ALAS-KA	Ruiperez-Valiente, Munoz-Merino, Kloos	Web-based	[15]
6	LATUX (Learning Awareness Tools – User eXperience)	Martinex-Maldonado, Pardo, Mirriahi, Yacef, Ky, Clayphan		[24]
7	Course signals	Arnold, Pistilli	Windows, Linux	[25]

7 SLA System's Prototype Developed

We developed a prototype of the SLA system for a smart university – the InterLabs SLA system. This system should eventually include the functionality of SLA for SmU as described in Tables 2 and 3. Graphic user interfaces of developed prototype of the InterLabs SLA system for various users are presented in Fig. 1 (student view), Fig. 2 (faculty/instructor view), and Fig. 3 (department chair or administrator view).

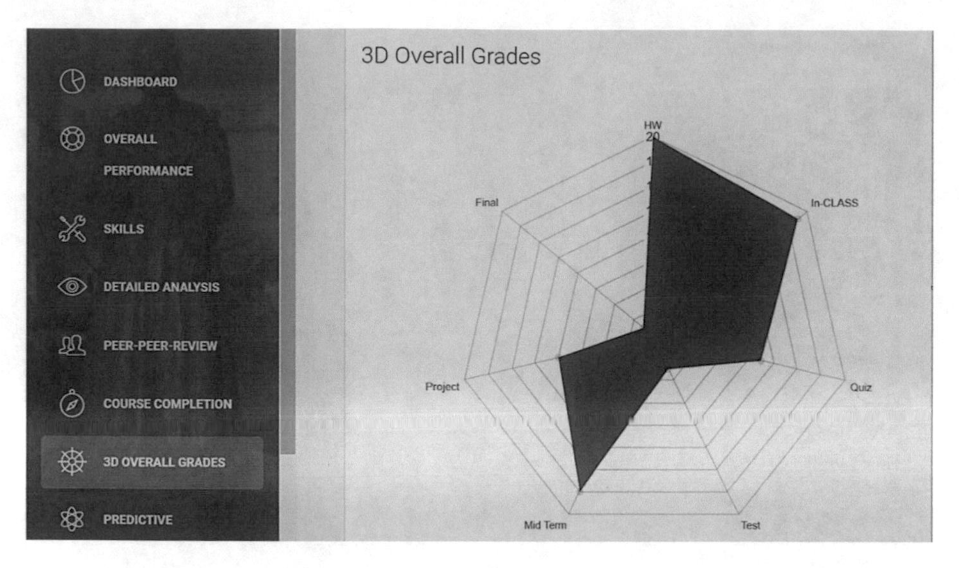

Fig. 1. A prototype of the InterLabs smart learning analytics system: student view

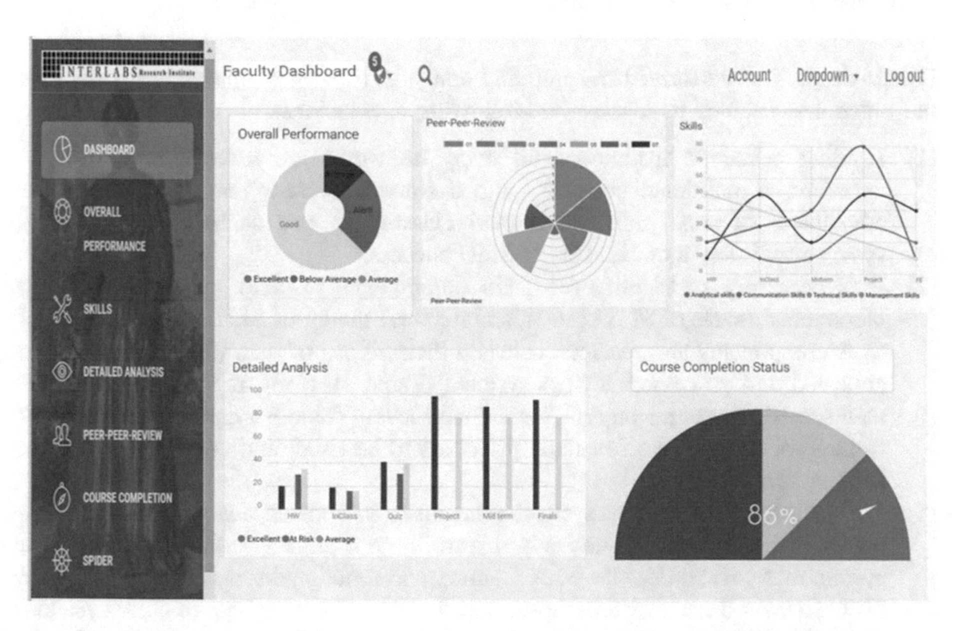

Fig. 2. A prototype of the InterLabs smart learning analytics system: faculty view

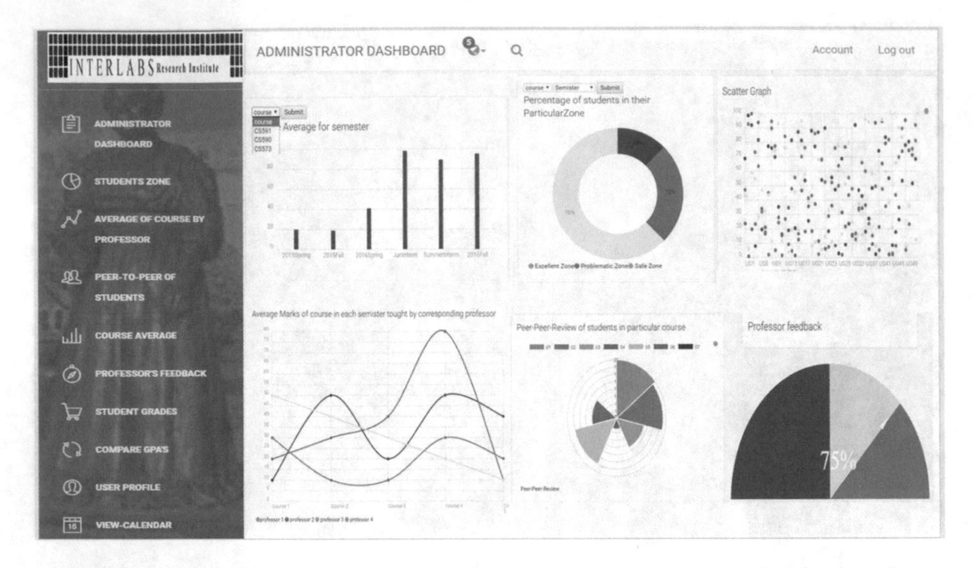

Fig. 3. A prototype of the InterLabs smart learning analytics system: administrator view

8 Conclusions. Next Steps

Conclusions. The performed research and analysis, as well as the obtained findings and outcomes, enabled us to make the following conclusions:

(1) Leading academic institutions all over the world are investigating ways to transform a traditional university into a smart university and benefit from the advantages of smart university, smart classrooms, and smart pedagogy. SLA systems will play a crucial role in SmU success.
(2) It is necessary to identify the main components of SLA systems, including hierarchical levels of SLA system, architectural model of SLA system with main SLA components and relations between them, types of data (data objects) to be collected and processed at SLA systems (Table 2), a set of SmU user requirements to SLA systems on each hierarchical level (Table 3), inputs to and outputs from SLA system, interfaces and protocols to be used, and constraints/limits of SLA system at SmU.
(3) We analyzed 30 + current systems relevant to various aspects of SLA; unfortunately, we could not identify any mature system among the existing ones – a system that will address the SLA features and functionality as required by SmU (Table 3). As a result, we designed and developed a prototype of SLA system - InterLabs SLA.

Next Steps. Based on the obtained research findings and outcomes, the next steps of this research project are to (a) use the Agile Software Engineering process model, (b) create a set of prototypes of SLA system for various types of users, and (c) involve various SmU users into the SLA design and development to ensure quality of final software system.

References

1. Uskov, V.L, Bakken, J.P., Pandey, A., Singh, U., Yalamanchili, M., Penumatsa, A.: Smart university taxonomy: features, components, systems. In: Uskov, V.L., Howlett, R.J., Jain, L. C. (eds.) Smart Education and e-Learning 2016, pp. 3–14. Springer, June 2016. 643 p., ISBN: 978-3-319-39689-7
2. Uskov, V., Bakken, J., Pandey, A. The ontology of next generation smart classrooms, In: Uskov, et al. (eds.) Smart Education and Smart e-Learning, pp. 3–14. Springer (2015). 510 p., ISBN 978-3-319-19874-3
3. Uskov, A., Sekar, B.: Smart gamification and smart serious games. In: Sharma, D., Jain, L., Favorskaya, M., Howlett, R. (eds.) Fusion of Smart, Multimedia and Computer Gaming Technologies, Intelligent Systems Reference Library, vol. 84, pp. 7–36. Springer (2015). doi:10.1007/978-3-319-14645-4_2, ISBN: 978-3-319-14644-7
4. Society for Learning Analytics Research. https://solaresearch.org/
5. Friesen, N.: Learning analytics: readiness and rewards. Can. J. Learn. Technol. **39**(4), 1–12 (2013). http://learningspaces.org/wordpress/wp-content/uploads/2013/05/Learning-Analytics1.pdf
6. Siemens, G.: The journal of learning analytics: supporting and promoting learning analytics research. J. Learn. Analyt. **1**(1), 3–4 (2014)
7. Suchithra, R., Vaidhehi, V., Iyer, N.E.: Survey of learning analytics based on purpose and techniques for improving student performance. Int. J. Comput. Appl. **111**(1), 22–26 (2015)
8. Borkar, S., Rajeswari, K.: Attributes selection for predicting students' academic performance using education data mining and artificial neural network. Int. J. Comput. Appl. **86**(10), 25–29 (2014)
9. Tempelaar, D.T., Cuypers, H., van de Vrie, E., Heck, A., van der Kooij, H.: Formative assessment and learning analytics. In: Proceedings of the 2013 International Conference on Learning Analytics and Knowledge (LAK), 8–12 April 2013, Leuven, Belgium (2013). https://pdfs.semanticscholar.org/29db/489b4532ee7e983a17f5653d65b31e9cb93d.pdf
10. Suchithra, R., Vaidhehi, V., Iyer, N.E.: Survey of learning analytics based on purpose and techniques for improving student performance. Int. J. Comput. Appl. **111**, 22–26 (2015). ISSN: 0975–8887
11. Mattingly, K.D., Rice, M.C.: Learning Analytics as a Tool for Closing the Assessment Loop in Higher Education. Knowl. Manag. E-Learning: An Int. J. **4**(3), 236–247 (2012). http://kmel-journal.org/ojs/index.php/online-publication/article/viewFile/196/148
12. Siemens, G., Dawson, S., Lynch, G.: Improving the quality and productivity of the higher education sector, Society for Learning Analytics Research (2013). https://sydney.edu.au/education-portfolio/ei/projects/SoLAR_Report_2014.pdf
13. Dietz-Uhler, B., Hurn, J.E.: Using learning analytics to predict (and improve) student success: a faculty perspective. J. Interact. Online Learn. **12**(1), 17–26 (2013). Spring. ISSN: 1541-4914. www.ncolr.org/jiol
14. Dyckhoff, A.L., Zielke, D., Bültmann, M., Chatti, M.A., Schroeder, U.: Design and implementation of a learning analytics toolkit for teachers. Educ. Technol. Soc. **15**(3), 58–76 (2012)
15. Ruiperez-Valiente, J.A., Munoz Merino, P.J., Delgado-Kloos, C.: An architecture for extending the learning analytics support in the Khan Academy framework. In: Proceedings of the First International Conference on Technological Ecosystem for Enhancing Multiculturality, Salamanca, Spain, pp. 277–284. ACM (2013). doi:10.1145/2536536.2536578

16. Siemens, G., Gasevic, D., Haythornthwaite. C., Dawson, S., Buckingham Shum, S., Ferguson, R., Duval, E., Verbert, K., Baker, R.S.J.d.: Open Learning Analytics: an integrated & modularized platform (2011). http://www.elearnspace.org/blog/wp-content/uploads/2016/02/ProposalLearningAnalyticsModel_SoLAR.pdf
17. Giannakos, M.N., Sampson, D.G., Kidziński, L.: Introduction to smart learning analytics: foundations and developments in video-based learning. Smart Learn. Environ. **3**(12) (2016). https://slejournal.springeropen.com/articles/10.1186/s40561-016-0034-2
18. Boulanger, D., Seanosky, J., Kumar, V., Kinshuk, Panneerselvam, K., Somasundaram, T.S.: Smart learning analytics. In: Chen, G., Kumar, V., Kinshuk, Huang, R., Kong, S. (eds.) Emerging Issues in Smart Learning. Lecture Notes in Educational Technology. Springer, Heidelberg (2015)
19. Smart Learning Analytics project at Bradley University – References. www.interlabs.bradleu.edu/SLA_project
20. Ogor, E.: Student academic performance monitoring and evaluation using data mining techniques. In: Electronics, Robotics and Automotive Mechanics Conference. IEEE (2007)
21. Verbert, K., Duval, E., Klerkx, J., Govaerts, S., Santos, J.L.: Learning Analytics Dashboard Applications. SAGE Publications (2013)
22. Jenny McDonald, et al.: Cross-institutional collaboration to support student engagement: SRES version 2 (2016)
23. Eleni Koulocheri, et al.: Applying learning analytics in an open personal learning environment: A quantitative approach. IEEE (2012)
24. Martinez-Maldonado, R., Pardo, A., Mirriahi, N., Yacef, K., Kay, J., Clayphan, A.: LATUX: an iterative workflow for designing validating, and deploying learning analytics visualizations. J. Learn. Analyt. **2**(3), 9–39 (2015)
25. Arnold, K.E., Pistilli, M.D.: Course signals at purdue: using learning analytics to increase student success. In: Proceedings of the Second International Conference on Learning Analytics and Knowledge LAK 2012, Vancouver, BC, Canada. ACM (2012). 978-1-4503-1111-3/12/04

A Half-Duplex Dual-Lingual Video Chat to Enhance Simultaneous Second Language Speaking Skill

Bui Ba Hoang Anh(✉) and Kazushi Nishimoto

Graduate School of Advanced Science and Technology, Japan Advanced Institute of Science and Technology, Nomi, Japan
s1620022@jaist.ac.jp, knishi@jaist.ac.jp

Abstract. This study presents a smart software system named BiTak which employs strict turn-taking to encourage learners to practice speaking skills via a half-duplex dual-lingual video chat using a recording function. BiTak aims to motivate the conversation using two different languages as well as to identify the mutual benefits through this kind of conversation for the need of improving simultaneous bilingual acquisition. The paper also examines the improvement of users using BiTak that is assessed objectively by certified language teachers following the scoring tool named rubric. It is revealed from the results of the experiments that the system brings about the sense of language learning and favorably boost students' speaking skills.

Keywords: Smart software system for education and learning · Smart video chat application · Simultaneous second language acquisition · Strict turn-taking · Dual-lingual communication

1 Introduction

Due to the growing influence of international mass media, global communication through Internet, it is crucial for many people to learn to speak other languages in addition to their first language. Apart from many ways of learning a second language, people tend to find chance to practice speaking second language with native speakers through informal communication to better improve their speaking skills. It is the most popular way to learn a language as well as effective way most people use when they live in the country where the language is spoken. However, it is not always easy to get an opportunity of speaking with native speakers face-to-face.

Besides, there are lots of difficulties to study languages. Some conventional ways of learning languages have been applied such as attending classes, group discussion and learning, self-study. Apparently, there is not the most appropriate method, people learn in different ways at different paces, and the most effective way may involve not one but a mixture of different techniques. Kimber, L. [1] recommends providing more opportunities for interactions between Japanese and internationals to help each other speak. This method may satisfy both of their requirements at the same time.

V.L. Uskov et al. (eds.), *Smart Education and e-Learning 2017*, Smart Innovation, Systems and Technologies 75, DOI 10.1007/978-3-319-59451-4_20

Being inspired by that recommendation and the attempt of encouraging learners to use smart software and hardware systems and smart technology for education and learning, our study suggests building an online environment for language learners to talk freely using two different languages. Specifically, we propose a video chat system named BiTak that employs strict turn-taking dual-lingual communication (that are described in detail in Sect. 2). In particular, this study is designed to examine the improvement of users using BiTak that is assessed objectively by certified language teachers following the scoring tool named rubric.

The rest of this paper is organized as the followings. Section 2 defines two important terms: dual-lingual communication and strict turn-taking. Section 3 reviews related works and correlates them with our proposed system. The description of our prototype system is mentioned in Sect. 4. Section 5 describes the experiments to estimate the proposed system as well as mentions its results which is assessed objectively by language teachers using rubric scoring scale. The effectiveness of the system is also discussed in Sect. 5 by comparing two experiment groups, one using "strict turn-taking" with the recording button and one without using it. Section 6 concludes the paper.

2 Term Definitions

Dual-Lingual Communication. The concept of Dual-lingual Communication in this research is defined as two languages being spoken in the conversation and understood by respective participating parties. For instance Japanese students will basically speak only English while American students will basically speak only Japanese though they can switch to his or her mother tongue, only if necessary. This is different from bi-/multilingual communication. Myers-Scotton [2] defines bi-/multilingual as "the ability to use two or more languages to sufficiently carry on a limited casual conversation". During the dual-lingual conversation, they firstly try to speak the second languages. However, if some problems are found, they will help each other correct speaking mistakes by using their native language. It will be a good opportunity for both parties to learn from each other to make comfortable communication.

Strict Turn-taking. Talking naturally without caring overlapping usually brings about the comfort of expressing ideas in an informal conversation. Smooth turn-taking is an essential aspect to coordinate one's communicative actions and interact successfully with others. However, it is not always good for learning a language. You may hardly recognize your speaking mistakes by yourself although the listeners can understand clearly. In many researches of second language learning, the fact that turn-taking in communication may affect the quality of group discussion between non-native and native speakers has been taken into consideration. According to Mynard, J. [3], non-native students seemed "to be overwhelmed and even lost in parallel and fast discussion, especially students who have slow keyboarding skills, slow reading/writing skills, or different cultural backgrounds." Hence, our system would like to strictly apply the turn-taking approach by using a recording function. In our system, users are asked to entirely obey the turn-taking rule and they are not allowed to overlap or interrupt another speaker. We suppose that the unfamiliar way of strict turn-taking will bring about unexpected but possible outcomes.

3 Related Works

3.1 Online Systems Support Language Learning

Recent researches have shown that computer-mediated communication tools are considered as potential source for students to enhance their language proficiency. In Freiermuth, M., and Jarrell, D. [4], their research of second language learning asserted that when compared with face-to-face communication, online text chatting provided a more comfortable environment for foreign students to make conversations. In spite of facing the pressure of immediacy that is typically expected by speakers in face-to-face communication, students found it less burden when communicating through text chat.

Besides, research in computer-mediated communication has also inferred that a student's willingness to communicate may be positively affected by computer. Specifically, Freiermuth [5, 6] claimed that when assigned a group task, group language learners seemed more eager to communicate using computer-mediated communication tool than using spoken language. They felt more freedom in expressing their ideas without being hindered from the teacher or other students or a plethora of other elements that might minimize the effect of the experience [7]. After making interviews about preference of media use of non-native speakers, Setlock, L.D., and Fussell, S.R. [8] also showed that non-native speakers preferred online chatting tools because these tools reduced the risk of misunderstandings that often caused by language problems.

The potential of computer-mediated communication tools in facilitating second language acquisition has been mentioned in various current researches. Angelova, M., and Zhao, Y. [9] conducted a collaborative online project between students from China and United States of America. They were paired up to communicate using the discussion board and e-mail tools for tutoring and learning different aspects of English grammar and developing culture awareness. The American students tried to correct mistakes of their Chinese partners in writing introduction essays or cultural lessons. The Chinese students used e-mail as well as Skype to communicate with their American partners. Apart from the benefits collected from different aspects, the study concluded that computer-mediated communication are used as a bridge to connect students from two different countries and two different programs to improve the teaching skills of the American as well as to enhance non-native speakers' language skills.

In addition, videoconferencing that has been called visual collaboration is becoming noticeable in the benefits of online language learning. Hampel, R., and Stickler, U. [10] conducted research about videoconferencing in supporting multimodal interaction in an online language classroom. The study concentrated on the use of videoconferencing in the context of a larger exploratory study to find out how language-learning interaction was influenced by the virtual learning environment. The findings demonstrated how an online videoconferencing environment can be applied in language teaching as well as how teachers and learners collaborate in online environment.

Nevertheless, few studies have aimed to utilize video chat applications for supporting simultaneous learning of multiple languages. Our study proposes a smart video chat system as a virtual turn-taking face-to-face environment for users to practice dual-lingual conversation. Instead of choosing one partner's language over the other, they practice "dual-lingual" pattern. It is a communication pattern in which each partner actively uses his or her second language and receives the partner's second language in response. There have not been a system that is designed for allowing people to communicate in a dual-lingual pattern.

3.2 Assessment of Second Language Speaking Proficiency

According to James E. Purpura [11], the term Language Assessment refers not only to formal tests like TOEFL or an end-of-chapter evaluation, but also to other methods of obtaining information about knowledge, skills, and ability of students such as observing second language performance during pair work or by asking learners to report their understandings and uncertainties. In this paper, we use rubric: a scoring guide used to evaluate the quality of students' constructed responses to assess their second language speaking proficiency. The usefulness of rubric has been recognized in the field of assessment for many decades [12]. When utilizing a rubric, evaluators use an analytic rating system whereby each component is scored individually or performance is rated holistically on the basis of an overall impression [13]. We create our rubric for Speaking Skill Test based on four criteria: "Relevance & Content", "Fluency", "Vocabulary & Word Choice", and "Interviews: Does interviewee understand question?"

4 BiTak System

We developed a web application called "BiTak" using the open source from WebRTC [14], which is a free, open project that provides browsers and mobile applications with real-time communications with simple APIs. Figure 1 shows the user interface of BiTak. The most prominent features of BiTak is following two functions: (1) a strict turn-taking function by discretely recording each utterance and (2) a text chat function related to each recorded utterance.

4.1 Strict Turn-Taking Function by Discretely Recording Each Utterance

In order to achieve the strict turn-taking, BiTak is equipped with a recording function to discretely record each utterance. When a person wants to talk, he/she just needs to click on the recording button then his/her voice will be automatically recorded. At the same time, others' microphone will be off; they can do nothing but listen to the speaker. After the speaker finishes talking, he/she clicks the recording button again, the blue recording link will appear in the right pane of the main window chat (See Fig. 1). The next person will take turn to talk by repeatedly clicking the recording button. Therefore, the communication style with using BiTak is in a half-duplex manner similar to that of a transceiver. As you can see, not only for achieving the strict turn-taking, the recording function can also be used to provide the users a chance to watch the video again to fully

Fig. 1. User interface of BiTak

understand the dual-lingual situation. Furthermore, the users can also download all the recording videos for further reference.

4.2 Text Chat Function Related to Each Recorded Utterance

The recording link will lead users to another tab where they can re-watch the video (See Fig. 2). Meanwhile, the main chat (where people are talking) will be still facilitated without any interruption. If, for example, an utterance in English from a Japanese participant includes some errors or unsuitable expressions, it should be corrected immediately. In order to readily achieve it, we provide a text chat function to each recording link, not to all recording links. The users can chat, ask or point out any unclear points by typing text in the chat bar right beside the recording video. This feature is designed not only for each recording link but also for the main chat to achieve deeper understanding.

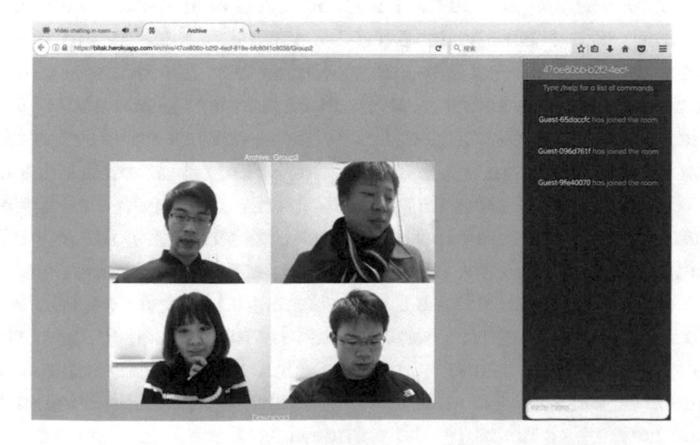

Fig. 2. An example of the recording link interface

5 Experiment and Discussion

5.1 Experiment Procedure

In order to investigate the effectiveness of BiTak, we compared the experiences of the two 4-member groups using all the functions of BiTak with who used just the interface of BiTak (i.e., a simple video conferencing system without the recording function). Meanwhile, both of the two groups will apply dual-lingual communication to discuss. Each group consists of 2 non-Japanese students who speak English fluently as native speakers and 2 Japanese students as specifically described in Fig. 3.

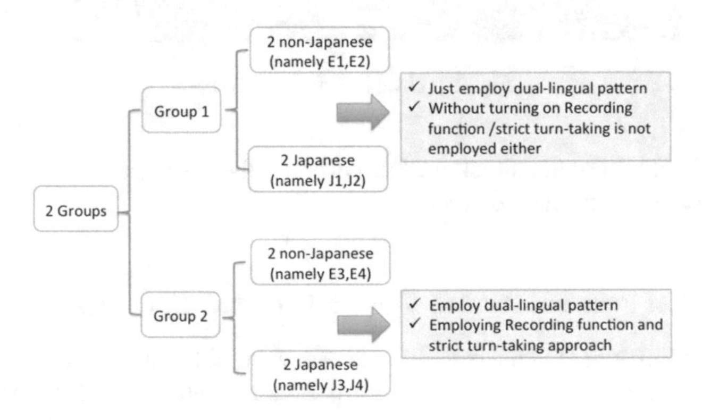

Fig. 3. Experiment description

To measure the improvement of speaking skill after using BiTak, all subjects were supposed to attend pre-experiment evaluation and post-experiment evaluation. Specifically, the Japanese students were interviewed their English speaking skills by a certified English teacher and the non-Japanese students were interviewed their Japanese speaking skills by a certified Japanese teacher. The interview questions during the two evaluations remain unchanged and their improvement is assessed followed a rubric for Testing Speaking Skill specially designed for the task. Each group was required to participate in a series of six experiments in which they could discuss intensively the topics given in the evaluation interview. Each experiment lasted about 90 min. While Group 1 held their discussion using BiTak without turning on the recording function which also means strict turn-taking is not employed either, Group 2's discussions used BiTak with employing the recording function and strict turn-taking approach. To ensure unbiased improvement, all subjects were requested not to use any other kinds of language learning tools during the period of experiments. In addition, each subject was asked to attend a 30-minutes individual semi-structured interview with the first author. The individual interview questions were guided by the general themes which aims to gain thinking about dual-lingual communication and strict turn-taking with the recording function. Besides, the questions were also open-ended enough for us to be able to pursue new topics raised by the participants. Each interview was recorded and transcribed to text then the transcripts were informally analyzed.

5.2 Result Analyses

Result of rubric. The speaking performance of all subjects in the evaluations were assessed by four criteria: "Relevance & Content", "Fluency", "Vocabulary & Word Choice", and "Interviews: Does interviewee understand question?" as summarized in Fig. 4.

	Group 1				Group 2			
	E1	E2	J1	J2	E3	E4	J3	J4
Relevance & Content	10	6.3	7.5	7.3	8.5	6.5	7.3	6.8
	9.7	6.9	10	8.2	8.5	7.7	7.8	7.7
Fluency	10	5.7	7.3	6.8	8	6.3	6.8	7.5
	10	7.0	9.7	8.2	8.2	7.3	7.7	7.7
Vocabulary & Word Choice	10	6.0	6.5	6.6	8	6.2	7.6	7.8
	10	6.7	10	7.8	8.3	6.8	7.5	7.0
Interview	9.8	6.5	9.3	10	8.7	7.7	10	9.5
	10	7.2	10	7.8	9.5	8.2	7.7	7.5

Note: - Scores of the 1st evaluation (pre-experiment) are indicated by numbers in black color
- Scores of the 2nd evaluation (post-experiment) are indicated by numbers in red color

Fig. 4. Result of rubric

As can be seen from the Fig. 4, all participants showed sufficient improvement between the pre-experiment and post-experiment evaluation. The distinctions also varied from small to big proportion in both groups. Interestingly, all subjects received positive feedbacks from the two examiners for their progress during experiment period and subjects in Group 2 slightly received more appreciation for their improvements. For Japanese students (J1, J3, J4), they were highly praised in gaining confidence of speaking. As most of them were seen to reluctant to answer the questions in the first evaluation, their attitude remarkably changed after the series of six experiments. The certified English teacher was amazed at their fluency in the second evaluation and all of them got better score in this criteria. Besides, the non-Japanese students (E3, E4) were considerably appreciated by the certified Japanese teacher about their changes in expressing ideas and choosing words. While they often answered in short phrases and simple words in the first interview, they managed to answer the same question in full sentences and more complicated phrases in the second one. It could be possibly explained that thanks to helping each other revising the mistakes they made (as they stated in the individual interview), Group 2 showed better enhancement in the second evaluation.

The unexpected direction was indicated in the fourth criteria. This criterion assesses the ability of understanding the interview questions without asking for repetition. While the majority of subjects showed their improvement in this measure, some of them (J2, J3, J4) unanticipatedly lost concentration and used the repetition clues as "Pardon, please", "Could you please repeat the question?". Thus, there was no wonder in the decrease of their score in this criteria in the second interview.

Result from individual semi-structured interviews. We held a 30-minutes individual semi-structured interview with every member to obtain an insight of their feeling during the time of using BiTak. The open-questions related to dual-lingual communication and functions of BiTak.

Group 1:

- Usability of BiTak (without the recording function)

All users reported to be able to use BiTak easily because they did not use any specific function. The feature of logging in without creating ID gave them the comfort of access when they sometimes just want to join the chat immediately and do not have to care about the nuisance of making nicknames. The text-chat is highly evaluated because it is convenient for them to type any words or sentences to make it more clearly for the others to understand.

- Dual-lingual communication

Although all the subjects found this kind of communication weird and hard at first, they gradually recognized it really helps people from beginner to intermediate level. The more familiar they get with BiTak, the more motivated they are to speak.

- Hesitate to correct friends' mistakes while in the middle of conversation

When being asked about helping to correct others' mistakes, most of subjects in the group revealed that they hesitated to do that due to they were in the middle of conversation. They sometimes recognized their friends' mistakes but neglected them to wait for the conversations to finish then unintentionally forgot the errors.

Group 2:

- Usability of BiTak (with the recording function)

For the first time, it is really difficult for this group to use strict turn-taking. Interestingly, they deliberately discuss the way to communicate in BiTak without any instructions of the authors to make the communication went smoothly: applying dual-lingual conversation with strict turn-taking for presentation, using recording link for realizing mistakes and normal conversation for correcting mistake and discussion.

The members steadily reported that this system aim to learn language, not merely for chatting. When they do the presentation in the recording part, only one person have to talk. They felt that it is a good challenge for them because they can do a lot of presentation to train their speaking skill.

- Dual-lingual communication

They all agreed with the idea of dual-lingual communication can help them learn languages. Japanese students normally do not have chance to speak English much and vice versa for non-Japanese students so it has mutual benefits. They can gain some new words and correct the mistakes they usually make before. In their opinion, this kind of communication may not be comfortable for chatting but effective for learning languages.

- Strict turn-taking

All group members pointed out that strict turn-taking feature give them time to think carefully before raising their voice. They consequently have confidence in expressing their ideas.

- Recording link

One more interesting point the subjects found is the recording link. They all felt this feature is really important because they can listen again their friend's presentation all the time to recognize and correct mistake for each other.

In summary, dual-lingual pattern proves its efficiency in both groups. However, it is easy to find out the difference in the performance of two groups. The communication in Group 1 is merely normal chatting; they do not care much about others' mistakes. It is alright as long as they understand then they gradually forget to correct mistakes for each other. It is not good for language learning. Meanwhile, members in Group 2 took all of recording link into serious consideration. They want to make sure their friends know their mistakes and willing to correct for them. All of them gradually have sense of learning, not simply gossiping on the account of the proposed features of BiTak.

6 Conclusion

In this paper, we proposed a smart video chat system named BiTak for simultaneous second language acquisition by employing dual-lingual communication and strict turn-taking based on the recording function, and recommended a method of assessing the improvement of users of BiTak for second language speaking. The learner's progress is positively evaluated by language teachers using a rubric scoring framework. Based on the experiments, it was suggested that BiTak has changed the notion of users from an ordinary video chat application to a language supporting system thanks to its two prominent features: strict turn-taking and the recording function.

Due to limitation of time and effort, we recognize that our observations come from a relatively small number of subjects. It is not appropriate to apply quantitative analyses for small samples such as this. A more extensive study would be needed for proving the solid efficiency of all characteristics we have mentioned with the ambition to create a smart language learning environment in further research.

Acknowledgements. The authors greatly thank all subjects who cooperated to our experiments. Especially, we would like to express our gratitude to Prof. Shungo Kawanishi and Ms. Masako Tsutsui who dedicatedly took part in student evaluation process. This work was supported by JSPS KAKENHI Grant Number JP26280126.

References

1. Kimber, L.: Attitudes and beliefs of students toward bi-/multilingualism at an international university in Japan. Ritsumeikan J. Asia Pac. Stud. **33**, 139–152 (2014)
2. Myers-Scotton, C.: Multiple Voices: An Introduction to Bilingualism. Blackwell Publishing, Malden (2006)
3. Mynard, J.: Introducing EFL students to chat rooms. Internet TESL J. **8**(2) (2002)
4. Freiermuth, M., Jarrell, D.: Willingness to communicate: can online chat help? Int. J. Appl. Linguist. **16**(2), 189–212 (2006)
5. Freiermuth, M.R.: Using a chat program to promote group equity. CAELL J. **8**(2), 16–24 (1998)
6. Freiermuth, M.R.: Features of electronic synchronous communication: a comparative analysis of online chat, spoken and written texts. ETD Collection for Oklahoma State University, AAI3021616 (2001)
7. Schwienhorst, K.: Evaluating tandem language learning in the MOO: discourse repair strategies in a bilingual internet project. Comput. Assist. Lang. Learn. **15**(2), 135–145 (2002)
8. Setlock, L.D., Fussell, S.R.: What's it worth to you?: the costs and affordances of CMC tools to asian and american users. In: Proceedings of the 2010 ACM Conference on Computer Supported Cooperative Work, pp. 341–350. ACM, February 2010
9. Angelova, M., Zhao, Y.: Using an online collaborative project between American and Chinese students to develop ESL teaching skills, cross-cultural awareness and language skills. Comput. Assist. Lang. Learn. **29**, 1–19 (2014). (ahead-of-print)
10. Hampel, R., Stickler, U.: The use of videoconferencing to support multimodal interaction in an online language classroom. ReCALL **24**(2), 116–137 (2012)
11. Purpura, J.E.: Second and Foreign language assessment. Mod. Lang. J. **100**(S1), 190–208 (2016)
12. Andrade, H.G.: Using rubrics to promote thinking and learning. Educ. Leadersh. **57**(5), 13–19 (2000)
13. Pomplun, M., Capps, L., Sundbye, N.: Criteria teachers use to score performance items. Educ. Assess. **5**(2), 95–110 (1998)
14. WebRTC, pp. 1235–1244 (1990). http://www.webrtc.org/ No.9

A Recommender System for Supporting Students in Programming Online Judges

Raciel Yera[1], Rosa M. Rodríguez[2](✉), Jorge Castro[2], and Luis Martínez[3]

[1] University of Ciego de Ávila, Ciego de Ávila, Cuba
ryera@unica.cu
[2] University of Granada, Granada, Spain
{rosam.rodriguez,jcastro}@decsai.ugr.es
[3] University of Jaén, Jaén, Spain
martin@ujaen.es

Abstract. Programming Online Judges (POJs) are tools that contain a large collection of programming problems to be solved by students as a component of their training and programming practices. This contribution presents a recommendation approach to suggest to students the more suitable problems to solve for increasing their performance and motivation in POJs. Some key features of the approach are the use of an enriched user-problem matrix that incorporates specific information related to the user performance in the POJ, and the development of a strategy for natural noise management in such a matrix. The experimental evaluation shows the improvements of the proposal as compared to previous works.

Keywords: Programming online judges · Problems recommendation · Natural noise management

1 Introduction

Programming online judges (POJs) are e-learning tools that enclose a set of programming problems to be solved by their users, which usually are students in computer science colleges, and include the automatic compilation and evaluation processes for the solutions proposed by students. Over the last few years, POJs have been successfully used for training students that participate in ACM-ICPC-like programming contests, and in the systematic practice of programming skills in universities and high schools [8,9,17].

Typical POJs present a sequential list of programming exercises that users can select and try to solve [8]. In these applications, it is considered that the users with better performance are those that solve the most problems, because it implies a high level of fulfilment. Such growing satisfaction would increase the learners' efforts and performance in problem solutions and programming subjects, because learners effort increases if they are more satisfied, and then users' learning process is enhanced.

V.L. Uskov et al. (eds.), *Smart Education and e-Learning 2017*, Smart Innovation, Systems and Technologies 75, DOI 10.1007/978-3-319-59451-4_21

In this way, it is important to remark that the learning process in current POJs, based on the previous rule *the more problems solved the better skills acquired* by the students, are far from other e-learning tools such as ITSs or ontology-supported systems. Additionally, in a similar way to most online information nowadays, novice users in POJs fail to solve problems, because they are not able to select suitable ones according to their level of knowledge, due to the information overload regarding the available exercises. This causes a discouragement in POJ participation, and users misplace the benefits that POJ could give them.

A typical solution used to mitigate the negative effect of information overload is the development of recommender systems [1]. Such systems are focused on providing users the information items that best fit their preferences and needs in a search scenario overloaded with possible options. Recommender systems are employed in diverse contexts such as e-commerce, e-learning, tourism, or libraries [1].

This contribution aims at introducing a memory-based collaborative filtering recommendation approach to support students by mitigating information overload in POJs. This is achieved identifying problems that they should solve regarding their skills to avoid failures and frustration. Such an approach, so-called Natural Noise Programming Judges, (NN-PJ), presents two novel features:

1. The use of an enriched user-problem matrix that facilitates the calculation of closer neighbourhoods and therefore a more accurate recommendation.
2. The incorporation of a data preprocessing step based on the concept of *natural noise management* [4,12,14] that detects and corrects anomalous users' behaviours that could affect the recommendation generation.

It is important to remark that, although the proposal is developed specifically for POJs, its innovative principles could additionally be used in other e-learning recommendation scenarios related to problem resolution. This proposal is evaluated through a case study on data associated to a real POJ.

The rest of the contribution is structured as follows. Section 2 revises the required background related to POJs and e-learning recommender systems for supporting programming subjects. Section 3 presents the proposal for recommending problems to solve in POJs. Section 4 shows an experimental evaluation of the proposal to compare its performance with previous related works. Finally, Sect. 5 concludes the contribution.

2 Related Works

Here, several concepts and previous works about POJs and systems for supporting the learning of programming are reviewed.

2.1 Programming Online Judges

Programming online judges (POJs) are systems that contain a list of programming problems to be solved by the users. Specifically, a usual interaction of the user in the POJ is composed of the following steps:

1. The user chooses a problem in the POJ and tries to solve it.
2. The user uploads his/her solution to the POJ.
3. The POJ evaluates this solution using predefined input-output data. If its output matches the expected one, then the solution is considered as correct and the problem solved. Otherwise, the user can either propose an alternative solution for the same problem, or choose a different problem.

The first use of the term POJs as systems for supporting the automatic evaluation of source codes developed by students as problems resolution was presented by Kurnia et al. [8]. Afterwards, a widely used system, Mooshak [9], supports the programming contest management and could also be used as an online judge. Verdu et al. [15] presented EduJudge, which is an effective platform that integrates Moodle and QUESTOURnament, a competitive e-learning environment. Finally, Wang et al. [17] presented OJPOT, a teaching approach that combines online judges and practice oriented programming teaching, which develops students' practical abilities through programming practices. Other works, although do not employ the term online judge, are centred on the automatic evaluation of programming problems solutions [3].

2.2 Supporting Programming Learning

Recommender systems have been applied in e-learning scenarios [7]. This section briefly reviews previous works related to the current e-learning context, which is the recommendation for supporting programming subjects.

In this way, De Oliveira et al. [10] presented a system for recommending "classes" of activities in programming practices, such as "Math operators", "data types", "if structure", or "while structure". This task is modelled as a multi-label classification approach, where the student's profile is associated with some of the mentioned "classes". Previously, Hsiao et al. [6] presented a system to guide students through the suitable questions in a Java programming course. More recently, Vesin et al. [16] described a recommendation module integrated with an intelligent tutoring system for learning programming. Such module focuses on recognizing the students' learning styles through the use of several domain ontologies for finding access sequences for the activities, which are later used to recommend relevant links and actions during the learning process. Similarly, Ruiz et al. [11] presented a knowledge-based strategy using semantic information represented by ontologies to recommend educational resources in a programming course to provide personalized access to such resources.

Summarizing, the referred works were conceived for specific predetermined scenarios and are highly content dependent beyond the user-learning resource interaction. However, in POJs only the users' performance when they aim at solving the problems is available. Therefore, the adaptation of most of the referred works to this context is not straightforward.

Recently, however, two researches [13,18] are the nearest antecedents to the current contribution, given that they are specifically focused on POJs. The proposal introduced in [13] presents a user-based collaborative filtering approach to suggest problems to solve in a POJ. Centred on the same purpose, the work presented in [18] suggests the application of an item-based collaborative filtering approach. However, both approaches are mainly focused on the direct application of a traditional collaborative filtering technique. The proposal uses techniques specifically focused on the POJs scenario.

3 A Recommendation Approach for Programming Online Judges

This section presents a novel approach for recommending problems to solve in POJs supported by the available information, which is represented by the users' performance when they try to solve the presented problems.

Such available information, which is the input for the proposal, is represented as a set of triplets $< u, p, j >$ that represent a user u attempt to solve a problem p with a judgement in the category j, which can be "accepted" if the problem is successfully solved or "not accepted" in any other case. To formalize the current proposal, this scenario will be represented as the relation $D : (u, p) \rightarrow (n, solved)$, that receives as input a user u and problem p, and returns the amount of times n that the user tries to resolve the problem, and whether at last it was solved.

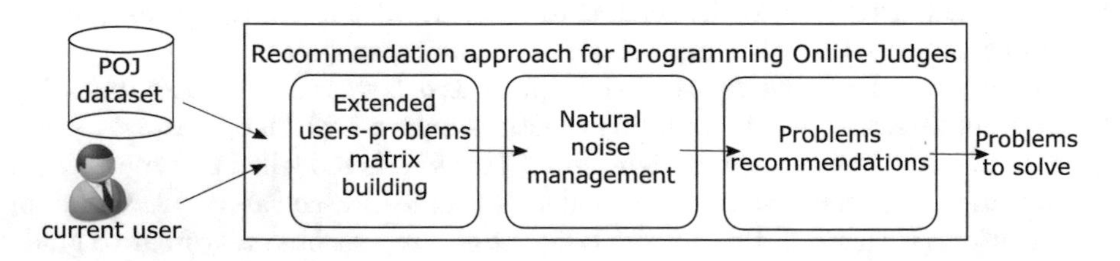

Fig. 1. The recommendation approach for programming online judges

Figure 1 presents an overview of the approach, which takes as input the relation D and the active user x to retrieve the suggested problems to be solved. It is composed of three main phases: extended user-problem matrix building (Sect. 3.1), extended user-problem matrix preprocessing for natural noise management (Sect. 3.2), and problems recommendation (Sect. 3.3). The following sections will describe these phases in further detail.

3.1 Extended User-Problem Matrix Building

In order to apply a memory-based collaborative filtering approach for recommending problems to solve in the online judge, firstly it is used the relation D to build a boolean user-problem matrix M, whose cells are filled with one if the corresponding user solves the corresponding item or zero in other case.

However, the dynamic of the users' performance in POJs makes that this basic matrix M is not enough to obtain a suitable user similarity that considers user knowledge and abilities. Therefore, instead of the analysis of a simple similarity regarding problems solved or not solved by the users, we propose to enrich matrix M by considering also the number of unsuccessful attempts ($D(u,p).n$) when the user u tries to solve the problem p. Such enrichment allows the construction of a new matrix M^* that leads to more accurate similarity calculation and, therefore, a better recommendation performance.

To implement this strategy, we propose to convert matrix M, which contains the binary user-problem interaction values, into a matrix M^* composed of different interaction values that consider the unsuccessful solution attempts:

- $M^*[u,p] = 0$, when the user u has never tried to solve the problem p.
- $M^*[u,p] = 1$, when the user u has not solved the problem p, and has tried it just few times.
- $M^*[u,p] = 2$, when the user u has not solved the problem p, and has quite a few previous failed attempts.
- $M^*[u,p] = 3$, when the user u solved the problem p, having failed many times beforehand.
- $M^*[u,p] = 4$, when the user u solves the problem p quickly.

Following, Eqs. (1) and (2) formalize such transformation, depending on the thresholds λ_{p1} and λ_{p2}, being the boundary between the few and the many attempts categories in not solved and solved judgements, respectively. In both cases the thresholds are initialized as the average number of failures associated with all users that have attempted to solve the problem p. Equations (3) and (4) formalize such initialization strategies, where NS_p and S_p are respectively the set of users that have not solved and solved the problem p, and v any user in the POJ.

$$M[u,p] = 0 \rightarrow M^*[u,p] = \begin{cases} \mathbf{0}, \text{if } D(u,p).n = 0 \ \ \text{(no solution attempts)} \\ \mathbf{1}, \text{if } D(u,p).n \geq \lambda_{p1} \ \ \text{(not solved, with many} \\ \text{solution attempts)} \\ \mathbf{2}, \text{if } D(u,p).n > 0 \text{ and } D(u,p).n < \lambda_{p1} \\ \text{(not solved, with few solution attempts)} \end{cases} \quad (1)$$

$$M[u,p] = 1 \rightarrow M^*[u,p] = \begin{cases} \mathbf{3}, \text{if } D(u,p).n \geq \lambda_{p2} \ \ \text{(solved with many} \\ \text{solution attempts)} \\ \mathbf{4}, \text{if } D(u,p).n < \lambda_{p2} \\ \text{(solved, with few solution attempts)} \end{cases} \quad (2)$$

$$\lambda_{p1} = \frac{\sum_{v \in NS_p} D(v,p).n}{|NS_p|} \quad (3)$$

$$\lambda_{p2} = \frac{\sum_{v \in S_p} D(v,p).n}{|S_p|} \quad (4)$$

3.2 Preprocessing Natural Noise in the Extended User-Problem Matrix

Once the enrichment of the binary user-problem matrix is performed, a new transformation step aims at preprocessing inconsistent data associated with the already transformed matrix M^* to improve the recommendations. This transformation is specifically focused on processing data erroneously related to the users' behaviour, i.e., natural noise [14].

Specifically in e-learning scenarios, it has been pointed out that sometimes the users do not know the solution of the activities, but supposed them correctly; and in contrast, sometimes they recognize their solutions, but make an error [7]. This referred work pointed out that this unexpected behaviour should be regarded in the management of user data in POJs. Considering this scenario and the features associated to natural noise [14], it can be established that these anomalies could be identified as natural noise.

Using such works as a basis, natural noise in the POJ context is characterized as the circumstances where the user does not solve an exercise using the expected predictable effort, which could be occasioned by causes such as plagiarism, guessing, or unexpected slip-ups.

To manage such inconsistent behaviours in the enriched matrix M^*, we suggest the grouping of the problems solved by user u, i.e., those belonging to the categories $M^*[u,p] = 3$ and $M^*[u,p] = 4$, in two sets:

1. Set of problems solved by the user u that needed too many attempts, $C1_u$: $C1_u = \{p \mid M^*[u,p] = 3\}$.
2. Set of problems solved by the user u that needed few attempts, $C2_u$: $C2_u = \{p \mid M^*[u,p] = 4\}$.

Similarly, the users that solved the problem p are grouped in two sets:

1. Set of users that solved the problem p that needed too many attempts, $C1_p$: $C1_p = \{u \mid M^*[u,p] = 3\}$.
2. Set of users that solved the problem p that needed few attempts, $C2_p$: $C2_p = \{u \mid M^*[u,p] = 4\}$.

Using such definitions, *noisy* interactions are detected considering that if the user and the problem being solved belong to homologous sets, then the associated interaction category in matrix M^* should also belong to the same group of homologous sets. In other case, such interaction is considered as noisy and, therefore, it should be replaced with the interaction associated to the group of the corresponding user and item.

For this assumption, two groups of homologous sets are considered: the first one composed of the set of users $C1_p$, the set of problems $C1_u$, and the interaction $M^*[u,p] = 3$; the second one by the set of users $C2_p$, the set of problems $C2_u$, and the interaction $M^*[u,p] = 4$. Equations 5 and 6 formalize how the matrix M^{**} is obtained, which is used afterwards to recommend.

$$M^*[u,p] = 3 \rightarrow M^{**}[u,p] = \begin{cases} \mathbf{4}, \text{if } |C2_u| \geq 2|C1_u| \text{ and } |C2_p| \geq 2|C1_p| \\ \mathbf{3}, \textit{in other cases} \end{cases} \tag{5}$$

$$M^{*}[u,p] = 4 \rightarrow M^{**}[u,p] = \begin{cases} \mathbf{3}, \text{if } |C1_u| \geq 2|C2_u| \text{ and } |C1_p| \geq 2|C2_p| \\ \mathbf{4}, \textit{in other cases} \end{cases} \quad (6)$$

3.3 Recommendation of Problems

This last step consists of the recommendation, which is supported by a traditional memory-based collaborative filtering paradigm [1]. This paradigm has been extended to work with matrix M^{**}. Let x be the active user, this step is composed of three phases: (1) Find x top-k similar users using the matrix M^{**}, (2) Predict a score for each exercise not solved yet by user x using his/her top-k similar users, and (3) Recommend exercises with the highest scores.

First, for each user x we build his/her profile, which is composed of his/her associated row in the matrix M^{**}. To find the top-k similar users regarding x, the proposal uses a modification of the Simple Matching Coefficient (SMC) [2] to discard the influence of the interaction category $M^{**}[u,p] = 0$ in its calculation. This consideration is done because, when two users are compared, the common presence of a problem in the no solution attempts category does not indicate closeness. This modification is explained as follows.

Let S_{xu} be the set of problems where users x and u have the same interaction category, which is different from the zero category: $S_{xu} = \{p \mid M^{**}[u,p] = M^{**}[x,p] \text{ and } M^{**}[x,p] \neq 0\}$. Let D_{xu} be the set of problems where x or u have at least some interaction that is different from the zero category: $D_{xu} = \{p \mid M^{**}[u,p] \neq 0 \text{ or } M^{**}[x,p] \neq 0\}$. With these sets, the similarity between x and u is formulated as $sim(x,u) = S_{xu}/D_{xu}$.

Once the similarity values between x and all users are calculated, the users with the top-k higher values are employed to calculate a score w_p for each problem p not solved by user x, as the sum of the similarities between x and each neighbour that solves p (Eq. 7).

$$w_p = \sum_{u \in \, top\_k(x)} sim(x,u) * M[u,p] \quad (7)$$

The top-n problems with the highest scores w_p, are recommended to user x.

4 Experimental Study

The evaluation of the proposal was performed by using a dataset obtained from the Caribbean Online Judge (2009–2010). This dataset is composed of 1910 users, 584 problems, and nearly 148000 attempts to solve problems. Training and test sets were created following the procedure stated in [5]. For each user, solved problems were split in two sets adding at least one of them to the training set and the other to the test set. This strategy guarantees that both sets contain data from all users. Additionally, the experiments were executed several times with different training-test partitions and their accuracy results were averaged. The proposal also uses the information related with the unsuccessful attempts, which is stored independently to be used as input for each evaluation.

As evaluation measure, F1 measure is used, which evaluates the quality of the top-n recommendation [5]. F1 is defined in terms of precision and recall. In this context, precision is the amount of recommended problems that were solved, recall is the amount of solved problems that were recommended.

The experimental procedure was developed as follows. For each user the proposal generated a list of recommended problems considering the data in the training set. Such list and the solved problems in the test set were then used to obtain the F1 value for the current user. Finally, all users F1 values were averaged to obtain an overall F1 value that characterizes the proposal. Such procedure was repeated 10 times with various train-test partitions and the values were averaged. The proposal used the value $k = 130$ as the number of nearest neighbours used for computing the recommendations. Such value was adjusted through previous experimentation.

To evaluate the performance of our proposal, NN-OJ, in relation to previous works, we compare it to the works referred in Sect. 2.2 that are directly related to the current proposal. Specifically, UCF-OJ [13] and ICF-OJ [18] were used in the comparison. Additionally, the Binary approach is added to represent a procedure similar to the proposal, but using only matrix M.

Table 1. F1 measure: NN-OJ vs. UCF-OJ, ICF-OJ and Binary

	5	10	15	20	25	30	35	40
NN-OJ	**0.3875**	**0.3933**	**0.3784**	**0.3582**	**0.3397**	**0.3220**	**0.3055**	**0.2916**
Binary	0.3614	0.3687	0.3540	0.3352	0.3189	0.3026	0.2895	0.2767
UCF-OJ	0.3833	0.3899	0.3736	0.3543	0.3367	0.3194	0.3035	0.2890
ICF-OJ	0.3602	0.3624	0.3494	0.3348	0.3191	0.3058	0.2932	0.2808

Table 1 presents the results of the approaches evaluated for different number of top-n recommended problems, which are given in the range [5, 40]. The proposal outperforms the previous works for all list sizes. These results show that the matrix enrichment and the natural noise correction imply an improvement of the recommendation accuracy in the POJ scenario.

5 Conclusions

This contribution presents an approach for recommending problems to solve for supporting students in programming online judges. Specifically, the proposal has two key steps: (i) extended user-problem matrix enrichment and (ii) natural noise management, that exploit the data particularities in POJs. This feature contrasts with previous works on POJs scenarios based on a direct application of typical recommendation techniques. The experiment confirms the superiority of the current approach in relation with such previous works regarding accuracy. Future works will be focused on managing the uncertainty inherent to the current scenario using fuzzy tools that allow a better representation of the information related to users behaviour.

Acknowledgements. This research work was partially supported by the Spanish National research project TIN2015-66524-P, the Spanish Ministry of Economy and Finance Postdoctoral Fellow (IJCI-2015-23715), the Spanish FPU fellowship (FPU13/01151) and ERDF.

References

1. Adomavicius, G., Tuzhilin, A.: Toward the next generation of recommender systems: a survey of the state-of-the-art and possible extensions. IEEE Trans. Knowl. Data Eng. **17**(6), 734–749 (2005)
2. Amatriain, X., Pujol, J.M.: Data mining methods for recommender systems. In: Recommender Systems Handbook, pp. 227–262. Springer (2015)
3. Caiza, J., Del Amo, J.: Programming assignments automatic grading: review of tools and implementations. In: Proceedings of INTED 2013, pp. 5691–5700 (2013)
4. Castro, J., Toledo, R.Y., Martínez, L.: An empirical study of natural noise management in group recommendation systems. Decis. Support Syst. **94**, 1–11 (2016)
5. Gunawardana, A., Shani, G.: A survey of accuracy evaluation metrics of recommendation tasks. J. Mach. Learn. Res. **10**, 2935–2962 (2009)
6. Hsiao, I.-H., Sosnovsky, S., Brusilovsky, P.: Guiding students to the right questions: adaptive navigation support in an e-learning system for java programming. J. Comput. Assist. Learn. **26**(4), 270–283 (2010)
7. Klašnja-Milićević, A., Ivanović, M., Nanopoulos, A.: Recommender systems in e-learning environments: a survey of the state-of-the-art and possible extensions. Artif. Intell. Rev. **44**(4), 571–604 (2015)
8. Kurnia, A., Lim, A., Cheang, B.: Online judge. Comput. Educ. **36**(4), 299–315 (2001)
9. Leal, J.P., Silva, F.: Mooshak: a web-based multi-site programming contest system. Softw. Pract. Experience **33**(6), 567–581 (2003)
10. De Oliveira, M.G., Ciarelli, M.P., Oliveira, E.: Recommendation of programming activities by multi-label classification for a formative assessment of students. Expert Syst. Appl. **40**(16), 6641–6651 (2013)
11. Ruiz-Iniesta, A., Jimenez-Diaz, G., Gomez-Albarran, M.: A semantically enriched context-aware OER recommendation strategy and its application to a computer science oer repository. IEEE Trans. Educ. **57**(4), 255–260 (2014)
12. Toledo, R.Y., Castro, J., Martínez, L.: A fuzzy model for managing natural noise in recommender systems. Appl. Soft Comput. **40**, 187–198 (2016)
13. Toledo, R.Y., Mota, Y.C.: An e-learning collaborative filtering approach to suggest problems to solve in programming online judges. Int. J. Distance Educ. Technol. **12**(2), 51–65 (2014)
14. Toledo, R.Y., Mota, Y.C., Martínez, L.: Correcting noisy ratings in collaborative recommender systems. Knowl. Based Syst. **76**, 96–108 (2015)
15. Verdú, E., Regueras, L.M., Verdú, M.J., Leal, J.P., de Castro, J.P., Queirós, R.: A distributed system for learning programming on-line. Comput. Educ. **58**(1), 1–10 (2012)
16. Vesin, B., Klašnja-Milićević, A., Ivanović, M., Budimac, Z.: Applying recommender systems and adaptive hypermedia for e-learning personalization. Comput. Inform. **32**(3), 629–659 (2013)

17. Wang, G.P., Chen, S.Y., Yang, X., Feng, R.: OJPOT: online judge & practice oriented teaching idea in programming courses. Eur. J. Eng. Educ. **41**(3), 304–319 (2016)
18. Yu, R., Cai, Z., Du, X., He, M., Wang, Z., Yang, B., Chang, P.: The research of the recommendation algorithm in online learning. Int. J. Multimedia Ubiquit. Eng. **10**(4), 71–80 (2015)

Assessment of Students' Programming Skills by Using Multi-Style Code Editor

Elena N. Cherepovskaya(✉), Ekaterina V. Gorshkova, and Andrey V. Lyamin

ITMO University, Saint Petersburg, Russia
{cherepovskaya, gorshkovakatya}@cde.ifmo.ru,
lyamin@mail.ifmo.ru

Abstract. e-Learning tools and environments, especially virtual laboratories are becoming of great interest nowadays as the reason of fast improvement of learning management systems and education itself. Virtual laboratories are often used for developing and assessing students' programming skills. However, these environments are commonly related to the specific programming language, so that some students may experience difficulties, while dealing with them. Hence, it is expedient to construct such learning tools that will not be connected to the specific language. This paper presents the results of evaluation of the developed virtual laboratory Multi-style code editor that corresponds with the stated rule.

Keywords: e-Learning · Computer science · Virtual laboratories · Remote Laboratory Control Protocol

1 Introduction

A tendency of universities related to advancing the quality of training graduating students makes a huge shift towards individual education and e-learning [1], as students' interest in massive open online courses is rapidly increasing. Massive open online course (MOOC) [2, 3] provides an opportunity to receive knowledge and improve skills in a particular sphere with a possibility of getting verified certificate of completing a course. In order to provide a certificate it is required to prove the level of skills a person has obtained during the course passage. For this purpose, different assessment procedures can be used, e.g. electronic tests. However, this type of assessing person's skills can be applied only for checking knowledge of theoretical information that a person had memorized throughout a course and the amount of correct answers is always countable and limited to a particular number. A verified certificate claims to be included in person's professional portfolio or applied during the recruitment procedures. Hence, it is necessary for MOOCs to include special assignments that can state the level of person's skills in a discipline. One of the most popular and the most efficient assignments of the described type is virtual laboratory (Lab) [4, 5].

A virtual laboratory is a complex environment that is used for developing, training and estimating person's professional skills in the sphere of the course, a Lab is related to. Labs are smart environments that are used for preparing a person for being able to deal with professional tasks. They are usually presented as specific programming tools

V.L. Uskov et al. (eds.), *Smart Education and e-Learning 2017*, Smart Innovation, Systems and Technologies 75, DOI 10.1007/978-3-319-59451-4_22

that are embedded in smart learning management systems (LMS) [6] or MOOC platforms. All virtual laboratories have different architecture and can be divided into three groups according to their characteristics. Each of the presented types has its advantages and disadvantages [5]. The list of the Labs' categories is following:

- Autonomous [7] or client-server [8] realizations of virtual laboratories;
- Virtual laboratories with an access to physical equipment [9] and Labs based on mathematical models [10];
- Virtual laboratories that provide automatic results assessment [4, 5].

One of the most essential merits, and, at same time, a demerit, of virtual laboratories is that they are mostly applied for complex tasks having an infinite number of possible answers. Hence, for the Labs that use automatic assessment of student's answers an evaluation on test sets stating an input and output set of values is applied. Virtual laboratories that are applied for information technology disciplines, e.g. used to assess student's programming skills, are often realized for only one programming language because of the complexity of their realization, as it is required to evaluate student's program in order to check its correctness that is usually implemented on the server side of the Lab. However, if the Lab is required to be used for a comprehensive discipline, e.g. computer science, it is necessary that all students will have an opportunity to pass the final assignment, even if they have experience in different programming languages. Hence, a Lab that comprises different programming languages should be developed in order to provide a tool for assessing programming skills of all students regardless the language they are using.

The paper presents a virtual laboratory called Multi-style code editor used for estimation of students' skills of algorithms' construction. The second section of the paper is related to the description of interaction scheme between a Lab and the system using Remote Laboratory Control Protocol. The third section describes the Multi-style code editor Lab itself. The fourth section describes the evaluation results and the last section contains concluding statements.

2 Interaction Between Virtual Laboratory and LMS

The Lab described in this paper was developed based on the rules formed in the learning management system of ITMO University called AcademicNT [3–5, 11, 12]. These rules are applied for almost all Labs developed in the University and specify the way of transferring data about student's answer and Lab's settings, regulations of the assessment procedures, etc. For this purpose, a Remote Laboratory Control Protocol (RLCP) [4, 5, 11] had been developed in order to provide all the necessary information as well as a convenient way of interaction between parts of the Lab and LMS.

The scheme of the structure of an RLCP-compatible virtual laboratory is presented in Fig. 1. The Lab consists of the following three parts interacting with each other and LMS by using RLCP and HTTP protocols:

- RLCP Stand (Stand)
- RLCP Client (Client)
- RLCP Server (Server)

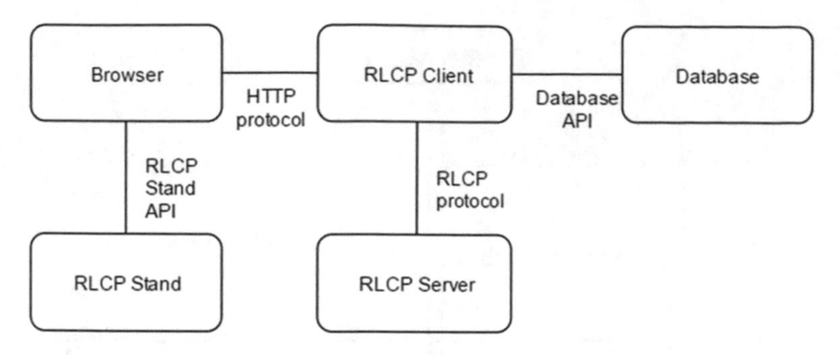

Fig. 1. Lab's structure

RLCP Stand is a virtual stand of the laboratory, which is realized based on Java, or during the last years due to the rejection of Java Applet technology – JavaScript, technology and is embedded in the Browser to be viewed by student inside LMS. Hence, a Stand represents a client-side application of the Lab. The server side consists of the RLCP Client and RLCP Server. A Lab's Stand is connecting to the related Client by HTTP protocol through Browser in order to request for Lab's variant, send student's solution for the evaluation, etc. Client is responsible to provide interaction between a Server and a Stand. It connects to the Server by RLCP protocol. If the Lab allows intermediate evaluation, a request is sent to the Server and the result is shown for a student after the evaluation was completed. The whole path of the request from a Stand to the Server containing student's final response consists of the following steps. A sequence diagram of a Stand interacting with the Server is presented in Fig. 2.

1. A request is sent from a Browser using HTTP protocol to the Client. The submitted response is included in the request.
2. Client processes the request and send it to the Server through RLCP protocol.
3. The student's solution is being evaluated. Evaluation results are sent to the Client through RLCP protocol.
4. Client stores the results in Database by accessing it using Database API and then constructs an update for browser's page.
5. The results of the evaluation are presented to a student.

Remote Laboratory Control Protocol realizes the interaction between RLCP Client and RLCP Server. When a student is sending a request to evaluate his or her solution, XML response containing a solution is being constructed and included in the request. The header of the request is separated from the body with a new line and contains information about the working mode (`Check` – for final assessment, `Calculate` – for intermediate assessment, `Generate` – for generating an assignment). The body contains necessary information about student's solution.

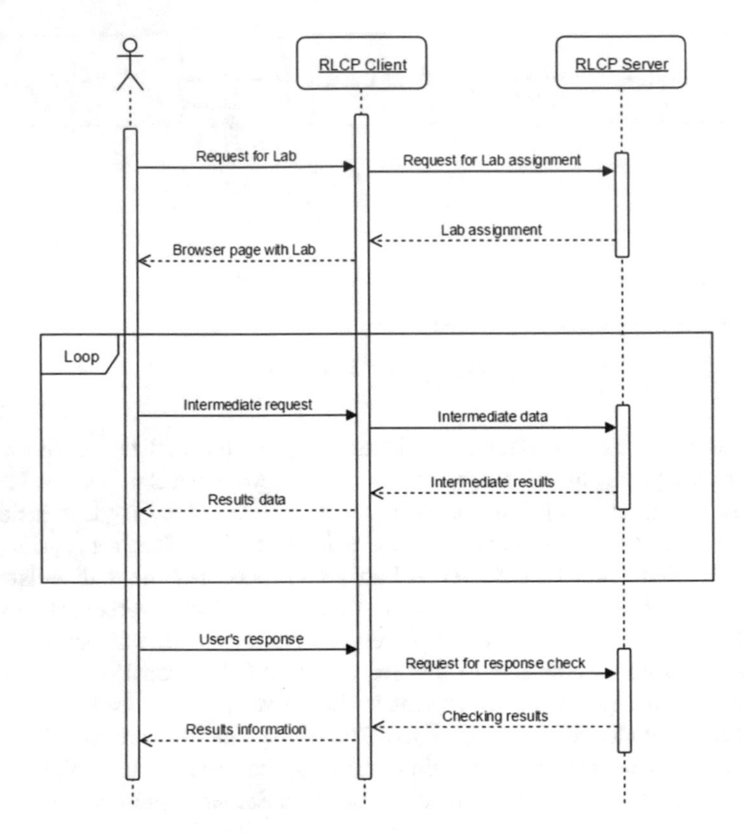

Fig. 2. RLCP sequence diagram

3 Multi-style Code Editor

The Multi-style code editor (MCE) virtual laboratory is a Lab developed in ITMO University that is embedded in University's LMS AcademicNT. It is successfully applied for almost 10 years as a laboratory exam task in computer science for the first-year bachelor students. Together with Post and Turing machines' interpreters, this virtual laboratory is used for the assessment of student's skills of constructing algorithms. However, contrary to the mentioned laboratories, MCE is used for more complex assessment as it is based on writing programs on one of the three programming languages that provides certain advantages of its application as an exam task. The final assignment results do not depend on the programming language that a student wants to use. Hence, students with the same level of building algorithms' skills will be assessed equally regardless the studied language. That also allows applying MCE in schools in order to develop skills of constructing algorithms after students were taught the basics of one of the common languages implemented in the Lab.

The interface of the Lab is presented in Fig. 3. It consists of five main areas: variables panel, code editor panel, the functions library, console area for viewing Stand's messages, e.g. errors and info-messages, and actions panel. The first area is used for displaying variables and is divided into three sections according to the type of variables, i.e. input, output and inner. Variables can be added or edited in the special dialog, which becomes available by clicking on the 'Edit'-button in the top right corner of the panel. The type of the variable is chosen from the following list: integer number, real number, integer array, real array and array of characters (double-byte Unicode character set). The functions library contains the list of operations and constructions for the selected programming language that can be inserted in the code editor by double-clicking on necessary operation's name. The last area is used for accessing the actions panel with a drop-down list for choosing programming language and a list of control buttons used to run program in different modes.

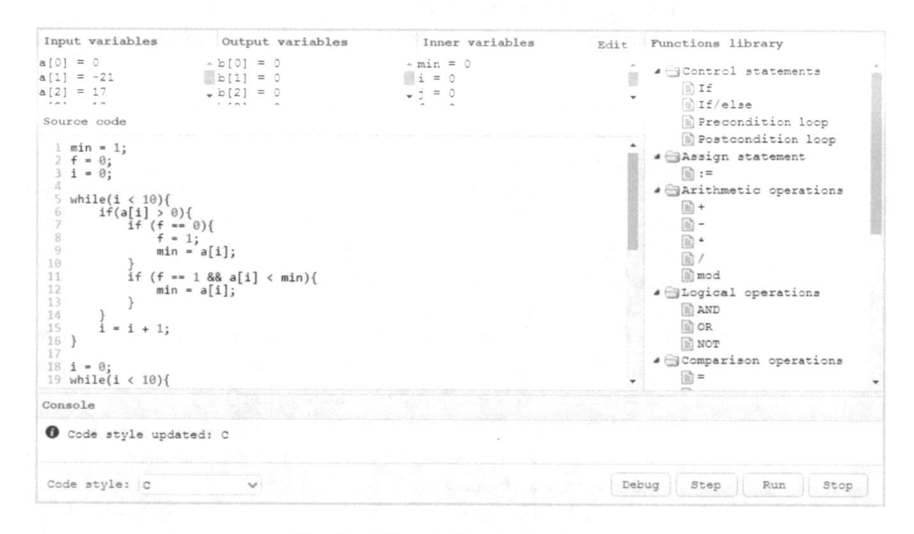

Fig. 3. Virtual Stand of MCE Lab

The virtual stand of MCE Lab contains an embedded interpreter that does not require intermediate evaluation on the server, as all necessary actions, e.g. running code, is realized on the client side in the stand itself. It was at first programmed using Java Applet technology, however, due to its rejection by most of the existing modern browsers, recently the stand was updated and renewed by applying JavaScript technology. The structure of the current version of MCE stand and description of the files is presented in Table 1.

Table 1. The structure of the MCE virtual stand

Filename	Description
Lab_mce_i18n.js	Contains translations for the UI of the Lab. Currently two different languages are supported: English and Russian.
Lab_mce_vlab.js	Contains main settings of the Lab, functions used to receive and process student's previous solution from server and current student's solution from the virtual stand. The following actions of processing

(*continued*)

Table 1. (*continued*)

Filename	Description
	programming code are presented: run program, run program in debug mode, stop program, and step (proceed) to the next line. This file also includes main settings for the allowed programming languages.
Lab_mce_ui.js	Contains functions of building user interface of the virtual stand and processing student's actions. It includes main virtual stand with a space that displays a list of existing variables according to their group (input, output or inner) and a code editor, dialog windows for entering and editing variables of different types together with dialogs used to display warnings and errors
Lab_mce_syntaxer.js	Contains functions for syntactic analysis of student's current solution for the Lab. The analysis of the programming code is based on the keywords specified for each of the presented languages
Lab_mce_tree_builder.js	Contains functions that are used for building a tree of constructions for student's solutions. The main function in this file gets a Syntaxer-object after the syntactic analysis of the programming code that allows iterating on the extracted keywords. The rules for the existing constructions are specified for each language in "Lab_mce_vlab.js" and are used in these functions in order to build the tree correctly. The list of functions includes the following methods among others: • build (The main function that receives a Syntaxer-object as an input and returns the formed tree of constructions) • buildCommand (The function is based on the analysis of a construction rule checking if the construction had been written correctly by a student) • commands (The function is used for processing a sequence of programming commands and returns a tree node that includes all the processed commands in a required order) • getExpression (The function is used for reading an expression and returns a tree node for the processed expression)
Lab_mce_interpreter.js	Contains an object that realizes unified interpreter for all the available programming languages. It receives a tree of constructions as an input and processes student's solution according to the specified mode. The list of objects includes the following objects among others: • mceExpression (The object is used for processing a tree node that realizes an expression; it includes checking the conformity of variables' types) • mceProgramProcessor (The object is used for processing the evaluation of student's solution, including pre- and postcondition loops and conditions, and realizes unified interpreter, which is based on sequencing tree nodes' identifiers and analysis of theirs parent nodes' ids) • mceErrorProcessor (The object is used for displaying errors and highlighting a line in the code editor according to the line number and error code that are specified as input arguments in the main object's function)

As the virtual laboratory is used for complex skills assessment, there are some issues of sending a request with student's solution of the assignment to Server. As mentioned earlier, student's solution is sent in XML format and due to the fact that programming code often contains special characters, such as '<', '&', etc., it is required to escape these characters in a specific order presented in Table 2.

Table 2. The order of escaping characters in XML request

#	Character	Entity
1	&	`&`
2	<	`<`
3	>	`>`
4	–	`& minus;`
5	"	`"`
6	Line breaks (single characters and sequences of characters with codes 10 и 13)	`< br/>`

Document Type Definition (DTD) for the aforementioned XML is following:

```
<!ELEMENT ProgramEnvironment (Variable+, Code)>
<!ELEMENT Variable (Name, Type, Value+)>
<!ATTLIST Variable
      input CDATA (true | false) "false"
      output CDATA (true | false) "false"
      inner CDATA (true | false) "true"
>
<!ELEMENT Type (#PCDATA)>
<!ELEMENT Value (#PCDATA)>
<!ATTLIST Value
      N CDATA #IMPLIED
>
<!ELEMENT Code (#PCDATA)>
<!ATTLIST Code
      style CDATA #REQUIRED
>
```

The root element is `ProgramEnvironment`. It contains two types of elements: `Variable`, number of these elements corresponds with the number of variables used in the program, and one `Code` element that contains student's solution. Each `Variable` is described by its input, output or inner type, `Type` of the value and `Value` itself presented as the related elements and attributes. The variables are included in order to apply them for server-side code evaluation on test sets.

4 Evaluation

The Lab is successfully applied in ITMO University for the first-year students as a practical task in the exam of computer science. Moreover, MCE Lab had been used during entry exams in the University. There are different variations of the assignment that are given to students during the last part that is of higher difficulty comparing to other tasks of an entry exam. Hence, it was decided to use a virtual laboratory for the exam as it was held in computer classes of the University in AcademicNT LMS. During the described evaluation, 677 students had been involved, 350 of whom submitted solutions on the Lab. The first part of the exam consisted of the basic tasks and tasks of higher complexity was presented as an electronic test. Most of the participants successfully finished the test. The basic problems included 8 tasks that had been finished with an average of 75.12% points. Problems of the higher complexity included 6 tasks that had been finished with an average of 60.71% points. The last part of the exam consisted of two assignments of the highest complexity. These assignments were passed in the presented virtual laboratory. The first problem called C1, which stated a capacity to write small program of processing an array, could result in 2 points for the overall exam score, and the second one – C2 related to writing programs used for solving problems of an average complexity – in 4 points. Students were allowed to choose one of the three available programming languages, i.e. C, Pascal and Basic, as these languages are the most popular to be taught in Russian schools. The results of the evaluation are presented in Table 3.

Table 3. The results of the lab evaluation for both types of the assignment

Points (%)	C1 (% of students)	C2 (% of students)
0	47	73.67
25		10.67
50	29.11	5.78
75		4.66
100	23.89	5.22

As presented in Table 3, most students found the tasks difficult, as they realize one of the most advanced learning outcomes, and did not achieve any points during the exam. However, a significant number of students passed tasks of the last part successfully that proves the possibility of using virtual laboratories during the exam, so that students could write programming code on their computers, while passing exams in computer classes, instead of passing them on paper. The Lab helped to assess one of the basic programming learning outcomes related to algorithms' construction. Moreover, the tool helped students not to be concentrated on the specific programming language and integrated development environments, so that their skills could be purely assessed.

5 Conclusion

e-Learning is rapidly growing nowadays and a giant number of universities are participating in different programs related to e-learning including massive open online courses or building learning management systems. As e-learning is aimed to provide knowledge to students and develop their professional skills, it is required from the developers of learning management systems to build special learning tools and environments that will fully accomplish the stated goal. Virtual laboratories are complex e-learning environments that are used to investigate real-world systems and processes, and, hence, develop students' skills and assess them by constructing specific assignments. These tools had shown their positive effect during the last few years, and Labs are now often used in MOOCs and learning management systems of different universities all over the world. Considering computer science sphere, it is necessary to develop skills of constructing algorithms as soon as a student started to learn the discipline. Therefore, a Lab that could provide all the required functions, i.e. a possibility of writing programming code and automatically assessing the results, should be used. However, most of the existing Labs are related to the specific programming language that limits the number of students it can be suitable for, so that using such Labs will not evidently develop the regarded skills for all the participating students. This paper presents a Lab called Multi-style code editor that is not related to a specific programming language. It is built for the three most common languages that are frequently used for algorithms' explanation. One of the main advantages of this Lab is that the syntactic analysis of the code and the way a Lab interprets and runs it is the same for all the presented languages. Hence, it provides a possibility to easily enhance the Lab by adding necessary languages so that more students will feel comfortable using it during their learning. Moreover, the evaluation of the Lab showed a possibility of its application for being included in the last part of entry exams to the University.

Acknowledgement. This paper is supported by Government of Russian Federation (grant 074-U01).

References

1. El Mhouti, A., Nasseh, A., Erradi, M.: Towards a collaborative e-learning platform based on a multi-agents system, In: Proceedings of the 2016 4th IEEE International Colloquium on Information Science and Technology (CiSt), pp. 511–516 (2016)
2. Zang, X.; Iqbal, S., Zhu, Y., Riaz, M.S., Abbas, G., Zhao, J.: Are MOOCs Advancing as Predicted by IEEE CS 2022 Report? In: International Conference on Proceedings of the Systems Informatics, Modelling and Simulation (SIMS), pp. 49–55 (2016)
3. Ivaniushin, D.A., Lyamin, A.V., Kopylov, D.S.: Assessment of outcomes in collaborative project-based learning in online courses. In: Smart Innovation, Systems and Technologies, vol. 59, pp. 351–361 (2015)

4. Chezhin, M.S., Efimchik, E.A., Lyamin, A.V.: Automation of variant preparation and solving estimation of algorithmic tasks for virtual laboratories based on automata model. In: Vincenti, G., Bucciero, A., Vaz de Carvalho, C. (eds.) LNICST, vol. 160, pp. 35–43, Springer (2016)
5. Efimchik, E.A., Cherepovskaya, E.N., Lyamin A.V.: RLCP-compatible virtual laboratories in computer science. In: Uskov, V., Howlett, R.J., Jain, L.C. (eds.) Smart Innovation, Systems and Technologies, Smart Education and E-Learning: 2016, pp. 303–314 (2016)
6. Ishak, M.K., Ahmad, N.B.: Enhancement of learning management system with adaptive features. In: Proceedings of the 2016 Fifth ICT International Student Project Conference (ICT-ISPC), pp. 37–40 (2016)
7. John, J.: Remote virtual laboratory on optical device characterization and fiber optic systems: experiences and challenges. In: 2012 IEEE International Conference on Proceedings of the Technology Enhanced Education (ICTEE), pp. 1–5 (2012)
8. Magyar, Z., Zakova, K.: Using SciLab for building of virtual lab. In: 2010 9th International Conference on Information Technology Based Higher Education and Training (ITHET), pp. 280–283 (2010)
9. Zhang, J., Li, J.: Design of a network virtual laboratory system for digital signal processing experiments. In: Proceedings of the International Conference on E-Learning, E-Business, Enterprise Information Systems, and E-Government (EEEE 2009), pp. 84–86 (2009)
10. Moritz, D., Willems, C., Goderbauer, M., Moeller, P., Meinel, C.: Enhancing a virtual security lab with a private cloud framework. In: 2013 IEEE International Conference on Teaching, Assessment and Learning for Engineering (TALE), pp. 314–320 (2013)
11. Lyamin, A.V., Efimchik, E.A.: RLCP-compatible virtual laboratories. In: Proceedings of the International Conference on E-Learning and E-Technologies in Education (ICEEE 2012), pp. 59–64 (2012)
12. Uskov, V., Lyamin, A., Lisitsyna, L., Sekar, B.: Smart e-Learning as a student-centered biotechnical system. In: Vincenti, G., Bucciero, A., Vaz de Carvalho, C. (eds.) LNICST, vol. 138, pp. 167–175. Springer (2014)

DILIGO Assessment Tool: A Smart and Gamified Approach for Preschool Children Assessment

Antonio Cerrato(✉), Fabrizio Ferrara, Michela Ponticorvo, Luigia Simona Sica, Andrea Di Ferdinando, and Orazio Miglino

Department of Humanistic Studies, University of Naples "Federico II", Naples, Italy
Antonio.Cerrato@unina.it

Abstract. This paper describes DILIGO, a tool for smart assessment of preschool children attitudes. Many tests for children assessment children are based on verbal materials that can result tiresome for pupils, taking a long time for the administration and also they are sensitive to biases. This paper illustrates a gamified approach to promote participation and involvement of preschoolers and to run an implicit and transparent assessment. This way, it is possible to evaluate the children preference for a certain activity, giving information for orienting in school and work context.

Keywords: Assessment · Involvement · Gamification · Preschool children · Technology enhanced learning

1 Introduction

Assessment plays a key role in every educational challenge as it offers the tools to understand where we are, where we have to get and how we can get there. Traditional assessment is run with a variety of methods, like paper-pencil questionnaires, interviews, observation. In the field of education, assessment is usually aimed at investigating students behaviors and attitudes or, for example, to individuate their favorite subjects and school preferences. Usually a lot of tests are thought to evaluate primary or secondary school children [22] and, often, some researchers do not focus in particular on the intrinsic nature of favorite activities, but on the external causes (e.g. gender effect) that influence children choices [3]. Instead, in this work, the target group is represented by kindergarten children (5–6 years old).

Innovations in recent years have opened new possibilities for assessment methodologies that try to overcome traditional assessment drawbacks, such as the social desirability bias [9] and the acquiescence [14]. The social desirability bias refers to the tendency of research participants to give socially desirable responses instead of choosing responses that really reflect their opinion, mood or feelings, whereas acquiescence is the tendency people to give tacit assent; agreement or consent by silence or without objection. Both biases are clearly elicited by the assessment situation so they can be reduced by hiding or dissimulating the setting connected to the assessment. In other words, the assessment must become implicit and non-evident.

V.L. Uskov et al. (eds.), *Smart Education and e-Learning 2017*, Smart Innovation, Systems and Technologies 75, DOI 10.1007/978-3-319-59451-4_23

Another problem, which is often connected to traditional assessment and is especially evident with time-demanding tools, is that participants can get tired during extremely long testing procedures and do not pay the same attention to the final items resulting in less accuracy.

At present, tests are increasingly exploiting the benefits deriving from digitalization, which allows benefiting of a wide range of features, such as the increased speed in administering tests, the more secure data collection and the possibility to quickly share this information.

Together with digitalization, another methodology which can be used to make test procedures, in the fields of education and psychology more manageable is represented by gamification.

The next subsection will describe in more detail what is meant by this approach.

1.1 Gamification of Assessment Tools: The Assessment Is Made Engaging and Implicit

Gamification is the act of applying game characteristics to a non-game situation [6] or implementing design concepts from games, in order to increase user interaction and participation [27].

Moreover, gamification is in relationship with an idea called flow [20] that is the mental state of operation in which the person performing the activity is completely immersed in it, feeling a deep involvement and enjoyment represented by the process of doing that specific activity.

More generally, gamification is referred to any process using games and game-like phenomena in non-leisure settings [11].

Gamification has been fruitfully applied to a wide variety of contexts, including tests. For example a psychological test can be gamified inserting some game-like features in the real situation/event. In some tests a type of gamification is contemplated in reference to the sentence "Imagine of…" in the participants' instructions which leads the participants in a fictional dimension to reply. Also in the educational field, some learning tools are ideated according to gamification strategies.

A stronger strategy consists in transforming the whole assessing situation in a game. An example is represented by digital Serious Games which, supported by emerging technologies, can use video games as assessing situation, instead of the traditional setting.

Several Serious Games, accessible on laptop, tablet or other electronic devices, aim at creating valid alternatives to traditional tests assessing cognitive abilities and personality traits [23, 24].

This technique allows more accurate data recording process; moreover, individuals performing a Serious Game are more absorbed in the session itself, due to higher engagement levels promoted by gamification [21].

And, Last but not least, the participant, while playing, is not aware of undergoing an assessment procedure, thus mitigating the mentioned biases.

One example is represented by the ENACT game [5, 12], developed during the ENACT project, funded under Lifelong learning programme and tested on more than

300 individuals. The ENACT Game starts from a traditional and validated psychological test to assess negotiation abilities, the ROCI-II (Rahim Organizational Conflict Inventory-II) [17, 18]. It is a role playing game where the user, interacting with an artificial avatar, employs her negotiation skills. This allows the system to evaluate the player negotiation skills without making her aware of undergoing an assessment procedure.

1.2 Smart Assessment

Using Serious Games and other non-standard assessment procedure, together with the fact that assessment is kept implicit, offers another relevant advantage, the assessment customization.

As happens with standard learning materials, standard assessing tools can not be enough flexible in order to fulfil assessment needs.

A certain paper-pencil test can be too long for children and this can impact its validity to investigate specific areas.

Instead, a smart assessment tool, relying on artificial intelligence techniques, can offer a customized assessment for each child, build individualized reports at different levels of granularity, adapt to child starting point and trajectories. In this manner, smart assessment can be thought as a prerequisite of smart technologies for education and learning in that the software can provide a unique learning experience for the students and also to provide useful information for teachers.

An example of artificial intelligence techniques applied to smart assessment and personalized learning are Intelligent Tutoring Systems [1, 8].

They incorporate computational models from cognitive sciences, pedagogical sciences, computational science, artificial intelligence, mathematics, and related fields that represent a prerequisite to transfer the assessment abilities from humans to artificial systems.

The act of analysing and inferring from the huge amount of data collected throughout the use of these smart systems gave rise to the fields of Educational Data Mining and Learning Analytics [2].

1.3 What DILIGO Is Meant to Assess

People are different also in the way they learn and many scholars in literature tried to understand individual differences in learning and their relation with motivation, cognitive styles and other variables, leading to the invention of different tools..

As an example, we can cite the LSI Learning styles inventory [10], dating back to 80's and improved along years, that identifies 9 typologies related to experiential learning (Initiating, Experiencing, Imagining, Reflecting, Analyzing, Thinking, Deciding, Acting and Balancing).

The idea of learning styles is very appealing because teachers and course designers pay attention to students' learning styles: by diagnosing them, by encouraging students to reflect on them and by designing teaching and learning interventions around them [4].

It is also interesting to look for connection between learning styles and other related cognitive concept such as cognitive style [19], multiple intelligences [7], individual or classroom profiles [26].

Another interest topic, especially active in Italian context, is *precocious orienting*.

The orienting towards work should take place long before the decision moment, starting from childhood. Orienting helps people to be self-aware, to progress in adapting studies to work, to fulfil their personal and professional expression.

In the field of life-long orienting, DILIGO offers a gamified and smart tool to assess activity preferences in very young children, a necessary starting point for every orienting challenge.

2 DILIGO Assessment Tool

Nowadays, people collect a large amount of data in order to increase the knowledge of many aspects of life. Also children have become the object of investigation of different surveys (longitudinal and not) with parents and teachers trying to achieve a better understanding of their skills, preferences and future attitudes. Moreover the adoption of new technologies in the educational field has a steady and continuous rise which permits an improvement of didactic methodologies and new forms of students participated interaction. Additionally, technology enhanced assessment plays a key role in the education, primarily because it represents a cheaper way to individuate a customized and unique pathway of students; moreover, it is helpful to provide an instant feedback about tested abilities, an informative schedule for teachers and a meaningful and positive learning experience for students [25].

On these lines, DILIGO has been built: *gamified and cartoonized* assessment software addressed to pre-school children aged 5 years. It has the purpose to investigate future trajectories and attitudes of kindergarten pupils among different activities, logic, artistic and literary activities, related to school subjects. The uniqueness of DILIGO represents the fact that it has been ideated for a specific audience (5 years children) that usually is not included in this kind of researches.

The main goal is to allow children to explore different type of activities trying to individuate the ones which they feel more confident with.

At the end of the test, kindergarten teachers read a unique profile for each child, selected among 13, depending on different choices picked by the pupils.

DILIGO, in particular, is a game taking place in a fictional astronomic space, with the leading character Ludo, a young alien girl (Fig. 1).

Ludo has the goal to give life to a new universe, by collecting different gems from the planets she visits. The gems collection during her journey is made possible by the player who performs activities and, in the meantime, undergoes the assessment.

The first step is to register the child, in order to allow the system to keep track of child interaction (Fig. 1). Then the game can begin.

Children have to accompany Ludo during the exploration of 5 different planets: in the first one, there is a tutorial that helps children understand the game functioning. Then, for each planet, kids can choose among three different activities selecting one of

Fig. 1. The DILIGO leading character Ludo

three characters leading to a specific activity: the painter, connected to the artistic activity, the reader, connected to the literary activity and the scientist, connected to logic activity (Fig. 2).

Fig. 2. This is an example of a single interaction between the user and a task of DILIGO: User performing the test can choose one of the three characters at the bottom of the page

These three exercises are available for each planet and are quite varied:

- the artistic activity consists in helping the painter color an animal, different for each planet, in three possible ways;
- the literary activity consists in helping the reader choose the conclusion for a certain story, different for each planet, among three ending scenes;
- the logic activity consists in helping the scientist free a caged animal, the same of artistic activity, completing a sequence of geometric shapes.

We have selected these 3 peculiar activities relying on our theoretical framework on the fact that Italian Ministry of Education divides high schools in three main branches: scientific/technical high schools, in which logic mold activities prevail, artistic high schools, where creative and artistic activities are pursued and humanistic high schools, in which literary activities are given priority. The idea behind DILIGO is the willingness to define a child exploratory profile rather than a constructive one: acquiring information about children tendencies, they can be encouraged to explore paths they are not so well accustomed to.

It is worth underlining that children can choose freely the option they prefer because a correct answer does not exist. This derives from the purpose of DILIGO, which has an attitudinal intent and not an evaluative one: there is no performance to be evaluated, only the children choices must be recorded. After the children choice and the activity execution, Ludo receives a gem for having accomplished the task and the story continues going to the following planet. At the end of the game, after an activity has been chosen for each planet, DILIGO ends with a rewarding video viewed by the child. The teacher can now insert a new child to be tested and the whole cycle can start again. Data about children interaction are recorded and become accessible only for teachers with a specific password. For each child, session data regarding the chosen activity during the test and the provided responses are collected.

Moreover, in the result page a profile is identified, among 13, depending on the type and amount of the activity chosen. For example if a user selects only logic activities to perform during the assessment, the obtained profile will describe "*The commitment to a specific field is obvious; However, to ensure that the positive attitude towards the environment is linked to the content rather than a tendency to repetition, it could support the exploration of other areas to stimulate curiosity, openness and interest.*" Conversely, if the child has performed DILIGO choosing 3 artistic activity, 1 logic and 1 literary the profile reveals "*Prevailing attitude towards the artistic activities, with some sporadic exploration of other types of activity*".

If a child is oriented to perform only a specific kind of activity, as in the example,, teachers can encourage him/her to explore also other activities, in order to guarantee a more comprehensive exploration of different tasks with which children can prove themselves. DILIGO, as every educational game, can be described in terms of different levels [15, 16], which refer to superficial and deep aspects.

These two levels are represented by coexisting interactive and non-interactive elements. The interactive elements refer to the assessment tool core and are related to an invisible dimension that is the main structure of the game, the engine.

Instead, the superficial dimension is connected with the narrative: for example, a plot may be present to keep children involved. The narrative dimension is useful to maintain the child in a fictional world and gives sense to the activities to be performed. This dimension includes non-interactive elements structured around interactive ones in the deep level.

In the next subsection, these different levels are introduced in more detail.

2.1 DILIGO Superficial Level

The superficial level of DILIGO is represented by the narration of the story itself, the graphic and the space environment created for the test (Fig. 3).

On the whole, the superficial level represents the set of elements (interactive and not) that are immediately visible to the user. For example, the tutorial starting at the beginning of the assessment is a non-interactive element that helps the child to gain confidence with the DILIGO functioning. Instead, another element of DILIGO, like the interaction presented above (Fig. 2), represents an interactive element requiring an "answer" from the child that is the selection of one of three characters to perform the corresponding activity.

Fig. 3. A screenshot of DILIGO representing an element of the superficial level

2.2 DILIGO Deep Level

The deep level is a structured graph (Fig. 4), which defines a pathway between nodes. The graph, actually, is the inner core of the DILIGO structure and represents the unique direction (from node 1 to 6) accessible to the user. To move towards nodes there are three possible and selectable alternatives that are the different characters and corresponding activities showed to the child in the superficial level. They all convey, without changes in the main path, from the precedent node to the following one.

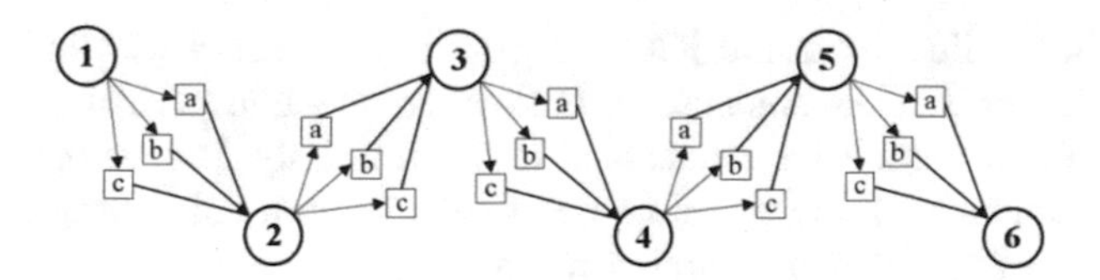

Fig. 4. Graph structure of the deep level

2.3 DILIGO Implementation

We ideated DILIGO using STELT (Smart Technologies to Enhance Learning and Teaching), an integrated software and hardware platform to build educational materials. It allows to design and implement prototypes based on a well-known artificial intelligence methodology (agent-based modeling) and tangible interfaces (usually concrete objects equipped with RFID sensors) as tools to support user-computer interactions [13]. STELT mainly consists of three parts/modules: Storyboarding (aimed at building personalized scenarios), Recording (to track users' data) and Adaptive Tutoring (that consists in on-time intelligent feedbacks).

It is important to underline that we can support the interaction game/child step by step, with immediate feedback. With the support of a technology like STELT it is possible to enhance educational materials possibilities.

2.4 DILIGO Learning Analytics

One of the goals of DILIGO is to collect data about children behavior during the game: to do this, DILIGO includes a module for learning analytics in which data are recorded and elaborated.

Starting from the choices and the activities performed it is possible to associate a specific session with one of the 13 possible profiles. Moreover, it is possible to inspect the whole frequency distribution related to children choices and enrich the analysis with diagrams and charts. In this manner, it becomes doable to analyze different levels of granularity about the data: the single user behavior during the entire session, his choices among the different interactions, the child in rapport to her class, the trend of her school or the whole population. Another aspect to take into account in the analysis about variables such as the children gender, the repetition of the assessment during time, the eventual age difference and so on.

Lastly, it is important to mention that the idea underlying this instrument is to reach a complete personalization of the user assessment, with the possibility to adapt the software requests to each specific user, for example taking into account her precedent choices, avoiding to demand the execution of the same type of activities and differentiating the sessions that came in succession.

2.5 DILIGO Validation

In regards to the validation of DILIGO, we work in partnership with Indire, the Italian National Institution for Documentation, Innovation and Educational Research, representing the reference point for educational research in Italy. It develops new teaching models, experiments with the use of new technologies in training programs, promotes the redefinition of the teaching-learning relationship.

The institute has recruited 5 public kindergarten school to administrate the DILIGO assessment tool in kindergarten classrooms, building a database composed of more than 100 children aged from 5 to 6 years.

3 Conclusions and Future Directions

In this paper we have described DILIGO, a technology enhanced assessment tool to examine future trajectories and attitudes of kindergarten pupils among different activities. Even if the assessment tool is yet to be tested and validated for children, it is possible to affirm that the software offers the appealing chance to ideate completely new assessing environment where gamification and involvement are central. Through a gamified approach of the assessment tool, a better involving and concentration of the children can be guaranteed in tasks that they are currently attending. The validation of DILIGO will be extensive and will be run on large samples.

The next steps for this research are twofold: on one side the goal is to propose new activities for different cognitive assessment, such as memory, categorization, decision making, etc.; on the other it is to validate this platform comparing the results obtained with traditional methods. Moreover in the next years the STELT platform could be integrated in other learning management systems such as Moodles and MOOCs.

Acknowledgments. The first version of DILIGO has been developed for activities connected to "*Costruire giocattoli con la stampante 3D*" project by INDIRE, Istituto Nazionale di Documentazione, Innovazione e Ricerca Educativa.

This project has been supported by INF@NZIA DIGI.tales, funded by Italian Ministry for Education, University and Research under PON-Smart Cities for Social Inclusion programme.

We would like to thank Jose Mangione and Maeca Garzia of INDIRE and Daniela Pacella of University of Plymouth for their precious help in developing this project.

References

1. Anderson, J.R., Boyle, C.F., Reiser, B.J.: Intelligent tutoring systems. Science(Washington) **228**(4698), 456–462 (1985)
2. Baker, R.S., Inventado, P.S.: Educational data mining and learning analytics. In: Learning Analytics, pp. 61–75. Springer, New York (2014)
3. Colley, A., Comber, C., Hargreaves, D.J.: Gender effects in school subject preferences: a research note. Educ. Stud. **20**(1), 13–18 (1994)
4. Coffield, F., Moseley, D., Hall, E., Ecclestone, K.: Learning styles and pedagogy in post-16 learning: a systematic and critical review (2004)
5. Dell'Aquila, E., Marocco, D., Ponticorvo, M., di Ferdinando, A., Schembri, M., Miglino, O.: ENACT: virtual experiences of negotiation. In: Educational Games for Soft-Skills Training in Digital Environments, pp. 89–103. Springer International Publishing (2017)
6. Deterding, S., Khaled, R., Nacke, L., Dixon, D.: Gamification: toward a definition. In: Proceedings of CHI 2011 Workshop on Gamification, 7–12 May, Vancouver, BC, pp. 12–15 (2011)
7. Gardner, H.: Frames of Mind: The Theory of Multiple Intelligences. Basic Books, New York (2011)
8. Graesser, A.C., Conley, M.W., Olney, A.: Intelligent tutoring systems (2012)
9. Grimm, P.: Social desirability bias. Wiley International Encyclopedia of Marketing (2010)
10. Kolb, D.A.: Learning styles inventory. The Power of the 2 × 2 Matrix, p. 267 (2000)
11. Lieberoth, A., Mrin, A.C., Møller, M.: Deep and shallow gamification - shaky evidence and the forgotten power of good games. In: Marti-Parreño, J., Ruiz-Mafé, C., Scribner, L.L. (eds.) Engaging Consumers through Branded Entertainment and Convergent Media. IGI-Global, Hershey (2014)
12. Marocco, D., Pacella, D., Dell'Aquila, E., Di Ferdinando, A.: Grounding serious game design on scientific findings: the case of ENACT on soft skills training and assessment. In: Design for Teaching and Learning in a Networked World, pp. 441–446. Springer International Publishing (2015)

13. Miglino, O., Di Ferdinando, A., Di Fuccio, R., Rega, A., Ricci, C.: Bridging digital and physical educational games using RFID/NFC technologies. J. e-Learn. Knowl. Soc. **10**(3), 89–106 (2014)
14. Paulhus, D.L.: Measurement and control of response bias (1991)
15. Ponticorvo, M., Di Ferdinando, A., Marocco, D., Miglino, O.: Bio-inspired computational algorithms in educational and serious games: some examples. In: European Conference on Technology Enhanced Learning, pp. 636–639. Springer International Publishing (2016)
16. Ponticorvo, M., Di Fuccio, R., Di Ferdinando, A., Miglino, O.: An agent based modelling approach to build up educational digital games for kindergarten and primary schools. Expert Syst. (2017)
17. Rahim, A., Bonoma, T.V.: Managing organizational conflict: a model for diagnosis and intervention. Psychol. Rep. **44**(3_suppl), 1323–1344 (1979)
18. Rahim, M.A.: A measure of styles of handling interpersonal conflict. Acad. Manag. J. **26**(2), 368–376 (1983)
19. Riding, R., Rayner, S.: Cognitive styles and learning strategies: understanding style differences in learning and behavior (2013)
20. Sillaots, M.: Achieving flow through gamification: a study on re-designing research methods courses. In: European Conference on Games Based Learning, vol. 2, p. 538. Academic Conferences International Limited, October 2014
21. Skalski, P., Dalisay, F., Kushin, M., Liu, Y.I.: Need for presence and other motivations for video game play across genres. In: Proceedings of the Presence Live (2012)
22. Stables, A., Wikeley, F.: Changes in preference for and perceptions of relative importance of subjects during a period of educational reform. Educ. Stud. **23**(3), 393–403 (1997)
23. Tong, T., Chignell, M., Tierney, M.C., Lee, J.: A serious game for clinical assessment of cognitive status: validation study. JMIR Serious Games **4**(1) (2016)
24. van Nimwegen, C., van Oostendorp, H., Modderman, J., Bas, M.: A test case for GameDNA: conceptualizing a serious game to measure personality traits. In: 2011 16th International Conference on Computer Games (CGAMES), pp. 217–222. IEEE, July 2011
25. Vander Ark, T.: Getting Smart: How Digital Learning is Changing the World. Wiley, New York (2011)
26. Zaina, L.A., Bressan, G.: Classification of learning profile based on categories of student preferences. In: 38th Annual Frontiers in Education Conference, 2008, FIE 2008, pp. F4E-1, October 2008
27. Zichermann, G., Linder, J.: The Gamification Revolution: How Leaders Leverage Game Mechanics to Crush the Competition. McGraw-Hill Education, New York (2013)

Eye Tracking Technology for Assessment of Electronic Hybrid Text Perception by Students

Kirill Zlokazov, Maria Voroshilova, Irina Pirozhkova, and Marina V. Lapenok(✉)

Ural State Pedagogical University, Yekaterinburg, Russia
zkirvit@gmail.com, shinkari@mail.ru, Irene22@live.ru, lapyonok@uspu.me

Abstract. This paper discusses some issues of e-learning, namely the problems of adaptations of learning materials to the individual opportunities and needs of students. The standardized content of e-learning doesn't take into account individual features of student's cognitive activity. It results in low motivation to study and poor progress. In this paper we describe an experimental research of hybrid text perception with the help of eye tracking technology. During the experiment we got data about behavior and emotions of students when completing learning tasks. A hybrid learning text consists of verbal text and several pictures. We believe that the visual component of a hybrid text influences efficiency of text perception because it may arouse different emotions: negative or positive.

Keywords: Eye tracking · e-learning · Hybrid text · Creolized text · Fixation · Eye movements · Positive modality pictures · Negative modality pictures

1 Introduction

We are facing a wide use of e-learning in Russia today. In 2014 the Ministry of Education and Science established Open Education Council which included eight federal and national research universities. In 2015 the Council was restructured into the "National Platform for Open Education" on the basis of which an on-line platform for e-learning programs was launched in 2015/2016. There are more than 300 e-learning programs on the basis of this platform. Their number is growing considerably year by year.

The perspective goal of e-learning is to include the basic subjects of the Bachelor's curriculum into the system of e-learning. This will help to improve the quality of higher and post-graduate education, as e-learning can make the leading universities of the country, and even of the world, easier of access.

To reach this goal, it is necessary to solve several problems, among which we find the following: development of the education content; use of hardware resources; development of the necessary software which can bring e-learning closer to learning in class; and introduction of the necessary changes and amendments into the existing regulations and standards of distance learning.

V.L. Uskov et al. (eds.), *Smart Education and e-Learning 2017*, Smart Innovation, Systems and Technologies 75, DOI 10.1007/978-3-319-59451-4_24

The experience of e-learning reveals the problems, which need to be solved to prove its worth and good prospects. Our paper discusses the problem of adaptation of educational content to the learning opportunities of students. Besides we describe the issues of identification of cognitive and emotional processes of students and the practice of verification of the students' success in learning.

The problem that motivated us to carry out this research is found in the standardized content of learning, which doesn't take into account individual features of student's cognitive activity: the tempo of information perception, the ability to interpret and master academic material [4].

In this aspect, learning in class has many advantages over e-learning [9]. The solution of this problem can be found in the use of eye tracking technology. It is widely used to adapt the content of education to the individual style of learning.

At the same time, some issues are beyond the interest of researchers, namely the analysis of perception of electronic didactic materials – textbooks, guidebooks and workbooks – by the students. Perception of the learning material may be connected with education quality: if the learning material is clear for the students, they get a motivation to study. The given paper focuses attention on this issue. The structure of our paper is as follows: after the introduction we give a theoretical overview of the use of eye tracking technology in practice of e-learning. The next part describes the experiment aimed at assessing the influence of the visual part of the hybrid learning texts on the comprehension of the information given in the text itself. The final part of the paper contains our conclusions on the problem under study.

2 Theory

The broad definition of e-learning says it is an educational technology giving a possibility of "asynchronous" learning. This asynchronous nature allows students to study the subject without any direct contact with the teacher on the basis of the didactic materials made in advance [7]. There is also a synchronous kind of learning but it is used mostly for the learning programs having time limits or it may be used from time to time either when the key issues of the subject are studied or for counseling or assessing the student's progress.

The experience of e-learning implementation reveals issues that cause students serious difficulties, but they are easy to solve in in-class learning when students can always consult their teachers. Among such problems are motivation of students, perception of learning materials and formation of practical skills. Taken together, these problems may raise the question of whether e-learning may compete with the other forms of education.

The solution of the problem may be found in a complex approach to it. First, it is necessary to choose the best model of learning: flexible, discreet and constructive; secondly, it is important to select the methods of teaching that might broaden communication between the student and the teacher or between students, in order they might solve the problems relevant to the academic subject – role-based and situation-based ones; finally it is essential to prepare didactic materials that allow integrating multimedia technologies into the process of teaching.

Worthy of mention is the fact that over the last decade the studies of the possibility of adaptation of learning materials to the individual needs of students became very important. In our opinion, this advantage of e-learning makes it possible to overcome such difficulties of asynchronous learning as predetermined tempo, prescribed structure and content of learning which may cause serious problems if the chosen models and/or methods are not optimal.

Adaptation of e-learning programs is carried out by means of biometric methods, the results of psychophysiological measurements and eye tracking technologies [14].

Such methods allow personalization of the learning process under the influence of the data about behavior and feelings of the student when completing training tasks. The contours of personalization are clearly identified in the project AdeLE (Adaptive e-Learning with Eyetracking). It analyzes eye movement reactions treated as representations of cognitive processes of a student and describes adaptation of the academic subject to the needs of every student [11]. It was noted that not only core parameters of the subject, such as the scope and content of academic information, may undergo adaptation but also micro-parameters, such as didactic material found on every page of e-textbook or e-practice workbook [10].

The advantages of eye tracking technology to assess psychological and psychophysiological parameters of students' cognitive and emotional reactions are as follows. Firstly, eye movements, fixation and saccades, as well as the pupil size in real time mode characterize cognitive and emotional processes of information perception [13]. At the same time, the possibility of correlation of these reactions with the student's movements allows one to immediately observe the changes. It is of great importance because it makes it possible to assess the parameters of cognitive and emotional processes which follow perception.

Secondly, non-intrusive eye-trackers do not cause students any discomfort as they are not observed and noticed by them. There is no need to attach any contact lens or other devices which might change the student's behavior during the experiment. This results in spontaneity of behavior and thus increases reliability of the data and conclusions and makes the measurements close to reality. Such advantages of eye tracking technology have stimulated its active use in researches of learning programs adaptation [3].

At the same time there are few scientific research works that describe the problems of perception of hybrid (creolized) texts [15]. Didactic materials offered for the students are difficult from the point of view of their content. They include verbal texts as well as visual and audio components. Their use produces a dual effect from the point of view of didactics. The advantage of visual and audio components is that they make the text clear, expand and elaborate it and make it easy to understand. The disadvantage of creolization is the heterogeneous nature of the text which requires the use of different strategies of perception in reduced attention and decline of effectiveness of the learning process [12]. This problem was discussed while planning the learning environment and learning strategies [2].

In the framework of this paper, we use the eye tracking technology in experimental research of hybrid text perception. The aim of the experiment is to reveal visual references of cognitive and emotional reactions which accompany the perception of learning texts. The main hypothesis is that the visual component of a hybrid text influences efficiency of didactic unit perception by emotional reactions of positive or negative modality.

3 Assessment of Hybrid Learning Text Perception

Research design. During the experiment we measured eye movements of students looking through the text prior to retelling it. All those tested were divided into two groups; both of them were offered a hybrid learning text. The first group was offered a text expressing positive modality, the second group worked with the text expressing negative modality. The order of reading and picture viewing was not specified; the participants looked through the page in the order they liked best (or randomly).

Method. The experiment involved careful looking through one page of an e-textbook that showed a hybrid learning text and was followed by a comprehension test.

To measure eye movements we used the technology of high-speed binocular remote eye-tracker SMIRED500. The software Experimental Center was used to plan the stimuli. The software BeGaze 2.x. was used to process the measurements.

We prepared a questionnaire to test the participants' knowledge that they got from the page of the e-textbook used in the experiment. The questionnaire included five questions. Each question had five variants of answers, only one of them being correct. The tested students were to choose only one variant out of five. The correct answer scored one point; all points were summed up. If all the five answers were correct, the participant got five points, if all the answers were wrong, the tested student got zero points.

Stimuli. The person tested looked at one page of a learning text. 30% of its square was a picture; 10% was a comment to the picture; 60% of the page was a text with didactic elements. The visual images on the page expressed positive or negative modality. Assessment of modality had been determined by the experts. The pictures showed children and adults. Positive modality images showed smiling people, negative modality images showed crying people.

Procedure. The participant of the experiment was sitting in front of the screen with the eye tracking devices under it. Before watching, the subject was instructed to memorize the information from the page.

Parameters. We measured and described the following parameters: (1) average time of fixation; (2) number of fixations; (3) average size of pupil by interest zones; (4) frequency of blinking; (5) "closed eyes" blink phase length. We singled out two zones of interest: (1) "picture" – an image that included didactic information, (2) "text" – a verbal text with didactic component.

Sample. Twenty people were chosen as subjects of our experiment (54% are men, the average age is 20 years, SD = 1.2 years).

Results. Assessment of the number of didactic units reproduced in the test did not reveal any significant differences in both groups, one group was shown a positive modality picture and the other was offered a negative modality picture (One-wayANOVA ($F(1, 18) = ,322$, $p < 0.846$)).

The differences were found in the parameters of participants' eye movements in the interest zones of "picture" and "text".

The interest zone "picture" differs from the interest zone "text" in the average time of gaze fixation, the number of fixations, the average size of pupil and "closed eyes" blinkphase length. In the interest zone "picture" the average fixation time on a positive image is less than the average time of fixation on the negative image (One-wayANOVA ($F(1, 18) = 2{,}565$, $p < 0.02$)).

The pictures expressing positive modality had 2–3 gaze fixations, while those expressing negative modality had 5–6 fixations. These differences are significant (one-wayANOVA ($F(1, 18) = 5{,}112$, $p < 0.01$)). The "closed eyes" blinkphase is longer with the testes students who are offered a positive image ($F(1, 18) = 3{,}461$, $p < 0.01$). The size of pupil is smaller with the tested students looking at a positive image, than with those looking at a negative image ($F(1, 18) = 2{,}337$, $p < 0.02$). The parameter "frequency of blinking" shows no statistical differences in both groups.

The interest zone "text" differs from the interest zone "picture" in the average time of fixation and the size of pupil.

The participants viewing a text expressing positive modality reveal shorter fixation phase than those viewing a negative modality text ($F(1, 18) = 2{,}347$, $p < 0.02$). As for the number of fixations, it is fewer with those participants who were offered a positive modality text: they fixed their gaze on different elements of the text 9–10 times; those participants who were reading the negative modality text fixed their gaze 11–12 times ($F(1, 18) = 3{,}971$, $p < 0.01$).

There are no statistical differences in such parameters as size of pupil and the length of "closed eyes" blinkphase.

Discussion. During the experiment we studied the influence of the visual element of the text on the subjects' perception and understanding of the verbal part of the text. Before the experiment we made a piece of training material of one page of e-textbook. It included both verbal text and pictures. The function of the pictures was not purely illustrative; they were rather visual representation of the studied phenomenon. In other words, the subjects of the experiment could get the main idea of the studied phenomenon from the pictures. We used two types of pictures: positive and negative, and we offered them to two different groups of students.

The results prove the absence of influence of modality of pictures on learning text memorizing. But eye movements show some peculiarities of perception of the pictures expressing negative modality, which, in our opinion, needs some consideration and explanation.

To sum up, we may say that the negative modality pictures attract the viewers' attention – they repeatedly return to these pictures and they spend additional time on their perception. Based on this we may conclude that negative modality texts arouse more interest than positive modality ones. But it is interesting to find out the nature of this interest. To reveal it we examined the size of pupil and the "closed eyes" blink phase.

The changes of the pupil size, if not caused by lighting condition, show psychophysiological processes of the subject. They show the work of the autonomic nervous system in the periods of physical and cognitive activity [1].

During our experiment we found out that when the subjects were looking at the negative modality pictures, their pupils grew bigger. It is argued that psychological stress caused by different sources makes the pupil get bigger [8]. In this research we believe the pupil size reflects the level of emotional tension of the subjects. The pupil is bigger in the group viewing negative modality pictures. While looking at the positive modality pictures, the pupil is of the smallest size, which may prove positive modality texts to be stressors of low potential.

The next parameter to be discussed is "closed eyes" blinkphase, which helps assess the level of importance of information. The research revealed that negative modality pictures caused shorter periods of "closed eyes" blink phase. Consequently viewing negative modality pictures stimulated the change of the neurophysiological process of blinking of the subjects of the experiment. The time of direct contact of the eyes with the picture was prolonged.

According J. Gross and R.W. Levenson's research, increased blinking and a long "closed eyes" blink phase can be considered a sign of greater arousal and it may serve to protect human's psyche from negative images [5]. The experiment conducted by C. Jackson and a group of psychologists had similar results [6]. Positive images led to smaller startle eyeblinks, while the negative images caused larger startle eyeblinks.

To sum up, the negative pictures had more impact than the positive ones. The subjects looked at the negative modality pictures more often, their pupils grew bigger and "closed eyes" blink phase was prolonged. Thus the negative image influences emotional state of the subject and causes more stress than the positive image.

Another question is if a negative modality picture influences verbal text perception. We believe that such influence exists although our experiment did not give us any physical evidence. Our assumption is based on several facts. First, the subjects looked at the texts accompanied by the picture of negative modality longer, secondly, they returned to such texts more often. It is obvious that this text (with the negative modality picture) required more will to focus and concentrate on it. But the emotional state of the subjects in both groups was almost the same, which is proved by the size of pupil and blink phase length.

In general, the research proves our assumption that the pictures expressing negative modality influence emotional state of students; it requires strong volition to focus on electronic learning text accompanied by negative modality pictures. The parameters of eye movements measured with the eye tracking technology prove this fact.

4 Conclusion

The development of e-learning technologies raises the question of their effectiveness assessment. The answer to this question depends on several factors, among which is the choice of methods and aids of e-learning. Our research is substantiated by the experience of eye tracking technology application to the study of electronic learning text perception and adaptation of learning materials to the individual needs of students.

The hypothesis of this research is the assumption that the hybrid learning text influences the efficiency of learning material perception depending on the emotional feelings it causes. The experiment results partially confirmed our hypothesis. The visual component of the text did not influence comprehension of its verbal part. But looking at the negative modality pictures the subjects of the experiment spent more effort to focus their attention, because those pictures caused them emotional discomfort.

In this connection, we suppose that further development and perfection of e-learning may be connected with the adaptation of learning texts based on the specificity of their perception.

Research disclaimer. We would like to attract the attention of the readers to the following facts. First, it is the parameters of the sample. We tested a group of adolescents and our conclusions are true only in reference to the group of people under examination. But at the same time, it is also important to study the peculiarities of electronic didactic material perception by children and teenagers. Such research may help to reveal the impact of a textbook on motivation – whether it increases or decreases motivation. Assuming that e-learning will develop and spread, such researches may be useful for the development of a new criterion of learning aids assessment. The other parameter of sampling is connected with the number of subjects of the experiment. As the groups were not numerous, we attribute these results only to the subjects of this experiment; we cannot prescribe the same results to the other groups of adolescents. But at the same time, these results prove that a more detailed and deep study based on the technology of eye tracking would be very promising.

Second, the learning text used in the experiment belongs to the Humanities and is not connected with Sciences. As we used only one text of a certain scientific sphere, we should note that the results of this experiment may refer only to social sciences.

Third, the experiment structure didn't include perception management and the choice of priorities prescription, although it is obvious that the students use different strategies of viewing. For example, some of them may skip the text and focus on the pictures, while others may focus on certain elements of the verbal text. We didn't analyze and assess perception style factor. Although in the light of the obtained results, it may play a certain role in students' motivation.

Fourth, we didn't take into account fatigue. This factor is crucial for ecological validity of the experiment, as it is important for assessment of the level of learning text reproduction. In other words, many adults do not study in the morning or afternoon, when they are supposed to, but they prefer to study in the evening. Thus fatigue may tell on the learning text perception. In this connection, we believe that the influence of pictures, video-images and audio-elements on verbal text perception may vary due to the level of fatigue.

So, the study of hybrid text perception has a great role for the assessment of such learning texts and for the assessment of reliability of e-learning technologies in general. The conclusions we make are broader than the hypothesis that stimulated this research. We believe that the further research in this field will support the ideas expressed in this paper.

References

1. Beatty, J.: The Pupillary System: Psychophysiology: Systems, Processes, and Applications, pp. 43–50. Guilford, New York (1986)
2. Bielikova, M., Moro, R., Simko, J., Tvarozek, J.: Adaptive web-based textbook utilizing gaze data. J. Eye Mov. Res. **8**(4), 252 (2015). (Special Issue ECEM)

3. Copeland, L., Gedeon, T.: What are you reading most: attention in eLearning. Procedia Comput. Sci. **39**, 67–74 (2014)
4. Fichten, C.S., Ferraro, V., Asuncion, J.V., Chwojka, C., Barile, M., Nguyen, M.N., Klomp, R., Wolforth, J.: Disabilities and e-Learning problems and solutions: an exploratory study. Educ. Technol. Soc. **12**(4), 241–256 (2009)
5. Gross, J., Levenson, W.: Emotional suppression: physiology, self-report, and expressive behavior. J. Pers. Soc. Psychol. **64**(6), 970–986 (1993)
6. Jackson, C., Malmstadt, R., Larson, L., Davidson, J.: Suppression and enhancement of emotional responses to unpleasant pictures. J. Psychophysiol. **37**, 515–522 (2000)
7. Nichols, M., Anderson, B.: Strategic e-learning implementation. J. Educ. Technol. Soc. **8**(4), 1–8 (2005)
8. Partala, T., Surakka, V.: Pupil size variation as an indicator of affective processing. Int. J. Hum Comput Stud. **59**, 185–198 (2003)
9. Pivec, M.: Knowledge transfer in on-line learning environments. Ph.D. Work at Graz University of Technology, Austria (2000)
10. Pivec, M., Pripfl, J., Trummer, C.: Adaptable e-learning by means of real-time eye-tracking. In: The ED-MEDIA, Montreal, Canada, 27 June–2 July (2005)
11. Pivec, M., Trummer, C., Pripfl, J.: Eye-tracking adaptable e-learning and content authoring support. Informatica **30**(1), 83–86 (2006)
12. Mayer, R.E., Moreno, R.: Nine ways to reduce cognitive load in multimedia learning. Educ. Psychol. **38**(1), 43–52 (2003). Pfeiffer
13. Rayner, K.: Eye movements in reading and information processing: 20 years of research. Psychol. Bull. **124**, 372–422 (1998)
14. Spada, D., Sánchez-Montañés, M., Paredes, P., Carro, R.M.: Towards inferring sequential-global dimension of learning styles from mouse movement patterns. In: Proceedings of the Adaptive Hypermedia and Adaptive Web-Based Systems, pp. 337–340. Springer, Heidelberg (2008)
15. Voroshilova, M.B.: Creolized text: aspects of research. In: Chudinov, A.P. (ed.) J. Political Linguistics, vol. 20, pp. 180–190. Ural State Pedagogical University, Yekaterinburg (2006) (in Russian)

The Smart MiniFab: An Industrial IoT Demonstrator Anywhere at Any Time

Tobias Schubert(✉), Benjamin Völker, Marc Pfeifer, and Bernd Becker

Chair of Computer Architecture, Faculty of Engineering,
Institute of Computer Science, Albert-Ludwigs-University Freiburg,
79110 Freiburg, Germany
{schubert,voelkerb,pfeiferm,becker}@informatik.uni-freiburg.de

Abstract. The fourth industrial revolution is predicted to be one major topic in the upcoming decades, affecting nearly every industrial facility. The key enabler for the successful integration of corresponding developments are well-trained engineers with elaborated knowledge in the area of the *Industrial Internet of Things* (IIoT).

This paper introduces a concept enabling engineering students to gain the required skills for the upcoming challenges connected with the fourth industrial revolution. Therefore, the *Smart MiniFab* was developed - a demonstrator of an industrial facility primed for the implementation and the analysis of different IIoT concepts on a small scale. Due to the modular structure of the system, students can use the demonstrator to programmatically implement the functionality of an entire facility from scratch or to test only certain new concepts. The hardware can be programmed via remote access from anywhere at any time, while live feedback is provided through a webcam and different other communication possibilities directly within the remote access application.

While the Smart MiniFab was already the HW environment of different student projects, a new lab course is currently under development which enables the access to the Smart MiniFab for a broad number of students.

Keywords: Industrial Internet · Internet of Things · Smart production demonstrator · Innovative lab course

1 Introduction

The intelligent connection of devices, machines, facilities and humans is regarded as the fourth industrial revolution. The main driving forces behind this revolution are known as the *Internet of Things* (IoT) and the *Industrial Internet* (II). While IoT describes the intelligent connection of usually small devices, II specifies the same intelligent connection combined with other smart improvements for larger scale production sites. Both concepts are combined in the *Industrial Internet of Things* (IIoT) which is based on so-called *Cyber-Physical-Systems* (CPS). These systems unite software and hardware to interact with their environment in

V.L. Uskov et al. (eds.), *Smart Education and e-Learning 2017*, Smart Innovation, Systems and Technologies 75, DOI 10.1007/978-3-319-59451-4_25

a smart way. Therefore, they comprise sensors, controllers and actuators on the hardware side and a corresponding controller firmware on the software side. Especially in IIoT scenarios different CPS are interconnected to dynamically control larger production sites so that the overall usage and throughput is optimized. The importance of IIoT for our future in nearly all industrial areas is indicated in different case studies. To give just one example, a survey conducted in 2016 claims that 82% of 173 interviewed companies consider the adaption of IIoT concepts as being critical to their future success [1]. To successfully employ the potential of IIoT, well-trained engineers are essential which not only develop innovative CPS, but also connect and maintain existing or newly created systems. This requires knowledge in the area of mechanics, electronics and computer science as well as a good glance on how to combine these disciplines into the context of IIoT. Especially for the latter this paper presents an innovative concept which aims at the effective training of future engineers with the help of a custom-made industrial facility demonstrator, called *Smart MiniFab*. It was developed at the Faculty of Engineering at the University of Freiburg, where it is used to train engineering students as well as professionals who join qualification seminars. The Smart MiniFab is comprised of different distributed production stations in a small scale, each controlled by a microcontroller and equipped with different communication interfaces. This flexible setup enables either the programming of the entire facility from scratch or the analysis of only certain new IIoT concepts.

To reach as many students as possible we created a web-based remote access to program and interface with the demonstrator. This allows for controlling the Smart MiniFab from anywhere at any time which perfectly adapts to the usually tight schedule of a contemporary student. By further integrating a webcam, different sensors and communication interfaces, this remote access provides an almost perfect live feedback and, therewith, the feeling of "standing right next to the fab". Based on continuous positive feedback received from students, who have already worked with the Smart MiniFab for team projects, a new lab course is currently under development which will be available to all students at the Faculty of Engineering. Participants of the course will learn the most important aspects connected with IIoT in a very practical way before they start to develop and implement their own project ideas with the Smart MiniFab.

The remainder of the paper is structured as follows: In Sect. 2 important concepts related to IIoT are introduced. The way they are implemented within the Smart MiniFab is shown in Sect. 3, together with a general overview of the demonstrator. Section 4 further highlights the individual hardware components of the SMF in more detail, while in Sect. 5 its remote access is described. Section 6 sketches how the intended lab course will be structured. Related work is presented in Sect. 7 before the work achieved so far is concluded and links to possible future projects are given in Sect. 8.

2 Concepts of the Industrial Internet of Things

While the interconnection between different machines and the centralized management of production sites is common today, IIoT introduces new concepts which aim to further boost the productivity and adjust the production to upcoming needs of the market. The most important concepts are briefly introduced in the following:

(a) **Decentral Decisions:** One key concept in IIoT is to switch from a centralized production setup to a more decentral model, in which different production machines are allowed to perform their own decisions based on their locally captured (sensor) data, while being at the same time connected to other machines for exchanging global information. When realized in an efficient manner the productivity of the overall production system can be increased. This design shift is also motivated by the fact that the manufacturing process of most products today has moved from one central to multiple distributed locations each performing some sub steps of the entire manufacturing process [2].
(b) **Smart Products:** Closely related to a decentral control flow is the so-called "smart product". The term describes the fact, that the product itself knows how it has to be manufactured and what components or features are needed. This is achieved by equipping a blank product or a product carrier with additional active or passive elements that communicate with the manufacturing machines. An example for a passive element is a QR-code that contains all manufacturing steps and production parameters required for a particular product.
(c) **Individual Products:** A common trend on the market heads towards individually designed products. While this is a neat feature for consumers it might be very hard to realize for a company since most existing production sites are typically optimized for high-volume instead of high variability. To enable the production down to "lot size one" IIoT concepts have to be applied such that different products can be manufactured by using the same production machines (in particular without any manual reconfiguration of the machines in between).

3 Overview of the SMF

The Smart MiniFab (also refered to as SMF) is a miniaturized production facility that can handle certain tasks in a smart way. It was build from scratch so that students get in touch with IIoT related hardware without the need to visit external companies or to carry home obtrusive hardware components. An overview of the SMF is presented in Fig. 1. It is capable of moving *Tetris* like bricks called tetrominos between different sized containers called palettes. The SMF as one entity is divided into currently three distributed stations **S1**, **S2** and **S3**. S1 handles tetrominos that are asymmetrical in shape since a rotation box allows it to rotate them to a given orientation (see also Fig. 3). S2 has no rotation box but features an assembly belt which can move palettes between S1, S2 and S3. Station S3 comprises a WiFi-module and is responsible for handling incoming orders. An order for a "product" can be placed in a web-shop and consists of several tetrominos of

Fig. 1. Overview of the Smart MiniFab.

different shape, color, and orientation. These are placed at different positions in different sized palettes. In the context of IIoT the palette can be seen as a blank product which can be equipped with different features - the tetrominos. Currently, eight different tetrominos exist in five different colors and by considering all possible orientations, up to 100 different features are possible. If the different possible positions of the tetrominos inside a palette are also considered, a huge variety of different "products" can be ordered. If such an order is placed, there is a high probability that it is the first time this specific configuration has to be produced. This has a strong relation to the mentioned production in lot size one. Furthermore, the SMF also provides the ability to highlight the concept of smart products since RFID tags with assemble information are attached to the bottom of each palette. Each station has access to the stored information and can start to work on a sub-step independently. To assemble palettes, S1 and S2 both feature an assembly robot which is highlighted in Fig. 2. This robot moves along the *x*- and *y-axis*, picking up tetrominos at one position and moving them to other positions where they are laid down. In order to do so, the robots incorporate different electrical components such as stepper motors, end-stops, vacuum pumps and valves together with their corresponding driver electronics.

4 Hardware Components of the SMF

Each station of the SMF features different hardware components that can also be found on large scale production machines. These components collaborate together to perform a given task and, if viewed from a larger scale, to achieve the IIoT concepts introduced in Sect. 2. An overview of the hardware components of station S1 is shown in Fig. 3. For precise movements of the robots along the *x*- and *y-axis* and for the assembly belt of S2, stepper motors are used.

Fig. 2. The robot arm of a station with its pneumatic system and stepper motors for movements in x- and y-direction.

Fig. 3. Overview of S1 showing all relevant HW components and the mechanical setup.

Mechanical end-stops determine the end of each axis and are attached to both sides. To pick up tetrominos, vacuum pumps and compressors connected to valves are used. The compressor lowers the robot arm while the pump produces vacuum to lift tetrominos. Valves are used to minimize the amount of residue pressure or vacuum in the pneumatic system, so that the procedure of picking up and laying down can be performed reasonably faster.

The brain of each station is a *Teensy++ 2.0* microcontroller featuring an *Atmel AT90USB1286 8-bit* core [3]. Besides a comparably large amount of program memory and SRAM, this controller also features enough pins to control the large amount of peripheral components. Furthermore, it can be programmed with the *Arduino* IDE [4] that has a large community and is said to be easy to use. Each station also possesses a *Xilinx Spartan-3A Field Programmable Gate Array* (FPGA) to protect the setup from erroneous code running on the microcontroller [5]. All stepper motors and the rotation box are connected to a protection circuit on the FPGA so that the SMF can not be damaged by uploading incorrect software onto the controllers. Further details about these circuits can be found in Sect. 5.1.

In order to enable a communication between the three stations (S1, S2, S3) and to use central and decentral information for making decisions, radio frequency modules (based on [6]) are used which operate at the 2.4 GHz ISM band. These modules offer both, unicast and broadcast message transmission to be able to send messages either to a specific station or to all stations at the same time. *Radio-Frequency Identification* (RFID) is used to mimic a "smart product" that tells the machine what to do. Therefore, a RFID with re-writable internal storage is mounted to each palette which is read or written by RFID readers installed at each station.

5 SMF Remote Access

Beside the hardware needed for the SMF's "production mode" additional components to enable remote access are added. Figure 4 gives an overview on how the remote access is realized and highlights special components introduced in the following.

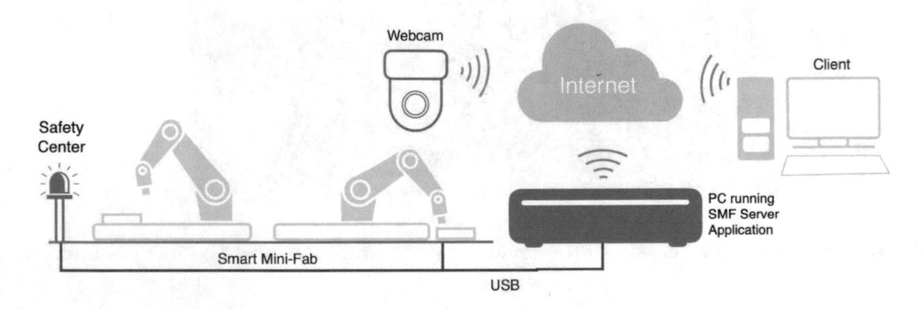

Fig. 4. Overview of the data flow needed for remote access. Additional modules are highlighted in dark orange.

5.1 Hardware

Since all components of one station of the SMF are controlled by a microcontroller, the entire functionality of the SMF is determined through the firmware running on the microcontrollers. To flash them USB connections to a development computer are established. This computer runs specialized software to enable the remote interconnection and the programming as discussed in Sect. 5.2 and is called "server" in the following.

As the aim of the work with the demonstrator is to perform a practical (production) task, optical feedback is essential. Especially, if the user is accessing the SMF remotely he or she needs to know whether the uploaded code leads to a correct operation of a station or not. Therefore, a webcam is installed directly above the Smart MiniFab.

A potential issue could arise if written code does not perform as expected or was uploaded incompletely. Since the user can not shutdown the hardware directly, unsafe conditions – e.g. if a robot arm continues to move towards its end-stop – can lead to damages. To prevent these situations, a *Safety Center* is connected to each station. This module is not directly accessible by the users. It consists of the already introduced Xilinx FPGA to which all end-stops and control signals are connected. The FPGA monitors these signals, detects potential dangerous situations and automatically performs emergency stops which disable all signals coming from the microcontroller. To provide programmers the possibility to properly react to end-stop presses a 100 ms delay is included in each safety center before an error is indicated; the signal, however, is blocked immediately. The safety center is directly connected to the server which is not only notified about the occurrence of a stop but also about its reason. Besides the self-stopping mechanism, users can also trigger emergency stops by themselves if they detect any failure via the webcam stream. Such an emergency request is forwarded to all safety centers which immediately block all outgoing signals.

Beside their safety functionality the safety centers also offer another teaching feature for students: One task could be to prove by formal verification methods that a bad state (a situation in which the SMF might be damaged) can not be reached, regardless of what has been uploaded into the stations' microcontrollers.

5.2 Software

The key element of the remote programming access for the SMF is the realized web application: It provides the possibility to input, develop, compile, and upload code to the SMF and, additionally, gives live feedback during the execution of this code. It is hosted on the server and is comprised of an individually developed server application and the client web application itself. The web application is executed on the user's computer as an interactive webpage based on HTML, *JavaScript*, and CSS. Figure 5 shows this webpage which can be easily accessed using standard web-browsers. The left side contains the programming section while on the right-hand side live feedback is given. The main part of the programming section is the code editor. Users can either upload their pre-written code or write a complete code from scratch in a C++ like language according to the Arduino [4] platform. All corresponding Arduino functions are available and proper code-highlighting and -indention is provided. Through a tab-based structure it is possible to maintain different code-files and to create classes with header-files. To store intermediate or final results, the code can be downloaded as a *ZIP*-file at any time. Furthermore, users can also compile and upload the written code. In this case, all files are sent to the server backend, compiled, and checked for compile-errors. The resulting hex-code is uploaded to the microcontroller which corresponds to the station the user has selected via a drop-down menu. The output of the compile and upload step is than sent back to the client and shown in the output field.

Live feedback is provided by a moveable and zoomable webcam at the upper right-hand side and an error output at the bottom on the right. If one or multiple safety centers, as introduced in the previous section, perform an emergency stop, the corresponding reasons are displayed. Emergency stops can also be triggered manually by pressing the stop button. The test-order field allows users to directly order pre-defined patterns of tetrominos for a quick test of their implementation. The order is received from the server and immediately forwarded to S3 of the SMF. A second way to place orders is provided through a button leading to a web-shop. This allows to configure and order arbitrary tetromino patterns.

To protect the web application from any unauthorized access a common user/password management is integrated. The access is currently limited to a single user at a time to avoid that two users programm the same station simultaneously, which may cause conflicts and confusion. In case of inactivity the user is automatically logged out after 20 min to prevent an accident or intentional blocking of the SMF.

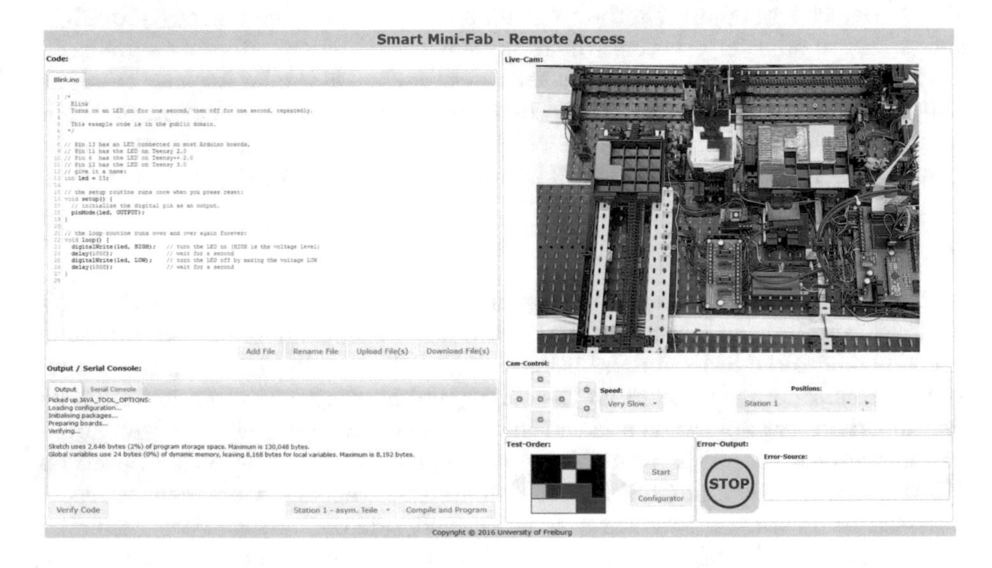

Fig. 5. Screenshot of the web application for the remote access to the SMF.

6 Lab Course

To give more students the opportunity to work with the SMF as the underlying HW environment, we are planning a new practical course in which students have to deal with IIoT related components like microcontrollers, RFID devices, and stepper motors and for which they have to provide software to control them. The lab course is divided into three parts:

1. **Software and Programming Introduction:** In the first part, the students get in touch with the web application and its visual feedback. To get familiar with the programming language lectures are given and the students will learn

how to read and write analog values or digital states of I/O pins, how to use timers, how to call functions and how to develop classes. To strengthen the newly acquired knowledge, the participants have to solve practical exercises by using the web application.
2. **SMF and IIoT concepts:** In the second part, the hardware modules of the SMF are treated in more detail, namely the stepper motors with their end-stops, the RFID devices, and the RF modules. This is achieved by practical exercises once again which have to be solved using the web application. In each exercise, a specific hardware module like a stepper motor with its end-stop is treated and the students have to develop code or, at best, a complete software module to control the hardware. Furthermore, the IIoT context of the used hardware is highlighted in more detail with the help of lectures.
3. **Project phase:** The last and major part of the lab course combines the knowledge of all previous exercises into one large project. The goal of the project is to solve a complex SMF task by dividing it into multiple subtasks that are executed by specific hardware components in a dynamic manner. In this part no strict exercise is provided as before, instead only the final goal is given to the students, meaning that the participants have to come up with a solution completely self-organized! Nevertheless, the organizers will support the students through the whole project phase by giving advice and feedback.

7 Related Work

The concept of a remote laboratory, shared hardware for interactive practical courses, and/or remote access for students has been of major interest since the upcoming of the internet. Different concepts exist in this area, depending on the hardware and the use-case under consideration. The general idea of a remote laboratory is compared to a competing hands-on laboratory in [7]. The authors performed a case study resulting in a comparable or even better performance of remote courses in contrast to corresponding on-site courses. A concrete example for a remote laboratory for hardware experiments with FPGAs, pattern generators, and measurement equipment such as logic analyzers is proposed in [8]. The presented system allows students to perform experiments with a FPGA attached to a PC from inside or outside the campus network. However, safety related regulation or protection is not mentioned at all and no visual feedback via webcam is provided. Video feedback has been added in a system developed at the University of Illinois [9] that allows users to access multiple servers with FPGAs attached. These can be programmed remotely while feedback is given by observing rudimentary LCD displays or LEDs attached to the FPGA with a webcam. However, the authors again do not provide information whether the hardware is protected. The solution proposed in this paper is, to our knowledge, the first that enables the remote access to IIoT related hardware. In contrast to the presented systems, it is possible to test complex software distributed to multiple subsystems from a single web application in which video feedback is given to observe moving parts of the SMF. On top of that, a protection circuit is added so that the system can not be damaged by uploading erroneous code.

8 Conclusion

This paper proposes a modular and expandable IIoT demonstrator called Smart MiniFab. It can be accessed from anywhere at any time via a web application which allows to remotely upload code to each of the microcontrollers placed in the SMF. Furthermore, the application provides visual feedback by a controllable webcam. Since code which is uploaded might be erroneous, a safety center was added to each station which checks for hazardous situations related to moving parts. If such a situation is recognized, the corresponding moving part is stopped and the user is informed.

As already mentioned, the SMF is divided into three stations. Another station is currently under construction and will feature a high-bay warehouse in which multiple palettes can be placed. Therewith, the demonstrator could be used to simulate even more complex scheduling and optimization tasks. Furthermore, a future version of the web application will feature a collaborate mode, in which students can work seamlessly on the same firmware version which will be managed by some kind of version control system such as *GIT*.

The positive feedback and enthusiasm received by students who actually worked with the demonstrator either for their complementary work or during practical sessions indicates that a remote access to interesting hardware should be integrated into the daily schedule of any student. Therefore, we are planning to use the SMF as the underlying HW environment in a new lab course to give a large number of engineering students the chance to perform experiments in the area of IIoT.

References

1. 2016 Reality Check: Transforming Industrial Businesses with the Internet of Things. Survey, IndustryWeek/Genpact (2016). http://www.genpact.com/downloadable-content/genpact-iw-report-final.pdf
2. Hermann, M., Pentek, T., Otto, B.: Design principles for industrie 4.0 scenarios. In: 49th Hawaii International Conference on System Sciences (HICSS), pp. 3928–3937. IEEE Press, Hawaii (2016)
3. PJRC - Teensy++ 2.0. https://www.pjrc.com/store/teensypp.html
4. Arduino - Official Website. https://www.arduino.cc/
5. Micronova - Mercury FPGA. http://www.micro-nova.com/mercury/
6. Nordic Semiconductor - nRF24L01+. https://www.nordicsemi.com/eng/Products/2.4GHz-RF/nRF24L01P
7. Corter, J.E., Nickerson, J.V., Esche, S.K., Chassapis, C.: Remote versus hands-on labs: a comparative study. In: 34th IEEE Frontiers in Education Conference, F1G, vol. 2, pp. 17–21. IEEE Press (2004)
8. Fujii, N., Koike, N.H.: A new remote laboratory for hardware experiment with shared resources and service management. In: 3rd IEEE International Conference on Information Technology and Applications (ICITA 2005), pp. 153–158. IEEE Press, Sydney (2005)
9. Hashemian, R., Pearson, T.R.: A low-cost server-client methodology for remote laboratory access for hardware design. In: 39th IEEE Frontiers in Education Conference, pp. 1–5. IEEE Press (2009)

"Semograph" Information System as a Framework for Network-Based Science and Education

Konstantin Belousov[1(✉)], Elena Erofeeva[1], Yuliya Leshchenko[2], and Dmitriy Baranov[3]

[1] Perm State University, Perm, Russia
genling.psu@gmail.com
[2] Perm State Humanitarian Pedagogical University, Perm, Russia
napo530@mail.ru
[3] Orenburg State University, Orenburg, Russia
baranov@semograph.com

Abstract. Present-day educational process requires interactivity from the teacher and the student, as well as employment of modern tools provided by the internet informational space and computer technologies. Network organization of the subjects' interaction in professional and educational activity is considered most effective; network-based software is currently used in many professional spheres. Educational activity in higher school requires resorting not only to proper "academic" network-based software solutions, but also to scientific ones, as far as educational tasks include mastering methods of obtaining new knowledge. The paper presents the "Semograph" Information System used for training specialists in linguistics while organizing both their educational and scientific activity. Particular instances of using the "Semograph" in teaching academic disciplines are introduced; examples of research projects used for teaching scientific activity are presented.

Keywords: Educational environment · Interactivity · Distance education · Network-based science · Principles · Organization · Higher school · Educational and research environment · The "semograph" information system

1 Introduction

Network structure of subjects' interaction in professional and educational activity is considered to be most effective [1, 2]; network software is widely used nowadays in different spheres of professional activity. In the educational sphere (mostly in higher educational institutions), program platforms Moodle, ILIAS etc., as well as educational portals Coursera, edX, Udacity, Russian INSTITUTE are used. Nevertheless, the tasks of higher educational institutions include not only organization of the academic process *per se*, but also involving students into scientific activity. The latter requires resorting to scientific software rather than to proper "academic" network programs.

At the same time, in the scientific sphere there are still few software products which realize the principle of the network-based science.

V.L. Uskov et al. (eds.), *Smart Education and e-Learning 2017*, Smart Innovation, Systems and Technologies 75, DOI 10.1007/978-3-319-59451-4_26

Network-based science is understood as a scientific process distributed in the real-time mode; this process implies organization of network interaction of its participants and the system of research activity control, using the identical information processing technologies and the generally accessible database that integrate every participants' results into a joint information space.

At present, in the internet open access there are a large number of sources giving information on research in different spheres of science and technology, as well as portals used for communication between researchers. Among them are web sites of scientific journals and scientists, knowledge bases and databases (including those which contain experimental results, such as DB Reaxys, e.g.), scientific social networks and sources created for perspective scientific research support, etc. Moreover, there exist systems of network organization of scientific research based on the Citizen science framework. These researches are carried out by volunteer groups in collaboration with or under the guidance of scientists and/or scientific organizations (the list of finished projects and projects in progress can be found in [3]). Still, implementation of the network-based science idea requires not only a multitude of sources, a great scope of scientific information, a set of specific formats for its representation, and a large number of participants. There also should exist a distributed analytical network space that gives an opportunity to realize online interaction of participants of the research process, and this process correction. Herewith, program products providing this environment should be directed at concrete scientific spheres and tasks (that can be quite broad, though). In the given paper the authors present the "Semograph" Information System [4] as a realization of network-based approach to organizing linguistic scientific research and text content analysis.

2 "Semograph": Its Purpose and Functionality

"Semograph" Information System is designed for analyzing text data, creating/tagging of corpora, carrying out and analyzing the data of psycholinguistic, sociolinguistic, etc. experiments, developing classifiers and thesauruses, creating models, and other research tasks that can arise in text content analysis.

The following principles are realized in the "Semograph":

- realization of the full research cycle from data collecting to creating models;
- network-based distribution of agents of the scientific process which implies the possibility of working out one project by a group of scientists geographically remote from each other;
- multi-users working pattern which provides online interaction of the project group members;
- methodological pluralism that enables the researchers with different theoretic and methodological opinions to use the "Semograph" IS.

The Information System includes the following modules:

- the search module (including a web crawler and a parser developed on the basis of the Python-framework Scrapy);

- data import (with the help of the search server on the basis of the Apache Solr and files with the table-type data organization);
- exporting the results into an external application, for example, into R (statistical data analysis software), Gephi (a tool for graph construction and analysis), as well as into table format;
- the organizing module (permitting to invite new users, to add the IS users to projects, to specify the system of access rights, to create open projects and communication systems);
- the research module (a wide range of tools for experimentation, analyzing language content, tagging corpora, classification, etc.).

The "Semograph" possesses API which can be used for working with external applications or developing applications on the basis of extending some functional aspect of the Information System.

3 Using "Semograph" for Training Linguists

Our experience of using the "Semograph" for training specialists in the sphere of linguistic analysis enables to single out two main situations.

3.1 Using "Semograph" in Teaching Linguistic Disciplines

The lecturer can resort to the Information System database for demonstrating linguistic data and connections between them. In the frameworks of executing laboratory or independent work the Information System can be used as a teaching platform with the opportunity of face-to-face and/or external guidance of students' work: formulating the task, depositing necessary materials, evaluating the results. Moreover, such work can be organized both as individual and group one, which is regulated by the provided access.

Work of this kind can be organized, for instance, while teaching semantics, lexis, morphology, syntax, and other subjects. An example of a concrete task proposed to students while teaching lexical semantics on the topic of semantic fields is demonstrated below. Figure 1 shows a screenshot of the "Semograph" window with the task of analyzing the "Water Surfaces" lexical-semantic group in the Russian language. In the right window concrete words students are working with can be seen: a river, a spring, an ocean, a sea, a lake, a swamp, etc. The students are asked to single out differential semantic features of the words belonging to the group in question and match them with the concrete words in the "Field analysis" window (a "Semograph" tool).

Using the Information System enables to engage a large scope of material (for example, the analysis of some close lexical-semantic groups and comparison between them) which helps students to form systematic (not fragmentary) knowledge of the structure of semantic space in Russian. The automated material processing gives a chance to quickly and adequately model the semantic space in form of graphs (task 3 asks students to construct and analyze graphs of words' connections and their differential features).

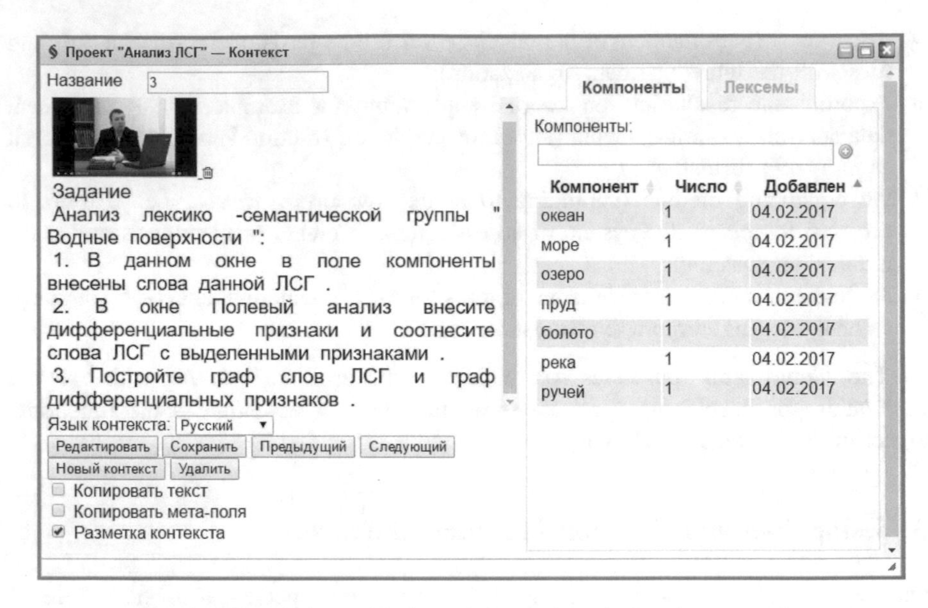

Fig. 1. The window with the task for students in Russian semantics

The teacher is provided with an access to any stage of students' work, so he/she can estimate the complexity of the task, see the stages where students have difficulties, direct and instruct them, comment on mistakes and, generally, estimate the performing of the task. Figure 1 demonstrates how in the instruction window a video file can be attached (in the particular case it is a file with the teacher's oral explanations about the task).

Similar tasks are possible to use not only for teaching theoretical subjects, but also while studying a foreign language, e.g., tagging a foreign text, classifying the text units into grammatical categories, etc.

3.2 Possibilities of the "Semograph" for Carrying Out Research Projects of Bachelor, Master and Postgraduate Students

While carrying out such work the teacher's control and his/her ability to influence the course of the work are extremely important.

In this case the "Semograph" maintains 3 functions: data collecting, its primary processing, and constructing models of the informants' verbal behavior.

3.2.1 Data Collection in the "Semograph" IS

Data collection in the Information System can be realized in several modes

1. Manual data entering. This mode can be used during online experiments when the informants enter the data while fulfilling the experimental task.

2. Data import with the help of a file loader. The mode is efficient in case the analyzed content had been represented in any table format (.xls, .xlsx, .ods, etc.) before the work in the Information System was begun.
3. Data import by the Solr (for these purposes both xml-files stored in the Solr and the results of work with the necessary web-content of the web crawler and the Scrapy parser transferred into the Solr can be used). In case graphic, audio or multimedia content is employed in the research, it can also be downloaded into the project.

Thus, the system provides the learner with the opportunity of working with the internet content and, at the same time, enables the instructor to control the quality of the content imported.

3.2.2 Data Processing in the "Semograph" IS

As it has been mentioned before, the work with research material in the "Semograph" can be realized by various linguistic methods chosen in accordance with the research tasks. Each project can be characterized with the help of a formalized description system which enables to reveal the research frame at the first stage of getting acquainted with the project. The basic descriptive parameters of the project are

- *Objects:* User, Project, Context, Component, Word, Lexeme, Field, Meta-field, Meta-type, Collocation, Fragment, Keyword.
- *Operators* (actions performed on the objects): Machine operators – Indexation, Excerpt; Data operators – Access Right, Tagging, Evaluation (subjective characteristics of the object based on using scales), Addition, Inclusion, Classification; Operators of Results – Classifier (or Thesaurus), Semantic Map, Semantic Graph, Semantic Distance, Transfer Matrices, Frequency Tables, Correspondence Tables.
- *Characteristics:* Meaning (e.g., a field has the X meaning), Frequency, Sequence, Localization, Type (text, integer, nonintegral, date, file), Volume (e.g., sample volume), Depth (e.g., depth of a thesaurus = a number of levels of semantic fields' hierarchy), Time Intervals.
- *Working mode:* online/offline.
- *The project status:* open/closed (in case of an open project its title is given), number of participants and their roles.
- *Meta-information*(the date of creating the project, the project author, the list of subjects who have access to the project, time of their work with the project, etc.).

Various combinations of the basic parameters enable to create different schemes of working with linguistic information that can be applied for a wide scope of research tasks.

3.2.3 Modeling

The Information System enables not only to collect linguistic databases and realize primary processing of linguistic material, but also to create models of the informants' verbal behavior. Modeling is realized in form of semantic tables and semantic graphs (for more detail see [5]).

4 Examples of Students' Projects in the "Semograph" IS

The authors of the paper have a solid experience of organizing students' research work with the help of the "Semograph" IS. Examples of three supervised student projects are presented as follows.

4.1 The Project "Terminology of Activity Theory"

There search task was to create a thesaurus of terminology system of activity theory basing on the material of scientific publications within the given subject domain.

While working on the project students learn to find the required terminological units in context and to analyze their meaning. The teacher controls the work in progress and is able to check both the list of units singled out from the context and their description (see Fig. 2).

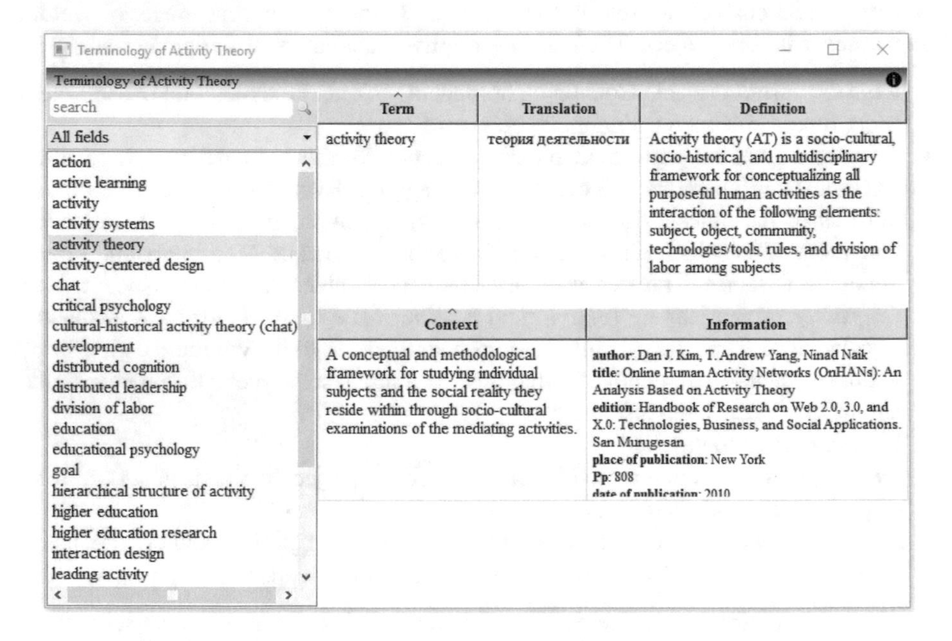

Fig. 2. The window of the project "Terminology of Activity Theory"

The description of the project is given below.

- *Project:* closed (within the information system); unindexed; 350 contexts (each context describes one term), 65 components (terms), 15 fields (classifying groups).
- *Users:* experts (two philology students and a teacher of philology).
- *Meta-fields of contexts:* term, translation (Ru), type of context, reference (author, title, edition). Meta-fields of contexts are available for samples.

- *Operators:* access rights are unrestricted for experts; adding components (terms of activity theory).
- *Working mode:* offline.
- *Fields:* Classification is based on the analyzed subject domain. Hierarchy depth: 1 – no hierarchy, the components belong to the semantic-fields, while the semantic-fields do not form a hierarchy.
- *Operators of result* (the results which can be received at the given stage of working with the project): analysis of meta-data (gender, age, education, etc.); the semantic map of components; the semantic graph of components; the semantic map of fields; the semantic graph of fields.
- *Meta-information:* The project was aimed at creating a thesaurus, so the project data (due to having the API) were exported into a separate desktop application.

4.2 The Project "Image of Profession"

The research task was to model a profession image characteristic for representatives of different specialties. The project is based on the material received in the directed chained associative test realized within the Information System (see Fig. 3).

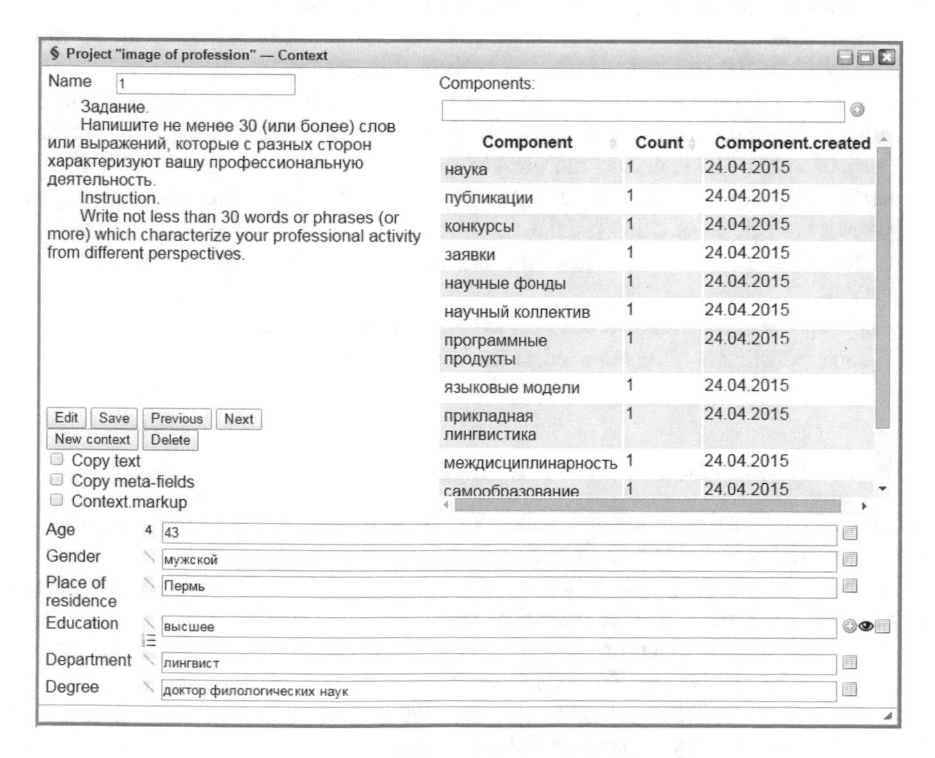

Fig. 3. The window of the project "Image of Profession"

While working on the project students learn to organize the excerpt of informants, develop classifiers of associative reactions, and model the semantic space of the received experimental data. In the course of linguistic data processing it is mostly important to overcome the subjective approach which is possible by organizing joint work. In particular, creating classifiers and classifying the data for the project work requires not less than two experts. Apart from the teacher who evaluates and controls the quality of expert work, learners interact with each other for working out the joint decision. The teacher has a possibility to correct the classifying system and classification results, as well as to help with interpretation of the constructed models.

The project description:

- *Project:* open; unindexed; 139 contexts, 4 540 components, 29 fields (in the project results of expert classification are available).
- *Users:* experts (3–2 students, 1 teacher), informants (139).
- *Meta-fields of contexts:* gender, age, education, specialty, place of residence, scientific degree. Meta-fields of contexts are available for samples.
- *Operators:* Access rights are restricted for the informants (the informants cannot see other informants' answers); adding components (associates received in the associative test).
- *Working mode:* online (the informants work within the project).
- *Fields:* Hierarchy depth: 1 – no hierarchy, the components belong to the semantic fields, while the semantic fields do not form a hierarchy.
- *perators of result* (the results which can be received at the given stage of working with the project): analysis of meta-data (gender, age, education, etc.); the semantic map of fields, the semantic map of components, the semantic map of words; the semantic graph of fields, the semantic graph of components, the semantic graph of words; the Matrix of shifts (fields), the Matrix of shifts (components); the Matrix of shifts (time intervals of fields); the Matrix of accordance (Meta-field: semantic fields).
- *Meta-information:* the directed chained associative test method was used; the task for the informants is represented in the context window.

4.3 The Project "Nominations of a Magician"

The project is aimed at semantic tagging of texts which are included into the corpus of mythological stories of the Perm region.

The aim of the project is to discern several micro-topics of mythological stories, and to find the nominations of magicians used in dialect speech.

Figure 4 shows a tagged text of a mythological story. Somewords in the text are tagged as belonging to certain lexical-thematic groups. While tagging all the words included into one and the same group are additionally detected by the same color. Text tagging is possible by means of adding meta-fields to words. Any number of meta-fields can be added to a word, though only one chosen meta-field is shown (the choice of the meta-field that will be shown is one of the IS tools).

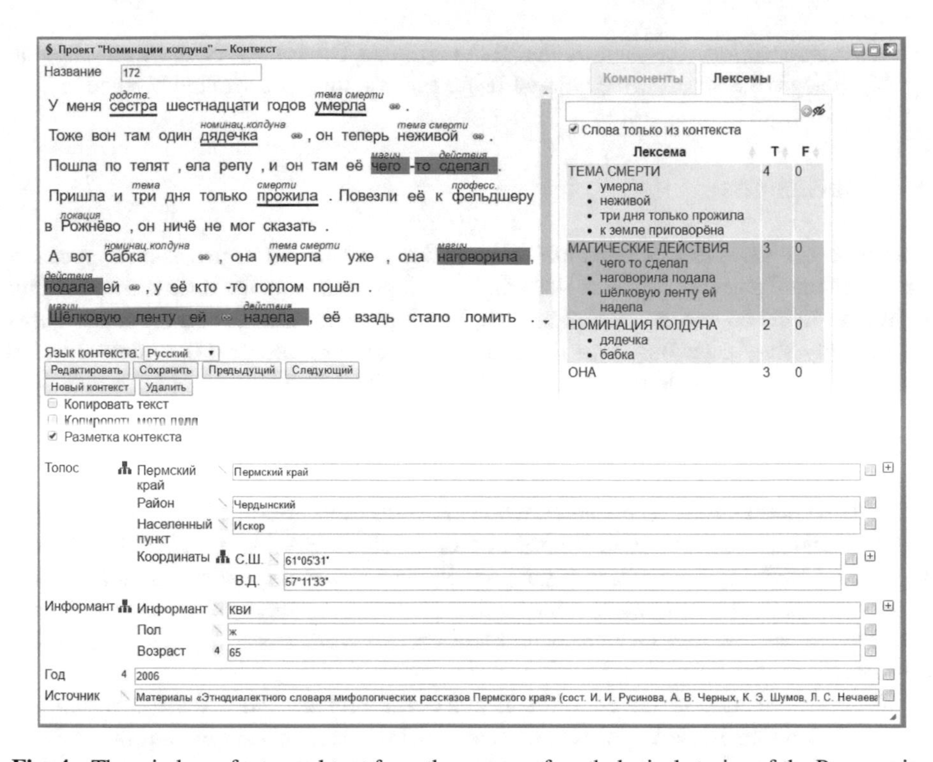

Fig. 4. The window of a tagged text from the corpus of mythological stories of the Perm region

In the lower part of the window content meta-fields which characterize the informants' residential place are shown. Meta-fields can differ for each particular project.

While working on the project students learn to find the necessary words in context, to ascribe them certain linguistic characteristics, to single out micro-topics of the texts. Besides, basing on the project data students can observe the difference in the structure of texts, as well as variability of nominations.

The project description:

- *Project:* closed; indexed; 257 contexts, 15 components (micro-topics), 0 fields.
- *Users:* experts (2 students, 1 teacher), informants (9).
- *Meta-fields of contexts:* Perm region, District, Residential place, Location, Informant (reference), Gender, Age, Year, Source. Meta-fields of contexts are available for samples.
- *Operators:* Access rights are unrestricted for experts; Adding components (micro-topics); Tagging the text (referring words/word combinations to a micro-topic).
- *Working mode:* any.
- *Fields*: no.
- *Operators of result* (the results which can be received at the given stage of working with the project): analysis of meta-data (age, gender, region, etc.); the semantic map of components, the semantic map of words; the frequency dictionary with the contexts of word usage.

- *Meta-information:* the data of the "Ethnographic Dictionary of the Perm Region Mythological Stories" were entered using the continuous sampling method.

5 Conclusions

The paper discusses the "Semograph" Information System and its use as an application for educational purposes. Present-day educational process requires interactivity (ability to interact in the course of academic activity) both from the teacher and the student; this interactivity should be based on using modern tools provided by the internet informational space and modern computer technologies [6].

Nowadays there are a great number of educational programs; nevertheless, their majority is either not oriented at interaction of the educational process participants, or can solve just a small number of specific tasks. Moreover, these programs are mostly aimed at secondary school education, while the systems of distance interaction between teachers and students used in higher school (such as SAKAI or MOODLE) only provide a possibility for students to get tasks in their personal accounts and for teachers - to check them. There is no real interaction in the frameworks of those remote educational systems, as well as there is no collaborative work wherein professional skills are developed.

A specific feature of the educational process in higher school consists in the fact that while training Bachelor, Master and Postgraduate students should not only master an academic discipline and acquire certain competence; what is more important, they should learn to obtain new knowledge on their own. That is why apart from proper educational activity it is necessary to create the scientific and research environment for higher school students. The "Semograph" Information System successfully performs these tasks while training specialists in linguistics.

Acknowledgments. The research is supported by the state grant of Perm State University 2017–2019, project № 34.1505.2017/PCh.

References

1. Castels, M.: Galaxy Internet: Reflections on the Internet, Business and Society. U-Factoria, Ekaterinburg (2004). (in Russian)
2. Purdehnad, D.: Open Innovations and social networks. Probl. Manage. Soc. Syst. **7**, 22–27 (2012). (in Russian)
3. Citizen science. https://en.wikipedia.org/wiki/List_of_citizen_science_projects
4. Information System "Semograph". http://semograph.com
5. Baranov, D.A., Belousov, K.I., Ichkineeva, D.A., Zelyanskaya, N.L.: The network organization of experimental research in linguistics: opportunities and prospects. Procedia Soc. Behav. Sci. **214**, 958–964 (2015)
6. Burbules, N.C., Torres, C.A.: Globalization and education: an introduction. In: Burbules, N., Torres, C. (eds.) Globalization and Education: Critical Perspectives, pp. 348–349. Routledge, New York (2000)

Social Interactions in a Virtual Learning Environment: Development and Validation of the Social Map Tool

Magalí Teresinha Longhi(✉), Ana Carolina Ribeiro Ribeiro, Fátima Weber Rosas, Leticia Rocha Machado, and Patricia Alejandra Behar

Digital Technology Center Applied to Education (NUTED), Federal University of Rio Grande do Sul (UFRGS), Avenue Paulo Gama 110, 12105 Building, Room 401, Porto Alegre, RS 90040-060, Brazil
magali.longhi@gmail.com, carol_ribeiro2@hotmail.com, fwrosas@gmail.com, leticiarmachado@gmail.com, pbehar@terra.com.br

Abstract. This article presents a sociometric social mapping tool and its application in the Virtual Learning Environment (VLE) ROODA (Cooperative Learning Network). The purpose of this tool is to map the social interactions that take place during the development of a classroom or Distance Learning activity. The maps are constructed based on selected categories (for example, participants and period of analysis) and visualized as sociograms. These graphs have been proven to be relevant in assisting teachers to both monitor and follow-up on their students during the teaching and learning process. In addition to providing punctual analysis of the relationships and forms of interaction among students, the availability of this sociometric tool in the VLE itself allows for the reconceptualization of educational practices and recreation of pedagogical strategies that can maximize not only virtual social interactions but also forms of collaboration and cooperation. Preliminary results were used to construct new research questions that enabled the collection of categories of social information and the construction of pedagogical strategies based on social aspects for Distance Learning.

Keywords: Virtual Learning Environment · Sociometric mapping · Sociograms

1 Introduction

Considering the number of students enrolled in Distance Learning (DL) courses, the information and communication technologies (ICT) incorporated in the Virtual Learning Environments (VLE) should be used for more than accessing and discussing content. They should also serve as a means of accompanying the construction of knowledge. This process involves the interactions between the individual subjects and the social relations established in the specific virtual space. It also encompasses the

V.L. Uskov et al. (eds.), *Smart Education and e-Learning 2017*, Smart Innovation, Systems and Technologies 75, DOI 10.1007/978-3-319-59451-4_27

conflicts, contradictions, arguments, and rhetoric as well as the recommendation of new subject matter and materials.

People have only just begun to study the social relationships in VLE. For the most part, studies focus on themes such as content management and monitoring of students, and do not give due importance to the interactions that occur during the process of knowledge construction in these environments [3, 4]. In fact, Piaget [1, 2] categorically stated that well-structured intellectual development presupposes due attention to affectivity and social interactions.

This work therefore presents the development and validation of a tool that presents sociometric graphs, called Social Map. This tool is linked to the AVA ROODA (Cooperative Learning Network, https://ead.ufrgs.br/rooda). It is meant to analyze the interactions of the subject in the learning environment and illustrate relationships in the form of sociograms.

Research was carried out based on the generated graphs which in turn enabled the improvement of the manner of visualizing the interactions and categories of social information as well as in the recommendation of pedagogical strategies in order to meet individual student and/or group needs in the process of knowledge construction. Thus, the Social Map tool is a digital service integrated into AVA ROODA that is exclusively for the teacher's use. It can transform the teaching experience, primarily in the E-Learning context. It is part of a group of smart solutions that have been developed in the Digital Technology Center Applied to Education (NUTED) at the Federal University of Rio Grande do Sul, Brazil (UFRGS).

In order to understand the importance of interactions in VLE, we present the theoretical foundation of the key terms of this study in Sect. 2 and the tool Social Map is outlined in Sect. 3. Section 4 presents the adopted methodology, while Sect. 5 describes the preliminary results that were obtained. Lastly, the final section presents new perspectives for the development of this research project.

2 Theoretical Structure

In order to understand the social relationships that can be formed in a virtual learning environment, a theoretical review is necessary, mainly regarding social interactions, VLE, and sociometry, as will be discussed below.

2.1 Social Interactions

Etymologically, the term interaction (inter + action) includes the concept of reciprocity, in which at least two elements (they do not need to be the same type of elements) are involved in an encounter that causes them to change. According to Piaget [1, 5], subjects construct knowledge through interactions. Thus [5], "social life is one of the essential factors in the formation and growth of knowledge" (p. 17).

Interactions may be intrapersonal or interpersonal. The first level of interaction occurs when one uses previous knowledge to reformulate or understand the new, for example the mental processes that occur during reading. At the second level, the relationship is considered S (subject) => O (object), to the extent that there is some reciprocity. That is, O can be another subject or any other element in the environment. Contact requires the intermediation of the physical environment (sound waves, electric waves, etc.) for interactions established at the interpersonal level in order for the subjectivity of each participant to be externalized [6].

In technological terms, interaction mediated by digital resources is understood from a technical perspective [7], with emphasis on the functioning of the computer system based on Information Theory and Behaviorism. According to Primo [7], the interaction can be defined as mutual and/or reactive. Mutual interaction is "characterized by interdependent relations and negotiation processes," whereas reactive interaction is "limited by deterministic relations of stimulus and response" (p. 57). The two types of interaction are not established exclusively. In some spaces, both can occur at the same time (e.g. while chatting one interacts with the application's interface, the mouse, and keyboard).

Therefore, the practice of conversation (and the social relations formed through it) supported by digital technologies can be confused with a flow of messages, where there is not necessarily social dialogue or interaction (often called interactivity). Yet, interactions in technological spaces are based on a dialogue that modifies the subject, the other, their messages, and their inter-relations [7].

2.2 Virtual Learning Environments (VLE)

Virtual learning environments are understood as the composition of a platform (programming and interface) and networks of relationships established through this platform (cognitive, symbolic, affective). Hence, a VLE are characterized as resources to support learning, communication, and collaboration, and can be used to support both face-to-face education as well as Distance Learning.

In this context, the digital technologies integrated in VLE support virtual spaces beyond their technological mechanisms, or beyond simply managing information. They are also formed by the subjects involved and their interactions. Therefore, interactions in VLE are important sources for understanding about not only cognitive, but also social, affective, symbolic and behavioral aspects [7]. Thus, a subject's behavior, relationships, and contributions, revealed through their texts and messages, can be indicators used by teacher to mediate the process of knowledge construction [7].

2.3 Sociometry

Sociometry [8] aims to implement an experimental technique based on quantitative methods of the mathematical study of the psycho-sociological properties of populations. According to Moreno [8], sociometry is a strategy for understanding the structure of a group.

One of its techniques is the application of sociometric tests that allow one to visualize the similarities and the differences between the individuals that compose a group. For example, the subjects express: (a) their choices of which partners they would like (or not like) help from to perform a particular activity; (b) which partners play a certain role in the group better (or worse). In making their choices, the reciprocal relationships of the subjects are identified and presented in the form of a graph, known as a sociogram. These can, according to Moreno, reveal even the "invisible."

A sociogram illustrates the position occupied by the individual in the group and the nucleus of relationships that are formed around them. This nucleus of relationships constitutes the smallest social structure, which Moreno defines as a social atom [8]. While certain social atoms are limited to the individuals participating in it, some of these individuals may be related to parts of other social atoms, and so on, forming complex chains of interrelationships.

Therefore, through a sociogram one can see the social position of each participant in a learning community and their relationship with the rest of the group. Their choices determine the most privileged member of the social atom and degrees of reciprocity. It also shows which individuals are rejected because they do not comply with reciprocity, and which ones are isolated because they do not show preferences.

Sociograms are graphical representations in the form of a network of relationships among a group of individuals. More than a method of presentation, sociograms constitute a method of exploration, because they allow for the identification of sociometric data and the structural analysis of a group.

Moreno [8] defined a set of symbols (which are geometric figures, such as circles, triangles with one or two outside edges as well as red or black straight lines with continuous or interrupted lines with or without arrows), manually drawn representing the gender of the subjects, their role in the group, their attractions and repulsions, indifferences, as well as uni or bilateral relationships.

In the 1960s, Moreno's sociogram incorporated the formalisms of graph theory that gave it a mathematical rigor [9] and was described by computational algorithms with graphic displays in various devices. Sociograms are now recognized as social network diagrams.

A network (or graph) is formed by a finite set of nodes that represent the actors (individuals, groups or organizations) and edges (or arcs) that reveal the connections between them. The main focus of analysis when reading the graph is on the pattern of connections, the distance, and the physical position of the nodes.

In this work, the analysis of the sociogram of the learning community makes it possible to verify, through the patterns of connection, the choices made and the reciprocity between individuals. The distance and physical position of the participants designate the those who are isolated, rejected, and leaders in the group as well as closed groups (or social atoms) [12].

Sociograms are valuable sources of data on relationships in a group of individuals. In this context, some research has been developed to explore social networks in VLE.

Most of the research presented in the literature extracts data from open source or proprietary software VLE databases (PAJEK, UCINET, MEERKAT-ED, GraphML and NETDRAW) to construct the matrices and their respective sociograms [11, 12]. On the other hand, there is the software created by the Social Networks Adapting Pedagogical Practice (SNAPP), which aims to reinterpret the interactions in postings in the forums of several VLE, such as MOODLE, BlackBoard, and WebCT. These interpretations are presented in the form of a social network diagram [13]. VLE TelEduc, developed by the State University of Campinas (São Paulo, Brazil), offers the InterMap tool [14], which is based on the interactions of the participants with communication tools (e-mail, discussion forums, and chats), exhibited in a graph or table. However, the model is restricted to quantitatively mapping the data.

Yet, sociograms must go beyond the quantitative data expressed in the form of graphs, to offer qualitative interpretations for the teachers to help them understand the relationships that are formed during the teaching and learning process.

3 The Social Map Tool

Social Map is a tool that enables the creation of sociograms, based on users' interactions with ROODA's communication tools (forum, chat, synchronous and asynchronous messages). It allows one to visually follow the relationships established in the VLE. The sociogram makes it possible to identify the links, influences, and preferences that exist within a group. This tool can only be accessed by the class' teacher.

On the initial Social Map screen, the teacher indicates the configuration options for the visualization (Fig. 1), such as:

A. Period of analysis: The teacher defines the interval of time that they want to visualize and the interactions that occur;
B. Participants' colors: The teacher indicates the colors for each user profile (monitor/tutor/TA, teacher, and student).
C. Analysis functions: The teacher chooses which tools and their relevance to analyze interactions in ROODA;
D. Analysis: The teacher indicates the members of the class that will participate in the analysis (everyone, only teachers, only students, only formal groups).

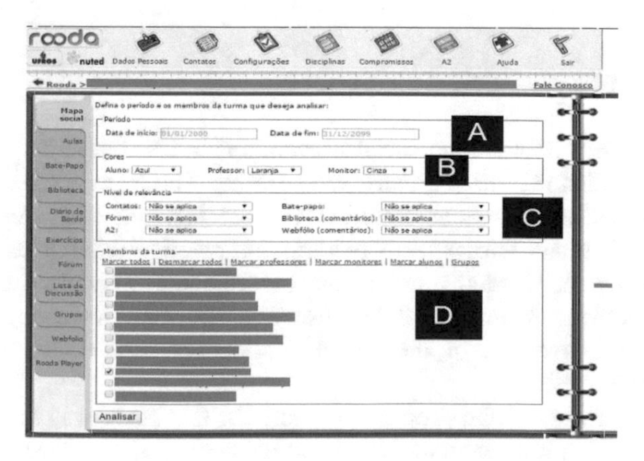

Fig. 1. Social map configurations. Source: ROODA. Available at: http://ead.ufrgs.br/rooda

After the teacher indicates the preference options, a graph is generated that shows the relationships that were established between the subjects and the tools that were used to collect the data (Fig. 2).

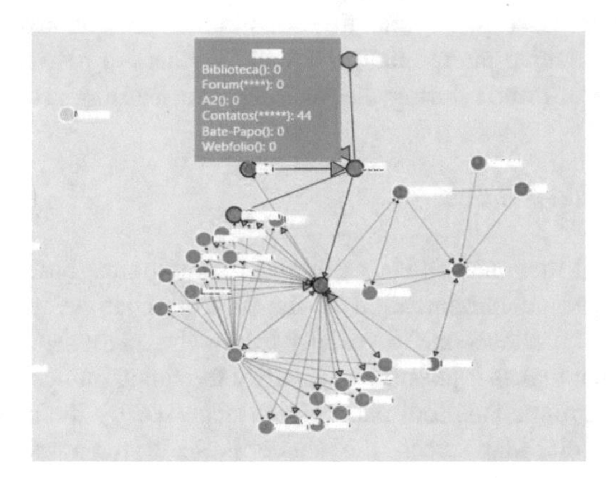

Fig. 2. Graph generated by the Social Map tool. Source: ROODA. Available at: http://ead.ufrgs.br/rooda

In the current version, the arrow indicating the direction of the connectors shows how the connections between two nodes (VLE participants) have been formed and the size indicates the number of interactions that occurred. The graphic visualization of this information allows the teacher to identify which students (and monitors/tutors/TAs) had higher (or lower) numbers of interactions during the selected period. They can also use the tooltip feature for each student. This feature numerically displays the number of exchanges that occurred for each VLE tool.

4 Methodology

This study addresses how to map social interactions in a virtual learning environment and presents them in the form of graphs (sociograms). These graphs enable the analysis of the subjects' social interactions. Based on these, it is possible to categorize the most frequent interactions and automatically recommend possible pedagogical strategies that can maximize the social interactions in the environment. Therefore, this research is characterized as theoretical-practical, because it is dedicated to the (re)construction of ideas and the improvement of theoretical foundations, especially those related to sociometric studies [8].

Thus, in order to meet the proposed objectives, this study was developed in four stages, which took place recursively: (1) construction of the theoretical reference regarding the topics addressed, primarily social aspects [7], sociometry [8], education [2–4], and Distance Learning [15]; (2) implementation of the Social Map tool, and interaction module that was built into the environment; (3) validation of the Social Map tool in undergraduate, graduate, and extension courses offered at or outside of the university; (4) consolidation of the social information categories based on the theoretical framework and the Social Map tool.

5 Preliminary Results: Validation of the Social Map Tool

Interactions in different contexts were analyzed in order to improve the Social Map tool in VLE ROOD. A difference was observed between the face-to-face and virtual the sociograms with undergraduate subjects. In fact, the interactions were far more intense in the Distance Learning modality. This is due to the fact that the teacher interacts more intensively and instigates the use of the VLE's communication tools.

This map was also applied in a music workshop with high school students. In addition to this tool, a sociometric test was also used with the students at the beginning and end of the course. The target audience consisted of 27 students in a public school, of which 10 were male and 17 were female, between 14 and 17 years old. The results of this test were compared with the sociograms generated by the Social Map tool in order to identify possible established relationships or social profiles of the VLE students. It was determined that Social Map can be an important tool to verify possible groupings between pairs of students, as well as the possibility of isolated or popular students. These mappings highlight the importance of developing smart technology for education and learning, especially for virtual learning environments.

By identifying the social profile of students based on this new technology, the teacher can then apply pedagogical strategies to mediate dialogues, investigations, interactions, and conflicts that occur not only in VLE, but also in face-to-face classrooms in high schools.

The Social Map tool was also used in an investigation of group formation. The results obtained were the basis of a Master's thesis [16] and signalled that the engagement in collaborative activities from the beginning of a course is fundamentally important for the constitution of a participatory group. Thus, it was possible to identify links, influences, and preferences that exist within the class. Moreover, the educator can diversify classes, as well as follow the interactions of students through the display of the sociograms constructed through the Social Map tool. The results of this research revealed that the Social Map could present the components of the groups with different colors in order to make it easier the teacher to identify the interactions between groups of students.

These applications and validations enabled the development of categories of social information in the map. The categories present a way of interpreting the most common relationships demonstrated in the map. It can also facilitate the teacher's actions by contributing to the identification of possible social profiles of their students. There were six categories of social information chosen: popularity, social distancing by the group, social distance from the group, absence, grouping, and collaboration. Based on this profile, the teacher can develop strategies to increase or diversify social interactions among students. These are important pedagogical strategies to qualify teaching, because the social aspects of each individual are intrinsically related to cognition [2]. In other words, they are always involved with learning.

It should be noted that the evaluation of the sociograms presented in this article did not take into account the pedagogical strategies used by the teachers, only the possibility of the Social Map as a source for teachers to examine or rethink their pedagogical practices. For the researchers, the visual data suggests several insights regarding the social interactions that occur in the education process and allows for new investigations and the possibilities of applying the map for other educational purposes.

Thus, the sociograms generated point to the need to carry out future studies that include qualitative and quantitative data in order to explore social relationships. The data also helped us in the development of a recommendation system that presents pedagogical strategies for teachers according to the social profile presented by the Social Map tool. Both the graphic presentation of the relationships and the recommendation of pedagogical strategies are part of the development of smart technology for education and learning.

In the current version, the recommendation tool for pedagogical strategies is linked directly to the Social Map tool. Teachers can choose the category of social information (collaboration, popularity, etc.) that they wish to examine in their students. Information on the number of interactions, the mean, the standard deviation, and the calculated degree in relation to the classmates is displayed for the each student and category selected. Therefore, pedagogical strategies, which are suggested educational actions that teachers can adopt to improve or maximize social interactions with thier students in the virtual learning environment, are recommended according to the social profile mapped by the tool. Yet, new perspectives point towards some improvements that could be made, that will be presented in the final considerations.

6 Final Considerations

This work presented the development, application, and validation of the Social Map tool for the mapping of the interactions that take place in the classroom and virtual teaching activities supported by VLE ROODA. Based on the data extracted from the Social Map tool, the teacher can analyze the collaboration, grouping, distancing of the subject by the group, the distance of the subject in relation to the group, popularity, and absence. Hence, the results obtained using the tool give the teacher important social information. Using these social indicators, the teachers can then focus their pedagogical practices on those students who require more help and can thus expand their range of communication.

The results of the sociograms made it possible to analyze and propose some modifications necessary to improve its development in the future:

- Adjustments in the visualization interface of the sociograms by social categories;
- Adjustments in the pedagogical strategies recommendation system based on the student's social profile in the VLE;
- Validation of social information categories in terms of establishing a relationship with the theoretical aspects raised by Moreno [8] and other theoretical authors.

In addition to the technological items that were discussed above, there are also educational aspects that are being discussed, evaluated, and validated. These aspects are related to the pedagogical strategies that may be recommended in the future to the teacher using the Social Map tool.

References

1. Piaget, J.: Les relations entre l'intelligence et l'affectivité dans le développement de l'enfant. Bulletin de Psychologie, VII, 143–150, 346–361, 522–535, 699–701 (1954)
2. Piaget, J.: Inteligencia y afectividad (Prólogo: Carretero, M), Buenos Aires, Aique Grupo Editor (2005)
3. Sacerdote, H.C.S., Fernandes, J.H.C.: Investigando as interações em um ambiente virtual de aprendizagem por meio da análise de redes sociais. Inf. e Doc. **4**(1), 22–33 (2013). Ribeirão Preto
4. Lima, L., Meirinhos, M.: Interacções em fóruns de discussão com alunos do ensino secundário: uma análise sociométrica. In: VII Conf. Internacional de TIC na Educação (2011)
5. Piaget, J.: Estudos Sociológicos. Forense, Rio de Janeiro (1973)
6. Longhi, M.T.: Mapeamento de aspectos afetivos em um ambiente virtual de aprendizagem. Tese de Doutorado, PPGIE/UFRGS, Porto Alegre, RS (2011)
7. Primo, A.: Interação mediada por computador: comunicação, cibercultura, cognição. Sulina, Porto Alegre (2008)
8. Moreno, J.L.: Fundamentos de la Sociometria. Paidós, Buenos Aires (1972)
9. Wasserman, S., Faust, K.: Social Network Analysis: Methods and Applications. Cambridge University Press, Cambridge (1994)
10. Rabbany, R., Takaffoli, M., Zaïane, O.: Analyzing participation of students in online courses using social network analysis techniques. In: Proceedings of Educational Data Mining (2011)
11. Willging, P.A.: Técnicas para el análisis y visualización de interacciones en ambientes virtuales. REDES- Revista hispana para el análisis de redes sociales **14**(6), 10–22 (2008)
12. Almeida, L.: Analisando Participação em Turmas de Ensino a Distância Usando Análise de Redes Sociais. Simpósios Nacionais de Tecnologia e Sociedade (2011)
13. Bakharia, A., Dawson, S.: SNAPP: a bird's-eye view of temporal participant interaction. In: Learning Analytics and Knowledge (2011)
14. Romani, L.: InterMap: Ferramenta para Visualização da Interação em Ambientes de Educação a Distância na Web, Dissertação de Metrado, IC/Unicamp (2000)
15. Behar, P.A.: Modelos Pedagógicos em Educação a Distância. Porto Alegre, Artmed (2009)
16. Ferreira, G.: Reconhecimento do comportamento sócio-afetivo do aluno: dicas pedagógicas para o professor, Dissertação de Mestrado, UFRGS (2015)

Education Information Interaction in a Group on the Basis of Smart Technologies

Natalya V. Gerova[1(✉)], Marina V. Lapenok[2], and Irina Sheina[1]

[1] Ryazan State University Named for S.A. Esenin, Ryazan, Russian Federation
nat.gerova@gmail.com, i.sheina@rsu.edu.ru
[2] Ural State Pedagogical University, Yekaterinburg, Russian Federation
lapyonok@uspu.ru

Abstract. The article deals with the issues of organization and management of education information interaction in a group of English language learners. The article looks at focus areas in using smart technologies for the purposes of foreign language acquisition and give an analysis of the most effective Internet resources for teaching and learning English. This research suggests the structure of education information interaction in a group in the processes of gathering, processing, transmission of information, exchange of messages among the members of the group in collaborative work with an electronic education resource and with the teacher in the local network. The research also suggests recommendations for the teacher concerning organization and management of education information interaction with the help of hard and software supporting an electronic education resource. The pedagogical experiment assessing students' knowledge and skills showed high level proficiency of the students in the field of education information interaction management on the basis of Smart Technologies.

Keywords: Competence · Education information interaction · Electronic education resource · Module · Smart technologies · Skills

1 Introduction

The goal of this research is to analyze the impact which the use Smart Technologies produces on the teaching and learning processes. The novelty of the research consists in creating opportunities for interactive information collaboration uniting teachers, students and the source of information. The experiment presented in the article demonstrates how use of education information resources in teaching a foreign language at the tertiary level ensures developing skills and assessing them [3, 4, 8, 10].

Relying on I.V. Robert's research, we can define education information interaction (EII) on the basis of electronic education resources (EER) which involves learner, groups of learners, electronic information resources functioning in the local network and ensures collaborative education information activities aimed at developing lexical and grammatical skills [8].

Interactive EER ensuring effective education information interaction in a group is viewed as electronic education information resources implementing immediate

V.L. Uskov et al. (eds.), *Smart Education and e-Learning 2017*, Smart Innovation, Systems and Technologies 75, DOI 10.1007/978-3-319-59451-4_28

feedback in the processes of gathering, processing, transmitting and producing education information by the members of the group.

2 Focus Areas in Using Smart Technologies for the Purposes of Foreign Language Acquisition

Having analyzed the potential of several education electronic means and of the instrumental systems responsible for developing applications in networks [2, 8], we identify the following features which ensure their implementation for education information interaction (EII) on the basis of electronic education resources (EER):

Data bases of audio-visual information provide the opportunity of interactive access to any scene, moving "into it" and "out of it", shift them.

Textual, graphic and audio-visual information can be not only presented but "manipulated" within the screen frame and also the previous and the following scree frames.

Real events can be demonstrated on-line as a video film.

Access to any textual, graphic audio and visual data placed in the system is available.

Assigning animation effect, video fragment, sound to textual objects is possible.

Audio-visual information can be differentiated according to selected features.

Smart Technologies make the use of the global Internet a most promising tool for teaching and learning foreign languages. The effective use of this tool depends on the following factors:

Effective information interaction among the learners, between the learners and the teachers, and between them all and the electronic education resource (EER). Interactivity is the key factor in the educational process based on the local networks:

- Quality of the teaching material;
- Effective transmission of the teaching material;
- Well-organized system of testing, assessment and motivation;
- Effective systemic computer-based feedback which is envisaged in the teaching material and which is implemented as communication with the teacher and partners in the network.

Effective foreign language education requires an effective system of teaching strategies. Computer technologies offer a wide range of opportunities for foreign language teaching methodology.

- The student has the opportunity of two-way communication with the teacher and the partners, with native speakers.
- The student has access to enormous an diverse amount of information, which is related to language acquisition and which is constantly being enriched and transmitted with the help of telecommunication technologies.

3 Education Information Resources of the Global Network

The development of education information resources of the Internet broadens the opportunities of learning a foreign language. Research shows a great number of teaching and assessing programs and materials placed in the Global Network. Let's take a closer look at some of them.

1. Internet resource – www.homeenglish.ru. This is a site for beginners and intermediate students. It contains:
   - lessons for developing pronunciation skills (isolated sounds, sound clusters, reading rules)
   - Texts of all kinds in English s (books, fairy tales, quotations, tongue-twisters, jokes, lyrics, proverbs and sayings);
   - Materials for students (topics, essays);
   - Reference materials (grammar, slang, abbreviation, idioms, buzz words, monetary units, etc.);
   - Miscellaneous (games, audio books, memory work, tests, dictionary, translators, programmes for learning English). strong and weak points.
2. Web Portal – www.bbc.co.uk/russian/learning_english. The site includes several sections where one can find materials in different fields, such as "Economics", "Science", "Great Britain", "Sports", etc. The site presents a wide scope of authentic materials for reading and listening, which is constantly renewed. Communicative competence is the main goal in language learning, but it is difficult to attain if the learner has no opportunity to communicate with native speakers. Blogs give learners a chance to practice their communicative skills in a real communicative situation.
3. Programme Speed Reader helps to develop fast reading skills in quick and comfortable way. Reading texts at a fast speed one gets a chance to get rid of pronouncing words in silence and the speed of reading increases 2–3 times. At the end of each page there questions with multiple choice to check the understanding. The programme has the function of widening the sight angle. A wider sight angle enables the reader to read without moving the eyes embracing whole paragraphs and even pages.
4. Internet resource - www.londononline.co.uk. The site presents up-to-date information about London: museums, exhibitions, places of interest, history, etc. The site is a guide which helps to find interesting facts and data being in any part of the world if one has access to the Internet. Anyone can know what is going on in the British capital.
5. Programme Laser CD-ROM B1+. Laser B1+ CD Rom offers a system of interactive exercises developing the necessary skills for passing the Unified State Exam and the Cambridge Exams PET and FCE. The system includes: exercises aimed at developing lexical and grammatical skills (the section Use of English) and tasks aimed at developing skills at listening, reading and writing (the sections Listening, Reading, Writing).

6. Information web portal – www.abc-english-grammar.com. This resource contains 75 sections, 4000 pages of useful information and more than 1200 files that can be downloaded and used to learn the English language. The site also presents radio programmers, interviews, dialogues, books, cartoons, teleshows, commercials, jokes.
7. Multimedia course English Discoveries. This is a multimedia education programmer for English language learners. It consists of several CD-ROMs divided into five levels:
   - Let's start (for beginners) – the first module for beginners (the alphabet, numerals, the most essential vocabulary)
   - Basic
   - Intermediate
   - Advanced
   - Executive. The programme covers all the four aspects of speech activities (reading, writing, speaking and listening) in communicative situations. It presents all grammar constructions and over 3 500 lexical units. Applications include an electronic dictionary, progress record, and the modules running the programmer. English Discoveries enables the teacher to control the learners' work. Using the opportunities offered by the system the teacher can monitor their progress, plan lessons and get reports showing the average scores, results of their tests and the time they spend at their computers.
8. Internet resource – www.lingualeo.ru. This site helps to enrich the personal vocabulary, watch video, do fascinating training and communicate with native speakers. The authors also use on-line games. Learners can watch videos, listen to audio texts, practice their pronunciation an do other tasks.

4 Opportunities Offered by Smart Technologies to Language Learners

One of the challenges the teacher faces using the Internet is selection of resources. There a lot low quality education resources, which do not meet the requirements of the education goals. The criteria for selection are determined by the mechanisms of language acquisition and the style of teacher-student interaction. The following guidelines can be used for selecting Internet resources: reliability of the resource, information about the author, content, topics, goals, correspondence of the grammar and vocabulary to the syllabus, up-to-dateness, comprehensibility. The design of the site can be assesses according to the following guidelines: clear structure, easy navigation, interactivity, absence of distracting visual elements, hyperlinks, functionality.

At the tertiary level teachers have to apply additional criteria related to university curricular and syllabi:

- Correspondence of the material presentation to the methods used by the teacher;
- Level division of the teaching materials;
- Means of assessment of learners' progress;

- Quality of the teaching materials including the design of the presentation;
- Help concerning each task which is available for the learners.

There are several approaches to intensification of language learning on the basis of information technologies. For example simulation games help to model situations of professional interactions in a foreign language. Profession-oriented simulation games enable learners to develop their professional skills and motivate them to polish up their English.

Designing a simulation game one has to take into account the number of the participants, the issues, the goal, the number of teaching hours, equipment, etc. The content is selected depending on the goals [4]. Smart Technologies are very promising for designing the tasks for such games.

5 How to Organize Education Information Interaction on the Basis of Electronic Education Resources

Research and practical work show that effective strategies of teaching and learning place the student into the center of the education processes being not the object, but the subject of education activities which ensures the development of intellectual abilities of the student [10]. This task can be carried out with the help of education information interaction on the basis of electronic education resources. Designing the structure of education information interaction for a group of students we relied on the theses put forward by prominent educators and psychologists [1, 6, 7, 9 and etc.].

The process of education presupposes transmission, perception and processing of information, that is education is always information interaction between the teacher and the learner organized by the teacher. Any education activity is designed along the same lines: orientation, fulfillment, assessment and correction.

Computer-based learning helps to organize this interaction within a group both among the learners and between them and an electronic education resource. In this case information is transmitted in real and virtual environment through a channel of direct link between an IT resource and the learners providing an immediate feedback [1]. The channel helps to transmit information about the learner's work with the education resource. The feedback helps to organize correction and transmit the necessary guidelines to the learner.

Feedback is an essential components of learning management. This is a mechanism that implements the system of interaction among the teacher, the learner and the education resource.

Information technologies functioning on the basis of the local network provide for goal-oriented management of the teaching and learning process developing foreign language students' skills in the field of education information interaction with the teacher and the electronic education resource [8].

The goal of education information interaction in a group on the basis of Smart Technologies is collaborative work in the teaching and learning process with the help of an electronic education resource. The objective can be defined in the following way:

Co-operative work of the group consists in gathering, processing, producing and transmitting information which increases their motivation and develops their skills of computer-based learning.

The structure of education information interaction on the basis of Smart Technologies depends on the teaching strategies chosen by the teacher, but the most essential components of these activities are the following: the group of learners, the organizer (the teacher), interactive electronic education resources (Fig. 1).

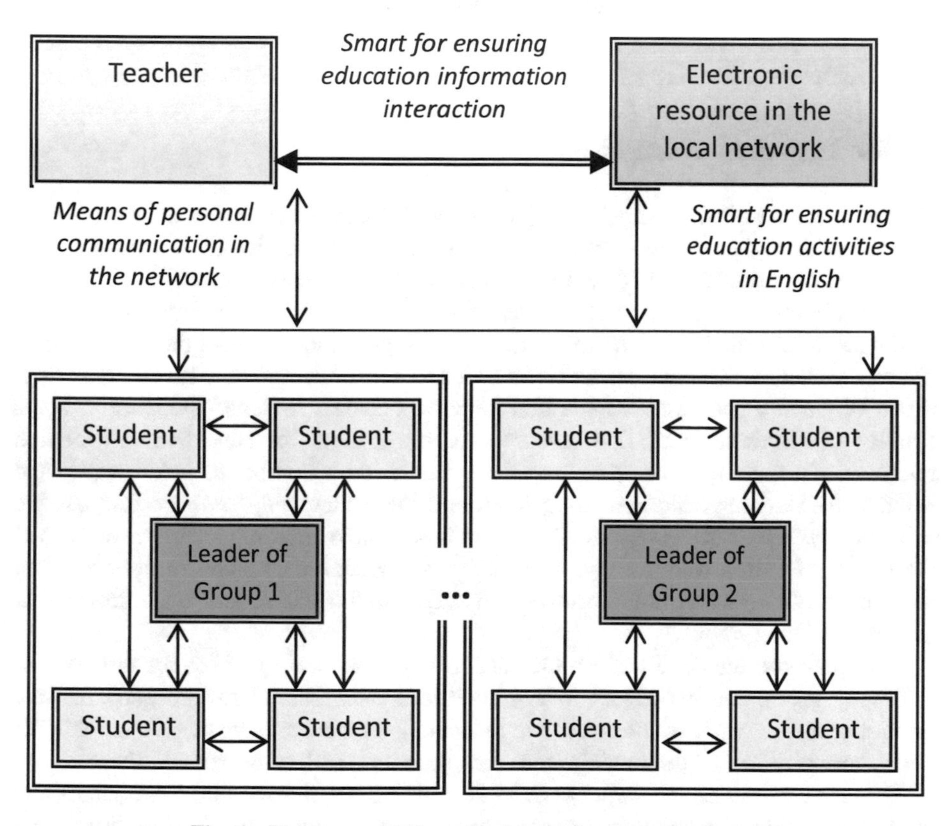

Fig. 1. Structure of EII on the basis of smart technologies

Organization of education information interaction on the basis of Smart Technologies goes through several stages:

- Interaction between students and the teacher is organized with the help of means of personal communication in the network as transmission of textual, graphic and audio-visual information by the teacher to study the parts of speech, tenses, syntactical constructions, variants of translation, etc. in the local network.

- Students interact in the group discussing the tasks with the help of means of message exchange. The teacher works out an electronic resource aimed at language acquisition. The resource is placed in the electronic means of the local network. Students can use the interactive electronic resource both in class and in their autonomous work. At this stage they learn how to use the resource doing the tasks and to correct each step in their activities to achieve the best results. The opportunity to use feedback makes their work more effective.
- The leader of the group receives the task in the form of a presentation, textual file or s test and distributes it among the members of the group, processes the results and sends them back to the teacher.

6 Guidelines for Organizing Education Information Interaction on the Basis of Smart Technologies for Learners of English

Before a teacher starts to organize education information interaction on the basis of Smart Technologies in the university local network he/she should be aware of the principles, opportunities and functions of the hardware and software.

Functioning of the electronic education resource has to correspond to the methodological principles of teaching a foreign language and to meet the requirements of appropriateness. Language acquisition relies to a great degree on vision and hearing, that's why audio and video information helps to enhance students' motivation. Interactivity of electronic education resources is the ground for education information interaction in a group, so in class a students has to do a number of tasks: watch and listen to the teaching materials, navigate around the content, copy information, use the reference system, answer question. A student has to have skills related to the use of the electronic education resource and to be able to organize interaction in the group, to receive, process and transmit information to and from the teacher and to and from other members of the group.

The dialogue interface enables the user to navigate easily looking for previous or following topics and get access to any reference materials. Hardware and software providing for the work of the electronic information resource have to give the teacher the opportunity to edit the modules and the materials, adding or deleting them.

The content management system CMS Joomla makes division into components and modules possible. The components are extensions adding more functions to the content management system. They make the electronic education resource fore flexible and they include forums, guest books, blogs, galleries, Internet shops, video conferences, data bases, etc.

The content of the electronic education resource has a variable module structure topics being the frame of the module which can be divided into micro-modules. The teacher has an opportunity to organize individual and varied learning depending on the teaching objectives.

Working out scenarios for language learners the teacher has to think of the order in which the material will be presented according to the methodological principles of consistency, coherence, comprehensibility, clarity, personal involvement, growing learning difficulties. For example, developing grammar skills presupposes the following order of activities: recognition of the grammar phenomena – understanding its meaning in word forms – understanding its difference from similar grammar phenomena – ability to choose it in appropriate contexts – automatic choice of this phenomenon constructing sentences.

To carry out the educational function an electronic education resource has to be able to pile up information: electronic teaching applications, manuals, etc. The content manager CMS Joomla is open software and it works with extensions and modules. It functionality can be increased and it can serve as a basis for creating a new unique resource.

The scenario also has to take into account the specific objectives for each stage of language acquisition and in the process of education interaction for some of the participants access to information is to be restricted depending on the degree to which his/her skills are developed.

Manuals for teachers who are planning to use electronic education resources to organize education information interaction in the university local network should contain: recommendations how to select content and structure to fill in the resource, descriptions showing how to work out scenarios of different types for different teaching objectives, recommendations how to choose software, ofrms and methods of organizing education information interaction on different stages of foreign language acquisition.

7 Results of the Experiment

To assess students' abilities to use electronic education resources and to perform collaborative activities in the process of education information interaction we identified the basic and the advanced levels for a group of 98 students.

On the basic level students are: to have the necessary knowledge of the structure of education information interaction, knowledge of the hardware and the software that supports this interaction, to have the ability to choose the algorithm of collaborative selecting, processing, producing and transmitting information in English in the educational environment, to be able to understand the instruction how to do the task placed in the electronic education resource, to participate in education information interaction under the teacher's guidance.

On the advanced level students are: to have knowledge of the structure of education information interaction in a group and the opportunities offered by hardware and software which support it, to be able to choose the operating mode with the electronic education resource to organize interaction, to carry out interactive activities using the hard- and software supporting the electronic education resource, to acquire skills necessary for operating with new hard- and software supporting education information interacting in English.

The pedagogical experiment had three stages: beginning, formative and final.

At the beginning stage all the second-year students learning the English language were divided into three groups consisting of 32, 34 and 32 people by random. the students' computer competence and ability to use electronic education resources were approximately on the same level.

The students were. administered a test consisting of 24 tasks. Each correct answer was assessed as one point. the measuring scale was divided into parts: one corresponding to the basic level (0–12) and the other corresponding to the advanced level (13–24) of students' proficiency at education information interaction management in a group.

The results of the test are presented in Table 1. The first line shows the number of points and the second line shows the number of students who got the corresponding number of points.

Table 1. Test results

Number of points	13	14	15	16	17	18	19	20	21	22	23	24
Number of students	8	13	17	14	11	8	6	6	5	4	4	2

The results were processed with the help of IBM SPSS Statistics as it is shown in Table 2.

Table 2. Test results processed with IBM SPSS statistics

	Group 1	Group 2	Group 3
α	0,121	0,114	0,132

The results of the test give an opportunity to bring them all together in one graph (Fig. 2).

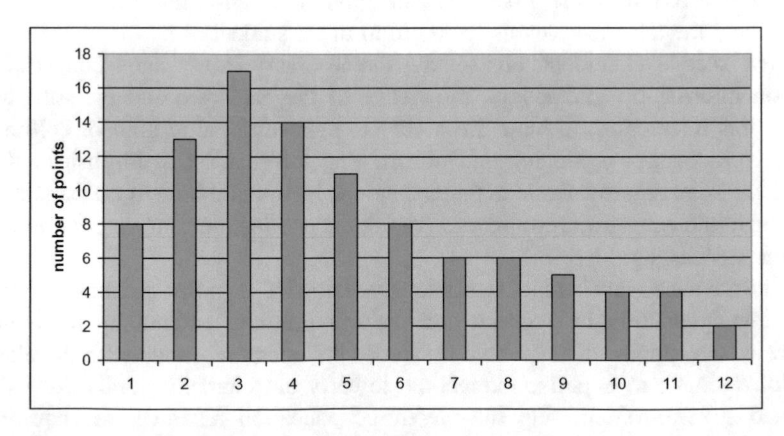

Fig. 2. Graph of test results distribution at the final stage of the experiment

The statistic analysis of the results shows that all the 98 students are highly proficient at education information interaction in a group. The average score of the students equals to 16,92 and standard deviation is 2,93. Thus, the results of the experiment show that the methodology applied in the course of the English language on the basis of Smart Technologies is effective.

8 Conclusion

Analysis of the scientific research and technological advances in the field of education information interaction shows the following technological approaches: use of personal communication means in the Internet, such as Skype, ICQ, Win Talk, Microsoft Chat, Net Forum etc.; use of distant education systems, such as Moodle, Training Ware Class, Caroline LMS, Tutor, Naumen Learning, etc.; use of multimedia authoring tools.

This research suggests the structure of education information interaction in a group in the processes of gathering, processing, transmission of information, exchange of messages among the members of the group in collaborative work with an electronic education resource and with the teacher in the local network.

The research also suggests recommendations for the teacher concerning organization and management of education information interaction with the help of hard and software supporting an electronic education resource.

The pedagogical experiment assessing students' knowledge and skills showed high level proficiency of the students in the field of education information interaction management on the basis of Smart Technologies.

References

1. Bespalko, V.P.: Pedagogy and Advanced Technologies in Education. Russian Academy of Education, Moscow, 336 p. (1995)
2. Yu, B.M.: Use of Telecommunications in Teaching Foreign Language at the Secondary Level (dissertation thesis), 198 p. (1994)
3. Vagramenko, Y.A.: Computerization of general education: results and prespects. Pedagogical Comput. Sci. **1**, 41–51 (1997)
4. Gerova, N.V.: Prospects of using flash technologies in humanitarian pedagogical education. In: Gerova, N.V., Vidov, S.V. (eds.) Proceedings of the Institute of Computer Education, Russian Academy of Education, vol. 9, pp. 163–166 (2003)
5. Yezhik, I.G.: Organization of education information interaction in a group on the IT basis for english language learners in military academy. In: Gerova, N.V., Yezhik, I.G. (eds.) Pedagogical Education in Russia, vol. 5, pp. 210–215 (2013)
6. Lerner, I.Y.: Instructional foundations of methods of teaching. Pedagogy, 185 p. (1981)
7. Pidkasistyi, P.I.: Computer technologies in the system of distant education. In: Pidkasistyi, P. I., Tyshenko, O.B. (eds.) Pedagogy **5**, 7–13 (2000)
8. Robert, I.V.: Computer technologies in science and education. In: Robert, I.V., Samoilenko, P.I. (eds.), 178 p. (1998)
9. Slastenin, V.A.: Pedagogy: textbook for students. In: Slastenin, V.A., Issayev, I.F., Shiyanov, E.N. (eds.) Academy, 576 p. (2003)
10. Grob, H.L.: Computer Assisted Learning (CAL). ZfB-Ergaenzungsheft **2**, 79–90 (1994)

Smart Evaluation of Instrument Scan Pattern Using State Transition Model During Flight Simulator Training

Kavyaganga Kilingaru[1], Zorica Nedic[1(✉)], Jeffrey Tweedale[2], and Steve Thatcher[3]

[1] University of South Australia, Adelaide, Australia
kavyaganga.kilingaru@mymail.unisa.edu.au, zorica.nedic@unisa.edu.au

[2] Defence Science and Technology Group, Adelaide, Australia

[3] Central Queensland University, Rockhampton, Australia

Abstract. Trainee pilots are expected to be thoroughly trained on both technical and nontechnical skills. Technical skills are relatively easy to evaluate. However, non-technical skills are hard to monitor and assess. This study investigates the feasibility of providing a smart evaluation technique that can generate feedback on trainee pilots' instrument scan behavior. The authors conducted gaze monitoring experiments with a number of trainee pilots while operating a flight simulator in order to isolate patterns of behavior associated with Situation Awareness (SA). Recorded data on eye movements are then processed and transitions are extracted using a state transition model. The sequence of transitions are then processed to retrieve repeated instrument scan pattern. The results are verified using chart visualizations.

Keywords: Pilot training · Flight simulator · Knowledge discovery · Situation awareness · Instrument scan pattern · Smart training

1 Introduction

Advances in modern technology over the past few decades have changed the air transport making it safer and more reliable. However, from the recent fatal air crash incidents it is evident that there is still a lot to be done in the aviation safety area. In addition to technical failures and other external factors, the level of pilot Situation Awareness (SA) have been found to be a main contributor to aviation accidents [1]. Therefore, it is important to provide an emphasized SA training to trainee pilots. In addition, it is important to provide accurate assessment of student performance.

A major part of flight training includes flight simulator training. This research focuses on an autonomous monitoring of a student pilot's instrument flight training and generating smart evaluation report of trainee pilot's instrument scan. Since gaze tracking is a proven nonintrusive method of assessment, eye movements are chosen as the physiological measures considered for analysis. The authors conducted experiments in order to verify that the gaze transitions can be captured and validated. Further the

V.L. Uskov et al. (eds.), *Smart Education and e-Learning 2017*, Smart Innovation, Systems and Technologies 75, DOI 10.1007/978-3-319-59451-4_29

captured transitions could be used to investigate the possibility of extracting patterns. The patterns are then used to determine if the scan indicate behavioral traits as defined in the study.

Using a number of aviation students the authors conducted simulator experiments and recorded their eye movements through each trial. The data collected was transformed using a Java program and the Azure machine learning toolbox. A Finite State Machine (FSM) model was created to track instrument transitions. Then a bubble chart visualization technique was used to verify the result collected. Further work needs to be done on an algorithm to classify the transition sequence into a meaningful feedback associated with level of SA.

2 Related Work

There are several SA evaluation techniques commonly used in aviation training [1]. Some of the most popular techniques are

- Freeze-probe techniques: where simulation of the task under analysis is paused and the participants are expected to answer SA queries based upon their understanding of the situation at that point in time [2].
- Real-time probe techniques: which involve the administration of SA related queries during task performance, but without pausing the current task under analysis [1].
- Self-rating techniques: where the performance is rated by participants themselves based on questionnaires.

However, the above techniques are highly intrusive and mostly rely on participants' qualitative opinion rather than an external entity monitoring quantitative performance. There are instances of using eye movement tracking and speech processing methods for SA training and evaluation. For example, Anders [3] conducted a research to analyze flight display and instrument panel scanning behavior. In his experiment, he used a head mounted eye tracker to record eye movements during a simulator training. The recorded values were used to determine good and bad attention distributions. However, the attention allocation computation was based only on average eye fixation time for a given Area of Interest (AOI).

Another experiment that relates eye movements to SA in pilots is experiment done by Van de Merwe, Dijk and Zon [4], where they injected failures in the simulator scenario and recorded eye movements of the pilots performing the scenarios. The pilots operating the simulator were asked to perform scenario with vague information based on the possibility of instrument failures. The visual search patterns of pilots who identified the failure with those who did not were then compared. It was found the result indicated a significant variation in visual search pattern for pilots who failed to identify faults.

Salmon et al. [5] used a method similar to freeze-probe technique in conjunction with eye tracking in their experiment. As such, the experiment was largely dependent on participants' qualitative feedback. The method did not look into extracting overall patterns; instead, it only focused on isolating patterns to detect failures.

In their experiments, Haslbeck and Bengler [6] studied pilot visual scanning strategies under visual restrictions and occlusion. The study analyzed the mean glance on primary flight display and other instruments. Frequencies of some of the preset scan paths using three instruments were computed during the experiment. The research was limited to statistical results related to the fixations on areas of interest and repeated occurrence of predefined scan path. Hayashi [7] used fixations on AOI and Hidden Markov Models (HMM) to identify instrument scan patterns. The research considered time spent on various instruments and tried to analyze the scan sequence using time series visualization. It was a challenge to manually analyze continuous stream of eye movement data using visual representation.

Obviously, there is a need for a smart pilot training system that can analyze and provide timely insight into student pilot performance with a minimum human intervention. In a previous study, the authors discussed the classification of flight operator behaviors using eye movements as misplaced attention, attention focusing and attention blurring [8]. This article introduces the next step towards the development of a suitable pilot training system. This new experiment promotes non-intrusive data collection to extract features from an eye-tracker that can be classified into relevant scan patterns during flight simulator trials.

3 Experiment

An experiment was conducted at the University of South Australia (UniSA) to record student pilots' eye movements during simulator operation. The experiment setup included a flight simulator with controls and an eye tracker running on a separate system.

1. Flight Simulator Setup: UniSA aviation department uses Lockheed Martin's Prepar3D flight simulator software to train aviation students. The simulator software has various options that can be selected, including an option for instrument flying with just panel view. It also provides the options to select various aircraft models, inject planned or random failures during a simulated flight scenario. Since the simulator is a proven training tool, the Prepare3D was chosen for our experiment. The Prepar3D flight simulator system was installed on a high-end computer and set up with main navigation controls such as pedals, yoke and gear.
2. Eye tracker Setup: EyeTribe tracker was used to record the gaze coordinates on the screen, time stamp, pupil size, good frame rates. Frame rate was set at 30 frames per second. Eye tracker was calibrated once for every participant before simulator scenarios are performed.

The experiment was conducted using the following steps:

1. Participant briefing: The participant is briefed about the scenario before every simulator session. For example, the participant is briefed about the departure airport and the landing airport, the weather condition settings. For the enforced scenarios participant is asked to perform some of the chosen crosschecks in a proper order.

The participant is then instructed to perform each scenario by switching gaze between instruments every 10 s on the researcher's instruction.

2. Gaze Calibration: Calibration is an important step prior to conducting any eye tracking experiment. The calibration involves software setup based on the participant's eye characteristics and lighting in the area for improved gaze estimate accuracy. Therefore, in the current experiment the student operator's eye movements had to be calibrated with the simulator screen coordinates prior to the first simulator operation. The calibration and verification step involved:

- Asking the participant (operator) to sit in a comfortable position in front of the simulator
- The eye tracker was adjusted so that the eyes of the operator are detected and well captured with both eyes almost at the center of the green area as shown in Fig. 1.
- The eye movements of the participant were calibrated using on screen calibration points on the simulator monitor. On successful calibration the eye tribe tracker shows calibration rating in stars as shown in Fig. 2. The calibration was then verified by asking the participant to look at the points and confirming the tracker is detecting the gaze correctly.

Fig. 1. Eyes tracked by the EyeTribe tracker

3. Simulator Configuration: Prepar3D simulator is configured to launch aircraft in Instrument Flying Rules (IFR) mode, with different departure and destination airports. Also, the participant is asked to do instrument flying using just the instrument panel. Weather conditions and failures are injected for different scenarios. For enforced scenarios the flight scenario was paused and the operator was asked to perform the instrument scan as per the researcher's instruction.
4. Gaze tracking: Gaze tracking is commenced from the EyeTribe tracker console, immediately after the scenario starts. The gaze records are saved into file named with the time stamp. The end result - crash or successful landing, and simulator configurations for each scenario is also recorded.
5. Analysis of the results: The data from various scenarios are preprocessed using azure machine learning studio. Then data mapping and transformation program are applied.

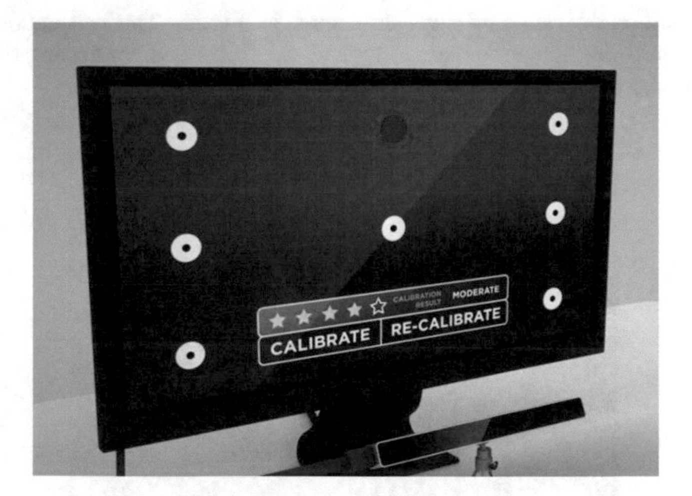

Fig. 2. Calibration points and calibration result: more stars = better calibration

Data collected during the crosscheck were run through the Java program written for this research and it was verified that transitions during the scan could actually be tracked.

4 Method

UniSA aviation students were asked to perform multiple scenarios with different sequence of instrument scan in a specific order. In this article, the authors discuss the radial crosscheck pattern scenario as a case study. The radial crosscheck pattern is the most common scan pattern recommended in flight training manuals. In this crosscheck method the AH is the primary instrument. While the pilot needs to check other

Fig. 3. Regular instrument crosscheck scenario. Instruments: Artificial Horizon (AH)– middle top position Air Speed Indicator (ASI) – left top, Altimeter (ALT) – right top, Navigation (NAV) – middle bottom row, Turn Coordinator (TC) – bottom left, Vertical Speed Indicator (VSI) – bottom right

instruments on the primary display the scan should always return to instrument AH in the middle of the top raw as shown in Fig. 3.

The calibration process was done as discussed in the previous section. The Prepar3D simulator was launched with instrument panel display for instrument flying. The participant was instructed to start the scan from AH, change to another instrument as per instructions during the briefing for radial crosscheck. The gaze coordinates on the simulator screen were recorded using the EyeTribe tracker while the participant was performing the instrument scan. The data from the eye tracker were in JSON format with parameters such as gaze coordinates, pupil size, frame number, time stamp, frame and state of the frame. The JSON data were then converted to comma separated file format for processing.

The raw data captured were preprocessed using a model prepared in azure machine learning studio. The elimination of invalid frames was done using SQL transformation scripts and missing values are cleaned by applying multiple imputation by chained equation based on average gaze coordinates from the left and right eyes and the pupil size. A state diagram was used to represent radial crosscheck pattern. A state diagram is defined as a directed graph:

$$G => (S, Z, T, S0, F) \tag{1}$$

Where:

- S represents a finite set of states. For the regular crosscheck, S = {S0, S1, S2,… Sn} represents a set of input symbols.
- Z represents output symbols. For the current model these are instruments such as Artificial Horizon (AH).
- T represents a set of transitions T0, T1… Tn; T => (Si)
- {S0} is the initial state.
- F is the final state.

Each transition from one state to another is triggered by an event mainly by the defined gaze changes from instrument to instrument or any other gaze point. Figure 4 shows various states and the change in instrument fixation as the events that trigger transition from one state to another. Further, the instrument scan for the whole scenario is transformed into a set of state transitions triggered on change of the area of interest.

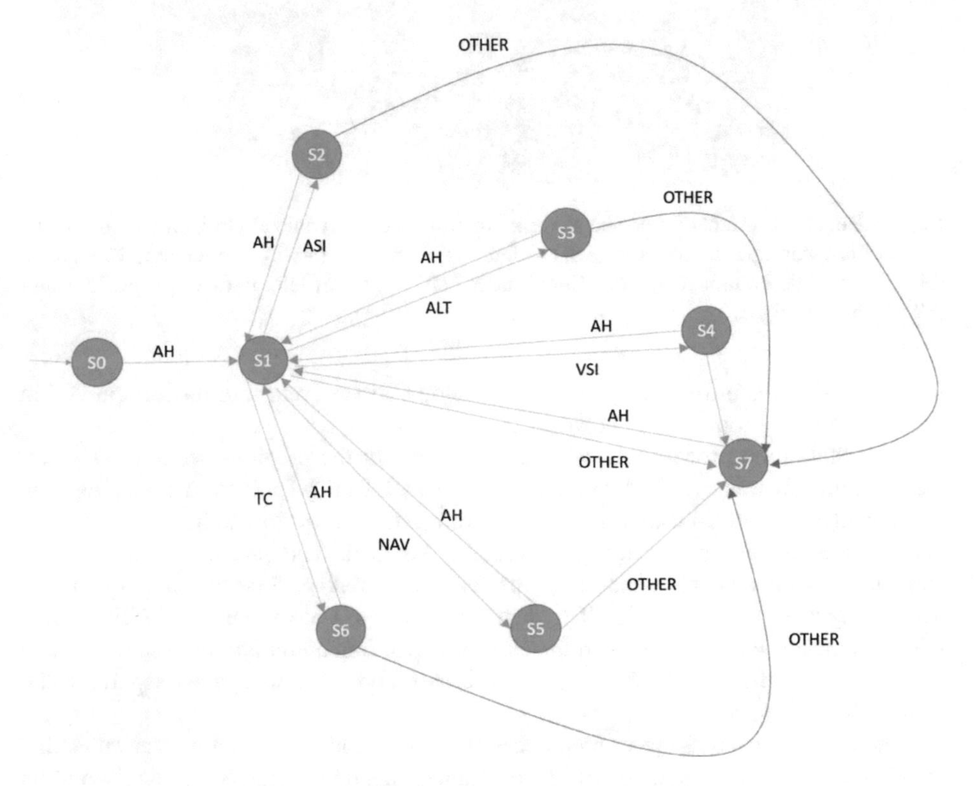

Fig. 4. Radial crosscheck finite state machine

5 Experiment Results

For the radial crosscheck it is expected that instrument AH was gazed for maximum amount of time because every scan starts and ends with the primary instrument AH. A sample set of transition records from the enforced scenario is shown in Fig. 5. The enforced scenario clearly showed transitions between the primary instrument AH and other instruments labeled in Fig. 4.

Hence, it has been verified that the program correctly identifies the transition between the area of interests (instruments). As an additional verification the aggregate visits are visualized as a bubble chart shown in Fig. 6. The figure shows fixation on each instrument tracked as area of interest. From the bubble chart it is evident that AH

T1 : S0 - AH -> S1

T5 : S1 - NAV -> S5

T2 : S1 - ASI -> S2

T8 : S2 - AH -> S1

T4 : S1 - VSI -> S4

T1 : S4 - AH -> S1

T3 : S1 - ALT -> S3

T9 : S3 - AH -> S1

T2 : S1 - ASI -> S2

T8 : S2 - AH -> S1

T4 : S1 - VSI -> S4

T8 : S2 - AH -> S1

T2 : S1 - ASI -> S2

T8 : S2 - AH -> S1

T3 : S1 - ALT -> S3

Fig. 5. A sample set of transition records from the enforced scenario

was confirmed as the instrument fixated for the maximum amount of time which is the expected result because any scan should return to, and pass through AH for radial cross check. The second highest fixated instrument is NAV which was not the expected result, so we observed the sequence of the instrument scan. It was noticed that NAV was fixated on for a prolonged time at the end of the scenario as the student was not instructed to transition back to AH before ending the collection of data stream.

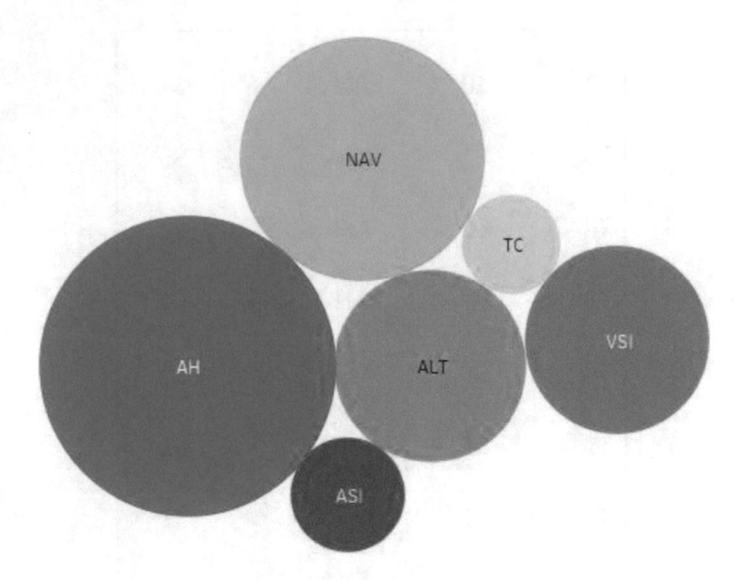

Fig. 6. Bubble chart

Finally the sequence of transitions recorded was processed by a dynamic program to extract repeated patterns. The recurring sequence P of transitions are extracted as:

$$\mathrm{Seq} = \mathrm{Sub}(\mathrm{T}_{1\,\mathrm{to}\,\mathrm{n}}, \mathrm{L}) \tag{2}$$

$$\mathrm{P} = \mathrm{MAX}(\mathrm{Seq}, \mathrm{N}, \mathrm{L}) \tag{3}$$

Where

- Seq is the subset of transitions T
- N is the number of occurrences of Seq with length $L > 2$.

The sequence with length L and maximum N is the most repeated pattern. Further research will look into association of the most repeated scan pattern to level of SA maintained by the trainee pilot.

6 Conclusion

This experiment takes advantage of instrument scan because it is the primary input for pilots during Instrument Flying (IF) when there are no visual cues. Failure to perceive instrument readings may lead to loss of SA and induce errors in decision making which in turn increases the risk of accidents. Therefore, students are extensively trained with various approach of instrument scan during IF. However, currently there are no suitable external systems that can non-intrusively monitor trainees' instrument scan behavior. This article we presented findings of smart instrument scan evaluation systems which can be used during flight simulator training. In particular, the authors presented an experiment focused on detecting trainee pilots' visual scan during simulator scenario through eye tracking and further processing of visual scan data to detect pattern using a state transition model to represent data and an algorithm developed as part of this research. The experiment verified that it is possible to detect the gaze transitions between the instruments. Further, scan patterns were detected from the sequence of transitions. Being able to monitor trainee pilots' gaze movements, identify the repeated scan pattern and verify the results is a significant milestone in the SA assessment. Future research will look into classifying the patterns into behaviors, mapping to cognition so that possible errors can be identified and relevant feedback/warnings could be timely generated.

Acknowledgment. This research is funded by the Australian Government Research Training Program (RTPd). The authors would like to thank Leonard Leano for his help with experiments.

References

1. Salmon, P., Stanton, N., Walker, G., Green, D.: Situation awareness measurement: a review of applicability for C4I environments. Appl. Ergon. **37**(2), 225–238 (2006)
2. Endsley, M.R.: Situation awareness global assessment technique (SAGAT). In: Proceedings of the IEEE 1988 National Aerospace and Electronics Conference, 1988, NAECON 1988, pp. 789–795. IEEE (1988)
3. Anders, G.: Pilot's attention allocation during approach and landing, eye and head-tracking research in a 330 full flight simulator. Focusing Attention Aviat. Saf. (2001)
4. van de Merwe, K., van Dijk, H., Zon, R.: Eye movements as an indicator of situation awareness in a flight simulator experiment. Int. J. Aviat. Psychol. **22**(1), 78–95 (2012)
5. Salmon, P.M., Stanton, N.A., Walker, G.H., Jenkins, D., Ladva, D., Rafferty, L., Young, M.: Measuring situation awareness in complex systems: comparison of measures study. Int. J. Ind. Ergon. **39**(3), 490–500 (2009)
6. Haslbeck, A., Bengler, K.: Pilots' gaze strategies and manual control performance using occlusion as a measurement technique during a simulated manual flight task. Cogn. Technol. Work **18**(3), 529–540 (2016)
7. Hayashi, M.: Hidden markov models to identify pilot instrument scanning and attention patterns. In: IEEE International Conference on Systems, Man and Cybernetics 2003, vol. 3, pp. 2889–2896. IEEE (2003)
8. Kilingaru, K., Tweedale, J.W., Thatcher, S., Jain, L.C.: Monitoring pilot situation awareness. J. Intell. Fuzzy Syst. **24**(3), 457–466 (2013). (2012)

Technologies for the Development of Interactive Training Courses Through the Example of LMS MOODLE

Leonid L. Khoroshko(✉), Maxim A. Vikulin, and Vladimir M. Kvashnin

Moscow Aviation Institute (National Research University), Moscow, Russia
{Khoroshko, Vikulinma, Kvashnin}@mati.ru

Abstract. Currently, all educational institutions and their teaching staff join their efforts for the purposes of joint educational activities through Internet on the basis of common learning standards and technologies. Hence, the significance of e-Smart learning has been steadily increasing and this necessitates wide application of distance learning systems. Today, e-learning can be used not only as a specific form for gaining knowledge based on the distance technologies, but as an additional opportunity to reinforce the material learnt or for an in-depth study of the scientific knowledge area.

Keywords: e-learning · Interactivity · Moodle LMS

1 Introduction

Reading material for smart learning should not be static. It is most likely that such approach will not spark students' interest and will not allow controlling the process of learning of the very material. Moreover, static resources fully entrust a student with organization of the process of learning of the provided material, which can lead to deterioration in information perception due to the possible failure to understand correct process of material learning. Better results can be achieved in the event if a lecturer to the fullest extent possible determines the process of learning for the student, whether it is a full-time study or further reinforcement of knowledge [1].

2 Education Course

Such opportunity in terms of e-smart learning can be implemented through the interactive reading material allowing breaking up lecture into separate parts with assessment of knowledge and limiting access for the learning of more complicated material till the essentials are not learnt (Fig. 1).

For example, the distance learning system Moodle this possibility is realized with the help of "Lecture" element, which is one of the tools of the main component of the system "training course". The course is a basic component used in the learning management system [2], under which it is possible to organize:

V.L. Uskov et al. (eds.), *Smart Education and e-Learning 2017*, Smart Innovation, Systems and Technologies 75, DOI 10.1007/978-3-319-59451-4_30

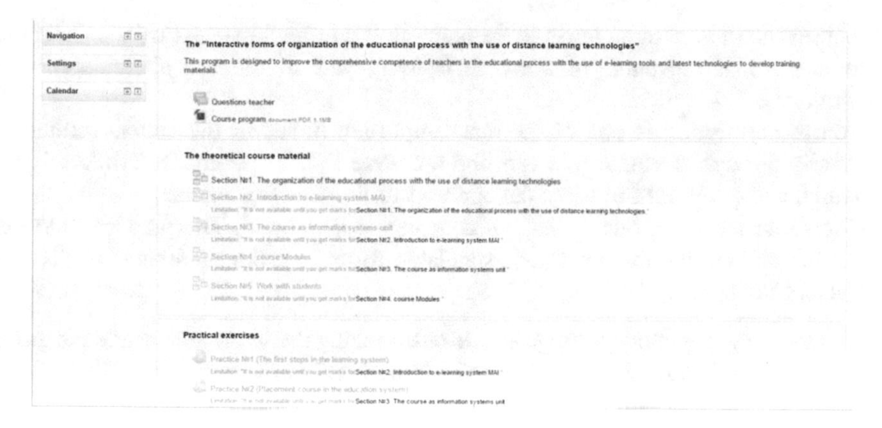

Fig. 1. Restricted access to training materials.

- Interaction between the students and feedback with a lecturer;
- Wide availability through the transfer of knowledge in electronic form;
- Individual and team work;
- Control over the gained knowledge.

First of all, it is necessary to form course structure and configure its main settings. Each training course has several mandatory parameters used to find courses by keywords and brief description (Fig. 2). Key parameters of the training courses' configuration and their descriptions are shown below:

- Category is location of a course in the training system. This parameter should be filled in on the basis of common rules for the location of courses in the specific system of distance learning;
- Full name of a course is the name that will be displayed on the user's general list of courses. As a rule, the name of a course corresponds to the area of scientific knowledge;

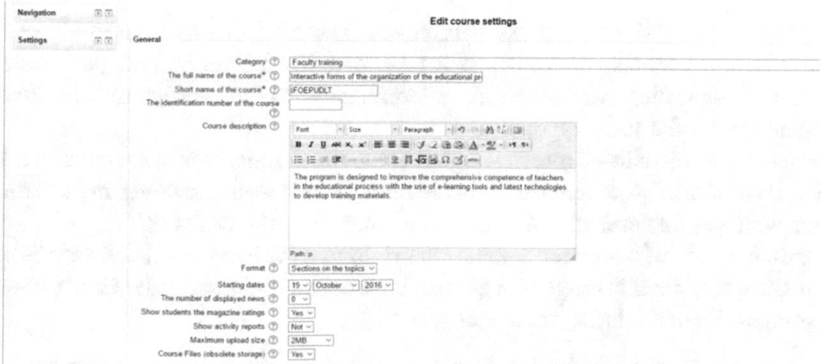

Fig. 2. Main parameters for course configuration

- Abbreviated name of a course is the name that will be displayed on the navigation toolbar, so, if the name of a course is long, you should use abbreviations and acronyms;
- Course description is one of the most important boxes in the course settings, as exactly through it search tool can find a course in the system. Description is also available for the outside users (as opposed to the course content);
- Course format – various formats of courses are supported. The practice of system use has shown that the most understandable format for the students is Sections by Subject Matters.

Tab for the group mode settings allows customizing the work with academic groups within a course. Group Mode parameter has three options:

- No groups – all participants are members of one big community;
- Isolated groups – members of each group work only within their group, other groups are not visible for them;
- Visible groups – members of each group work only within their group, but can see other groups.

Group mode, determined at the level of a course, is the default mode for all elements created within a course. You can specify individual group mode for each element that supports group mode. If a despotic group mode is set for a course, group mode settings for each element of a course are skipped.

3 Roles in Course

In addition to configuration of the very course, it is necessary to establish access procedure. For the purpose of access sharing, roles are used in Moodle distance learning system. Role is a set of pre-defined rights that are set for specific user and allow limiting his/her user options. Each role within the system has certain context constituting a page where the user can perform certain actions [3].

In particular, a training course can be a context. The main roles within the course are as follows:

- Student is the role of a course participant. The student can review all available (unhidden) materials, perform tasks, take part in recitations and pass tests. All student's activities are assessed automatically or manually by the lecturer, depending on the tasks type;
- Lecturer is a role allowing reviewing all course materials, assess students' work and give them marks. A lecturer can also review formed tests, however marks from the user with the Lecturer role will not be displayed in the register;
- Expert is a role of a lecturer who has the right to edit. Expert is a course editor who can change general settings of a course, create, edit and delete elements (modules): resources, lectures, tasks, tests. etc.

Hence, you must have the Expert role to create, edit and update information of a training course. But to give lessons according to the already developed training course and prepared materials, it is sufficient to have the role of a lecturer. Furthermore, there is a possibility to appoint an unlimited number of experts for a course which provides for joint creation of a smart learning content.

Users having the roles of Lecturer and Expert within a course are displayed when searching a course and on the My Courses section on the main page. If necessary, it is possible to change names of the roles within a course. The corresponding section of course settings (Roles Renaming) can be used for this purpose.

Course expert can enroll users in a course and assign specific roles to them (Student as a default). To enroll the user in the course, it is necessary to go on the Enrolled Users page, find the user by surname and enroll him/her in a course, indicating the role.

4 Lecture Elements

When a course has been completely configured, you can start to fill it. Reading materials that students should learn during the course of training is a basis of a training course. The most convenient option for placing materials is built-in format Lecture. Actually, lecture is a set of pages (logically complete semantic content fragments) with intermediate questions and complicated transition logic for organization of the students' logic of work with learning material. There are two main types of pages in the module:

- Rubricator Card (Section) – the page containing material and button(-s) for unconditional transitions to other pages of a lecture;
- Question – the page containing a question, options of answers, comments for the options of answers and transitions for each option of an answer.

In addition to the main page types, there are special pages that do not contain material or questions and serve to manage a lecture:

- Cluster Heading, End of Cluster – clusters, designed to cluster pages with questions in logical group. Cluster begins with the cluster heading and ends either with the end of the cluster or, if it is not defined, with the end of a lecture. In most cases, cluster is used to choose random questions from it.
- End of Section – section starts with the rubricator card and ends with the end of section or, if it is not set, with the end of a lecture. Sections unite any pages (both with questions and material). The following special transitions can be made within the sections: unanswered question from the section, random question from the section, random rubricator card.

Transition is the next important notion of the Lecture activity element. This notion is available almost on all pages of a lecture, both main and special. The transition determines which page will be displayed next for the student. Unconditional transitions are carried out by all pages except for the pages with questions. Conditional transitions are carried out by the pages with questions.

You have to create the lecture element in the course and configure it to work with it. For the lecture, you have to enter the name, which will be displayed in the course materials. You can limit the lecture in minutes. In this case, the student will see the timer and each lecture view will be time-limited.

You must specify the maximum number of answers/transitions in the general settings. By default, the lectures are set to 4. It means that after pages with the text and in questions you can create no more than 4 transitions.

In addition, it is possible to protect the lecture with a password. That is the students, who do not know the password, cannot complete this material. Such possibility is almost not used in practice, since there are other options for determining the logic of studying the course materials.

The next group of parameters defines how the lecture will be graded. It is possible to create a training lecture. In such lecture, students will be able to correct their answers, therefore, it won't be displayed in the records of grades.

Points for each answer – it allows defining the grade for each answer in a question. Answers can have negative or positive grade values. For the imported questions, 1 will be automatically set for the correct answers and 0 – for wrong, but it can be changed after import.

It is also possible to configure the additional element parameters that allow defining the lecture progress:

- To allow students to take the course again that is to allow or prohibit students entering to the lecture again;
- To provide an opportunity to answer a question again – if this parameter is enabled, the student will be asked after each wrong answer either to try to answer the question again (already without getting points) or just to continue the lecture;
- Maximum number of attempts – this parameter determines the maximum allowed number of attempts to answer each question. If the answer is repeatedly wrong and the maximum is reached, the next page of the lecture is displayed;
- Action after the correct answer – after the correct answer to the question there are 3 options when proceeding to the next page: ordinary – according to the lecture path; show not viewed pages – pages are displayed at random, previously displayed pages are not displayed; show the unanswered questions – pages are displayed at random, the pages containing the unanswered questions are displayed repeatedly. In practice, it is better to set transition parameters in logic of the lecture;
- Display the comment by default – if the parameter is enabled, when choosing the specific answer (if there is no comment to it), the comment by default will be displayed – "It is the correct answer" or "It is the wrong answer". In practice, it is better to make comments when describing the logic of the lecture;
- Display the progress indicator – if this parameter is enabled, the progress indicator, showing the approximate percent of the lecture completion, will be displayed at the bottom of the lecture pages;
- Display the lecture menu – if this parameter is enabled, the lecture menu (the list of pages) will be displayed on the block at the left. When you enable this parameter, it is recommended that to set the next parameter value of more than 50%;
- Display the menu at the left, only if grade is bigger than – this parameter defines whether the student should get a certain grade to see the lecture menu. This parameter can be used to ensure that the student has passed the lecture at the first

attempt, and then, after getting the necessary grade, he could use the menu at repeated viewing and look through material in the convenient sequence;

- Number of the displayed pages – this parameter defines the number of pages displayed in the lecture. It is only applicable for lectures with the pages displayed at random (when the "Action after the correct answer" parameter is set to "Show the not viewed page" or "Show the page with the missed answer"). If parameter is equal to zero, then all pages are displayed;
- Transition to the course element – in order to offer the student a transition to another element of this course after the lecture being completed, it is necessary to choose this course element in this list.

After the lecture is created and configured, the system will redirect the user to the page of editing the lecture logic. Since no component of the lecture has been created yet, the page displays the suggestions, in the form of links, to import questions, add the information page, cluster and page with questions.

Before creating the lecture, it is necessary to think over logic of the student's movement on a training material. In this module, you can configure any logic of the movement. The simplest option is implementation of the linear logic of mastering the training material. Pages with educational information will alternate with questions, which, in case of the wrong answer, will return the student to reading of the corresponding page, and pass to the subsequent materials of the lecture, if the answer is right. For a start, questions will not be generated at random, and will be strict and predetermine [4].

For the page with material, it is necessary to enter its heading – it will be displayed at the top of the screen when viewing the lecture. Contents of the lecture are entered into the Page Contents field. After the main contents, it is possible to set the transit buttons arrangement (horizontally or vertically) and turn on or off the display of this page in the lecture menu (if corresponding option is enabled in the lecture settings).

You must use the groups of Contents 1, Contents 2 and so forth to set the transition logic. Each group allows creating one transit button with the text, which will be entered into the Description field, and the corresponding transition. Next Page transition should be chosen for the linear logic of the lecture.

In the same way, you can add the other pages with the lecture contents. After that, you can begin adding questions, which will determine the logic of the students' movement on the created lecture [5].

The system supports several types of questions. Multiple choice questions (testing) are most often used. Such type of a question in system is called Multiple Choice.

For a multiple choice question, it is necessary to enter a heading – it will be displayed at the top of the page when viewing by students. The text of a question is entered into the Page Contents field. If the question has more than one answer (it is necessary to select several answers at once), then it is necessary to select the Multiple Answer check box.

Answers are entered into the corresponding groups. It is necessary to enter the correct answer, for which you must set Next Page in Transition field. Points for the correct answer must be greater than zero and established according to the chosen grading system.

For the wrong answers, the transition is set to Previous Page, and points for the answer are set to 0. Thus, the linear logic of passing the lecture is implemented.

Another option for forming the lecture logic is adding a cluster of questions with the random choice. If you pay attention to the previous logic of study, you can note that it has a major shortcoming, as each student always gets the same set of questions. Control elements – the beginning and the end of a cluster, were added to the lecture module to avoid this shortcoming. They allow organizing the random choice from a set of questions on a specific algorithm upon transition between information pages [6].

Just enter Add a Cluster before the first question, and add End of Cluster after the last question to form a cluster of questions (Fig. 3). The system will automatically set up a cluster to display not viewed question in it. It is necessary to change transitions in questions in order that the student can see only one random question. They must switch the user to the next section or the end of a lecture, if the answer is correct, and return to the information card in front of the cluster, if the answer is wrong.

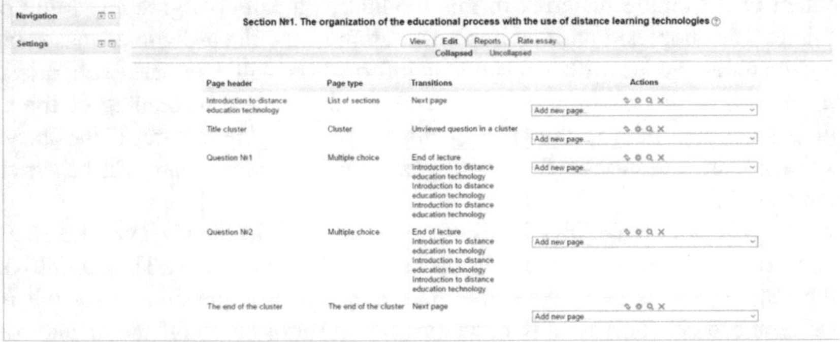

Fig. 3. Use of the questions cluster

The lecture module allows creating even more difficult logic of studying the material. After preparing the lecture, it needs to be tested from the student role, the View tab is used for this purpose. Typical errors: current page transition, which does not allow moving on, formation of the cluster without questions (an error is given, since the system does not understand what question is to be presented to the student), the transition logic does not allow passing on (return in any case) [7].

5 Conclusion

Thus, Moodle distance learning system allows implementing Smart learning of students in the interactive educational environment using electronic educational content regardless of the study mode: full-time, distance or mixed. By means of the main component of the system – a training course – and the lecture element, it is possible to organize the independent flexible study of materials by students both as the main occupation, and for additional consolidation of the material provided by the teacher.

References

1. Siluyanov, A.V., Khoroshko, L.L.: e-Learn expo Moscow 2010. In: Proceedings of 7th Moscow International Exhibition and Conference on e-Learning. IEC "Expocentre", Moscow (2010)
2. Ukhov, P.A., Kostykova, O.S., Khoroshko, L.L.: Using distance learning technologies for training young professionals. In: Proceedings of Moscow International Conference RosNOU, Moscow (2011)
3. Official site of moodle community (2017). http://www.moodle.org. Accessed 20 Jan 2017
4. Khoroshko, L.L., Sukhova, T.S.: The use of virtual and remote laboratory exercises in engineering education with e-Learning. In: Proceedings of 15th International Conference on Interactive Collaborative Learning and 41st International Conference on Engineering Pedagogy, Austria, Villach. IEEE Catalog Number: CFP1223R-USB (2012). ISBN 978-1-4673-24267
5. Andreev, A.V., Andreeva, C.V., Dotsenko, I.B.: Practice of e-learning using Moodle, 146 p. Publishing House TTI SFD, Taganrog (2008)
6. Pastuscha, T.N., Sokolov, S.S., Ryabova, A.A.: Creating e-learning course. Lection in SDL MOODLE: Teaching Aid, 44 p. SPSUWC, Saint Petersburg (2012)
7. Ukhov, P.A., Lomakin, A.L.: Distant Learning Technology in College: Monography. Moscow Humanistic Technical Academy, 180 p. Publishing House of MHTA, Mumbai (2010)

Analysis of Big Data in Colonoscopy to Determine Whether Inverval Lesions are De Novo, Missed or Incompletely Removed

Piet C. de Groen[1(✉)], Wallapak Tavanapong[2], JungHwan Oh[3], and Johnny Wong[2]

[1] University of Minnesota, Minneapolis, MN, USA
degroen@umn.edu
[2] Iowa State University, Ames, IA, USA
[3] University of North Texas, Denton, TX, USA

Abstract. We have created a system that automatically records the inside-the-patient images of each colonoscopy in de-identified fashion. At present this "big data" database contains around 100 TB of de-identified endoscopy data. Interval colorectal cancers (CRCs) are CRCs that develop despite periodic colonoscopy and are due to de novo tumor growth, a missed lesion or incomplete lesion removal. Using a combination of location, date, time and image information we were able to find a video file within our de-identified big data from a prior colonoscopy that belonged to a patient with a recently diagnosed large interval lesion. Analysis of the video file showed that a large lesion was incompletely removed. Analysis of big endoscopy datasets has the potential to resolve the cause of most if not all interval lesions and CRCs and can provide specific, focused education to endoscopists related to their individual limitations.

Keywords: Video stream analysis · Quality features · Colonoscopy · Interval colorectal cancer · Education

1 Introduction

Colorectal cancer (CRC) is a preventable cancer that is diagnosed in around 135–140,000 people each year in the US. Despite the fact that it can be prevented, about 45,000 patients die from CRC annually. [1] There is wide-spread consensus that the current preventive strategies in place in the US should drastically reduce the incidence and mortality of CRC, yet for a multitude of reasons this has yet to happen. [2, 3] Of all the methods to prevent death from CRC, colonoscopy holds most promise; it is a technique that allows detailed inspection of the entire colon and at the same time removal of all premalignant lesions. [4] The latter is commonly performed during the

Piet de Groen has a financial interest in EndoMetric INC, a company that analyzes colonoscopy video streams for features of quality. Wallapak Tavanapong, JungHwan Oh and Johnny Wong have a financial interest, hold an equity position and hold a position in EndoMetric INC.

V.L. Uskov et al. (eds.), *Smart Education and e-Learning 2017*, Smart Innovation, Systems and Technologies 75, DOI 10.1007/978-3-319-59451-4_31

withdrawal phase of the procedure. Colonoscopy is also readily available in most geographical areas of the US with wide-spread coverage of the procedure by payors.

The main problem with colonoscopy is the relatively limited CRC protective effect it currently provides. In theory, most CRCs should be preventable; interval CRCs are defined as "colorectal cancer diagnosed after a screening or surveillance exam in which no cancer is detected and before the date of the next recommended exam." [5] Several studies have shown limited or even total absence of a protective effect (i.e., in the right colon) against mortality of CRC, especially outside carefully controlled trials. [2, 3] More recent studies have shown a definite protective effect, in particular for CRC of the left colon. [6, 7] CRC of the right colon appears to be more difficult to prevent and numerous explanations for the relative failure of colonoscopy have been proposed. In general, the explanations for interval CRCs can be divided into two sets. One set focuses on patient and biology related factors; these include a poor preparation, an inability of the patient to cooperate during the procedure, an abnormal anatomy, flat polyp morphology or an unfavorable polyp or tumor biology. Indeed, these factors all may be present more in right-sided CRC: frequently bile and small bowel content covers the right colon, the deep folds of the right colon make inspection difficult, and flat, more rapidly progressing tumor biology (CIMP pathway) is much more likely in neoplasia of the right colon. The other set of explanations focuses on procedure and endoscopist related factors: suboptimal equipment, no removal of remaining debris, not reaching the cecum, fast withdrawal, no effort at inspection of areas behind folds and angulations, and inadequate polyp removal technique.

A key study supporting this concept was published in 2014 and shows that for every 1% increase in adenoma detection rate (ADR), there was a 3% decrease in interval CRCs; the lowest interval CRC rate was observed among endoscopists with an ADR > 33.5%. [8] Proponents of the first set of factors may point to the patient responsibility for a clean colon, the type of preparation and patient compliance, outline the benefits of propofol sedation and believe that interval CRC is a result of rapid growth. Proponents of the second set of factors are of the opinion that gastroenterology-trained endoscopists provide better quality than other endoscopists, believe that removal of debris, complete inspection and total removal of all neoplasia can be achieved and should lead to nearly complete protection against CRC if screening and surveillance guidelines are followed; interval cancers are considered a result of missed lesions (polyp or small cancer) or incomplete resection of identified lesions at prior colonoscopy. To conclude, interval CRCs therefore are either the result of (1) de novo rapid growing tumors, (2) missed lesions that were already or developed into CRC, or (3) a pre-existing lesions that were not completely removed and developed into a CRC.

We are in the process of developing a system that automatically will link all video files of endoscopic procedures of an individual patient over time. Such system would allow both manual and automated assessment of quality of mucosal inspection and completeness of removal of all identified lesions by comparison of subsequent colonoscopies. Indeed, such a system would allow in almost all cases a definitive explanation of the cause for interval CRCs as either caused by a de novo, a missed or an incompletely removed lesion. Here we show that this is possible using manual analysis of a very large dataset of de-identified video files for two subsequent video files of a patient with a large interval lesion. The implications of such a smart system are numerous; most importantly, based on the objective data that can be derived from subsequent colonoscopy video files we will be able to propose specific education to endoscopists related to their individual limitations.

2 Related Work

2.1 Indicators of Endoscopic Quality and Methods of Feedback to Endoscopists

Currently there are three somewhat objective measures that report on the combination of endoscopic behavior and skill set: cecal intubation rate, withdrawal time and ADR. [9] Most endoscopists achieve a high cecal intubation rate, i.e., they traverse the entire colon from anus till cecum in >90% of patients, therefore it has little discriminatory value. Withdrawal time has lost some of its luster yet the fact that time, i.e. 6–10 min during withdrawal from cecum to anus, was spent remains important; the problem is that it does not provide information about what was done. That explains why ADR currently is the single, dominant determinant of quality of colonoscopy [8, 10].

The problem is that these three indicators of colonoscopy quality are averages and provide no information about a single procedure. [10] Instead, these quality parameters are summary data that reflect a group of procedures performed by an individual endoscopist or a group of endoscopists over a specific time period. An inherent feature of summary data is that a few really poor procedures combined with a larger set of higher quality procedures will result in acceptable overall quality scores. A fourth indicator of quality that is the most important one, interval CRC, is difficult to measure as it requires careful follow up of patients after colonoscopy for many years in order to determine whether the patient develops CRC or not. [5] Another parameter of quality is the absence of complications – these too are more difficult to determine as these may occur hours (i.e., evidence of perforation) or days and weeks (i.e., a polypectomy related bleeding) after the procedure, are not universally preventable, and may or may not be related to the endoscopic behavior or skill set of the endoscopist.

Therefore, most feedback to endoscopists consists of relatively easy to measure parameters that can be measured during endoscopy or within a period of hours or days after endoscopy: cecal intubation rate, withdrawal time, ADR, immediate complications [10].

2.2 Continuing Medical Education Related to Endoscopy

It is mandatory for all US physicians, who obtained board certification after the early 1990s, to participate in Continuing Medical Education (CME), repeat certifying examinations (e.g., every 10 years for Gastroenterology) and recently Maintenance of Certification (MOC). Yet all of these programs are theoretical: they test knowledge and the theory of what should happen during procedures. Indeed, once a gastroenterologist finishes formal training and enters clinical practice, there is not any expectation of further hands-on training related to existing or new procedures as those develop in the next 30–35 years of practice. Indeed, the absence of any form of intermittent testing of practical skills or ability to learn new skills is a recognized, growing problem [11].

2.3 Endoscopic Multimedia Information System (EMIS)

Intuitively it does not make sense to set as goal a specific withdrawal time (several specific times have been proposed) or a specific number of cases in which polyps should be detected. Instead, it would make sense to measure features that directly define quality of each procedure.

Since 2003 our group has worked on creating an automated system to capture, analyze and summarize video files representing an entire endoscopic procedure. [12] We have called our system EMIS for Endoscopic Multimedia Information System. We have focused our efforts on colonoscopy. Our work has shown that our manual EMIS annotation technique is reproducible among annotators with fair to good inter-operator agreement; inter-operator agreement is best for very low and very high quality procedures, but varies when quality is average. Our automated EMIS technology results correlate with our manual annotation results, and both manual and automated annotations correlate with ADR – the most widely accepted main determinant of colonoscopy quality – for a set of video files representing the work of a single endoscopist or an endoscopy group. [13]

2.4 Dedicated Video Browser

In order to rapidly review EMIS-derived, MPEG2 formatted video files, we developed a dedicated video browser that allows fast video file scrolling at 2- to 32-fold speed. [14] This is critical when dealing with a very large video file corpus and performing manual searches for scenes of interest.

3 Methods

3.1 Test Case Identification

A trainee informed the authors about a large polyp that appeared to have been missed during a recent colonoscopy and was discovered during a follow up colonoscopy in a teaching room. At the request of the authors, the trainee provided the dates of both procedures (summer and fall of 2015), and the endoscopy location where the procedures were done – no patient or endoscopist information was provided; however, each time different staff and trainee endoscopists were involved (i.e., 4 endoscopists were involved) in the two procedures. Lastly, de-identified versions of the endoscopy and pathology reports were provided as well.

3.2 Identification of Video Files

First we manually examined all video files present in the big dataset of about 100 TB containing approximately 100,000 endoscopy video files from multiple institutions for video files with a date stamp and endoscopy location identical to that in the report of the second colonoscopy; eventually, we identified a single colonoscopy that contained a large polyp at the reported anatomical location. Next we examined the big data set for a video file with a date stamp and endoscopy location identical to the first date. We were unable to find a video file that matched the endoscopy report, but did find a video file that showed a polyp nearly twice the size in the same location as in the second colonoscopy.

4 Results

We found two video files that provided details of two colonoscopies performed on the same patient. Figure 1 shows the timeline and data we identified. Both video files were carefully examined for polyps, polypectomy attempts and visible completeness of polypectomy or lack thereof. All data were de-identified for patient and endoscopists.

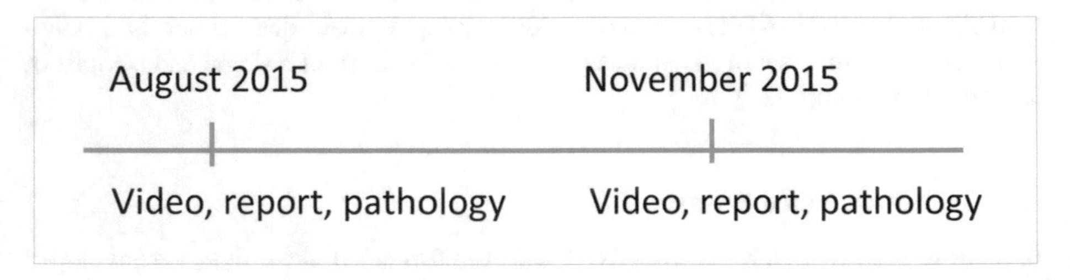

Fig. 1. Timeline of patient with interval lesion

4.1 First Colonoscopy

In the first colonoscopy, a large polyp was only partially removed just distal of the hepatic flexure. After removal of part of the polyp, the endoscopists retrieved the polyp specimen by removing the endoscope with the specimen hanging onto the tip. The endoscopists then reinserted the endoscope but never far enough to reach the location of the remainder of the polyp. In fact, it appears that the endoscopists were somewhat lost in a very long and redundant colon. The formal endoscopy report erroneously reads that a large polyp was completely removed from the splenic flexure whereas in reality a partial polypectomy was done closer to the hepatic flexure.

4.2 Second Colonoscopy

In the second colonoscopy, an asymmetric polyp is found in the area just distal of the hepatic flexure; the aspect of the polyp facing the camera appears scarred from a previous attempt at polypectomy. The rest of the polyp is more spherical, typical for the shape of an untouched polyp. The endoscopists remove this polyp fragment in piecemeal fashion and by the end of the polypectomy most of the polyp appears to be removed. As during the first colonoscopy, the endoscopists remove the main parts of the specimen by removing the endoscope with the specimen hanging onto the tip. Upon reinsertion of the endoscope the second two endoscopists also appear lost in this long colon. However, after several back and forth movements through the entire left colon, these endoscopists eventually find the polypectomy site just distal of the hepatic flexure in the right colon, confirm near complete or complete resection, and mark the location with a tattoo.

4.3 Comparison of Colonoscopies

The video images strongly suggest that the two colonoscopies represent the same patient. First, the colon is exceptionally long; this appears to confuse all four endoscopists as all appear to have great difficulty in returning to the original polypectomy site. In fact, only the second two endoscopists succeed in finding the original polypectomy site back.

Second, the location of both polyps in the two procedures is identical: just distal of the hepatic flexure in a very long colon with numerous curves and a very redundant transverse colon.

Third, the second polyp looks like the remnants of the first polyp after partial polypectomy (see Fig. 2). Indeed, the position on the colon wall, the round features remaining on the uncut part of the polyp and the size of the remaining part in the second colonoscopy all strongly suggest that there only was one polyp that was not completely removed during the first colonoscopy. Furthermore, the cut surface of the polyp in the second colonoscopy is along the plane of the snare placed in the first colonoscopy.

Fourth, careful examination of the anatomy immediately adjacent to the polyp reveals similar anatomical structures as shown in Fig. 3.

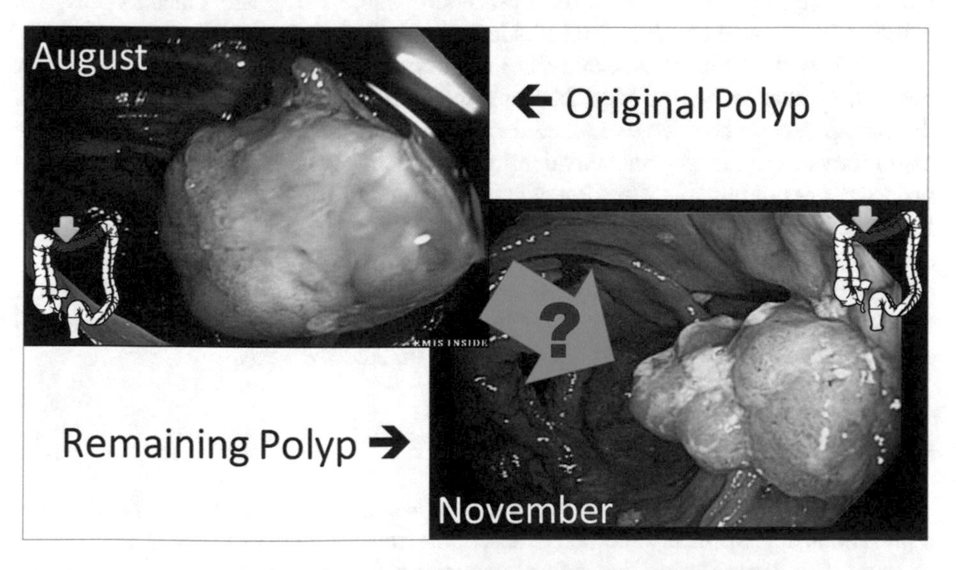

Fig. 2. Representative images from the same polyp at different dates; the colon schema shows in red the colon segment where the polyp is located. The green arrow shows the approximate location within the colon segment where the polyp was found

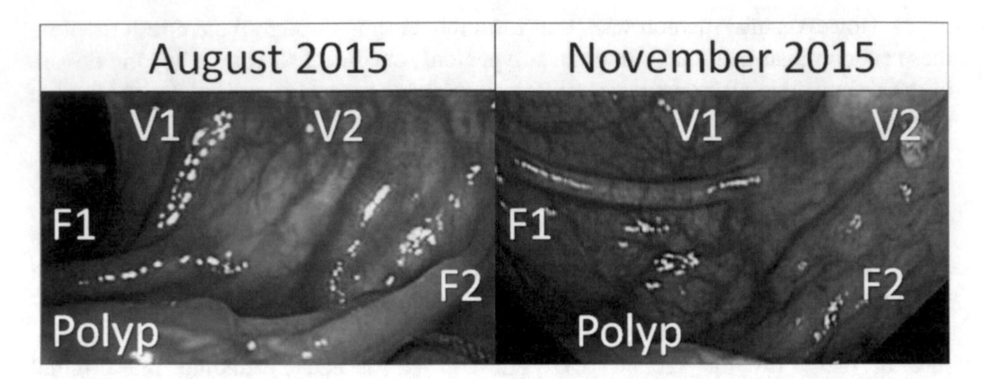

Fig. 3. Peri-polyp anatomy. V1 = Vein 1, V2 = Vein 2, F1 = Fold 1, F2 = Fold 2. The polyp is located on F2

Fifth, the pathology report of both polyp specimens is identical: tubulovillous polyp with low grade dysplasia.

5 Educational Analysis

Colonoscopy is a very operator-dependent procedure; this explains the wide variations in withdrawal time and adenoma detection rate. Indeed, there is a continuous search for better parameters of quality as just spending more time inside the colon does not equate to better quality. Similarly, as ADR is defined by the frequency where partial or complete removal of at least one polyp occurs during a single colonoscopy, it does not really define quality: the ADR is equal for a procedure where 1 of 1 adenomatous polyps is completely removed or where 1 of 10 adenomatous polyps is partially removed.

The best determinant of quality is the interval CRC rate: it reflects true cancer prevention. Ideally, we provide feedback to an endoscopist when a CRC develops during the interval between two scheduled colonoscopies. However, until now we only could provide the fact that an interval lesion was found; providing information why this occurred was not possible. Creation of a de-identified big data archive with anonymous linking of video file, endoscopy report and pathology data using a key representing a specific patient will allow specific interval CRC feedback to endoscopists by providing the endoscopist the opportunity to review his/her own colonoscopy that did not prevent the interval cancer. This is truly smart and individualized learning as only those procedures that lacked the required quality can be marked for study.

6 Discussion

We created a very large endoscopy video file dataset: a big data repository of endoscopy representing approximately 100,000 endoscopies from multiple institutions and 100 TB of data. This large dataset has served numerous functions including gold standard baseline for algorithm development, data source for chronological quality studies and the data source for studies of variation of individual endoscopist performance. Here we tested using a manual method whether we can use endoscopic big data to determine the reason for interval CRCs and found that this indeed is possible. Even better, comparison of two video files of procedures performed on the same patient revealed that several errors were made during the first procedure: the polyp was not completely removed, the location of the polyp was incorrectly reported, and the endoscopists appeared lost in a very long colon. Yet despite all these errors, we were able to conclude with great certainty that this interval lesion was the result of an incompletely removed polyp. This case supports the current opinion that most interval lesions are due to endoscopist error, not tumor biology.

Although this first report is based on manual analysis of our big data, we know we can completely automate the identification of multiple endoscopic procedures performed on a single patient, perform extensive analysis of each video file for degree of inspection and completeness of polypectomy, and link all video files with all endoscopy and pathology reports. Automated analysis of video information and linking this to other, structured and unstructured information for automated tracking of quality currently does not exist; yet such a system now is possible and would allow targeted feedback about procedural performance of individual endoscopists. Furthermore, by using a de-identified key that is automatically assigned to the video files and the patient data, we can perform long-term prospective studies without the need for formal patient consent. The latter is practically not possible given the large volume of endoscopic procedures (i.e., all) that need to be analyzed in order to find all interval lesions.

Finally, our method allows implementation of very specific, individualized, continued medical education. By comparison of video files, endoscopy reports and pathology results for individual patients prospectively over time, we can find patients where colonoscopy does not prevent large polyps or CRC. This will allow us to determine the reason for the interval lesions. For endoscopists without any interval lesions, no remedial education is required. For endoscopists with missed lesions, education will be focused on improving inspection. And for those with incompletely resected lesions, educational efforts will emphasize the need to perform complete polypectomies with clear tissue margins. Thus by combining big video data volume, compute-intense automated video file analysis, procedure reports and pathology results in an automated pipeline that links individual patient and endoscopist data, we can create the foundation for a truly smart learning and safety environment where endoscopists will automatically be informed about opportunities for improvement. Such a smart learning and safety environment should benefit patients, endoscopists, and the medical field in general.

References

1. Siegel, R.L., et al.: Cancer statistics, 2016. CA Cancer J. Clin. **66**, 7–30 (2016)
2. Brenner, H., et al.: Protection from right- and left-sided colorectal neoplasms after colonoscopy: population-based study. J. Natl. Cancer Inst. **102**, 89–95 (2010)
3. Baxter, N.N., et al.: Association of colonoscopy and death from colorectal cancer. Ann. Intern. Med. **150**, 1–8 (2009)

4. De Groen, P., et al.: Real-Time feedback during colonoscopy to improve quality: how often to improve inspection? In: Uskov, V.L., et al. (eds.) Smart Education and Smart e-Learning, vol. 41. pp. 501–512. Springer (2015)
5. Adler, J., Robertson, D.J.: Interval colorectal cancer after colonoscopy: exploring explanations and solutions. Am. J. Gastroenterol. **110**, 1657–1664 (2015). quiz 1665
6. Zauber, A.G., et al.: Colonoscopic polypectomy and long-term prevention of colorectal-cancer deaths. N. Engl. J. Med. **366**, 687–696 (2012)
7. Nishihara, R., et al.: Long-term colorectal-cancer incidence and mortality after lower endoscopy. N. Engl. J. Med. **369**, 1095–1105 (2013)
8. Corley, D.A., et al.: Adenoma detection rate and risk of colorectal cancer and death. New Engl. J. Med. **370**, 1298–1306 (2014)
9. De Groen, P.C.: Polyps, pain and propofol: is water exchange the panacea for all? Am. J. Gastroenterol. **112**, 578–580 (2017)
10. Rex, D.K., et al.: Quality indicators for colonoscopy. Am. J. Gastroenterol. **110**, 72–90 (2015)
11. Aggarwal, R. et al.: Simulation research in gastrointestinal and urologic care – challenges and opportunities: summary of a national institute of diabetes and digestive and kidney diseases and national institute of biomedical imaging and bioengineering workshop. Ann. Surg. (2017). In press
12. Oh, J., et al.: Measuring objective quality of colonoscopy. IEEE Trans. Biomed. Eng. **56**, 2190–2196 (2009)
13. Thackeray, E., et al.: Quality of colonoscopy depends on the quality of the endoscopist. Am. J. Gastroentero **106**, S520 (2011)
14. Liu, D., et al.: Arthemis: annotation software in an integrated capturing and analysis system for colonoscopy. Comput. Methods Programs Biomed. **88**, 152–163 (2007)

Smart Teaching

Smart Teacher

Blanka Klimova(✉)

University of Hradec Kralove, Rokitanskeho 62,
Hradec Kralove, Czech Republic
blanka.klimova@uhk.cz

Abstract. The arrival of modern information and communication technologies (ICT) and their widespread use in all spheres of human activities definitely has influenced teaching and learning processes, which is manifested in student centered learning, exploiting creative potential of a student, integration of different teaching plans, or diversity of teaching methods and strategies. Consequently, the traditional, in most cases the authoritative role of the teacher in the classroom is changing, too. Nevertheless, the teacher still has his/her role in student's learning. S/he is responsible for careful planning and running of the whole course, setting the learning aims and objectives, exploiting different teaching approaches and strategies, and determining whether and how to apply ICT in his/her teaching. Therefore the purpose of this article is to discuss a new, challenging role of the teacher in the new environment, the so-called smart learning environment enhanced by ICT. This is done on the basis of literature review of available studies on the research topic in the world's acknowledged databases Web of Science, Scopus, Springer, ScienceDirect and PubMed in the period of 2010 till October 2016. The results of this review study indicate that teachers have a positive attitude to the use of ICT in their teaching. Nevertheless, they also reveal that teachers are still not able to fully integrate ICT in their teaching. Therefore they should be made aware of the role of ICT in educational processes and provided with a chance to be trained in the new competences which are necessary for teaching in the new smart environment.

Keywords: Teacher · Tutor · ICT · Smart learning environment · Changes

1 Introduction

With the penetration of modern information and communication technologies (ICT) in education, the position of the teacher has significantly altered and s/he had to inevitably acquire new roles such as a role of tutor. There are many definitions of tutor. In Anglophone countries s/he is perceived as a person charged with the instruction and guidance of another as a private teacher or a teacher in a British university who gives individual instruction to undergraduates [1]. However, with the arrival of ICT and emergence of distance education, the word tutor acquired another new meaning. In this sense, the tutor is seen as an advisor, facilitator of student's learning in distance education, s/he individually works with his/her students, s/he is available as a consultant, evaluates his written assignments, tests, helps him/her in exam preparation, provides feedback and motivates him/her in his/her further studies [2]. In this study

V.L. Uskov et al. (eds.), *Smart Education and e-Learning 2017*, Smart Innovation, Systems and Technologies 75, DOI 10.1007/978-3-319-59451-4_32

tutor is understood as a smart teacher who in the new environment enhanced by ICT is responsible for the preparation, design and running of the whole course, as well as guiding and stimulating his/her students in their learning process. The new smart learning environment is then perceived as a technology-supported learning environment that can make adaptations and provide appropriate support (e.g., guidance, feedback, hints or tools) in the right places and at the right time based on individual learners' needs, which might be determined via analyzing their learning behaviors, performance and the online and real-world contexts in which they are situated [3]. In this smart environment teachers must foster collaboration and cooperation, identify and help struggling students, encourage students to learn, and provide feedback to develop students' confidence and satisfaction [4, 5].

2 Methods

The methodology of this study is based on [6]. The topic of the role of smart teacher/tutor in the new environment enhanced by ICT is based on literature search of available sources describing this issue in the world's acknowledged databases Web of Science, Scopus, Springer, ScienceDirect and PubMed in the period of 2010 till October 2016. The study was included if it was written in English; covered the designated period, focused on the research topic, i.e., *smart teacher and ICT*, or *smart tutor and ICT*, and if it was an original study, not the review or any other theoretical study. In addition, the author of this study used a method of comparison and evaluation of the findings in the selected studies on the researched topic. Furthermore, other relevant studies were reviewed on the basis of the reference lists of the research articles from the searched databases.

Most of the articles based on the selected keywords were found in the database ScienceDirect, their numbers are as follows:

- smart teacher and ICT (532)
- smart tutor and ICT (146)

As the numbers indicate, there are surprisingly not so many articles on the research topic. Nevertheless, their number is gradually rising as Fig. 1 below shows.

3 Findings

On the basis of literature review, altogether ten research studies were eventually included in this study. The study was involved if it focused on the selection period, i.e., from 2010 till October 2016. The reason is that the studies on the research topic started to occur after the year of 2010. All studies had to be written in English and they had to explore the issue of the teacher's role in the environment enhanced by ICT. The main outcome measures were mainly based on questionnaire surveys. The findings of the selected studies are summarized in Table 1 below. The studies are listed in the alphabetical order of their first author.

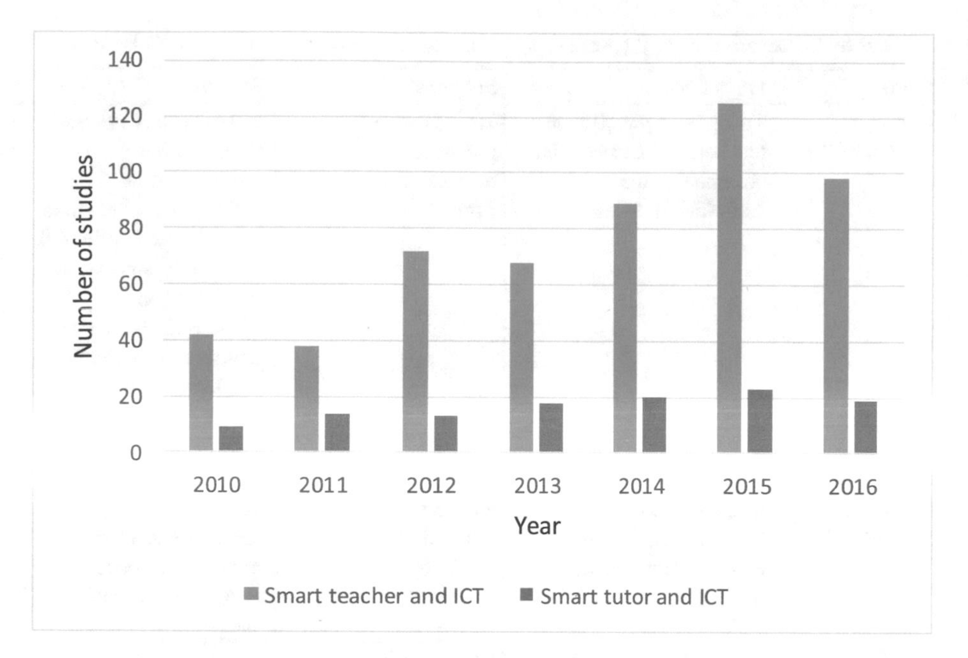

Fig. 1. Number of articles found on the research topic in the database ScienceDirect from 2010 till October 2016 (author's own processing based on ScienceDirect [7])

4 Discussion

The results of the studies presented in Table 1 indicate that teachers have a positive attitude to the use of ICT in their teaching [8–17]. Nevertheless, the findings also reveal that teachers are still not able to fully integrate ICT in their teaching [13, 15, 16]. Therefore they must be made aware of the role of ICT in educational processes. Gorghiu et al. [11] claims that the pedagogical use of the ICT leads to the easier understanding of the discipline content due to the using of ICT; the increasing of the quality of the learning process; the feeling that the use of ICT tools needs to be extended to the teaching of the whole palette of disciplines; and to the great attractiveness of the new teaching methods that combine ICT with traditional ones.

In addition, teachers in the new environment enhanced by ICT should possess advanced ICT competences [17]. Gorghiu et al. [10] suggest that teachers should be able to introduce specific discussion frames based on real reasons, reflective activities, guidance and explanations in order to promote computer supported collaborative learning. Nazarenko [14] adds critical thinking and processing information skills. Bednarikova [18] in her study on the role of tutor in distance education and in e-learning distinguishes the following competences:

- *Educational competence*, which presupposes expertise knowledge of teacher's professional field of interest, but also pedagogical-psychological and andragonic knowledge and skills. An inseparable part of the educational competence should also be communicative and reflective competence.

Table 1. Description of the teacher's role in the new environment enhanced by ICT

Study	Objective	Subjects	Findings
Demir, Yurdugul [8]	To explore prospective teachers' ICT usage and communication self-efficacy levels	1693 volunteer prospective teachers from Turkey	Male teachers possess higher online communication self-efficacy levels than female teachers; there is statistically significant positive strong relationship between prospective teachers' ICT usage and communication self-efficacy levels
Frydrychova Klimova, Poulova [9]	To examine the role of a teacher in the new teaching and learning environment enhanced by ICT	38 university teachers from different disciplines	Teachers acquire new roles in the new teaching and learning environment enhanced by ICT such as technical roles
Gorhiu et al. [10]	To explore interaction in a computer supported collaborative learning environment; to examine issues related to pedagogic use of ICT	57 participants from 15 European countries	The tutors are the key agents in delivering an on-line course and supporting the learners; they must possess relevant competences such as being able to introduce specific discussion frames based on real reasons, reflective activities, guidance and explanations in order to promote computer supported collaborative learning
Gorghiu et al. [11]	To discover opinions of both teachers and pupils about the use of ICT in the teaching and learning of different scientific topics	245 teachers; 2100 students from five European countries	Higher awareness of the importance of technology in teaching process; both teachers and pupils' answers proved that the pedagogical use of the ICT leads to: (a) the easier understanding of the discipline content due to the using of ICT;

(*continued*)

Table 1. (*continued*)

Study	Objective	Subjects	Findings
			(b) the increasing of the quality of the learning process; c) the feeling that the use of ICT tools needs to be extended to the teaching of the whole palette of disciplines; (d) the great attractiveness of the new teaching methods that combine ICT with traditional ones
Gorghiu et al. [12]	To improve teachers' skills and attitudes by developing and delivering a training program which is aimed at the integration of ICT tools into the educational process	500 in-service teachers from all the educational levels of Romanian schools	Positive responses from the teachers; in addition students' motivation was increased
Liu et al. [13]	To explore e-tutors' teaching readiness in distance learning	44 undergraduate students who worked as e-tutors	Job readiness was improved, however, the psychological readiness was low. Therefore it should be designed in the training workshop to support the e-tutors
Nazarenko [14]	To examine the role of teachers in blended learning; and students' feedback on experience in the blended learning (BL) environment	62 students	The findings show that competences such as ICT skills, critical thinking and processing information skills are necessary for BL; teachers must motivate their learners to use ICT; academic policy and administration should support the use of ICT in education
Sanchez et al. [15]	To investigate teachers' attitudes towards the use of ICT in the classroom	170 in-service teachers ranging from kindergartens to high schools	The results show that teachers' attitudes towards ICT are highly positive but the use of them in class is scarce and it is subjected to

(*continued*)

Table 1. (*continued*)

Study	Objective	Subjects	Findings
			innovative processes. New training is needed such as collaborative projects
Umar, Hassan [16]	To assess Malaysian teachers' levels of ICT integration and its impact on teaching and learning	2661 Malaysian teachers	The results indicate that the teachers' level of ICT integration is still at the low level although in general they responded that ICT positively affected their teaching practice and their students' learning. More training in the integration of ICT skills is required
Umar, Yusoff [17]	To investigate the levels of Malaysian teachers' ICT skills (basic, advanced, Internet skills for seeking and sharing information and Internet skills for communication	2661 Malaysian teachers	The findings show that the teachers have very good basic ICT skills as well as the Internet skills for seeking and sharing information and communication, however, they have only moderate advanced ICT skills. In addition, there is no correlation between the service years and the impact of ICT on teaching and students learning as well as teachers' computer experience and their teaching. Male teachers seem to use ICT in their classrooms more than their female colleagues

Source: author's own processing.

- *Counselling competence*, which involves tutor's ability to flexibly, operatively and constructively solve study problems of participants.
- *Managerial competence*, which is connected with directional and controlling activities of the tutor, as well as with his/her conceptual and evaluation activities.
- *Administrative competence*, which requires from the tutor to be reliable, systematic and consistent.

- *Technical competence*, which includes an ability to use didactic and communication technologies, including work on PC from the basic skills up to the running of an online course.

Probably the main constraint for the teacher who is not used to exploiting ICT in his/her teaching is the technical competence [9]. Therefore more attention should be paid to ample teacher trainings focused on the integration of ICT into educational processes [13–17]. Ruiz, Puig, and Soler [19] say that the teacher should be able to use ICT in the following areas of education: web pages, emails, videoconferences, telephone, chats, news, cell phone messages, and software. In fact, it is a must since students are now digital natives who want to interact (communicate, share, exchange), create, meet, coordinate, evaluate, learn, search, analyse, report, socialize and evolve differently [20–23]. As the results of this review study show, male teachers are more efficient in using ICT in their teaching than their female colleagues [8, 17].

Thus, in this new environment the teachers should act as experts who facilitate students' learning processes and create favourable conditions and stimulating opportunities to unlock students' potential [24, 25]. Jorda [26] states that teacher/tutor activity focuses on favouring a permanent maturation process which enables students succeed in acquiring and processing correct information about themselves and their environment, within intentional proposals of reasoned decision making. And the academic policies should support them in this process [14].

Future research should further explore the role of the teacher/tutor in the new smart environment and concentrate on the efficacy of teaching with the use of ICT in meeting students' needs. In addition, more research should be conducted in the area of teacher's competences in the educational process, both at national and international level. The researchers should also strive for larger sample sizes to be able to make objective generalizations. Moreover, a variety of methods examining these issues should be applied, apart from the questionnaire surveys.

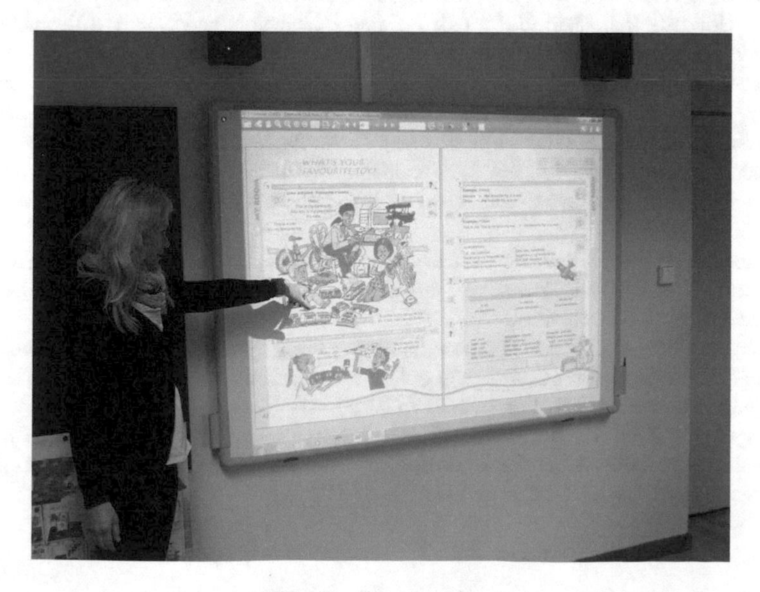

Fig. 2. Smart teacher

5 Conclusion

The traditional role of the teacher has radically changed in the new smart environment; smart teacher now has many more competences which require him/her to flexibly react to students' learning needs and purposefully run the whole learning process with the help of ICT (Fig. 2). In addition, as Sulcic and Sulcic [27] state, *it is necessary to choose individuals with a set of existing skills that can be developed through careful prepared training programs. Only well-trained tutors will be able to satisfy student expectation about the quantity, frequency and quality of learning supporting activities.*

The results of this review study indicate that teachers have a positive attitude to the use of ICT in their teaching. Nevertheless, they also reveal that teachers are still not able to fully integrate ICT in their teaching. Therefore they should be made aware of the role of ICT in educational processes and provided with a chance to be trained in the new competences which are necessary for teaching in the new smart environment.

Acknowledgments. The paper is supported by the SPEV project (2017) at the Faculty of Informatics and Management of the University of Hradec Kralove, Czech Republic. The author also thanks Jiří Forman for his help with data collection.

References

1. Merriam Webster Dictionary (2015). http://www.merriam-webster.com/dictionary/tutor
2. Palan, Z.: Vykladovy slovnik vzdelavani dospelych. [Expository Dictionary of Adult Education]. DAHA, Praha (1997)
3. Lin, Y.C., Liu, T.C., Kinshuk: Research on teachers' needs when using e-textbook in teaching. Smart Learn. Environ. **2**, 1 (2015)
4. Klimova, B.: Assessment in smart learning environment – a case study approach. In: Uskov, V.L., Howlett, R.J., Jain, L.C. (eds.) Smart Education and Smart e-Learning. Smart Innovation, Systems and Technologies, vol. 41, pp. 15–24 (2015)
5. Klimova, B., Simonova, I.: Study materials in smart leasing environment – a comparative study. In: Uskov, V.L., Howlett, R.J., Jain, L.C. (eds.) Smart Education and Smart e-Learning. Smart Innovation, Systems and Technologies, vol. 41, pp. 81–92 (2015)
6. Moher, D., Liberati, A., Tetzlaff, J., Altman, D.G.: The PRISMA group. Preferred reporting items for systematic review and meta-analysis: the PRISMA statement. PLoS Med. **6**, e1000097 (2009)
7. ScienceDirect (2016). http://www.sciencedirect.com/science?_ob=ArticleListURL&_method=list&_ArticleListID=-956424470&_sort=r&_st=13&view=c&md5=519dafd1af9ed84a6cb84372debb9604&searchtype=a
8. Demir, O., Yurdugul, H.: The examination of prospective teachers' information and communication technology usage and online communication self-efficacy levels in Turkey. Procedia Soc. Behav. Sci. **176**, 371–377 (2015)
9. Frydrychova Klimova, B., Poulova, P.: Theory versus field experience. In: Recent Researches in Educational Technologies, pp. 68–73. World Scientific and Engineering Academy and Society, Athens (2011)

10. Gorhiu, G., Lindfors, E., Gorghiu, L.M., Hamalainen, T.: Acting as tutors in the ECSUT on-line course – how to promote interaction in a computer supported collaborative learning environment? Procedia Comput. Sci. **3**, 579–583 (2011)
11. Gorghiu, L.M., Gorghiu, G., Dumbrescu, C., Olteanu, R.L., Glava, A.E.: Integrating ICT in traditional training-reactions of teachers and pupils involved in FISTE project activities. Procedia Soc. Behav. Sci. **30**, 1142–1146 (2011)
12. Gorghiu, G., Gorghiu, L.M., Brezeanu, I., Suduc, A.M., Bizoi, M.: Promoting the effective use of ICT in Romanian primary and secondary education – steps made in the frame of EDUTIC project. Procedia Soc. Behav. Sci. **46**, 4136–4140 (2012)
13. Liu, E.Z.F., Lin, C.H., Lin, Y.H.: E-tutors' teaching readiness in distance learning companion project in Taiwan. Procedia Soc. Behav. Sci. **176**, 386–389 (2015)
14. Nazarenko, A.L.: Blended learning vs traditional learning: what works? (A case study research). Procedia Soc. Behav. Sci. **200**, 77–82 (2015)
15. Sanchez, A.B., Marcos, J.J.M., Gonzalez, M., GuanLin, H.: In service teachers' attitudes towards the use of ICT in the classroom. Procedia Soc. Behav. Sci. **46**, 1358–1364 (2012)
16. Umar, I.N., Hassan, A.S.A.: Malaysian teachers' levels of ICT integration and its perceived impact on teaching and learning. Procedia Soc. Behav. Sci. **197**, 2015–2021 (2015)
17. Umar, I.N., Yusoff, M.T.M.: A study on Malaysian teachers' level of ICT skills and practices, and its impact on teaching and learning. Procedia Soc. Behav. Sci. **116**, 979–984 (2014)
18. Bednarikova, I.: Tutor a jeho role v distancnim vzdelavani a v e-learningu. [Tutor and his role in distance education and in e-learning]. Olomouc, PF (2013)
19. Sanchiz Ruiz, M.L., Martí Puig, M., Cremades Soler, I.: Orientación e intervención educativa, retos para los orientadores del Siglo XXI. Tirant Lo Blanch, Valencia, España (2011)
20. Prensky, M.: The Emerging Online Life of the Digital Native: What they do differently because of technology and how they do it (2004). http://www.bu.edu/ssw/files/pdf/Prensky-The_Emerging_Online_Life_of_the_Digital_Native-033.pdf
21. Reid, E.: Multimedia used in development of intercultural competences (ICC). In: CALL and Foreign Language Education: E-Textbook for Foreign Language Teachers, pp. 83–92. Nitra, UKF (2014)
22. Horvathova, B.: Implementing language learning strategies into a series of second foreign language learning textbooks. J. Lang. Cult. Educ. **2**, 60–94 (2014)
23. Vesela, K.: Teaching ESP in new Environments. Nitra, ASPA (2012)
24. Klimova, B.: Teacher's role in a smart learning environment – a review study. In: Uskov, V. L., Howlett, R.J., Jain, L.C. (eds.) Smart Education and Smart e-Learning. Smart Innovation, Systems and Technologies, vol. 59, pp. 51–59 (2016)
25. Semradova, I.: Nejvyznamnejsi konstanty pojeti ucitelske profese, [The most important constants in teacher's profession]. In: Pelcova, N., Semradova, I. (eds.) Fenomen vychovy a etika ucitelskeho povolani. Karolinum, Praha (2014)
26. Jorda, J.M.M.: The academic tutoring at the university level: development and promotion methodology through project work. Procedia Soc. Behav. Sci. **106**, 2594–2601 (2013)
27. Sulcic, V., Sulcic, A.: Can online tutors improve the quality of e-learning? http://proceedings.informingscience.org/InSITE2007/IISITv4p201-210Sulc388.pdf

Search, Exchange and Design of Learning Objects in Learning Objects Repositories

Daina Gudoniene[1,2](✉), Danguole Rutkauskiene[2](✉), and Valentina Dagiene[1](✉)

[1] Vilnius University, Vilnius, Lithuania
{daina.gudoniene,valentina.dagiene}@mii.vu.lt
[2] Kaunas University of Technology, Kaunas, Lithuania
danguole.rutkauskiene@ktu.lt

Abstract. The growing number of open educational resources (OER) causes the problem of search and sharing, design and reusability of learning objects in the semantic web according to a chosen subject. The authors have worked on the learning objects (LO) design model based on semantic technologies and now intend to present the integration of the model into a learning objects repository. The aim of the paper is to present the role of semantic web technologies in LO design, exchange and store in learning objects repository (LOR).

Keywords: LOR · Semantic · Learning objects · Metadata · Ontologies

1 Introduction

New educational technologies and semantic web bring new challenges for the designers of educational content. Metadata and ontologies play an essential role in the design of learning objects based on many metadata standards that exist and help in retrieval as well as reusability and adaptation of these LOs. Learning objects designed by means of new technologies and applications and featuring such characteristics as reusability, interoperability and manageability have better chances to be involved into the study process and to be stored at LOR. The aim of the paper is to present the role of semantic web technologies in LO design, exchange and store at learning objects repository (LOR). To achieve the aim, the analysis of literature has been carried out and the discussion on the architecture for LOR design with regard to the possibilities of exchange of LO based on semantic web technologies is presented. The LO standards, metadata, ontological approach, learning objects repositories are overviewed and the principle schema on LO design, search and exchange is presented.

2 Review of Existing Practises

2.1 Overview of LO Standards and Metadata

According to Riley et al. e-learning object repository 2.0 (LOR 2.0) requires a strict metadata and data standards [1]. The authors also provide the list of different types of

V.L. Uskov et al. (eds.), *Smart Education and e-Learning 2017*, Smart Innovation, Systems and Technologies 75, DOI 10.1007/978-3-319-59451-4_33

technologies such as automated LO metadata generation [20, 21], automated categorization of LOs, empirical evaluation of LOs, algorithms to math students and LOs and also algorithms to diagnose learner behaviour patterns. There are also some possible disadvantages mentioned by Riley et al.: "(1) current learning object metadata standards, (2) the absence of a learner information profile, and (3) the limitations of the current learner interaction tracking capabilities". Nikolopoulos et al. [2] and Rodrigez et al. [3] present LO meta data profile for distance learning from the point of view of ontological approach saying that the use of a single metadata standard is not a recommended solution since each application has its special features.

The standard IEEE LOM (IEEE Learning Object Metadata) is intended to describe learning material and learning resources. Dublin Core Metadata Initiative developed with the aim to share web resources and has no particular focus on education [4]. The second version, the Qualified Dublin Core (QDC), is very important for the present research as it is directly related with the idea of the paper regarding the exchange of semantic learning objects, i.e. it comes to extend the previous schema by importing 7 new elements. At the same time, QDC includes a group of qualifiers specifying the semantics of the elements in such a way that they may be reused by searching in the LOR and exchanging between external learning objects repositories [5]. However, SCORM describes how learning objects are developed, designed, and linked with learning management systems (LMS). LOM has an extension possibility and every institution can extend the list of necessary new elements.

The Dublin Core standard is directly related with the LO design, exchange and sharing in different LOR [6]. Conley [7] presents the new Common Core State Standard which allows educators to share a common language about what they want students to learn and enables the development of high-quality materials that correspond to the standards [4]. The metadata of these three standards are analysed in the Table 1.

Table 1. Main elements of the learning object metadata standards

IEEE LOM	Can Core	LOM Dublin Core metadata
1. General	1. General	1. Contributor
2. Life cycle	2. Life cycle	2. Coverage
3. Metadata	3. Metadata	3. Creator
4. Technical	4. Technical	4. Date
5. Educational	5. Educational	5. Description
6. Rights	6. Rights	6. Format
7. Relations	7. Relations	7. Identifier
8. Annotation	8. Annotation	8. Language
9. Classification	9. Classification	9. Publisher
		10. Relation
		11. Rights
		12. Source
		13. Subject
		14. Title
		15. Type

However LOs that are finally chosen should present some main properties in the LOR: they should be machine-readable, machine-processable and machine understandable; they should provide XML or other kind binding possibilities to import and export both full and partial student profiles, to allow for the merging of partial records into a comprehensive record. It should be designed with privacy concerns in mind [1]. The implementation of a database for storage, import, and export of these student models should be mandatory for SCORM 2.0-conformant LMSs.

2.2 Overview of the Ontological Approach

Nikolopoulos et al. [2] have analysed the ontological approach and claimed that even though RDF is intended for representing knowledge, it lacks reasoning abilities; RDF does not support making inferences or deductions. Therefore, a much more expressive framework is required, so that metadata can be meaningfully encoded. Ontologies, expressed in OWL (most widely-used ontology language) are the pillar of Semantic Web and provide the ability to represent any domain of interest in a more structured way.

The importance of search in semantic web requires ontological expression of LO metadata that could convert them into machine-understandable information and metadata of ontological models in OWL is even more enhanced by richer properties applied in LOR. Gasevic et al. [8] analyse ontologies for reusing LO content. Mohan and Brooks [9] discuss about the necessity for each learning object to specify exactly how that learning object is related to concepts in a particular domain and the kinds of learning outcomes that are possible in that domain, i.e. the ontology of concepts in a domain. There are important ontologies that describe learning objects: (a) metadata ontology (MO), (b) domain ontologies (DOi), and (c) content structure ontology (CSO).

Dhuria and Chawla [10] analysed Semantic Web as a content-aware intelligent web where Semantic Web technologies will influence the next generation of e-learning systems and applications. The importance of structuration and meta descriptions of learning objects and their material is growing and ontology is a key constituent in the structural design of the Semantic Web. Ontology is a formal specification of a particular domain that describes set of objects, properties that objects can have and various ways how these objects are related to each other.

The aim is to use semantic web for intelligent discovery and reusability of learning objects. By using shared ontologies, it is possible for software agents to perform most of the processing required in discovering and assembling learning objects.

2.3 Overview of Learning Objects Repositories

Currently learning objects repositories are operating online by providing to users a lot of different collections of LO by covering different educational levels and topics and also developed by variety of technologies and having different metadata described with the aim to classify LOs [11].
There are a lot of different well known learning objects repositories:

- CLOE Co-operative Learning Object exchange (Login required)

- La main à la pate, Activity-based science teaching for primary and secondary schools.
- MIT Open Courseware (OCW). A free and open educational resource for educators, students, and self-learners around the world.
- VCILT Learning objects repository, a multi-purpose repository from our friends at VCILT, University of Mauritius.
- Ariadne A European Association open to the World, for Knowledge Sharing and Reuse.
- CAREO, a project supported by Alberta Learning and CANARIE that has as its primary goal the creation of a searchable, Web-based collection of multidisciplinary teaching materials for educators across the province and beyond.
- Carnegie Mellon's Open Learning Initiative (OLI), A collection of "cognitively informed," openly available and free online courses and course materials that enact instruction for an entire course in an online format.
- Commonwealth of learning, Learning Object Repository, An online database of learning content that provides software to Commonweath countries.
- Edclicks
- Encore (Community for Open Resource Exchange based on tribes, resources and collaborations, ie. small private wikis). This is new (Jan 2007).
- Educational Object Economy Lots of Java applets
- GEM Gateway to Education Materials (in particular lesson plans)
- LOLA Learning Objects and Learning Activities (all kinds, but focus on information literacy)
- MERLOT (specialized in microworlds)
- NDMA (www.oer.ndma.lt/lor/ repository for LO and different educational activities design and store).

Most of these LOR follow IEEE-LOM metadata standard and metadata annotation is done manually. However, the learning objects repositories used not only for storing but also for sharing, reusing and LO design.

Mohan and Brooks [9] discuss a situation where learning objects are embedded within learning systems. It is analogous to the situation with mainframe systems in the past. The learning systems of today are essentially centralized, with learning objects (data) managed at a single place by a single system.

However, with the growth of learning object repositories on the Semantic Web, it is necessary to find out the model for the LO search, exchange and design where a learning object may also contain links to other courses or content packages where it has been used before to support searches by software and human agents on the Semantic Web.

3 Discussion on Semantical Learning Objects Exchange

The Semantic Web provides distributed information with well-defined meaning, understandable for humans as well as computers. Learning objects should contain metadata – information to classify, identify learning objects in the learning objects repositories. Authors present a variety of technological aspects like metadata, ontologies, for LOR design and architecture, where search, exchange, design and store of LO in the LOR can be carried out.

3.1 Meta Data Role for LO Search, Exchange and Design

Metadata will be part of learning object and will be visible to object users in the designed repository [12]. However, learning object metadata will be used to generate semantical relations to similar learning objects hosted outside repository. It is planned to have several types of metadata (Fig. 1).

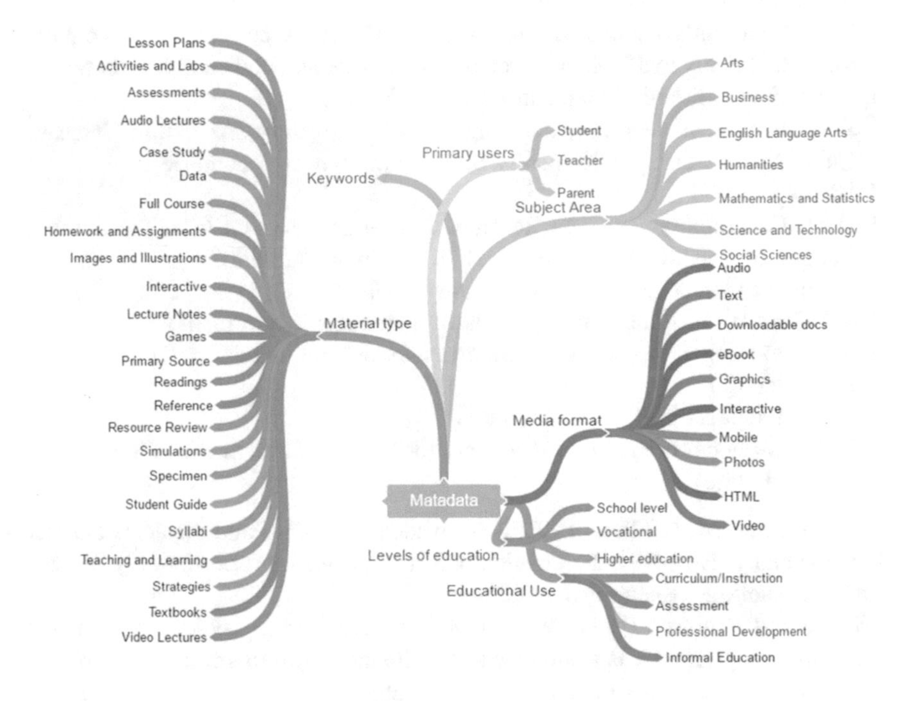

Fig. 1. Metadata map for LO description at LOR

The principal schema of learning objects relations has the following parts: Educational (learning material and activities); life circle; rights information; technical (relations with existing LO); relations; annotation; relations; classification (Fig. 2).

For preparation to develop a new LO design model there were used different (Fig. 2) modelling languages. "By using the evaluation and (re)design framework proposed here, we have managed to demonstrate cases in which each of the properties of a suitable representation mapping is lacking in the current UML meta-model when interpreted in terms of our reference ontology. In other words, as a conceptual modelling and ontology representation language, the UML meta-model can be shown to contain cases of construct incompleteness, overload, redundancy and excess" [13] (Fig. 3).

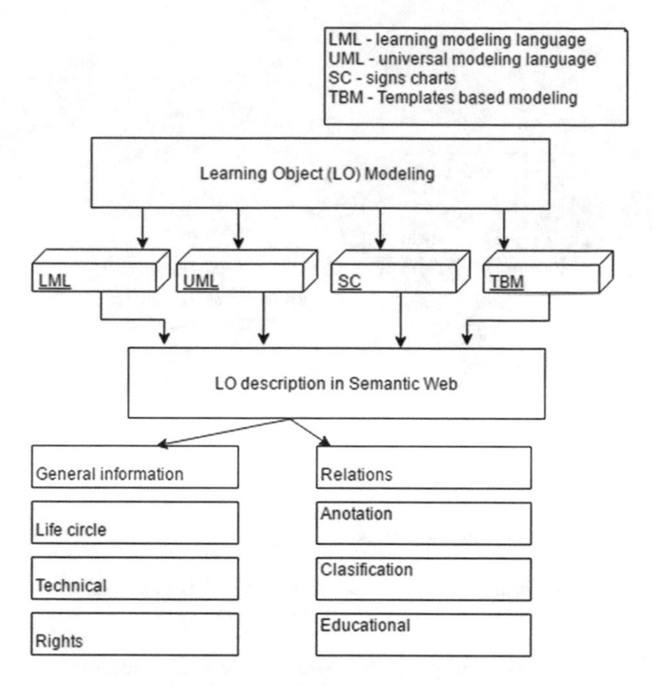

Fig. 2. Principal schema of learning objects relations

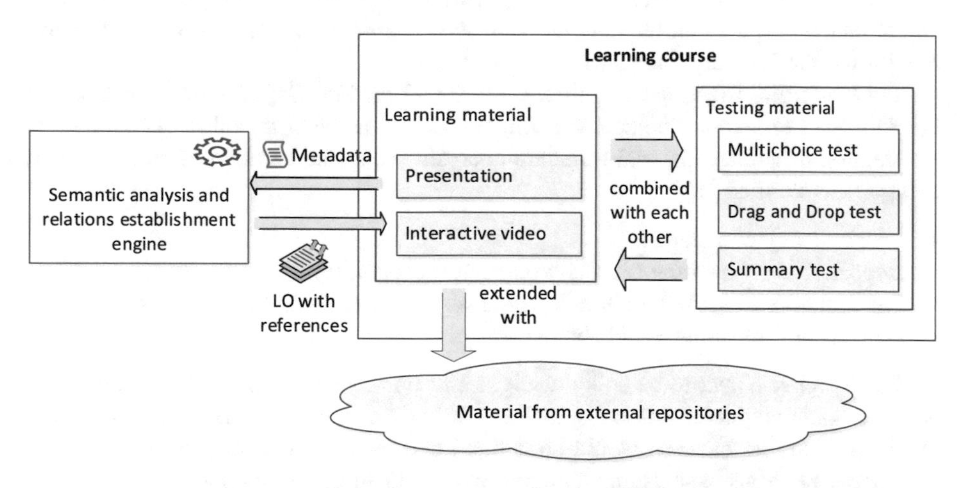

Fig. 3. Principle schema of the LO design and improvement (created by Gudoniene)

A learning course consists of blended learning material with tests. Learning material enriched with metadata that enables semantic analysis and relations establishment engine to build-in external references to learning objects [9].

Course presentations enable to design and deliver course material directly in the browser. Course presentations contain slides where it is possible to add various multimedia- and interactive elements to engage the learner. Presentation type material

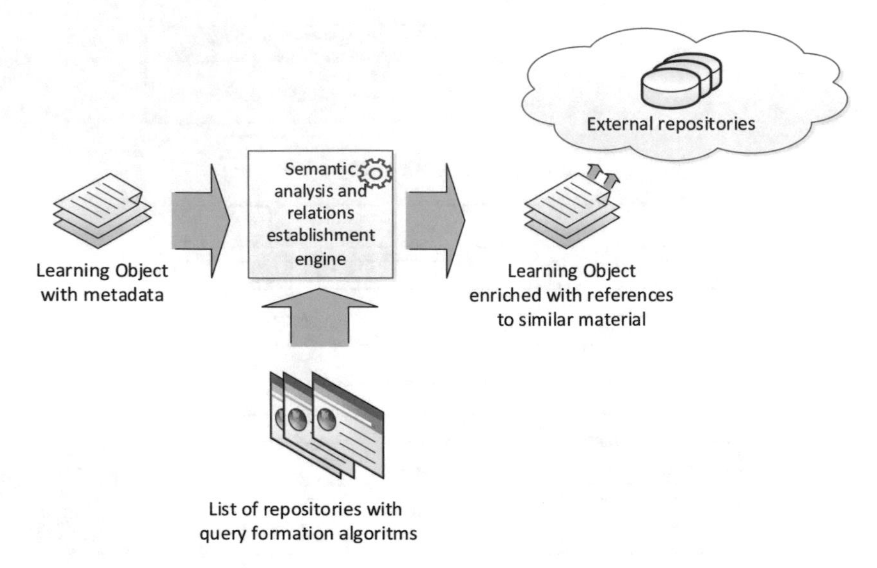

Fig. 4. Relations to similar material establishment schema

could be used when you want to package a piece of learning content in a structured and interactive format. Learners swipe through slides to experience the learning material, while solving various quizzes or watching videos along the way. Presentation learning object can be used as an alternative to present learning content as Power Point presentations, PDFs or text-based web pages (Fig. 4).

For copyright, Creative Commons might be referred to. Copyright information will be embedded to learning objects in a similar manner as other metadata [14]. To make the relation to similar material semantic analysis should be carried out and necessary components verified [15]:

- Learning object metadata;
- Repositories of learning objects with search capability;
- Functionality to set relations to learning object. We can call this part as "Semantic analysis and relations establishment engine".

For learning object to be as lightweight and portable as possible it is necessary to have semantic engine on authoring tool and, after learning object has been completed, relations to similar material would be defined and included into the object [16].
In general, relations establishment engine will work in the following way:

- engine will have flexible functionality to add new repositories and search engines [17],
- call to repository or search engine will be formed using construction algorithms responsible to form queries using strings, variables and placeholders [18, 19].

To create flexible engine, the ability to add variables and define variable-value relations with search query formation rules will be created [20]. From the beginning, relations establishment engine will be prepared to work with the following repositories and search engines:

- www.oercommons.org
- www.youtube.com
- www.google.com
- www.scholar.google.com

According to user interface, semantic analysis and relations establishment engine will be seamlessly integrated into learning object construction functionality and will work autonomously [21]. No user intervention is required – learning object creator has to define an object's metadata and relations establishment engine will form queries to know repositories and search engines.

4 Conclusions

The novelty of the LO modelling plays the central role in achieving unified authoring support in the process-awareness of authoring tools which should reflect semantic evolution of LOR.

It has been designated that an ontology oriented approach for e-learning improvement is very important and can offer a perfect technology for individualized learning based on interactive learning objects not only for teachers but for the learners as well as they can be uniquely identified, the content can be specifically personalized, and learner progress can be monitored, supported and assessed.

The growing importance of certain LO repository features, such as LO design functions and storing, remain the object for further research for the authors of the present paper.

References

1. Riley, S., Soh, K.L., Samal, A.: On the importance of metadata and learner interaction data in SCORM 2.0. LETSI SCORM 2.0 White papers (2008)
2. Nikolopoulos, G., Kalou, A., Pierrakeas, C., Kameas, A.: Creating a LO metadata profile for distance learning: an ontological approach. In: Dodero, J.M., Palomo-Duarte, M., Karampiperis, P. (eds.) Metadata and Semantics Research, vol. 343, pp. 37–48. Springer, Heidelberg (2012). ISBN 978-3-642-35233
3. Rodriguez, V., Ayala, G.: Adaptability and adaptability of learning objects interface. Int. J. Comput. Appl. (0975 – 8887) **37**(1), 6–12 (2012)
4. Dimitrova, V.: Semantic social scaffolding for capturing and sharing dissertation experience. IEEE Trans. Learn. Technol. **4**(1), 74–87 (2011)
5. Ermalai, I., Dragulescu, B., Ternauciuc, A., Vasiu, R.: Building a module for inserting microformats into moodle. Adv. Electr. Comput. Eng. **13**(3), 23–26 (2013). Timisoara, Romania
6. Sakarkar, G., Deshpande, S.P., Thakare, V.M.: Intelligent online e-learning systems: a comparative study. Int. J. Comput. Appl. **56**(4), 21–25 (2012)

7. Conley, T.D.: Common Core State Standards. CCSSO (2014)
8. Gaševic D., Jovanovic J., Devedžic V., Boškovic M.: Ontologies for reusing learning object content. In: Advanced Learning Technologies (2005). doi:10.1109/ICALT.2005.215
9. Mohan, P., Brooks, C.: Learning objects on the semantic web. In: Proceedings of the 3rd IEEE International Conference on Advanced Learning Technologies (ICALT 2003) (2003). 0-7695-1967-9/03
10. Dhuria, S., Chawla, S.: Ontologies for personalized e-learning in the semantic web. Int. J. Adv. Eng. Nano Technol. (IJAENT) **1**(4), 13–18 (2014). ISSN 2347-6389
11. Magnisalis, I., Demetriadis, S., Karakostas, A.: Adaptive and intelligent systems for collaborative learning support: a review of the field. IEEE Trans. Learn. Technol. **4**(1), 5–20 (2011)
12. Garrido, T., Onaindia, E.: Assembling learning objects for personalized learning: an AI planning perspective. In: IEEE Intelligent Systems, pp. 64–73 (2013)
13. Guizzardi, G.: Ontological foundations for structural conceptual models. ISSN 1381-3617 (2005)
14. Ermalai, I., Mocofan, M., Onita, M., Vasiu, R.: Adding semantics to online learning environments. In: Proceedings of 5th International Symposium on Applied Computational Intelligence and Informatics – SACI2009, pp. 569–573. Timisoara (2009)
15. Aroyo, L., Dicheva, D.: The new challenges for e-learning. Educ. Semant. Web Educ. Technol. Soc. **7**(4), 59–69 (2013)
16. Rutkauskienė, D., Mark, R., Kubiliūnas, R., Gudonienė, D.: Functional architecture of a service-oriented integrated learning environment. In: Proceedings of ECEL 2013 - 12th European Conference on e-Learning, pp. 431–439. Sophia Antipolis (2013)
17. Dragulescu, B.: Semantic web technologies used in education, Ph.D. thesis, Politehnica University of Timisoara (2012)
18. Alsultanny, Y.A.: e-Learning system overview based on semantic web. Electron. J. e-Learn. **4**(2), 111–118 (2010). Amman, Jordan. www.ejel.org
19. Aroyo, L., Dicheva, D.: The new challenges for e-learning. Educ. Semant. Web Educ. Technol. Soc. **7**(4), 59–69 (2014)
20. Arimoto, M.M., Barroca, L., Barbosa, E.F.: AM-OER: an agile method for the development of open educational resources. Inf. Educ. **15**(2), 205–233 (2016)
21. Gudoniene, D., Bartkute, R., Rutkauskiene, D., Blazauskas, T.: Technologichal aspects of the gamification model for e-learning participant's engagement. Baltic J. Mod. Comput. **12** (1) (2016)

Technology-Enhanced CLIL: Quality Indicators for the Analysis of an on-Line CLIL Course

Michele Della Ventura(✉)

Department of Music Technology,
Music Academy "Studio Musica", Treviso, Italy
dellaventura.michele@tin.it

Abstract. This article is aimed at identifying a set of quality indicators to analyze an on-line learning process referred to one of the methodologies that is currently considered among the most effective to promote language learning in formal contexts, i.e. the CLIL (Content and Language Integrated Learning). This methodology calls for various activities (speaking, listening, reading and writing) supported by the ICT (Information and Communication Technologies). The main objective of this study is to evaluate the quality of a CLIL practical activity (*task*) carried out in e-learning mode, using Technologies to support the teacher's/tutor's activity and boost students' learning of the basic aspects of a certain subject matter and their competence in discussing and debating. The indicators are used to analyze cognitive, meta-cognitive and relational aspects, drawing on a content analysis methodology. The model appears to have a wide range of possible applications in other online courses.

Keywords: CLIL · Learning indicators · e-learning process · On-line CLIL · Serendipity

1 Introduction

CLIL is an acronym for Content and Language Integrated Learning. It consists of teaching a curricular subject through the medium of a language other than that which is normally used, In CLIL courses, learners gain knowledge of the curriculum subject while simultaneously learning and using the foreign language. All of this entails the designing of a completely new learning environment, in which the teacher adopts didactic strategies that make the content to be transferred accessible [1] and methodological choices that offer learners new opportunities of reflection and cognitive stimulation as well as spaces for linguistic production and interaction [2]. Within this context, the ICTs (Information and Communication Technologies) are an important tool to support the learning process.

The constant refining of the ICTs is proposing an ever richer and more various range of opportunities for training in the classroom and/or on the web. E-learning is a new and important opportunity to build knowledge, that often, is not considered valid for certain contexts: for instance the CLIL, where rare and partial applications have been seen so far (Blended Learning).

V.L. Uskov et al. (eds.), *Smart Education and e-Learning 2017*, Smart Innovation, Systems and Technologies 75, DOI 10.1007/978-3-319-59451-4_34

However, the development of the ICTs fully satisfies some fundamental principles of the CLIL: socialization, interaction, motivation and knowledge sharing [2].

ICT development allowed for the creation of actual learning [3] and practicum [4] communities that collaborate in virtual environments to build and to share knowledge or to reach a common goal (solution to a problem or creation of a product). Within this framework, on-line interaction among the attendees assumes crucial importance not only as a defining status of this kind of communities, but also and most of all for the "survival" of the community itself and for the knowledge building process that is meant to be activated through it [5–7].

This definition undoubtedly highlights the dialogical, social and cultural nature of this process, aimed at "jointly creating and elaborating meanings", where the single, inasmuch as part of a group, receives support and motivation within the virtual community. This is how the type of language used (formal/informal) may become yet another tool for learning. It is interesting to note, as Coonan [8] highlights, how the comprehensibility of the contents transmitted by the teacher and the possibility to produce language that the students have (by returning knowledge and skills) entail positive relapses on both of the plans that CLIL must guarantee: content learning and language competence development.

The study presented in this article is based on a "Digital School" pilot-project that used a (temporary) on-line platform opened to students and teachers (from all over the world) of various disciplines. The main objective was to realize a CLIL on-line course using disciplines that also involved practicum activities based on one "*problem*" (PBL – Problem Based Learning), such as for instance Computer Science (with the realization of algorithms), digital synthesis of sound (with the realization of patches), analysis and composition theory (with the harmonization of the bass line or a given chant)… The entire process (which lasted for about 3 months) involved teachers and students of secondary schools. By analyzing the results of a series of questionnaires proposed to the students (before, during and after), in addition to the analysis of the data derived directly from the use of the platform (see paragraph 4) a draft was made of a series of indicators (for the e-learning) useful for the analysis and the assessment of a learning process: a standardized way was looked for so as to plan a general e-learning project, but also for the assessment of the results and of its efficiency.

This paper is organized as follows. Section 2 describes the CLIL Learning Environment. Section 3 describes the Task concept in CLIL. Section 4 describes the Interaction concept in CLIL. Section 5 describes the Monologue concept. Section 6 describes the Quality Indicators. Finally, conclusions are drawn in Sect. 7.

2 The CLIL Learning Environment

The learning environment is (generally) given by the totality of situations and designed activities that involve teachers and pupils.

It is based on the need to pass

- from the "teaching" concept to the "teaching" concept;

- from a vision centered on teaching and on what to teach to a perspective focused on the processes of the subject who is learning, paying meticulous and thorough attention to how to facilitate, how to guide, how to accompany the students in the construction of their knowledge and, therefore, which situations to organize so as to favor learning.

In a CLIL-type learning environment the methodological strategies acquire a significant importance inasmuch as in CLIL (as highlighted by the very name of the methodology: Content and Language *Integrated* Learning) the teacher, besides transmitting the contents of the discipline (for instance biology) by using a foreign language (for instance English), must also guarantee the learning of that foreign language. All of this entails the designing of a completely new learning environment, in which the teacher adopts didactic strategies that make the content to be transferred accessible [1] and methodological choices that offer learners new opportunities of reflection and cognitive stimulation as well as spaces for linguistic production and interaction. The teacher is therefore the architect of this learning environment, which on the one hand must support the learning processes and on the other hand must throw into crisis its weak points. In this dynamic process language plays a vital role: the curiosities, the researches, the questions, the reflections, the criticism, the negotiations of meaning give life to a multi-voiced discourse, a sort of inter-individual discourse [9], which stimulates and develops the interiorization and the personalization.

To allow students to interiorize the contents of a discipline, the teacher must naturally act in such a way as to render the contents clear and comprehensible. But the comprehensibility of the knowledge to transmit will also have positive relapses on the development of the students' linguistic competence [10]. Analogously, offering students spaces and opportunities to produce output will have a positive impact both on the learning of contents and on language learning [11]: on the one hand, in fact, a real and long-lasting interiorization of knowledge only occurs after a re-processing of the same knowledge by the learners, on the other hand the hypothesis of the comprehensible output postulates that production is a fundamental and indispensable element to learn a foreign language [12], given that it serves as an "impulse" that compels the learner to pay attention to the linguistic elements that he needs in order to successfully transmit his own thoughts (the concept of *noticing*) [13].

3 CLIL and Task

CLIL is an extremely innovative yet challenging learning environment. Based on all the considerations described above it can be inferred that cooperative learning, collaborative learning and, consequently, interaction [14] become three fundamental "ingredients" of a successful graduation of an on-line CLIL course, especially when carrying out a practical activity (Task).

Practical activity must try to "captivate" the student by leveraging on his cognitive involvement, by proposing activities devised as "problems" to solve (Problem-Based Learning): this requires the student to use the foreign language as if it were his mother tongue (as a means to discuss and solve the "problem") both because it is intrinsically interesting (and therefore more engaging), by virtue of the challenge within [8] and because it helps to consolidate a personal vocabulary of terms related to the non-linguistic discipline [15].

To have the student engaged to a maximum level the practical activity must have:

- An *Information gap* and/or a *required information exchange* that create the circumstances for interaction, for negotiation and for the information exchange [16];
- The *closed/open feature.* A closed activity, where participants know that there is only one possible outcome (as opposed to an open activity where such delimitations do not exist), entails more interaction and negotiation of information [17]: in the absence of a defined outcome the participants may opt for any outcome abandoning the field if they are in trouble [16].
- The *convergent/divergent* feature: in a closed task the participants must collaborate to reach the goal, they work in synergy, in a convergent manner, to this end [14, 17]. Convergence causes a very frequent use of interaction by means of confirmations given and received, clarifications, reformulations, corrections, repetitions and so on [8].

On-line practical activity may by more stimulating (motivating) for the student as compared to a mere exercise for the application of analyzed/studied concepts. At the same time, for some students, it may actually be a hindrance in the learning process, because a sense of abandonment and of disorientation, discouraging them and leading them to abandonment. One of the elements taken into consideration in the drafting of the indicators refers to the monitoring of the student so as to prevent this inconvenience.

The CLIL learning environment must therefore permit experiences capable of favoring the learning of knowledge together with the learning of doing: the doing that generates learning is never separated from knowledge. In these cases, group work and working in a group may facilitate the solution to a problem, and therefore reaching a certain result, thanks to sharing the personal knowledge/experiences of every student or to the individuals researches carried out. Group work and the necessity for the student to give his own contribution to the community may lead the student to carry out researches (even online), also facilitating "the learning by positive randomness" [18]: or, in other words, serendipity, i.e. the possibility to randomly make unexpected lucky discoveries while looking for something else.

Group work implies, therefore, the concept of *interaction* among the various members of the same group.

In the following paragraphs the attention will be focused on the concept of interaction rather than on the concept of cooperative learning or of collaborative learning, for which the literature has already provided important and exhaustive indications: interaction as an essential requirement of a CLIL learning process.

4 The Interaction

In the transmissive teaching model the student does not interact with the teacher, hence communication essentially goes only one way, i.e. from the teacher to the student [19]. The student only answers the questions he is asked. His role is therefore reactive.

Instead, the student must have a proactive role so as to insure the possibility of interaction: it is necessary to allow the student to ask questions or to propose new themes or subjects for discussion [20].

The value of interaction as far as language production is concerned is threefold:

- it permits, through negotiation, interventions tailored for individual problems associated with the comprehension of the input [21];
- it provides more opportunities for the student to use the language and, therefore, to expand and automatize it [1];
- it allows the student to manage himself his own learning [21];

An interactive-type didactic organization that also entails couple or group activities, it allows the set-up of collaborative learning forms where students interact to work together and help one another. The interaction, which is made possible through collaborative learning, has two fundamental values for linguistic production [22]:

- a quantitative value: more opportunities to use the language are provided. The student-to-student interaction allows the formation of groups, with the possibility for all to simultaneously and actively become involved in an activity:
- a qualitative value: from the standpoint of the promotion of the foreign language, the qualitative aspect lies in the possibility that the interaction gives the students to use the foreign language to convey meanings for authentic purposes and, most of all, to negotiate the meaning of the message if necessary. Through negotiation the students struggle themselves to understand and be understood using the language.

Hence, the interaction does not automatically produce learning, but it is facilitating factor inasmuch as it powers up the capacity to notice certain aspects of the language and to set in motion an intentional process of production of 'comprehensible *output*', the comprehensibility of which is measured not only in terms of communicative efficiency (*meaning-based output*), but also in terms of syntactic validity [23]. In the *output* production process the student has the opportunity to record a grade, the lack of a piece of knowledge, to seize the distance between his own production and that of his interlocutor (especially if – together with the feedback – amore appropriate linguistic model is offered) [1, 8]. This capacity to"notice" peculiar aspects of the language may lead either to an immediate reassessment and correction of his own production or to a process of collection of further information (request for information from the teacher, use of grammar manuals, consultation of dictionaries or data banks) that will enable, even subsequently, further learning.

5 The Monologue

Whereas interaction sharpens dialogical competence we must also take into consideration the fact that it may be necessary to promote monologue competence. While in the interaction statements [24] are typically brief and, at times, unfinished because they are interrupted by the interlocutor, in the monologue statements [24] are usually long because they are produced to transmit information. To make statements the student must have the capacity to produce elaborated. long sentences, connected to one another for cohesion and coherence [25]. It is a capacity that resembles the ability to write with the subsequent difficulty of having to do everything in real time.

6 Quality Indicators

Difficult as though it may be to sketch a shared model of measurement of the effects produced by an on-line CLIL course, it is however possible to try to define a set of indicators able to favor the assessment process aimed at understanding if the community works well, if it manages to reach its own objectives, if it is facilitating knowledge sharing and the realization of a shared inventory of best practices.

The model of analysis proposed here (Table 1), which took into consideration all the points described in the paragraphs above, consists of a series of indicators divided on the basis of different aspects or elements that characterize a CLIL course in e-learning modality: this way it is easier to differentiate the strong points from the weak points, the opportunities from the menaces.

Table 1. Quality Indicators for the Analysis of an On-Line CLIL Course

Indicators referring to the internal process	• total number of visits a learner has made • intensity of the cooperation among the members • speeding up of problem-solving (reduction of time needed to find the suitable solutions) • creation of a shared inventory of standards, methods, best practices • error risk reduction • number of ideas and new products • higher qualitative level of products and processes • capacity of optimal use of resources
Indicators referring to the learning and growth process	• searches done by learner to discover ideas added by peers • the websites added (link to websites, forums, blogs….) by learner • the number of ideas that were generated • the documents added by learner • amount of products realized • amount of common activities carried out to learn together • quality of the products and of the resources realized together • development of new key competences • growth of the capacity to reuse resources and knowledge • identification of explicit information (referential questions) • identification of implicit information (inferential questions) • idea connections added by learner

(*continued*)

Table 1. (*continued*)

	• idea connections with peer-generated ideas • identification of alternative methods/different than the ones studied to reach the objectives (serendipity)
Indicators referring to the interaction process	• the number and relevance of the interventions for every single participant • the number of messages sent by every single participant to another student on a subject relating to the course themes; • the number of messages sent by every single participant to a group of students on a subject relating to the course themes; • the number of messages sent by every single participant to the teacher/tutor on a subject relating to the course themes; • the number of questions (referential and inferential) asked • the number of answers given • the number of monologues held • change of perspective (signaled/not signaled) • the kind of language used (formal/informal)
Indicators referring to the user's perspective	• User's satisfaction degree • Reduction of the number of complaints from the user • Capacity to entertain the users • Activity sharing behavior • Increase of activity load

The model consists in:

- indicators referring to the internal process, which analyze the aspects related to the efficiency of the process;
- indicators referring to the learning and growth process, which analyze instead the cognitive and metacognitive aspects related to the construction and the sharing of knowledge, to the modality of elaboration of the learner's own experiences and the experiences of the others etc.
- indicators referring to the interaction process, which analyze the relationships associated with communication
- indicators referring to the user's perspective, which analyze the perception of the student.

7 Discussion and Conclusions

The CLIL methodology, which is already propagating at all the educational levels (from primary school to higher education), has proven efficient in a process of integrated learning of a Foreign Language (FL) and of a Non Linguistic Discipline (NLD).

In recent years this methodology has been applied to various disciplines (History, Mathematics, Biology…) obtaining positive results that encouraged its application to other branches of knowledge. The use of ICT allows carrying out various activities (speaking, listening, reading and writing) to develop different competences. In general they are activities that though using ICTs are carried out in the classroom and only partially (and not always) through the use of the WEB.

In this study we took into consideration the potential application of the CLIL methodology to a discipline, through the articulation of a course to be carried out exclusively in e-Learning modality. The attention was focused on certain disciplines that entail the necessity to develop practical competences (besides the ones indicated here above). The main idea was to introduce the CLIL methodology to the secondary schools through the identification of indicators aimed at guiding and supporting the teacher/tutor in the design and realization of the CLIL modules against the background of the most recent developments in the didactic and technological innovation ambit.

The indicators were identified without considering a specific discipline, but rather looking for a standardized way to plan an e-learning project in general and also to assess the results and its efficiency.

For their identification we took into consideration different aspects related to the learning process in general (where the serendipity concept was also considered) and to the interaction process: students develop communicative and interpersonal strategies to give helpful qualitative feedback to their peers.

It is highlighted that the validity of such indicators is fully expressed when they are used to make an *in itinere* (in-progress) assessment, because it is especially this way that interventions may be designed to insure the efficiency of the tutor's intervention and of the proposed training event.

It is an empirical methodology that provides results expressed in numbers that may be analyzed and then used in order to compare the effectiveness and the efficiency of an e-learning process. Finally, it is a method that may be improved by greater experience, that is to say, by applying it to a large range of educational institutions and contexts.

References

1. Coonan, M.C.: I principi di base del CLIL. Fare CLIL, I Quaderni della Ricerca, Loescher, Torino (2014)
2. Coyle, D.: Content and language integrated learning: motivating learners and teachers. Scott. Lang. Rev. **13**, 1–18 (2006)
3. Brown, A., Campione, J.: Communities of learning and thinking or a context by other name. Contrib. Hum. Dev. **21**, 108–126 (1990)
4. Wenger, E.: Communities of Practice – Learning, Meaning and Identity. Cambridge University Press, New York (1998)
5. Talamo, A.: Apprendere con le nuove tecnologie. La Nuova Italia, Firenze (1998)
6. Calvani, A., Rotta, M.: Comunicazione e apprendimento in Internet. Erickson, Trento (1999)
7. Varisco, B.M.: Nuove tecnologie per l'apprendimento. Garamond, Roma (1998)

8. Coonan, C.M.: La metodologia task-based e CLIL. In: Ricci Garotti, F. (ed.) Il futuro si chiama CLIL. Una ricerca interregionale sull'insegnamento veicolare, Provincia Autonoma di Trento (2006)
9. Varisco, B.M.: Costruttivismo socio-culturale. Carocci, Roma (2002)
10. Swain, M.: Communicative competence: some roles of comprehensible input and comprehensible output in its development. In: Gass, S., Maden, C. (a cura) Input in second language acquisition. Newbury House, Rowley (1985)
11. Swain, M.: Three functions of output in second language learning. In: Cook, G., Seidlhofer, B. (a cura) Principle and Practice in Applied Linguistics. Oxford University Press, Oxford (1995)
12. Swain, M.: The Output Hypothesis: Theory and Research. In Hinkel, E. (a cura di) Handbook of Research in Second Language Teaching and Learning. Lawrence Erlbaum Associates, Mahwah, pp. 471–483 (2005)
13. Johnson, D., Johnson, R.T., Smith, K.A.: Active Learning: Cooperation in the College Classroom. Interaction Book Company, Edina (1991)
14. Oxford, R.L.: Cooperative learning, collaborative learning and interaction: three communicative strands in the language classroom. Modern Lang. J. **81**, 4 (1997)
15. Della Ventura, M.: Problem-based learning and e-learning in sound recording. Int. J. Inf. Educ. Technol. **4**(5) (2014)
16. Ellis, R.: Task-based language learning and teaching. Oxford University Press, Oxford (2003)
17. Della Ventura, M.: Process, project and problem based learning as a strategy for knowledge building in music technology. In: Proceedings of the Multidiscilinary Academic Conference in Education, Teaching and e-Learning (MAC-Etel 2014), Prague, Czech Republic (2014)
18. Trentin, G.: Apprendimento in rete e condivisione delle conoscenze. Ruolo, dinamiche e tecnologie delle comunità professionali on-line. Angeli editore, Franco (2004)
19. Panselinas, G., Komis, V., Scaffolding through talk in groupwork learning. Thinking Skills and Creativity **4**(2) (2009)
20. Nikula, T.: The IRF pattern and space for interaction: comparing CLIL and EFL classrooms. In: Dalton-Puffer, C., Smit, U. (eds.) Empirical Perspective on CLIL Classroom Discourse. Frankfurt Am Mein, Peter Lang (2008)
21. Malamah-Thomas, A.: Classroom Interaction. Oxford University Press, Oxford (1987)
22. Kolloff, M.: Strategies for effective student/student interaction in online courses. In: 17th Annual Conference on Distance Teaching and Learning (2011)
23. Guazzieri, A.: Clil e apprendimento cooperativo. Studi di Glottodidattica **2**, 48–72 (2009). Università Ca' Foscari, Venezia
24. Brown, G., Yule, G.: Discourse Analysis. Cambridge University Press, Cambridge (1983)
25. Menegale, M.M.: Tipi di domande utilizzate durante la lezione frontale partecipata e output degli student. In: Coonan, C.M. (a cura di) La produzione orale in ambito CLIL, sezione monografica di Rassegna Italiana di linguistica applicata

Motivation of Students and Young Scientists in Robotics

Dmitry Bazylev(✉), Denis Ibraev, Alexey Margun, Konstantin Zimenko, and Artem Kremlev

ITMO University, Kronverkskiy av. 49, Saint Petersburg 197101, Russia
{bazylevd,kremlev_artem}@mail.ru, ibray1522@gmail.com, alexeimargun@gmail.com, kostyazimenko@gmail.com

Abstract. In this paper we consider development of interest among master students and young scientists involved in robotics. One of the key factors influencing a human involvement in learning and work processes is an ability to absorb knowledge in his/her usual way.

If a teacher presents a new theme in a way, not peculiar to student thinking, then the understanding process breaks and, as a consequence, the interest falls down. Similar problems can arise in communication of young scientists being a part of an international laboratory.

To "speak the same language" it is important to determine individual features of thinking of students and laboratory staff and organize exchange of information appropriately. To solve this problem we propose two approaches based on usage of meta programs (MP) and person's needs. The first method refers to increase of student interest in acquiring new knowledge. The second one is devoted to motivation young scientists and organization of their work in the international laboratory "Nonlinear Adaptive Control Systems". It is shown that the use of proposed approaches increases the number of students interested in self-realization in the field of robotics.

Keywords: Motivation · Meta program · Master student · Research laboratory · Robotics

1 Introduction

Standard educational programs are mainly focused on a structured presentation of information and a practical application of given knowledge for the sake of student interest. In each lecture, the teacher tries to explain theoretical material in a logical sequence when reporting new facts, statements, etc. Student motivation is taken from the need of given knowledge for a future profession and supported by practical lessons on laboratory benches with tasks related to situations occurring in real life.

Disadvantage of such training systems is in poor consideration of *human* features. It is assumed, that person's ability to acquire and use knowledge is determined by only intelligence, talent and motivation. The latter mostly depends on

V.L. Uskov et al. (eds.), *Smart Education and e-Learning 2017*, Smart Innovation, Systems and Technologies 75, DOI 10.1007/978-3-319-59451-4_35

his/her desire to be successful in life. Such description does not account unique way of thinking and person's needs, the use of which can significantly improve understanding and increase interest.

There are many works devoted to the problem of students' motivation in learning different subjects. However, the number of studies addressed to the development of young scientists' interest, is significantly lower. In [1] a theoretical and empirical approach to students' motivation is presented. The method is based on meta program modelling of knowledge and building models that help students understand and use teaching models in practice. In [2] authors underline the influence of computer gaming aspects of learning to motivate students in learning. Some authors pay attention to self-motivation in students [3] using the innovative method that includes computer tracking, protocols of thinking aloud, training diaries, direct observation and microanalysis. Studies of various effects on the students' ability to get new information and the factors influencing them are described in [4,5,7,8].

In [4] an internal dynamics of engagement metrics in the training and methods of organization of work in the classroom are investigated. Authors take into account the influence of a stimulating effects on the motivation and indicators of student work throughout the school year. The paper [5] shows an impact on the learning of factors such as personality traits, motivators, self-regulating students' learning strategies. Inspiration of motivation incentives on the students' attention to learning is described in [7]. [8] demonstrates the influence of mood and emotions on learning, creativity and interest itself.

The another way, resulting in gaining the student motivation, is active learning. Application of an active learning method to specific technical course "The integrated design and control systems" is proposed in [9,10]. See also [6] where the similar problem is studied.

In this paper we are interested in studying of person-to-person communication process being applied to either a teacher with a group of students or young scientists in the international laboratory. Two methods of motivation with a smooth transition between them are proposed – the first one is referred to master students at university and the second one is for laboratory employees. The basis of both methods is MP analysis and assessment of the person needs [11].

The remain of the paper is organised as follows. Section 2 shows the main features of the influence of MP profile on communication and dependence on the person needs. In Sect. 3 the detailed description of MP and person's needs is given. The algorithms of motivation for students and young scientists are proposed in Sect. 4. Section 5 presents application of the developed approaches in ITMO University. Finally, the paper is wrapped up with concluding remarks given in Sect. 6.

2 Meta Programs and Person Needs

MPs have a significant impact on the focus of human attention, his thought process and thus, behavior. They determine individual preferences of information reception, sorting and filtering for a person – what he is interested in, how he makes decisions, manages time, motivates himself, his effectiveness in solving problems, etc. In order to maintain a high level of interest and ensure involvement in a workflow we first identify individual MPs of students and young scientists. In this study, we distinguish the following MPs:

- Toward and Away From motivation;
- Proactive and Reactive;
- Internal and External Reference Frame;
- In-Time and Thru-Time;
- Specific and General;
- Induction, Deduction and Analogy;
- Past, Present or Future;
- Sameness, Sameness with Exception and Difference.
- Primary Interest.

Let's consider *Toward* and *Away From* motivation. This MP determines a person's training incentive. People with Toward motivation focus on what they want, they are interested in the ability to achieve something, they want to get closer to the goal and derive results from their work. A distinguishing feature of ones, who have Away From MP, is a strategy of avoidance. Such people pay much attention to the things they don't like and want to avoid, their stimulus is the problem to be solved.

Active – Proactive. People with MP Active are easy-going, they prefer to do first and then think. Such students and researches set themselves deadlines and often self-motivate themselves to take action. Proactive people like to ponder everything well before acting, they are heavy-going, it is difficult for them to take action, they do not like to work in a strict time frame.

Reference Frame MP. There are three types of references: internal, external people-oriented and external focused on the situation. People with dominant internal reference are free to decide and do as they see fit. External reference makes a person to be dependent on the opinions of others. A man with people-oriented reference takes a decision typically after he discusses the situation with others and consults with close friends or relatives. A person with reference, focused on the situation, relies on opinion of majority and prefer to do as accepted, in a common way.

In-Time and *Thru-Time.* This MP determines how a person perceives the situation – he is either immersed into it, or looks at it from the outside.

The next MP refers to the size of a layout, which can be *Specific* or *General.* A person utilizing General manner of thinking sees the information as a single piece and doesn't like details. Specific MP forces its owner to perceive new material in details and focus on sequential steps.

Three ways of thinking – *Induction, Deduction* and *Analogy* – determine how a person processes information and on what basis he makes conclusions.

Time orientation MP displays a part of experience, to which the principal value is paid. If the *Past* directs action of a person, then he is a conservative, focuses on the tradition and builds causal relationships from the past to the present. People with *Present* MP focus on the current moment, store events separately and do not build causal links. Planning and expectations are very important for a man with *Future* orientation.

Comparison preferences – *Sameness, Sameness with Exception* and *Difference*. People with Sameness MP want their world to stay the same for a very long time, Sameness with Exception enables slow change over time and Difference requires fast and major changes occurring continuously.

The last MP we take into account is a *Primary Interest* which denotes to what specific things we direct our attention and what of them we ignore – people, place, time, thing, activity or information.

In the organization of work and education process a special attention should be paid to the needs of group and particular person.

In accordance with the well-known Maslow's hierarchy of needs [11] a human has the following needs: psychological, safety, love and belonging, esteem, cognition, aesthetic and self-actualization. Physiological needs include metabolism requirements (air, water, food), protection from the weather (clothing and shelter) and sexual instinct for birth rate maintenance. Safety needs are mainly given by personal security, financial security, health and safety net against accidents. The next stage in the hierarchy is love and belonging which can be generally expressed through the need to form and keep such important relations as friendship, intimacy and family. Esteem is presented by respect and self-respect, being a "low level" and a "high level", respectively. Cognition needs include knowledges, skills, understanding and analysis. Essential components of aesthetic needs are harmony, order and beauty. Finally, the need of a person in self-actualization is connected with realization of his goals, abilities and self-development.

Many research centres care only about four needs – cognition needs, money (food, clothes, housing), safety (financial security, well-being and accident insurance), self-actualization. Our main focus in the science lab is directed to use and satisfaction of belonging, aesthetic needs, esteem and self-actualization.

3 Motivation Aspects and Communication

Most researchers consider the motivation of the student through the alignment of the education system, the relevance of acquired knowledge and creation of opportunities for the realization of student potential. However, many training programs do insufficiently or does not take into account at all what is going on in the human mind – how he thinks. How to make a decision that pushes him to action that draws attention and how he intends to work and learning process. Let's consider this questions through such incentive tools as MP profile and usage of a person needs. These instruments are useful in building a workflow and relationships in the team, taking into account both individual and overall team motivation.

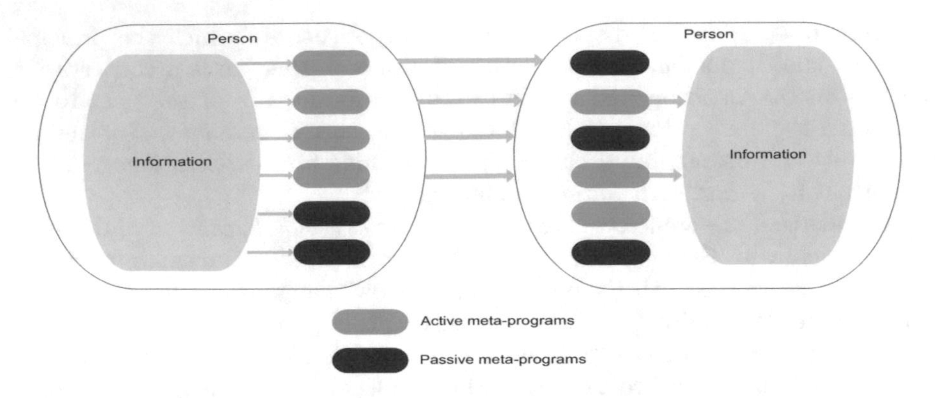

Fig. 1. Communication with not fully matched MPs

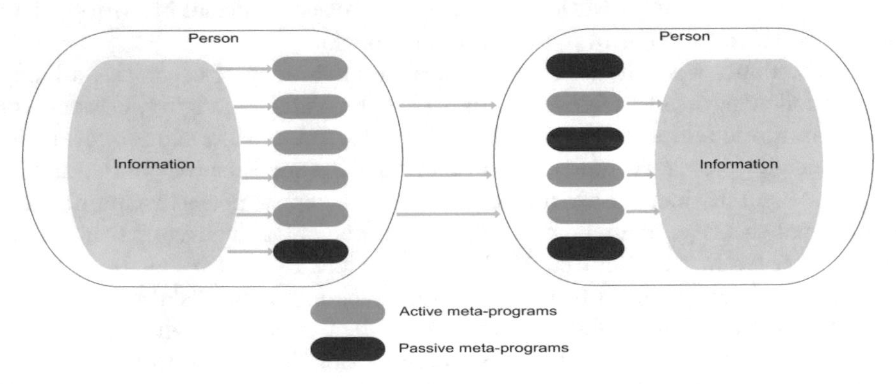

Fig. 2. Communication with matched MPs

MP profile defines filters of human perception and helps to identify the student's focus of attention, motivation, decision making, thinking, planning. The focus of attention determines where the human attention is, what is important to him and what kind of information it is easier to absorb and process. Motivation MPs indicate types of motivation, values and ability to self-motivation, which are individual for each person. One can say that meta-programs are individual "language" of thinking and communicating each person. Every man thinks and speaks in his usual meta-programs. Information obtained from other MPs is worse digestible or not perceived. Illustration of the relevant communication process is shown in Fig. 1. In the case with the known MP profile of the interlocutor, it is possible to maintain a dialogue and present information in a familiar way for him. The understanding of information and interest in a new material increases when people speak "the same language". The latter is demonstrated in Fig. 2. The another important factor of the person's interest is meeting his needs. Satisfaction of the needs directly affects on the motivation [11], and increases the interest of students and researchers in the work and development in the field of robotics. In [11] it is shown that the more satisfied the needs, the more people

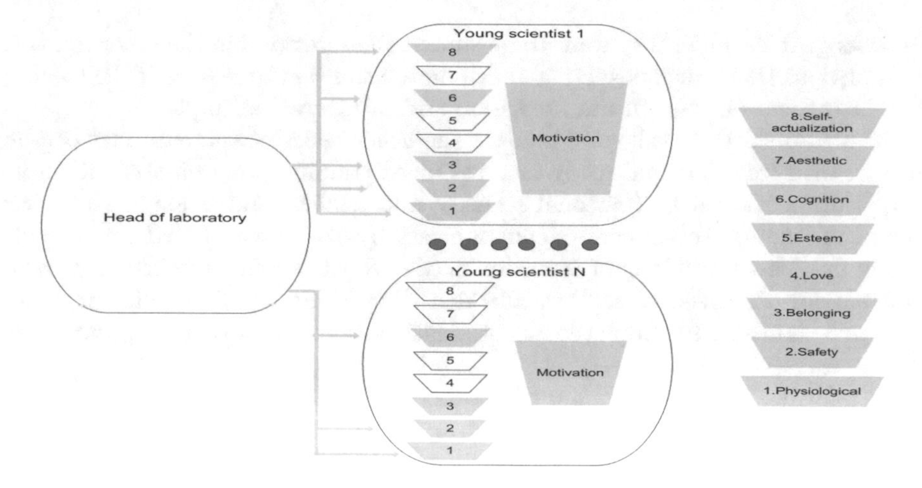

Fig. 3. Chief influence on young scientists motivation level

are satisfied with life and motivated for the realization of high-level needs. The purpose of education and the work of the scientific team lies in the need for self-realization and cognition, which are at high levels in the hierarchy of needs.

The motivation of employees in the science lab mainly lies on the head. Figure 3 illustrates the impact of the needs realization for young scientists on their motivation level.

4 Motivation Approaches

4.1 Master Students

Proposed approach for student motivation consists in the following steps. At first, a teacher learns the theory of MPs and needs. It is important for him to get practical experience in this area in order to clearly understand correlation between MPs, needs and different types of lessons.

The second step is an evaluation of students' MPs. Motivating students during a lecture depends on the teacher's personality, as well as the method of presenting information. In this study, we assume that we can not influence the personality of the teacher, but he can make lectures more informative for everyone. In general, the lecturer should use as much as possible MPs in his speech. Thus, the larger number of students will "hear" the teacher. However, being more precise for a current group he can make it easier for himself either without the loss of efficiency. To determine MPs one can use different questionnaires that show familiar to human ways of thinking - see, for example [12]. Knowing typical filtering preferences of a group of students he can significantly improve communication process and facilitate understanding for them.

Next, lessons are enriched with mentioning and underlining the student needs. Particularly, students expect to see practical application of academic knowledge in real life, they want to be sure in their demand in the labor market. It is obvious that learners will pay much more attention to lessons if the teacher regularly mention their future profession and brings real examples.

It is also worth to take into account the other needs of students that can be met in their free time and study. In a standard training program students meet cognition needs, safety (personal security and health) and belonging (weaker than the others). Adjustment of communication with respect to MPs will result firstly in a better fulfilling of cognition needs. Words about professional growth will employ the need of self-actualization. The other needs – belonging and esteem – can also be employed for student motivation through specific types of lessons.

Finally, we complete the education process by "Practical lesson with debates and disputes", "Lesson with planned errors" and "Case training". Interconnection between lesson types, MPs and needs is given in Table 1. The positive impact of such lessons is in additional usage of MPs which are rarely enabled in standard lectures.

Table 1. Relation between lesson types, MPs and needs

Lesson	Meta program	Need
Classic	Primary Interest: information Specific and General Sameness and Difference	Cognition
Practical with debates and disputes	Toward and Away From Primary Interest: people Internal and External Reference Sameness and Difference	Cognition Esteem Belonging
With planned errors	Toward and Away From Induction and Deduction Proactive and Reactive Primary Interest: things	Cognition Self-actualization
Case training	Toward Specific and General	Cognition

Practical lessons with debates and disputes. Students are divided into groups, each group chooses a topic for debate, which takes place after presentation of the topic to other groups. The main task of such lesson is to define the scope of application of each topic and find the pros and cons for it. These lessons involve the following MPs: Toward - to give own point of view to the audience, Away From – to avoid criticism, Interest in people - assessment of a particular student, Reference defence of own opinion and agreements with the other members, Sameness and Difference – whether the topic is familiar or not.

Lessons with planned errors are aimed at the development of critical thinking and in-depth understanding of the topic. During each session there are tasks that already contain errors in the condition problem or there is an incorrect solution. Students have to find mistakes and to offer the right solution to the problem.

MPs are: Away From – to find the error, Toward – to find right solution, Induction and Deduction – the manner of search, Primary Interest in things – to take into account specific details (initial conditions, applied algorithms), Proactive and Reactive – identify the error immediately or consider carefully.

Case training includes formulation of technical problem and search for solutions with argumentation. This lesson involves mainly motivation Toward – the desire to propose efficient solution. Specific and General MPs are used during the search process. It should be mentioned, that the other MPs should be used mainly by the teacher.

Though we assume we do not affect the lecturer's personality, he still can motivate students by personal example telling about famous scientists and well-known businesspeople in the corresponding field of activity. The same occurs during the practical lessons in the laboratory, when in the classroom the teacher describes the projects in the lab, invites students on presentation of projects and present successfully completed or on-going projects. The main idea is that students see with their own eyes how the work goes in real projects. Eventually, many of them are interested in completing some part of the research and in joining the project team.

4.2 Young Scientists in the International Laboratory

Motivation of researchers and young scientists have a strong impact on the work of the group, its unity and a desire to work for everyone.

Proposed algorithm consists of the following stages: the distribution of roles in the team and presentation of information, creation of a working atmosphere and organization of leisure and team-building activities.

Distribution of roles in project teams must take into account member's MPs, things he drew attention to, what motives prevail and needs should be satisfied. Basing on the results of tests determining MPs, the most suitable for every person role can be chosen – head, organizer or executor. MPs corresponding to each role are shown in Table 2.

Translating information about the current challenges for team members should occur through preferred MPs. Thus, understanding of problem formulation can be significantly improved as well as a number of redundant questions can reduce.

It should be mentioned that not all needs can be met during working hours. Thus, a special attention should be paid to leisure activities, joint events, exercises for trust development and activities on certain days.

Working atmosphere mainly depends on the need of esteem and belonging. It can be improved by setting rules governing the communication - an appeal to colleagues in a respectful manner, a censorship of jokes related to the personality and an agreement on mutual support of colleagues. Organization of leisure meet the needs of belonging, aesthetic and esteem, which can be satisfied within joint trips to different event, talks on interesting topics not related to work, etc. Special attention should be paid to team building – a few exercises per month can essentially unite the group and raise trust between its members.

Table 2. Motivation in laboratory

Level	Meta programs	Needs
Roles in team	Head - Proactive, Internal reference, Future, Primary Interest: activity Executor -External reference, Primary Interest: people, information Organizer - Future, Primary Interest: people	Belonging Esteem Self-actualization
Communication	Toward and Away From Sameness and Difference Specific and General Induction and Deduction	Esteem Belonging Cognition
Team building	Toward and Away From Primary Interest: people	Belonging Esteem
Leisure	Toward and Away From Primary Interest: people	Belonging Aesthetic Esteem

5 Implementation in ITMO University

In this section we present numerical indicators achieved due to realization of proposed approaches in ITMO University.

Motivation approach was applied for master students enrolled in programs "Mechatronics and Robotics", "Control in Technical Systems" and "System Analysis and Control". The total number of students studying in these areas is 69. After the first year of application a survey of students demonstrated significant improvement in motivation: 62 students were interviewed and 47 students pointed out the increase of interest in learning.

International laboratory "Nonlinear Adaptive Control Systems" was founded in 2013 and on the basis of ITMO University. The heads of the lab are Romeo Ortega and Alexey Bobtsov. The total number of scientists engaged in research is 71. Many scientists are working part-time, some of them are foreigners often working abroad. In view of this, not all of researchers were able to participate actively in proposed motivation program.

As a result of the incentive program, the number of students involved in international lab has increased from 19 to 35 people, which is almost 2 times more than last year. A survey of these 35 students and 18 more young scientists showed most of them willing to continue the program: 43 voted for and 10 abstained. 16 new students in the lab joined the existing projects or engaged in the implementation of their own research.

6 Conclusion

The problem of many education systems is in treating the student as a "black box" with a poor consideration of the individual needs and person characteristics. In this paper, we presented two approaches devoted to gaining interest of master students and young scientists. Introduced methods are based on the use of meta programs and person's needs. It is demonstrated that meta programs have a major impact on communication between young researchers, teacher and students. Giving information through preferred MPs significantly facilitates understanding of a new material. Consequently, motivation can be increased combining MP analysis and attraction of needs. Proposed approach was implemented in ITMO University. Numerical indicators have shown good results in three robotic education programs as well as in International Laboratory "Nonlinear Adaptive Control Systems".

Acknowledgments. This work was supported by the Ministry of Education and Science of Russian Federation (Project 14.Z50.31.0031) and by Government of Russian Federation, Grant 074-U01.

References

1. Schwarz, C.V., Reiser, B.J., Davis, E.A., Kenyon, L., Acher, A., Fortus, D., Shwartz, Y., Hug, B., Krajcik, J.: Developing a learning progression for scientific modeling: making scientific modeling accessible and meaningful for learners. J. Res. Sci. Teach. **46**(6), 632–654 (2009)
2. Papastergiou, M.: Digital game-based learning in high school computer science education: impact on educational effectiveness and student motivation. Comput. Educ. **52**, 1–12 (2009)
3. Zimmerman, B.J.: Investigating self-regulation and motivation: historical background, methodological developments, and future prospects. Am. Educ. Res. J. Manth. **45**(1), 166–183 (2008)
4. Skinner, E., Marchand, G., Furrer, C., Kindermann, T.: Engagement and disaffection in the classroom: part of a larger motivational dynamic. J. Educ. Psychol. **100**(4), 765–781 (2008)
5. Richardson, M., Abraham, C., Bond, R.: Investigating self-regulation and motivation: historical background, methodological developments, and future prospects. Psychol. Bull. **138**(2), 353–387 (2012)
6. Zimenko, K., Bazylev, D., Margun, A., Kremlev, A., Application of innovative mechatronic systems in automation and robotics learning. In: Proceedings of the 16th International Conference on Mechatronics, pp. 437–441 (2014)
7. Gable, P.A., Harmon-Jones, E.: Approach-motivated positive affect reduces breadth of attention. Assoc. Psychol. Sci. **19**(5), 476–482 (2008)
8. Baas, M., De Dreu, C.K.W., Nijstad, B.A.: A meta-analysis of 25 years of mood-creativity research: hedonic tone, activation, or regulatory focus. Psychol. Bull. Am. Psychol. Assoc. **134**(6), 779–806 (2008)
9. Bazylev, D., Shchukin, A., Margun, A., Zimenko, K., Kremlev, A., Titov, A.: Applications of innovative "Active Learning" strategy in "Control Systems" curriculum. Smart Innovation, Systems and Technologies, vol. 59, pp. 485–494 (2016)
10. Bazylev, D., Margun, A., Zimenko, K., Shchukin, A., Kremlev, A.: Active learning method in 'System Analysis and Control' area. In: Proceedings of the 44th Frontiers in Education (FIE) Conference, pp. 2113–2117, February 2014. Article no 7044339
11. Maslow, A.H.: Motivation and Personality, 3rd edn. Longman, New York (1987)
12. Meta Programs Survey. www.nlpaustralia.com.au/nlp/LifeSet/index.php

Attention Retention: Ensuring Your Educational Content Is Engaging Your Students

Mariia Kravchenko(✉) and Andrew K. Cass

Department of Energy and Environment, University College Nordjylland, Aalborg, Denmark
mark@ucn.dk

Abstract. As teachers look for ways to improve practice and enhance student engagement, referral to the literature leads to a dichotomy between specific activity and heavy academic research on metadata and learning analytics. This paper is intended to tread the pathway between the two so that teachers can, using the findings, introduce video for flipped classes, or online teaching. Taking 5 iterations of an action research approach the authors present the techniques and principles for making an interactive video and clear examples of lesson design. The research was conducted at a Danish higher education institution in technical subjects, but the methods are applicable to most teaching situations. The outcomes are, that paying attention to the process of recording existing lecture content leads to interactive video. This new content can be embedded in a structured lesson design and can greatly improve educational outcomes and student engagement. The researchers' goal was to create and sustain smart, rich and active learning environments in both online and in classroom environments, and utilize existing content across both platforms.

Keywords: Smart learning · Rich and active learning environments · Online learning · Flipped classroom · Interactive video · Lesson design

1 Introduction

Tertiary education is undergoing a substantial change driven by a need to adapt to the digital age. This is signified by a drive to fit into society that is influenced and driven by information and communication technologies (ICT), with widespread social media services and portable electronic devices [1, 2]. In response, Higher Educational Institutions (HEI) across the world are transitioning from traditional "on campus" classes into partially (blended) or fully online classes with an extensive utilization of multimedia materials and heavy reliance on ICT [3, 4]. Such a combination of technology services and web-based rich multimedia learning enables HEI to provide a smart education and create smart and rich learning environments. Indeed, many online courses rely on multimedia content, usually available in format of a video lecture that is remotely accessible by learners [5]. As Chandler and Sweller [6] note that use of videos in pedagogy should reduce cognitive load and optimize working memory. However, recent research asserts that integrating multimedia and pedagogical practices can enrich learning experience and improve students' performance [4, 7].

V.L. Uskov et al. (eds.), *Smart Education and e-Learning 2017*, Smart Innovation, Systems and Technologies 75, DOI 10.1007/978-3-319-59451-4_36

Importantly, incorporating videos into courses is not sufficient to create an engaging and active learning environment and improve learning [8]. Many studies on online learning and Massive Open Online Courses (MOOCs) conclude that relatively high rate of early stage drop out and low performance among students are caused by unclear course design and lack of interactivity in the course [9, 10]. Simply put educators often underestimate the importance of a course structure and the necessity of facilitating interaction in a systematic and organized way. This paper addresses the importance for educators involved in blended or online teaching to produce an interactive video and ways to embed such in the course so a more active, smart and engaging learning environment is created.

2 Background

Education has been constantly adapting new forms of teaching and learning in order to be able to satisfy the demand for knowledge in a society in which usage of products of development, such as ICT, is omnipresent and ubiquitous [11, 12]. According to Tom van Weert [11], such a society places three demands upon its citizens and workers, namely lifelong learning, knowledge development through practical research and knowledge sharing.

HEI are often asked to carry all the responsibility for preparing people to be productive workers and members of society [13]. In order to meet the society's demand as according to van Weert, the educators have to ensure that students develop critical-thinking skills instead of developing memorizing or information absorbing skills.

One educational strategy that can promote critical thinking skills and create active learning is use of Rich Environments for Active Learning (REALs). An education in a REAL is based on constructivist values as collaboration, reflectivity and engagement [14]. Constructivism learning environments place learner in the center of the learning process, which enables students to feel autonomy and responsibility for their own learning [15]. Moreover REALs promote critical and reflective thinking due to problem based nature of instruction [14, 16, 45].

There are several principles for designing REALs. First, in such environments teachers become facilitators of thinking process, not presenters of knowledge. For example, students should work with lesson plans and objectives trying to solve a problem and discover knowledge rather than be given that knowledge directly by the instructor [13, 17]. Secondly, learning has to occur by interaction. For example, students should work cooperatively in teams. Team activities make students analyse their own and others' knowledge and reflect upon others' interpretations. According to Vygotsky [18] most important learning occurs through social interaction. Moreover, this interaction is not only limited to interaction with peers in the class, but also to interaction with instructor and interaction with educational content [19]. Thus, the role of instructor is to encourage interaction, so higher order thinking such as analysis, synthesis and evaluation is involved [20]. Thirdly, instructors have to create a room for critical and reflective thinking, which is dialogic in nature [8, 13, 21]. Instructors can ask such questions as: "What methods did you use? What worked? What did not?

Are there any other methods you would use next time". According to Paas, Renkl and Sweller [22] and Jong [23], such instructional interventions as reinforcement, cooperative learning, tutoring, feedback and adaptive instruction have the highest effect on learning.

2.1 Online Education and Multimedia Learning

It is argued that constructivist learning environments can be replicated to an online course to create smart, active and rich learning environments [8, 24]. Hwang [25] defines smart learning environments as "...technology-supported learning environments that make adaptations and provide appropriate support (e.g., guidance, feedback, hints or tools) in the right places and at the right time based on individual learners' needs..." As recent research by Uskov et al. [26] points, universities can become smart by introducing smart educational features such as adoption of new educational styles (online learning, flipped classroom), new technical platforms, innovative use of software/hardware (smartboards, web-lecturing, lecture recording), etc. Additionally, smart e-learning environments should be developed on vision of "empowering the students' learning ability as well as empowering the teacher for smarter course preparation and delivery" [27]. Therefore, educators must remember that online environment and traditional classroom environment are different and faculty members need training and support in designing, delivering and maintaining online course and content [28]. Notably literature on retention rates and students' perceived effectiveness of distance education shows lack of clear course structure and content. Lack of interaction with and feedback from instructor are major factors of low retention rates in MOOCs and online courses [9, 10, 29]. Also many [3, 30] highlight that it is not the actual course content that affects the level of student's engagement, but as much the instructor's "presence", support and facilitation during the whole course period. Students need to be provided with a clear course structure, outcome expectations and be systematically guided through the course. As Berge [8], Paloff and Pratt [24] and Rienties and Toetenel [31] state, course design is one of the most important factors in online course retention and higher student engagement. Beetham [32] states that the educators must pay a special attention to a course design in a technology driven environment so to avoid frustration and discouragement of learners.

A successful course planning is a balance between three elements: learning goals, learning activities and feedback and evaluation [8] and course designing should consist of five main phases: course content design, course development, course implementation, course evaluation, and course revision [24]. When it comes to learning activities and course content in online education, educational content is usually available in format of a video lecture [5]. Video instruction is seen as an important element in both e-learning and blended learning environments, as it enriches students learning experience and gives flexible management of learning [4, 7]. Video helps students to create a sense of autonomy for own learning and better manage working memory as students can freely self-pace recorded lectures and analyse the content [33].

Although there exist various video lecture formats, such as lecture capture, voice-over, picture-in-picture or Khan-style video, there is no conventional standard to create a video lecture. Also, none of the video lecture formats is more superior to another due to the fact that they benefit various learner types (visualizers and verbalizers) differently [5]. Although videos containing a combination of animation (moving images and/or direct handwriting) and narration (text and/or audio recording) show more positive effect on students attention, learning experience overall [34]. Thus interactive videos, containing both animations and narrations as opposed to videos containing either of those, significantly improve students' performance and show higher learning satisfaction and engagement [5, 34].

3 Methodology

This research is based on analysis of authors' direct participation in creating video content for an online course, video integration in the course and course design in LMS. As authors were directly involved in the process themselves, this study resides on an action research approach. In an active research, it is crucial that researchers collaborate with other researchers/participants and engage in the situation, while focusing on problem solving via practice followed by critical and reflective learning [35, 36]. As a result, researchers contribute both to the practical concern in a given situation and provide an increase of knowledge.

The data is derived from the metadata available for viewing videos online. One of the advantages of the ICT environments is the ability to collect metadata and use it for statistical analysis. A simple comparative analysis is carried out based on recorded viewing times. In total, the researchers were involved in 10 classes spread over four different courses. The courses were a mix of flipped on campus classes and blended e-learning classes. The classes are of mixed size with between 4 and 15 students enrolled, which is not enough for a quantitative analysis. The authors used metadata retrieved from the LMS to identify potential action research interventions. The total number of videos created for individual lessons is 58 and vary in length from 5 min to 42 min. The videos are presented via several multimodal methods as direct downloads and embedded in the LMS. No metadata is available for downloaded videos other than the students feedback.

In addition to the video metadata, a qualitative analysis is carried out based on informal interviews and discussions and results from student support meetings (studerendes udviklingssamtale, originally in Danish – SUS). SUS are quality assurance meetings held regularly throughout the semester. They comprise of a short questionnaire related to course functioning and a chance for an open dialogue with the students. Using the action research approach, the researchers take feedback form class SUS and reflect of the responses. SUS meetings were held twice during the semester. By far the bulk of the feedback was from informal class feedback meetings. While several authors indicate that Learning Analytics (LA) should take a social LA perspective [37], traditional theory acknowledges the power of communication and collaboration [38].

The output of this paper are techniques that can be utilized by educators in order to create a "better" interactive video as well as principles to be considered when designing a course for a smart learning environment.

4 Results

The first iteration of videos produced by the researchers was a simple voiceover presentation of a slide show. The video comprised of 30 slides equivalent to the material the researcher had used for a two hour lecture period. The video was embedded in a flipped class style lecture where the first period was spent watching the video and the second part was doing an exercise based on the content. The feedback from the class was not particularly good. The video was 42 min long and the researcher was compelled on several occasions to stop the video and add commentary. After 30 min, the researcher stopped the video and reverted to the slide presentation due to the lack of attention from the class. It was observed that while the researcher could lecture for long periods, the class lost interest without the two way interaction of the lecture style. The researchers discussed the implications, and based on research of best practices, agreed that videos should be limited to one specific topic and exercises should be dispersed throughout the lesson block.

The second iteration used three videos between 7 and 12 min in length. The shorter videos focused on one topic. The first minute of each video was spent explaining how that video related to the other two. Additionally, every video would begin with researcher introducing him/herself and the course the video relates to. Each video was followed up by a quiz or an assignment where students were asked to research some specific aspect of the topic and send their findings to the researchers within the LMS platform. The findings from the second iteration were that students tended to skip first video slides, as they knew same introduction was coming, often missing the topic presentation and it's key points consequently (Fig. 1). Notably, the students enjoyed video follow up assignments; however, their research results were variable due to lack of understanding about research techniques and evaluating sources, since the researcher did not provide a written guidance on such.

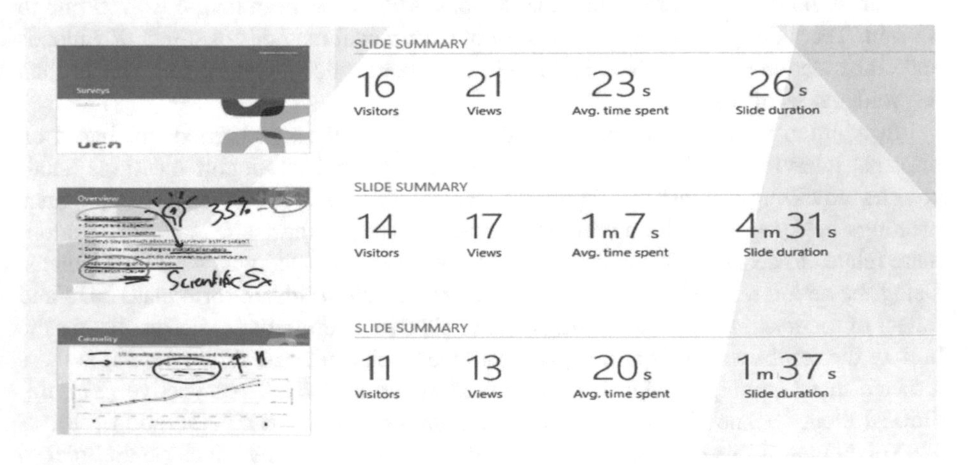

Fig. 1. Metadata slides showing standard presentation slideshow and showing low retention figures.

The remaining iterations are summarised in the Table 1 below.

Table 1. Results of the action research iterations showing the accumulation of techniques

Iteration	Video features	Feedback
1	1x 42 m, VoP	Long and boring lacked engagement
2	3x 7–12 m, VoP with A	Long introduction, poor learning outcomes (Fig. 1)
3	7x 5 m, AV	Great introduction, Well integrated
4	25x 5–17 m, AV, IV, FL	Improved class atmosphere
5	22x 5–12 m, AV, IV, OL	Good outcomes, better engagement

Key: x – Number of videos m – Minutes duration, VoP – Voiceover Presentation, A – Annotations, AV – Animated Video, IV – Interactive Video, FL – Flipped Class (video watched outside the classroom), OL – Online class

The results from the metadata survey show that students are watching the videos more completely after the third iteration (Fig. 2). In the last iteration, the students engaged with about 90% of the slide timings. The result almost never got higher than that number. The class represented in Fig. 1, had 10 students and the number of viewings is higher due to review viewings.

Data extracted from course analytics show that on the week ending 28th February 2016 students were less engaged with the course. The light grey bar shows page views only, whereas the dark bars shows students participation, i.e. interaction via educational content in form of assignments and quizzes (Fig. 3).

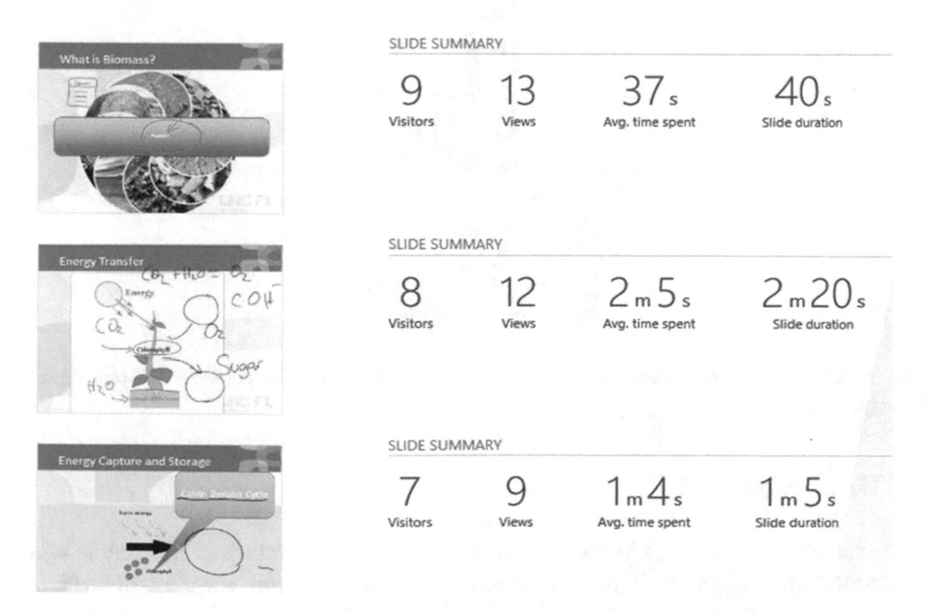

Fig. 2. Metadata showing slides without text, introduction and heavily annotated multi layered slides and showing very high rates of viewer retention.

After the third intervention with integration of educational content into LMS and researcher's guidance and facilitation, the students' participation rate was high and constant (Fig. 4).

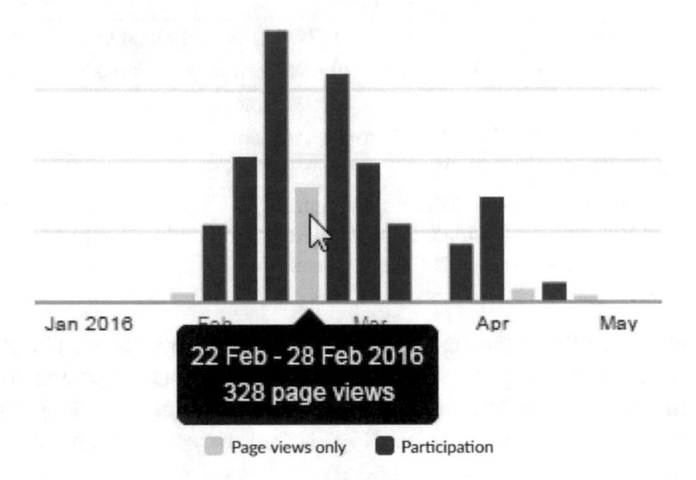

Fig. 3. Course analytics showing page views and participations.

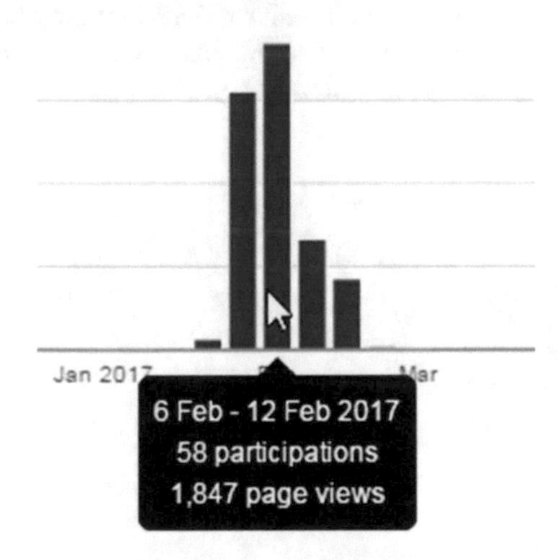

Fig. 4. Course analytics showing page views and participations when being constantly guided by educator.

5 Discussion

The researchers' goal was to analyse videos and their integration in the course within an institutional LMS so it boosts students' engagement and learning interest. The researchers had to define what would make a "better video" through the action research methods. The learnings from the first iteration were instituted as making the video shorter. Long videos were too monotonous as often the visual content was the same as the spoken content, so the students quickly lost interest. Thus, a general guideline was adopted that presentations that are in excess of 10 slides should be divided into multiple videos. Also long videos become unwieldy (large file sizes not shareable by email). Many articles on online education and online video design suggest that teachers should consult literature on the aspects of making a video and designing an online course prior implementation [3] and this is strongly supported by our findings.

The second iteration reinforced the findings of the first and the importance that videos have to be kept short. As Raths [39] and Zappe [40] suggest videos can be as short as 4 min and preferably no longer than 15 min depending on the students grade. Another finding was that the material should be honed prior to recording to remove any superfluous content. This requires teachers to review the existing material and identify the important slides that can be converted/adapted to a video. This process is time consuming, but gives a long term advantage as the content can be reused next time [41]. Often a short video was followed immediately by a research style investigative exercise that added context to the point, but this was student driven and then shared peer to peer. These lessons were appreciated by the students and scored highly on the SUS reports as they operated autonomously and were actively engaged in the topic. Thus providing evidence for the researchers that a smart and rich learning environment had been created. Additionally the results are supported by Fernandez et al. [42] and their research of video education in several engineering course with 487 students. The results summarized a survey data of students' evaluation of low-cost educational videos, where students gave high scores for such video features as interest, interactivity and convenience that influence students' motivation level. Also such video features as specificity, appeal, interactivity and real time were highly scored as the ones that increase students' attention level.

Another outcome was that researchers introduced annotations using interactive whiteboard system. This made video content versatile and made students maintain their attention to the whole video.

The third, fourth and fifth iterations focused on the point that video should be interactive: contain a mixture of text, pictures and animations. As Bergmann and Sams, [43] and Evans and Mathur [44] indicate that every video should excite students, so there is no harm in including humour (a comment or a picture). The researchers took this aspect to the highest level and moved from presentation style slides to animated videos, where there was almost no text. The ability to annotate a picture and have multiple pictures fade in and out created animated videos with a very high attention retention where all students would watch the entire videos.

The results showed that students would watch the videos multiple times, and the average time per slide would go below 100% only when students watched for a second time. It was surmised that the students were looking for a specific item in the video, but interestingly, the average time per slide was consistently higher towards the end of the video indicating that they would skip a slide or two, but would continue to watch to the end once started.

The presentation teachers normally showed in class commenced with a title slide, followed by the learning goals or an agenda. When transformed into a video however, the metadata showed that students tended to skip past this section. Iteration three in particular was implemented after the researchers went looking for inspiration on YouTube. Namely, the videos, that have high viewing numbers, are often those which start with the main point, like a teaser. The researchers applied this practice, and instead of presenting an introductory slide containing who is talking in every video, the authors stated main points of the topic/lecture within the first few seconds. It is noted that the teachers were reluctant to change this practice departing from their already prepared lecture. However, this practice is only applicable if the video is embedded in a comprehensively prepared LMS where the student is locked into a learning pathway, unlocking videos by completing previous sections. A critical element here is how the video is embedded within the LMS. The student comes to the lecture already knowing the subject and the context in which it is being applied. This enabled the teacher to focus on a key point, present it, then follow up with an activity.

Lecture plans and teaching material are made available for students via the institutional LMS platform, Canvas by Instructure. In reflecting on the LMS, you can see here (Fig. 5) how the educational content fits into a structured lecture and activity lesson plan.

Fig. 5. Embedding educational content into a LMS in the first iteration.

In the course analytics, as shown in Fig. 3, the low ranking for engagement related to the lesson design as shown in Fig. 5. The researchers inferred from the results of interviews and feedback that the lesson design was not able to keep the students on track due to a lack of guidance as to what was required. This could be interpreted that the students forgot or constructively dismissed the activity. However, during the reflection phase of the iteration, the researchers realised that there was no space created for the students to participate actively and this was a fundamental flaw in this lesson design.

Figure 6 shows learning design menu structure the embedded videos, followed by activities to complete the lesson. This lesson relates to the course analytics as shown in Fig. 4. The results show high and constant activity throughout the lessons. The researchers inferred that the increased participation was due to the improved lesson design where activities are integrated with educational material. This activity can, for instance, be a short quiz, which can, first, encourage students to watch the video as they know the teacher will follow up on the quiz. Second, the quiz is an excellent way to take the main point made in the video and add sub-context or get the students to extend

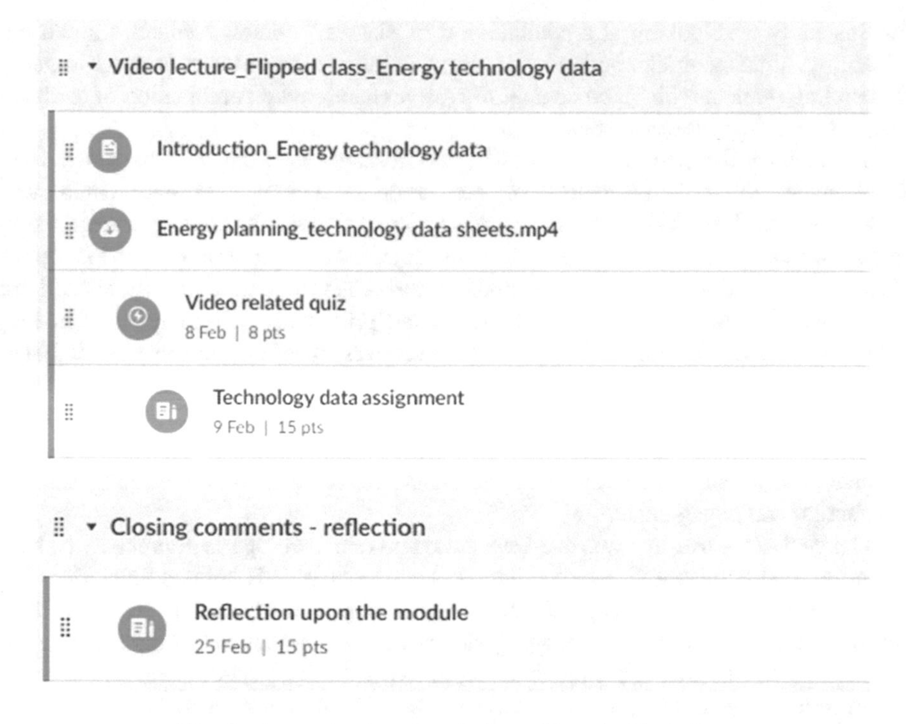

Fig. 6. Interactive learning design with embedded video.

their knowledge. The activities can as well be a discussion panel or an assignment. It was found that the students preferred activity in class and the classroom became a more engaging place with students devolving into working groups to work collaboratively. At the end of the topic or course, there is an afterword page, where instructions for reflections or follow up activities are issued. For example, the teacher can give some indicative questions to be considered once students are asked to reflect upon the course or topic. There, as well, can be an advice to consider applying knowledge from the course in a project. This supports the findings in the literature review, and was the key to the success of creating interactive learning environments, where peers, instructors and educational content combine for reflective practice based learning.

6 Conclusions

The action research method was ideal for this research methodology as the researchers were able to make multiple interventions of normal teaching practice and gain immediate feedback about the results. The literature is rich in discussions about video in an educational context, however the desire for a smart education system involves making a REAL and the process of moving to REALs utilising video is a complex one. It brings with it many technological innovations that can have mixed results because it is more than simply adding technology to the classroom [34]. The point of departure for this study was looking at a combination of classes, courses, blended, flipped and e-learning, utilizing similar action research interventions across each. In order to gauge student engagement with video content the researchers used a combination of feedback systems including viewer metadata.

Throughout the period of the study, the researchers were able to see a rapid development in the engagement of the students that was complimented with researchers' skill at producing videos. A key breakthrough in making students fully engage with a video was to eliminate the introductions and get straight to the key point of the video. Additionally, the educational material had to be thoughtfully revised and all the superfluous content had to be eliminated prior video recording, thus making videos short and relevant. Using shorter videos was beneficial to the overall lesson because more time could be spent with activity based learning exercises. Short videos are preferable in the learning environment [39], and this is supported by our findings. Also a combination of narration, animation and annotation ensured the videos were as interactive as possible, thus improving students' performance and increasing learning satisfaction and engagement.

The goal of the research was to create smart and rich learning environments that have the properties of collaboration, reflectivity and engagement [14]. Embedding videos in a course does not inherently engender these properties. The researchers return to these properties and ensured that the engaging videos were accompanied by collaborative and reflective exercises thus creating rich and active learning environment. Importantly, lesson design played a major role in the success of introducing video into the blended and online environment, and this is supported by major quantitative studies [31].

References

1. Lawless, K.A., Pellegrino, J.W.: Professional development in integrating technology into teaching and learning: knows, unknowns, and ways to pursue better questions and answers. Rev. Educ. Res. **77**, 575–614 (2007)
2. Kukulska-Hulme, A.: How should the higher education workforce adapt to advancements in technology for teaching and learning. Internet High. Educ. **15**, 247–254 (2012)
3. Keengwe, J., Kidd, T.T.: Towards best practices in online learning and teaching in higher education. MERLOT J. Online Learn. Teach. **6**(2), 533–541 (2010)
4. Mayer, R.: Multimedia Learning. Cambridge University Press, New York (2001)
5. Chen, C.-M., Wu, C.-H.: Effects of different video lecture types on sustained attention, emotion, cognitive load, and learning performance. Comput. Educ. **80**, 108–121 (2015)
6. Chandler, P., Sweller, J.: Cognitive load theory and the format of instruction. Cogn. Instr. **8**, 293–332 (1991)
7. Webb, M.: Affordances of ICT in science learning: implications for an integrated pedagogy. Int. J. Sci. Educ. **27**, 705–735 (2012)
8. Berge, Z.L.: Active, interactive, and reflective e-learning. Q. Rev. Distance Educ. **3**(2), 181–190 (2002)
9. Adamopoulos, P.: What makes a great MOOC? An interdisciplinary analysis of student retention in online courses. In: Thirty Fourth International Conference on Information Systems, Milan (2013)

10. Hone, K.S., El Said, G.R.: Exploring the factors affecting MOOC retention: a survey study. Comput. Educ. **98**, 157–168 (2016)
11. van Weert, T.: Education and Knowledge Society. WFEO/IFIP Kluwer Academic, Boston (2005)
12. Watson, D.: Understanding the relationship between ICT and education means exploring innovation and change. Educ. Inf. Technol. **11**, 199–216 (2006)
13. Dunlap, J.C., Grabinger, R.S.: Rich environments for active learning in the higher education classroom. In: Constructivist Learning Environments: Case Studies in Instructional Design, p. 252. Educational Technology Publications, New Jersey (1996)
14. Lebow, D.: Constructivist values for systems design: five principles toward a new mindset. Educ. Technol. Res. Dev. **41**, 4–16 (1993)
15. Piaget, J.: The Construction of Reality in the Child, vol. 82. Routledge, New York (2013)
16. Kemp, S.: Constructivism and problem-based learning. Temasek Polytechnic, Learning Academy, Singapore (2011)
17. Fetherston, T.: Becoming an Effective Teacher. Cengage Learning, Australia (2006)
18. Vygotsky, L.: Mind in Society. Harvard University Press, Cambridge (1978)
19. Moore, M.G.: Three types of interaction. Am. J. Distance Educ. **3**(2), 1–6 (1989)
20. Bloom, B., Krathwohl, D., Masia, B.: Taxonomy of Educational Goals. David Makay Company Inc., New York (1956)
21. Mulcare, D.M., Shwedel, A.: Transforming Bloom's taxonomy into classroom practice: a practical yet comprehensive approach to promote critical reading and student participation. J. Political Sci. Educ. **4**, 1–17 (2016)
22. Paas, F., Renkl, A., Sweller, J.: Cognitive load theory and instructional design: recent developments. Educ. Psychologists **38**, 1–4 (2003)
23. Jong, T.: Cognitive load theory, educational research, and instructional design: some food for thought. Instrum. Sci **38**, 105–134 (2010)
24. Paloff, R., Pratt, K.: Lessons from the cyberspace classroom: the realities of online teaching (2001)
25. Hwang, G.J.: Definition, framework and research issues of smart learning environments - a context-aware ubiquitous learning perspective. Smart Learn. Environ. – Springer Open J. **1**, 4 (2014)
26. Uskov, V.L., Bakken, J.P., Singh, U., Yalamanchili, M., Penumatsa, A.: Smart University taxonomy: features, components, systems. In: Smart Education and e-Learning. Smart Innovation, Systems and Technologies, vol. 59. Springer, Cham (2016)
27. Shehab, A.G.-D.: Smart e-learning: a greater perspective; from the fourth to the fifth generation of e-learning. Egypt. Inform. J. **11**, 39–48 (2010)
28. Olapiriyakul, K., Scher, J.M.: A guide to establishing hybrid learning courses: employing information technology to create a new learning experience, and a case study. Internet High. Educ. **9**(4), 287–311 (2006)
29. Dron, J., Ostashewski, N.: Seeking connectivist freedom and instructivist safety in a MOOC. Educ. XXI **18**(2), 51–76 (2015)
30. Halili, S.H., Zainuddin, Z.: Flipping the classroom: what we know and what we don't. Online J. Distance Educ. e-Learning **3**(1), 28–35 (2015)
31. Rienties, B., Toetenel, L.: The impact of learning design on student behaviour, satisfaction and performance: a cross-institutional comparison across 151 modules. Comput. Hum. Behav. **60**, 333–341 (2016)
32. Beetham, H.: Designing for active learning in technology-rich contexts. In: Rethinking Pedagogy for a Digital Age, p. 283. Taylor & Francis, New York (2013)
33. Abeysekera, L., Dawson, P.: Motivation and cognitive load in the flipped classroom: definition, rationale and a call for research. High. Educ. Res. Dev. **34**, 1–14 (2015)

34. Zhang, D., Zhou, L., Briggs, R.O., Nunamaker, J.F.J.: Instructional video in e-learning: assessing the impact of interactive video on learning effectiveness. Inf. Manag. **43**, 15–27 (2006)
35. Agyris, C., Putnam, R., MacLain-Smith, D.: Concepts, Methods and Skills for Research and Intervention. Action Science, p. 451. Jossey-Bass, San Francisco (1982)
36. Checkland, P., Holwell, S.: Action research: it's nature and validity. Syst. Pract. Action Res. **11**, 9–21 (1998)
37. Ferguson, R., Buckingham Shum, S.: Social learning analytics: five approaches. In: Proceedings of the 2nd International Conference on Learning Analytics and Knowledge, Vancouver (2012)
38. Vygotsky, L.: Mind in Society. Harvard University Press, Cambridge (1980)
39. Raths, D.: Nine video tips for a better flipped classroom. Educ. Dig. **79**(6), 15–21 (2014)
40. Zappe, S., Leicht, R., Messner, J., Litzinger, T., Lee, H.W.: Flipping the classroom to explore active learning in a large undergraduate course. In: Proceedings of the 2009 ASEE Conference. American Society for Engineering Education (2009)
41. Matamoros, A.B.: Answering the call: Flipping the Classroom to Prepare Practice-Ready Attorenys. Capital University Law Review, CUNY (2015)
42. Fernandez, V., Simo, P., Algaba, I., Albareda-Sambola, M., Salan, N., Amante, B., Sune, A., Garcia-Almin, D., Rajadell, M., Garriga, F.: Low-cost educational videos for engineering students: a new concept based on video streaming and Youtube channels. Int. J. Eng. Educ. **27**(3), 1–10 (2011)
43. Bergmann, J., Sams, A.: Flipped learning: maximizing face time. Training Dev. **68**, 28–31 (2012)
44. Evans, J.R., Mathur, A.: The value of online surveys. Internet Res. **15**, 195–219 (2005)
45. Savery, J.R., Duffy, T.M.: Problem based learning: an instructional model and its constructivist framework. In: Constructivist Learning Environments: Case Studies in Instructional Design, p. 135. Educational Technology Publications, Englewood Cliffs (1996)

Utilization and Expected Potential of Selected Social Applications in University Setting – a Case Study

Miloslava Černá(✉) and Libuše Svobodová(✉)

Department of Applied Science, Department of Economics,
Faculty of Informatics and Management, University of Hradec Králové,
Rokitanského 62, 500 03 Hradec Králové, Czech Republic
{Miloslava.cerna,Libuse.svobodova}@uhk.cz

Abstract. Education and use of technology have undergone through a turbulent development over the last two decades in the Czech Republic. Students widely utilize latest advanced technologies on the daily basis for their private purposes as well as for study purposes at majority of universities where they have access to online study materials in virtual platforms. The Faculty of Informatics and Management, University of Hradec Králové is one of the major institutions that has co-formed and influenced the local Czech educational scene in e-learning. Both education and common life have been closely connected with social software applications. Real world is being enriched with the virtual space. It is valued for creativity and innovations it enables and fosters. The paper contributes to the exploration of utilization of Web 2.0 phenomenon in the university setting with students of financial and information management. The goal of this paper is to map real situation relating to utilization of social applications with focus on students' expectation in potential of selected social software applications. Three hypotheses which assume that students who can see potential for educational purposes in YouTube, Facebook and a learning management system, already use these applications either for private or study purposes or both.

Keywords: Potential · Social software application · Statistics · Utilization

1 Introduction

Virtual space has been giving our real world new dimension; creativity and innovations are valued characteristics. Social software applications (SSA) fit knowledge management as they open the door to new ways of communication, enable development, editing, sharing and storing materials on the virtual platform. All aforementioned predetermines new options brought by utilization of virtual space to their new role in the process of education. Technical innovations possess substantive drive in young people hence university students as well; their strength lies in enabling users participate and interact in web-space which also represents a valuable contribution for educational purposes.

V.L. Uskov et al. (eds.), *Smart Education and e-Learning 2017*, Smart Innovation, Systems and Technologies 75, DOI 10.1007/978-3-319-59451-4_37

The paper contributes to the exploration of utilization of Web 2.0 phenomenon by university students of the second year of their Applied Informatics, Financial and Information Management studies within Bachelor Study Programs. Utilization of Web 2.0 is followed at two levels which often mingle: at the level of common use and at the level for study purposes. When students use social software applications for study purposes it is obvious that they use them in everyday life as well. If we look at that from a higher perspective, interconnection with wide areas can be explored as this may be reflected in advertising, product sales, track users etc. Another interconnection might be identified in the fact that students who get familiar with these applications will realize their benefits not only for private life, but also will be able to apply them in the business sphere. They will not be afraid to use these networks, they will see the potential in them or not, they will know how to move them, behave etc.

Web 2.0 concept was defined by Tim O'Reilly in MediaLive International in 2004 as a designation of the new generation of the Web. Reilyho definition of Web 2.0 is as follows: "Web 2.0 is the business revolution in the computer industry, which is caused by deflection in the understanding of the Web as a platform. Key among those rules is this: build applications that will be better and better due to the network effect with an increasing number of people." [1] With the concept of Web 2.0 it is an extension of the so-called classical 2nd generation web. Briefly Web 2.0 concept can be described as a set of principles, where the first and fundamental element of the web as a platform. It is obvious that without this platform Web 2.0 could never exist. Tim O'Reilly [1] defined the main differences between classical Web site and a new generation web. In terms of software development Web 2.0 is characterized as a shift from centralized processing and services to decentralization. Second generation web gave users the ability to handle their website and use social networks to converge with other users and attract potential customers (commercial use). Web 2.0 is a term for applications where the user affects the content or communicates with other users. These applications then become social networks. It is necessary to add that the concept of internet web 2.0 is not related to graphic design or web-design.

The discussed topic of the role of social applications in the tertiary education with the focus on collaboration and cooperation as their key features have been highlighted and analysed in a wide range of highly professionally oriented papers, e.g. [2–4]. They were also sources of inspiration for our research and evaluation of findings. Papers repeatedly underlined the magic words of pedagogical effort which are "engagement" and "motivation" of students into learning process; how to get students engaged, how to motivate them and at the same time tailor the way of the educational process so that it could reflect current trends and foster the demands of the digital age of university students. A role and responsibility of every instructor is to facilitate and to strengthen student's engagement in learning, however that is a very delicate process [2].

The philosophical core of the long-term research which has been conducted at the Faculty of Informatics and Management within the frames of national and specific educational projects is "interest in students' learning engagement". Researchers are trying to reveal how students utilize and perceive the new possibilities of social software applications like social nets, learning management systems (LMS), communication apps, etc. See more on partial and summarizing outcomes which bring the insight into the utilization of social software applications (SSA) for private and study purposes [5, 6].

This broad concept was narrowed last year when we focused on the cooperative and collaborative feature of SSA [8]. The issue of cooperation and collaboration has been discussed by academia [8, 9]. The paper brings current data on the issue of real active involvement of students into utilization social software applications.

The organization of the paper is as follows: firstly a theoretical background with a review of the literature is provided, then research methodology is described, the key part brings results from survey which was run with 67 full time students from the Faculty of Informatics and Management, University of Hradec Králové this academic year. Examined areas related to the general use of social software applications for everyday life and for study purposes. The highlighted analysed area was students' utilization and expectation in the potential SSA. Graphs visualize the findings. Conclusion summarizes the findings and raises questions calling for further discussion.

2 Literature Review

Information and communication technologies form a natural part of our lives where education is no exclusion; ICT is a topic which never stagnates. Researchers approach the issue of social applications in higher education from various perspectives. For example, Valtonen discusses social networking in the paper and brings implications for learning which are based on the findings from a conducted survey with more than one thousand students on their ICT skills and use of social software but from the perspective of a tutor [8]. Tess examines the role of social media and their potential as a facilitator and enhancer of learning, his literature review explores social media in higher education where he takes into consideration both real and virtual worlds [10]. Deep insight into the issue of Web 2.0 and practice of teachers in Higher Education was done by Greener; she reviewed peer-reviewed journal articles since 2006 [11]. As for characteristics of Web 2.0 and its appropriateness for teaching-learning purposes it encourages sharing and construction of information; it is participative [11]. The most underlined potential is seen in cooperation and collaboration. Findings gained from the review prove strong belief in real active cooperation and sharing knowledge via social media. But our findings from previous surveys where mapping of awareness and utilization of selected social applications in higher education didn't indicate students' expectations of future utilization of SSA [7].

2.1 Social Media

Social media are applications which become social networks via users' use.

Social media are a form of electronic communication (social networking and microblogging networks), through which users create online communities to share information, ideas, personal messages and other content (videos). [14] Social media can be divided on the basis of their focus, and also according to marketing tactics. Social media according to the focus divided into:

- Social networking - blogs, videos, audio, photos, chats, discussion, etc.
- Business Networking - connecting people to the business.
- Bookmarking - information sharing, mainly articles through public bookmarks.
- Sites on content quality voting.
- News - web sites, where news is displayed.

Contrary to this division the division according to the marketing tactics is considered to be more transparent. Surveys on utilization of social media are more frequently conducted on the basis of marketing media division:

- Social networks (Facebook, MySpace, LinkedIn).
- Blogs, video blogs, microblogs (Twitter).
- Discussion forums, Q & A portals (Yahoo Answers).
- Wikis (Wikipedia, Google Knol).
- Bookmarking systems (Digg, Delicious, Jagg).
- Shared multimedia (YouTube, Flickr).
- Virtual worlds (Second Life, The Sims). [15]

2.2 Social Networks

A sociologist J. A. Barnes (1954) is considered to be the author of the definition of the social network. The term social network is associated with the Internet and directly with the social networks on the Internet. But this is not the only connection there is a need to deal with the concept of social network from the sociological point of view, therefore, the social network can be described as a "map of the area near and distant surroundings where relationships of involved people are defined. The definition of social networks in sociology is that it is a set of social actors interconnected with relations. Entities are nodal points of the network, relationships which are the traces of points." [16]

Social network on the Internet is considered to be a group of people who communicate and share documents and information on users. The Social Network concept is described and defined by Boyd and Ellison: "Social networking is defined as a web service that allows individuals to create a public or semi-public profile within the bounded system, create a group of users with whom they share a connection, and browse the list of own connections and that created by other users of the system. The nature and terminology of these connections may be different network from the network." [17]

In case of Web 2.0, social networking means any system that enables creation and maintenance of a group of interconnected contacts and friends. Even systems where creation of contacts isn't their primary mission but only one of supported functions can be included into social networking. Each user of social networking system defines his/her characteristics and attributes, which are publicly available for other users. People within the system can mutually seek each other within the system and create virtual "community". The more advanced form is a search on social networks - access to the lists of friends in sense of searching friends of friends at other levels. On the above mentioned function other properties of the software are based, like the possibility to publish various information, post photos and albums, create diaries, etc. [6].

2.3 Social Networks Geography

Geographical view of the distribution of social networks reflects a cultural difference in nations; it is possible to visually monitor the distribution of the world map. The largest share of the world map belongs to Facebook, which is located mainly in North and South America, Europe, Australia, India and North Africa. Social network Vkontakte.ru is a popular throughout Russia and Russian speaking countries, social network Qzone is popular in China. Both social networks Vkontakte and Qzone serve as analogues to Facebook, including almost identical functions. Social network Facebook has such a wide representation in the world thanks to translations into 60 languages.

Social nets map [18] shows that in Janury 2016 Facebok was the most utilized social net. Instagram was the second most utilized one. The most utilized networks in the Czech Republic are Facebook and Twitter. Neither Google + or Youtube was not included into the evaluation.

3 Research Methodology

The study problem of the research and concise history of projects, research objectives, the research tool, the accessible sample and the way of data processing are discussed in this chapter. Development of trends in acceptance, utilization and satisfaction with social software applications in higher education on a limited scale of local university was monitored for five years within national educational projects. 'Evaluation of the modern technologies contributing towards forming and development of university students' competences' and within a follow up Excellence project. 'The ICT reflection within cognitive process development', see more [6]. This current study is conducted within the frame of a new grant project. "The influence of social media and mobile technologies on formation of optimal model of teaching." The paper is a logical continuation of the research dealing with social media landscape which concerned individual types of social software applications (SSA) selected on the basis of their various missions fitting various goals. This time the aspect of real cooperation or even collaboration is examined.

The study problem deals with social media in tertiary education. Two main levels are monitored: students' readiness to cooperate or even collaborate in the virtual space of social applications and their opinion in sense of expectations on potential of individual social software applications.

Collaboration is understood as an active involvement of more people pursuing one goal whereas cooperation might be even passive and is more about personal-individual goals. Villiers states that all players must collaborate; players have to work together to produce a result. Collaboration is described, as a higher order skill, demanding more than cooperation [12].

The main objective of the paper is to accept or reject three following hypotheses:

- "When the student sees the potential in Youtube, he/she already uses it".
- "When the student sees the potential in Facebook, he/she already uses it.
- "When students see potential in the learning management system (LMS), he/she already uses it."

The paper is based on the survey followed by semi-directed discussions which were conducted with 67 sophomore students of Applied Informatics, Information and Financial Management within the Professional English classes this winter semester. Findings were widely discussed with students during the following language hours of instruction. They served as proven activating factor because social applications represent the topic which is "their" topic, which is an essential part of their life.

As for the research tool, the survey with a modified questionnaire was applied. The survey stems from a long term survey on utilization of software apps; this survey is based on a short list of 8 relevant applications. Five key areas were examined. There were two main determinants influencing creation of a SSA shortlist: findings from previous surveys and appropriateness of SSA for cooperation and collaboration activities. The shortlist contained following applications: Google+, Facebook, You Tube, Wiki platform which enables developing common knowledge, Skype as exclusively social communication application, Blog as a form of online reflective diaries, Social-bookmarking and a learning management system (LMS). Rather old fashioned method was applied; 67 hard copies of adapted questionnaires were distributed and collected. It took about 10 min to fill in the questionnaire. A follow-up discussion with open questions on most beneficial ideas and proved applications utilized for studying foreign language based on students' own experience took about 15 to 20 min. Both teacher and students made notes. Return of the questionnaires was 100%. Research accessible sample consisted of 67 full-time students from the University of Hradec Králové from the second year of their study.

4 Findings

All students who formed the research sample were of the same background; field of study, age, time spent at the university. Firstly findings from the quantitative research are presented. This section on utilization of social applications brings a general view of the issue and then a focused view on utilization of social applications for study purposes and expectations of potential are solved.

Part of the Findings chapter deals with the core research problem. Findings on students' utilization of SSA in real life, utilization for education purposes and expectations related to potential in selected social applications are revealed and illustrated. The evaluation is described on three most often utilized SSA by involved students.

It is apparent from the findings and Fig. 1 that all respondents use Youtube, 94% use Facebook and 93% LMS.

Upon further analysis of data focused on SSA and their use in education, it was found that out of the 67 students, 91% of students used Youtube for education, 88% used LMS and only 57% of respondents used Facebook for study purposes. (Fig. 2)

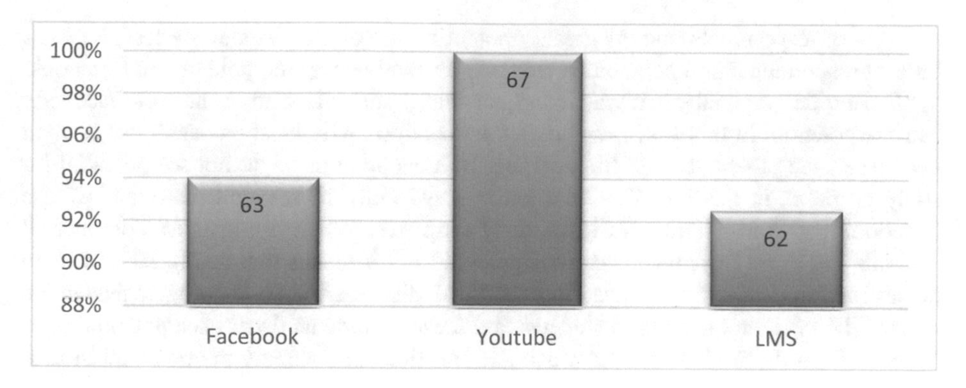

Fig. 1. Utilization of selected SSA by students of FIM, UHK

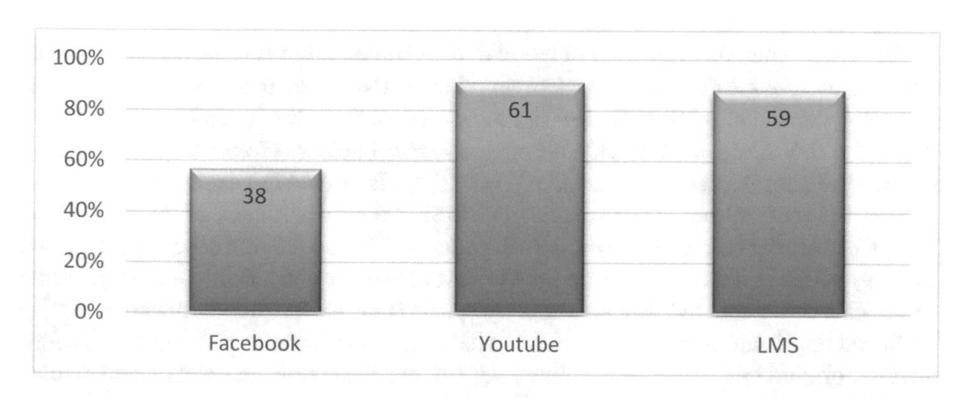

Fig. 2. Utilization of selected SSA by students of FIM, UHK for study purposes

The last graph (Fig. 3) shows the expectations of students in the potential of selected SSA.

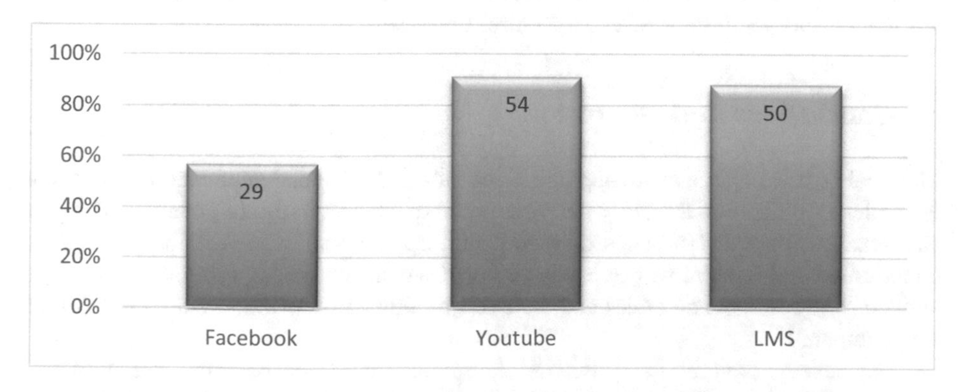

Fig. 3. Expectation of potential of selected SSA by students of FIM, UHK

81% of respondents find the greatest potential in Youtube. About 3% less, it means 88% of respondents find potential in the LMS and only 43% find potential in Facebook.

During deeper analysis it was found out that 3 students who do not use Facebook can see potential in this app. Here arises the question: Why he/she doesn't use it when he/she can see potential in it? But contrary, 35 Facebook users do not see potential for study purposes in this app. The question is why? Only 26 respondents out of 67 use Facebook and find potential in it for study purposes. When it comes to Youtube, all students use it, 54 respondents can see potential but it means that nearly 20% of them do not see potential in it. Regarding the third discussed app, learning management system (LMS) is not used by 5 students and these 5 students do not see potential in it, as well. Then there 12 respondents using it, but does not see any the potential in it. It can be assumed that in this case, some students are made to use LMS due to its link to their classes, where work in e-courses is mandatory. LMS is used by more than three quarters of students, 50 respondents use it and at the same time can see potential in this crucial teaching/learning platform.

In the case that the evaluation of use and of the potential of selected social software applications in education is conducted via other statistical methods, we can draw the following conclusions. Statistically significant consensus was found at LMS (Cramer's V = 0.487, sig = 0.000). On the contrary, Facebook was found to be nearly zero and insignificant consensus (Cramer's V = 0.093, sig = 0.447) which means that the answers are inconsistent. The results derive from the total numbers of the 67 respondents where Facebook is employed by 63 students, 38 respondents use it to study, but only 29, which is less than half find the potential for learning. For Youtube the consensus cannot be counted, because nobody stated 0 in the Youtube utilization.

Social applications were evaluated by students in the follow-up discussion. A few citations of students' statements follow. As for the Facebook one respondent wrote: "My First priority is Facebook (1) Solving Problems, (2) first response, (3) upload/download materials". YouTube was most widely discussed and appreciated application. One student highlighted visual power in the learning process "In my opinion YouTube is an excellent thing; thanks to videos we can comprehend everything much better than from theorems. I definitely can see future in YouTube". Another student perceives YouTube as kind of pieces of mosaic of the world: "YouTube – those are videos and solutions – science, culture, the world."

5 Conclusions and Discussion

The goal of this paper was to map utilization of selected social applications with focus on students' utilization for study purposes and expectation in future potential. The goal was reached; current landscape of used social applications was described from both quantitative and qualitative perspectives. For the data processing cross tabulation was applied. Via contingency tables and their results utilization of SSA and their potential was compared.

The survey brought the following findings: all 67 respondents' use Youtube, Facebook is used by 63 and LMS is used by 62 students. Regarding the use of the SSA for educational purposes, this area is still dominated by Youtube with 61 users, followed by 59 LMS users and only 38 students using Facebook for educational purposes. Regarding the potential of SSA for educational purposes, 80% of students find potential in Youtube. Almost ¾ of respondents find potential in the LMS. Despite the fact that 94% of respondents Facebook used in everyday life, for study purposes it is used only 56% and only 43% of them find the potential to study in the future.

The results show that students, who see potential in Youtube, use this application. As for the LMS, some students do not find potential for education in future but still use it. This result might be influenced requirements of some teachers to submit mandatory work in this platform. 28 students find potential and use Facebook. One can see potential but does not use this app. Other 35 respondents use Facebook but no not see potential. Based on the findings, all three of the hypotheses can be accepted.

- "When the student sees the potential in Youtube, he/she already uses it".
- "When the student sees the potential in Facebook, he/she already uses it."
- "When students see potential in the learning management system (LMS), he/she already uses it."

In the case that hypotheses had been set differently, they would have had to be rejected. Specifically, if it had been stated that: "students who use individual applications will find potential for study purposes" this statement couldn't have been accepted on the basis of gained results. Youtube is a winner as the most used application for both private and educational purposes as well as the app with the highest seen potential for future learning purposes followed by LMS and Facebook. The results show that students prefer learning in content-based SSA to profile-oriented SSA.

Future direction of the research might be conducted in the sphere of identifying the benefits of content-oriented networks compared to profile-oriented networks, which can lead to greater individualization in providing information and resources.

Acknowledgements. This study is supported by internal research project No. 2103 Investment under concept Industry 4.0 at Faculty of Informatics and Management, University of Hradec Kralove, Czech Republic. We would like to thank student Lenka Kutikova for cooperation in the processing of the article.

References

1. O'Reilly, T.: What is Web 2.0: Design Patterns and Business Models for the Next Generation of Software. O'Reilly website. O'Reilly Media Inc. http://www.oreillynet.com/pub/a/oreilly/tim/news/2005/09/30/what-is-web-20.html. Accessed 10 Dec 2016
2. Silapachote, P., Srisuphab, A.: Gaining and maintaining student attention through competitive activities in cooperative learning, a well-received experience in an undergraduate introductory Artificial Intelligence course. In: IEEE, Global Engineering Education Conference (EDUCON) 2014, Istanbul, Turkey, p. 295 (2014)
3. Jovanov, M., Gusev, M., Mihova, M.: The users' evaluation of an on-line collaborative activity for building ontology, In: IEEE, Global Engineering Education Conference (EDUCON) 2014, Istanbul, Turkey (2014)

4. Claros, T I., Garmendía, A.: Towards a collaborative pedagogical model in MOOCs. In: IEEE Global Engineering Education Conference (EDUCON), Istanbul, Turkey (2014)
5. Černá, M., Poulová, P.: Social software applications and their role in the process of education from the perspective of university students, In: Proceedings of the 11th European Conference on e-Learning, (ECEL 2012), Groningen, pp. 87–96 (2012)
6. Černá, M.: Trends in acceptance of social software applications in higher education from the perspective of university students - case study. In: Proceedings of the 13th European Conference on e-Learning, ECEL 2014, Copenhagen (2014)
7. Černá, M., Svobodová, L.: Utilization of social media and their potential in tertiary education – myths and reality, In: Proceedings of the 11th International Conference - Efficiency and Responsibility in Education, (ERIE 2014), Prague (2014)
8. Valtonen, T., et al.: Net generation at social software: challenging assumptions, clarifying relationships and raising implications for learning. Int. J. Educ. Res. **2010**(49), 210–219 (2011). Elsevier B.V
9. Weller, K. et al.: Social Software in Academia: Three Studies on Users' Acceptance of Web 2.0 Services; Web Science Conference 2010. Raleigh, NC. http://journal.webscience.org/360/2/websci10_submission_62.pdf. Accessed 10 Dec 2016
10. Tess, P.A.: The role of social media in higher education classes (real and virtual) – a literature review. Comput. Hum. Behav. **29**, A60–A68 (2013)
11. Greener, S.: How are Web 2.0 technologies affecting the academic roles in higher education? A view from the literature. In: Proceedings of the 11th European Conference on e-Learning, ECEL 2012, Groningen, pp. 124–132 (2012)
12. Villiers, A.: What is the difference? Cooperate and collaborate (2015). http://www.selectioncriteria.com.au/a-cooperate.html. Accessed 10 Dec 2016
13. Schroeder, A., Minocha, S., Schneider, C.: The strengths, weaknesses, opportunities and threats of using so social software in higher and further education teaching and learning. J. Comput. Assist. Learn. **26**, 159–174 (2010)
14. Social media – Definition and More from the Free Merriam. Webster Dictionary. http://www.merriam-webster.com/dictionary/social%20media. Accessed 10 Dec 2016
15. Janouch, V.: Internetový Marketing Prosaďte se na Webu a Sociálních Sítích. Computer Press, Brno (2010). 216 p.
16. Barnes, J.: Class and committees in a Norwegian Island Parish. Hum. Relat. **7**, 39–58 (1954)
17. Boyd, D.M., Ellison, N.B.: Social Network Sites: Definition, History, and Scholarship. J. Comput. Mediat. Commun. **13**(1), 210–230 (2007)
18. World map of social networks (2016). http://vincos.it/wp-content/uploads/2016/02/WMSN0116_1029.png

Electronic Mind Maps as a Method for Creation of Multidimensional Didactic Tools

Marina Mamontova(✉), Boris Starichenko, Sergey Novoselov, Kirill Zlokazov, and Marina V. Lapenok

Ural State Pedagogical University, Yekaterinburg, Russia
{mari-mamontova, zkirvit}@yandex.ru,
{bes, lapyonok}@uspu.ru, inobr@list.ru

Abstract. The article addresses the issue of tool support for teacher activity and learning activity of students. Synthesis of the technology of multidimensional didactic tools (hereafter MDT) and the smart technology for creation of electronic mind maps is suggested for presentation and analysis of the content and structure of educational material. Multidimensional didactic tools are universal logical and semantic models for the presentation and analysis of knowledge in a natural language possessing the properties of multi-dimensionality, fractality and solarity (have a radial and ray structure). Electronic mind maps provide technological and substantial flexibility in the development and use of MDT. Synthesis of technologies makes it possible to create verbal and image models for presentation of the educational material at various structural levels - from a separate element of knowledge to interdisciplinary links. Multidimensional didactic tools serve as a systemic orientation basis in the course of studying educational material and enable students to control knowledge quality and manage the development of their own knowledge. The synthesis of technologies is demonstrated using structuring and presentation of the content of the academic discipline "Testology".

Keywords: Professional training · Smart training · Smart technologies · Mind map · Electronic mind map · Structure of educational material · Multidimensional didactic tools

1 Introduction

The modern man lives in a dynamically developing information and technology environment. Development of technologies creates prospects for growth of human intelligence through brain-machine integration. It is expected that people with any level of ability will be able to quickly and efficiently acquire necessary knowledge and skills. Radical expansion of human intellectual activities will be ensured convergent development of nano-, bio-, informational and cognitive sciences and technologies [1].

However, the technologization of the intellectual activity has a reverse side. Psychological research and educational practice evidence significant changes in the natural human intellect due to the use of modern information and communication technologies.

V.L. Uskov et al. (eds.), *Smart Education and e-Learning 2017*, Smart Innovation, Systems and Technologies 75, DOI 10.1007/978-3-319-59451-4_38

People are increasingly focused on external storage devices which results in weakening of their ability to memorize and prevents formation of logical thinking. Experts are speaking about the development of the so-called “mosaic thinking”. The relatively easy access to information makes the need of learners to be able to create knowledge independently unnecessary. These phenomena are particularly dangerous in the course of specialist training when it is necessary, in a relatively short time, to assimilate and learn to apply large amounts of specialized knowledge.

Solution to the problem of individual knowledge and technologies lies in the area of interdisciplinary research. This article looks into one aspect of the problem associated with instrumental support of teacher activity and learning activity of students. The article is focused on the use of smart technologies for presentation and analysis of the content and structure of educational material. According to V.E. Steinberg, “the instrumental basis of traditional and emerging technologies needs to meet the requirements of universality, multidimensionality and geneticity (evolutionary development). The first requirement is determined by the fact that the language of instruction is a natural language, the second is driven by the development of didactics towards multidimensionality which is the most common characteristic of the reflected reality, and the third requirement is based on the principle of conformity with natural laws)” [2].

Multidimensional didactic tools are used for coding educational information to facilitate its memorization and increase the volume of involved content. Didactic multidimensional tools constitute “universal image and concept models for multidimensional knowledge presentation and analysis in a natural language in the external and internal planes of educational activities” [2].

The modern information technology market offers many services for knowledge presentation. To develop multidimensional didactic tools computer programs are necessary which on the one hand, make it possible to present the content and structure of the subject material and, on the other hand, enable the student to display and adjust the system of individual knowledge emerging in the course of studying the educational materials. Such tools can include computer programs (services) supporting creation and use of electronic mind maps in the course of learning. A mind map is presentation of information in graphic form reflecting links (semantic, associative, causal-and-effect and others) between concepts and parts of the studied subject area [3].

Knowledge representation using mind maps makes it possible to transfer from linear structures of educational material with their typical small portions that are studied spread in time and space which impedes the development of a systemic view and analysis of knowledge – to compaction and consolidation of the educational material and its presentation in interrelations using multidimensional nonlinear structures. Electronic mind maps provide flexibility in the use of multidimensional didactic tools, enable the students to control the quality of knowledge and manage the development of own knowledge in the process of studying the educational material.

The synthesis of the technology for development of multidimensional didactic tools with the technology for creation of electronic mind maps is shown on the example of structuring and presentation as a set of logical and semantic models of the content of the academic discipline “Testology”. Testology is taught when training master students of the Ural State Pedagogical University.

2 Knowledge and Its Types

The term "knowledge" means both "public (fixed, objectified using words, signs and symbols)" and individual "images of things, properties, processes, relationships of the material reality arising due to consolidation and generalization of the objective content of mental formations and stored in memory in the form of views, concepts and judgments" [4].

The following types of knowledge are classified in didactics - basic concepts and terms, facts, laws of science, theorics, ideas and knowledge about methods of activity [5]. The terms and concepts denote different objects or bodies of knowledge. Knowledge of facts reflects the reality and forms the basis for other types of knowledge. Laws reflect substantial links between specific sets of facts. Theories reflect sets of laws.

V.S. Tsetlin [6] proposes division of all body of knowledge to be assimilated in the course of studying an academic discipline into knowledge about the world and knowledge about methods of activity. Knowledge about the world is subdivided into theoretical and factual knowledge (material). The units of theoretical knowledge are "concepts of varying degrees of generality and systems of concepts and abstractions as well as theories, hypotheses, laws and methods of science." Factual knowledge is represented by individual concepts. Knowledge about methods of activity is presented as regulations (rules) and algorithms.

Knowledge is the basis of the educational content. Student knowledge is the result of the process of acquisition (remembering, understanding, application) of the subject matter. Knowledge is acquired by students if they show understanding of the system of the concept attributes and the system of concepts, the knowledge of the methods of activity and are able to store them in their memory ready to be operated in familiar and unfamiliar situations, within complex activity and within individual skills. The criterion for description of knowledge acquired by students is conformity of this knowledge with the content of subject matter specified in the state educational standards and educational programs in the form of requirements.

"Requirements to knowledge and skills is description of planned learning outcomes which provides understanding what and how the students should acquire and in what activities should knowledge and skills be manifested" [7]. The basis of the requirements to the outcomes of learning a discipline is a system of knowledge about characteristics of the studied objects and methods of activity.

3 Approaches to Knowledge Presentation and Analysis

3.1 Systemic Structural Approach

"Knowledge quality is identified as a result of multidimensional analysis of assimilation and application of knowledge in various activities. In education, the notion of "knowledge quality" includes correlation of the types of knowledge (laws, theories or applied, methodological, evaluation knowledge) with elements of the educational content and, thus, with levels of assimilation. Such correlation is necessary because any knowledge is potentially associated with its application, can be included in a creative process and acquire one meaning or the other" [8].

In the works of I.Ya. Lerner [8], M.N. Skatkin and V.V. Krajevsky [9], L.Ya. Zorina [10] a systemic structural approach to the description of knowledge quality is suggested and substantiated. Knowledge quality of students is considered at three levels - subject-content (reproduction of certain aspects of the educational content, reproduction of links between different content objects), content-activity (results of knowledge consolidation and actualization, its adjustment and application) and content-personal (results of knowledge application in independent extracurricular and academic activities using varied educational material).

The characteristics of individual knowledge at the subject-content level is its completeness, generalization and consistency. Completeness of knowledge reflects the result of reproduction of the features of the studied objects known to students that are necessary and sufficient for explanation of the essence of these objects. Generalization of knowledge characterizes the result of reproduction and explanation of the object essence based on the links between its features. "The essence of the study object directly depends on the level and method of its generalization. It can be a link between basic, major and marginal, additional features. At the stage of defining the concept this relationship acts as a relation between generic features, common for the class of objects, and features, that are specific for this type of objects. At the stage of judgment the essence is manifested as an idea of the belonging of an object to a class of objects, at the stage of conclusion – as a conclusion about belonging of an object of one class to another" [7]. When controlling knowledge it is necessary to take into account the level of generality of the educational material and correctness of its reproduction.

Systemacity of knowledge characterizes the result of reproduction by students of the essence of links and relationships of two or more objects of study. While assessing the knowledge systemacity it is important to identify critical links and relationships between the studied objects by their purpose (functional links), by origin (genetic links), by structure and interaction (subordination, inclusion links, etc.).

3.2 Systemic Evolutionary Approach

N.V. Timofeev-Resovsky connected the idea of systems with their evolution: "… any definition which we tried to form for the concept of a system, must necessarily include time, history, continuity; otherwise all is meaningless, and the concept of a system is entirely identified with the concepts of a structure" [11]. The systemic evolutionary approach is based on the idea of systemogenesis – harmonization of the idea of evolution and the principle of consistency. The system of individual knowledge of a student when studying the educational discipline undergoes various stages in the course of its development - from fragmented state to an integral system.

The system of individual knowledge of a student at various stages of studing the educational material can be correlated with a specific condition. The products of various stages of evolutionary development of the system of individual knowledge can be documented and analyzed in the control process. The connection of the condition of the system (the level of knowledge system development) of individual knowledge with its completeness and structure is reflected in SOLO-taxonomy (the structure of observed learning outcomes). In taxonomy the development levels of individual knowledge are identified - prestructural, monostructural, polystructural, systemic and relational. The system of individual knowledge at various levels of development is described using the completeness of the elements of its content and the links between these elements [12].

In the course of teaching it is important to consider both approaches - correlate (compare) the gradually evolving system of individual knowledge with the system of knowledge which was originally preset and reflected in the content and structure of the educational material.

4 Knowledge Presentation and Analysis Tools in Learning

4.1 Traditional Visual Teaching Aids

Human knowledge is based on assimilation of achievements in the cognitive activity of the previous generations. Transfer of experience is carried out using various educational techniques (tools). As a rule, in the capacity of such aids material objects are typically used to support the substantive activity of students, and formal languages and sign and symbol models for theoretical presentation of studied objects and phenomena. Traditional teaching methods are characterized by fragmentation of the educational material in small portions and by one-dimensional linear mechanisms for knowledge transfer. The used visual aids to present knowledge reflect small logically closed portions of the educational material. Multidimensional educational material is distorted, making it difficult to internalize knowledge to form multidimensional images in the internal plane of cognitive activity of students.

4.2 Multidimensional Didactic Tools

To support cognitive activity of students in verbal form Steinberg suggested the use of multidimensional didactic tools which make it possible to present various logical links between the elements of knowledge, to condense and minimize information, "to transfer from nonalgorithmic operations to algorithm-like structures of thinking and activity" [2].

The advantages of the suggested tools include the fact that they are based on the patterns of thinking and knowledge presentation and "are used for adequate knowledge explication and representation, knowledge operation, assigning materialized nature to knowledge, programming and control of their processing and assimilation" [2]. Multidimensionality is regarded as a particular property of knowledge visualization provided by combining meaningful properties of the studied object or phenomenon, on the one hand, and, on the other hand, consistent with the morphological features of the human brain.

Didactic multidimensional tools are developed on the basis of the principles of objectivity, systemacity, development, conflict, variability, integrity and multidimensionality. Along with these principles the principle of splitting and combination of elements is used (splitting of the multidimensional knowledge space into semantic groups and their consolidation in a system), the principle of coordination of the verbal and image dialogue in internal and external plane, the principle of the multidimensionality of knowledge presentation and analysis, the bi-channel activity principle (division of the information presentation and perception into visual and verbal channels), division of the design channel into the channel for designing educational models and the reverse channel for comparative and assessment activity).

5 Creation of Didactic Multidimensional Tools for an Academic Discipline

5.1 Basic Structures of Logical and Semantic Models

Creation of multidimensional didactic tools is based on the concept of multidimensional semantic spaces. Multidimensional didactic tools are measurer of such spaces with built-in multi-axial support and node frames with minimized information. Logical and semantic models for knowledge presentation and analysis contain two components – logical (the order of placement of coordinates and nodes), which is represented by the coordinate and matrix graphics, and semantic content - in the form of the content of coordinates and nodes (represented by the keywords).

Three basic types of frames (structures) have been developed: a support and node coordinate system, a support and node coordinate and matrix system, a support and node linking matrix as a part of the coordinate system. A detailed description of frames is presented in the work of [2]. Frames (structures) have the properties of multidimensionality, fractality (self-similarity) and solarity (radial ray structure). The structures can be considered as universal schemes for presentation of knowledge on an academic discipline. The concept of multidimensional semantic spaces is consistent with the idea of knowledge presentation using mind maps (radiant structure and verbal elements) reflecting non-linear structures formed by neurons in the brain during learning (neural networks).

5.2 Algorithm for Creation of Didactic Multidimensional Tools for an Academic Discipline

For the verbal and image presentation of educational material it is necessary to perform such operations as coordination of multidimensional semantic space, semantic granulation (separation of semantic information nodes), semantic coordination (arrangement of semantic coordinates from of semantic granules), semantic intersection (identification of semantic links between semantic granules located at different coordinates).

The algorithm for creation of multidimensional didactic tools:

- in the primary unstructured information “power information lines” are identified – semantic coordinates that are placed on a plane in the form of radial ray structure (identification of major sections in the content of an academic discipline);

- the initial information in accordance with the set of coordinates is divided into heterogeneous semantic groups with identified content elements arranged along the coordinates on a particular base (themes are specified within sections);
- Identification of semantic links between the nodal elements within themes and their location in intracoordinate spaces (or matrices).

The converted semantic space presents a semantically linked system where information bits acquire semantic valence (connectivity) features. The content of an academic discipline presented in this way provides a framework for systematic orientation of students in the studies material, for gradual development and updating of their knowledge. The system of individual knowledge emerging in the course of training acquires the features of the presented system.

5.3 Electronic Mind Maps as a Tool for Creation of Multidimensional Didactic Tools

Software programs for building mind maps is a convenient way to create multidimensional didactic tools. Among the many services we decided on XMIND software program [13].

In comparison with other services this program has a number of advantages: an opportunity to create free of charge an unlimited number of mind maps, to repeatedly refer to the created maps and develop them, to copy maps and transfer them to other users. The program contains various templates for future mind maps. There are templates with the structure necessary for creation of multidimensional didactic tools (support and node coordinate systems, support and node coordinate and matrix systems). Created maps can be used in two modes - online and offline. The online mode makes it possible to use sources of information in various formats and to embed in the map hyperlinks to these sources. When working offline, you can refer to files in the computer memory including to mind maps created and stored on this computer.

Therefore, the use of XMIND service makes it possible to create verbal and image models that reflect the broad logical and semantic space showing the links between the elements of the studied educational material at various structural levels - presentation of individual elements of knowledge, the intradisciplinary integration of the elements of various levels (theme, section) and the interdisciplinary integration.

6 Multidimensional Didactic Tools for Presentation and Analysis of the Content of the Academic Discipline "Testology"

Let us demonstrate the use of the synthesis of the multidimensional didactic tools technology and the technology for building electronic mind maps for creation of logical and semantic model of the structure and content of the academic discipline "Testology".

In the content of the discipline “Testology” several “power information lines” (major sections) are identified that are presented by using radial and ray structure (semantic space coordination). The content structuring in sections is shown in Fig. 1.

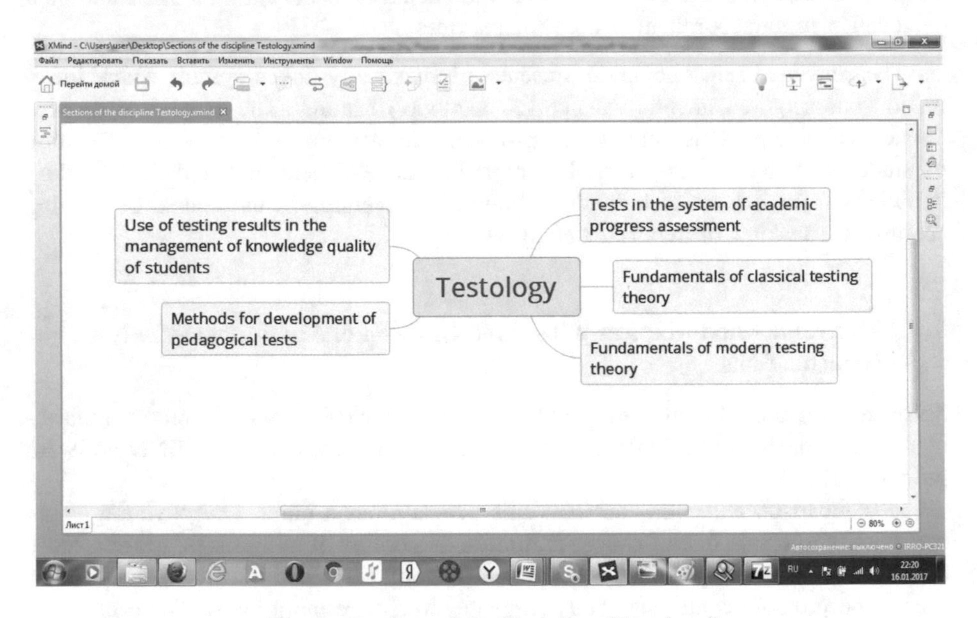

Fig. 1. Sections of the discipline “Testology”

The next structural level is identification of themes within sections (semantic granulation). Figure 2 shows a fragment of a mind map with highlighted themes in one of the sections of the discipline “Foundations of the classical testing theory”. Within each theme semantic nodes (elements of the content of the educational material) are identified - separate concepts, facts, laws and objects studied within the theme). The nodes can be described and presented using various formats (verbal description in the form of notes, a chart, a drawing, a separate diagram, an audio file, a video segment, etc.).

To create a single semantic space links are established between the selected nodes. Links between elements are presented in a mind map using various tools (floating sections, matrices and arrows located in intercoordinate intervals). Figure 3 presents links between knowledge elements of two themes.

It is obviously not possible to present on a sheet of paper a complete structured system of knowledge on a discipline. The function of electronic mind maps is to expand and minimize separate coordinate lines and nodes and they have virtually no space limitations.

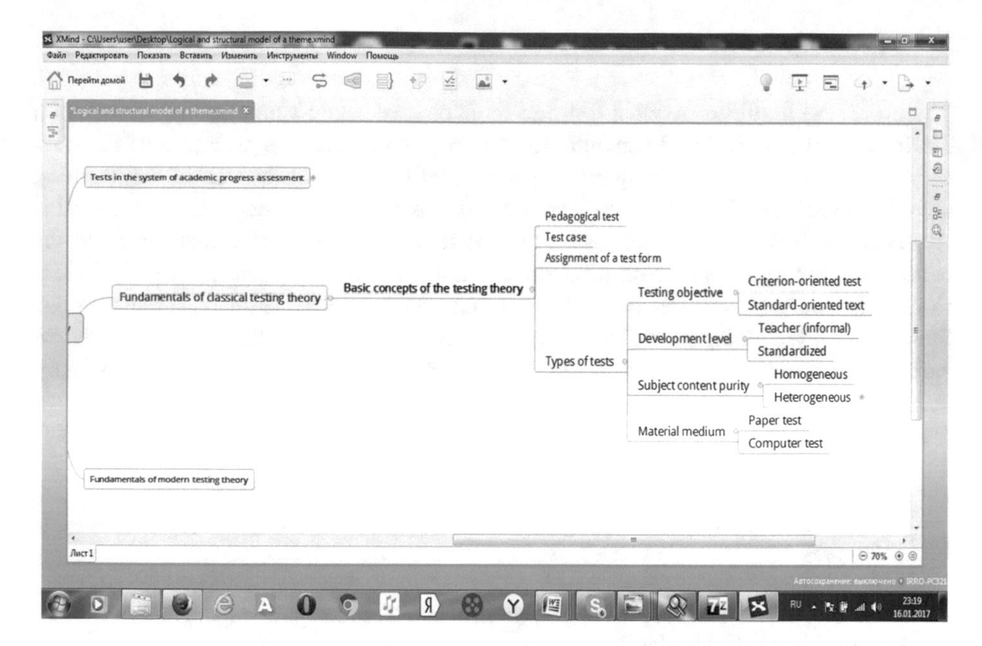

Fig. 2. Logical and structural model of a theme

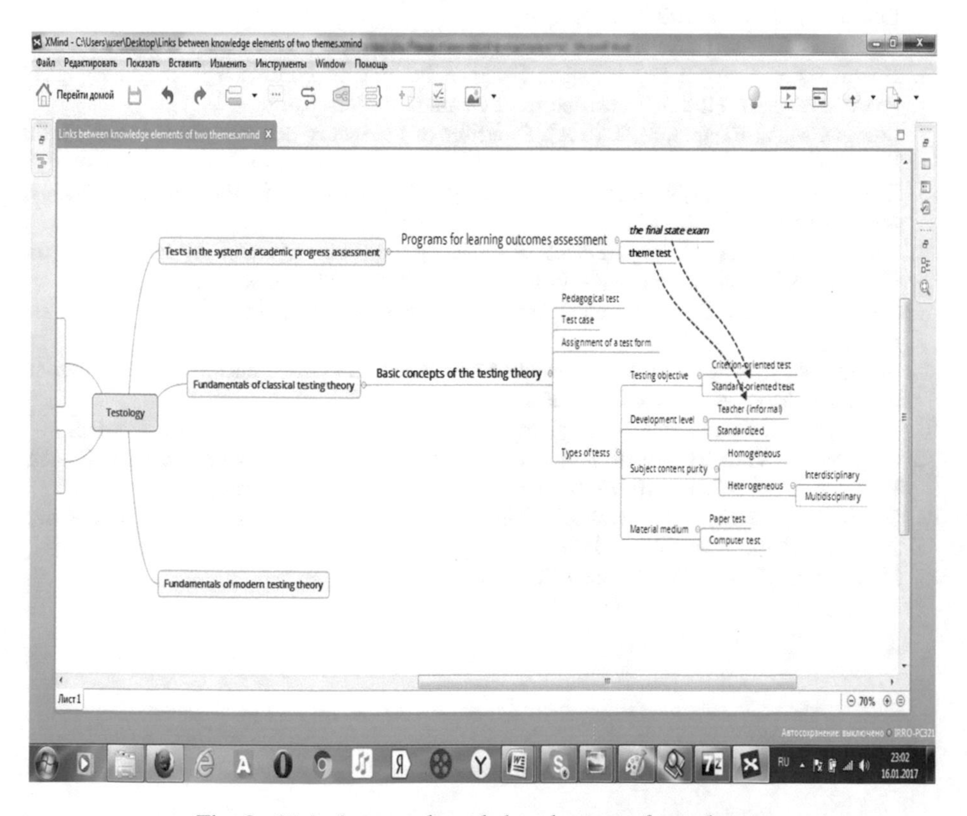

Fig. 3. Links between knowledge elements of two themes

7 Conclusions

Therefore, the multidimensional didactic tools created using electronic mind maps on the basis of the logical and semantic models of the content and structure of the educational material, serve as a systemic orientation basis in the course of the learning activity of students. The structure of verbal and image models of educational material is consistent with the structure of neural links that are formed in the human brain in the course of learning. Application of smart technologies to development of multidimensional didactic tools ensures technological and conceptual flexibility of created tools. Synthesis of technologies makes it possible to present the educational material at various structural levels - from a separate element of knowledge to interdisciplinary links. Multidimensional didactic tools enable students to manage the development of their own knowledge.

References

1. Alekseeva, I.Yu., Nikitina, E.A.: Intelligence and Technologies: Monograph. Prospect, Moscow (2016). (in Russian)
2. Steinberg, V.E.: Educational Multidimensional Tools: Theory, Methods, Practice. National Education, Moscow (2002). (in Russian)
3. Buzan, B., Buzan, T.: Super-Thinking. Potpourri, Moscow (2003). (in Russian)
4. Kossakovski, A. (ed.): Psychological Basis of Personality Formation in the Pedagogical Process, Moscow (1981). Translated from German. (in Russian)
5. Danilov, M.A., Skatkin, M.N. (eds.): Didactics of Secondary School, Moscow (1975). (in Russian)
6. Tsetlin, V.S.: Underachievement of Secondary School Students Its Prevention. Pedagogy, Moscow (1977). (in Russian)
7. Kuznetsova, A.A. (ed.): Requirements to knowledge and skills of secondary school students: Didactic and methodical analysis. Pedagogy, Moscow (1987). (in Russian)
8. Lerner, I.Ya.: Quality of students' knowledge. What should it be? Moscow (1978). (in Russian)
9. Skatkin, M.N., Krajevsky, V.V.: Quality of students' knowledge and ways to improve it. Pedagogy, Moscow (1978). (in Russian)
10. Zorina, L.Ya.: Consistency is the quality of knowledge, Moscow (1976). (in Russian)
11. Malinovski, A.A.: Theory of Structures and Its Place in the System Analysis, System Research, Yearbook, pp. 10–79. Science, Moscow (1970). (in Russian)
12. Biggs, J.B., Collis, K.F.: Evaluating the Quality of Learning: The SOLO Taxonomy. Academic Press, New York (1982)
13. http://soft.mydiv.net/win/download-XMind.html

M-learning and the Elderly: Construction of Inclusive Pedagogies

Leticia Rocha Machado(✉), Jozelina Silva da Silva Mendes, Deyse Cristina Frizzo Sampaio, Tássia Priscila Fagundes Grande, and Patricia Alejandra Behar

Digital Technology Center Applied to Education (NUTED), Federal University of Rio Grande do Sul, Federal University (UFRGS), Avenue Paulo Gama, 110, 12105 Building, Room 401, Porto Alegre, RS 90040-060, Brazil
leticiarmachado@gmail.com, jozelinasilvadasilva@gmail.com, deysefrizzo@gmail.com, tpri.fagundes@hotmail.com, pbehar@terra.com.br

Abstract. The objective of this study was to identify possible pedagogical strategies for m-learning with the elderly in order to develop inclusive pedagogical practices. Each year new digital technologies, such as smartphones and tablets, are being developed and consumed by different age groups. At the same time, the number of people 60 years of age and older continues to increase. Thus, it is important to investigate possible pedagogical strategies for m-learning, aimed to include the elderly in the digital and mobile world. Therefore, this work has used a qualitative and quantitative research methodology with a case study. Five steps were adopted in the methodology to address the research objective. The study was carried out with a group of 23 elderly people that had a mean age of 70 years of age. These subjects participated in a digital inclusion course at the Federal University of Rio Grande do Sul (UFRGS), Brazil. As a pedagogical practice, the research adopted the construction of applications in order to promote authorship for the elderly, as well as to encourage them to use of their smartphones and tablets as a possible way of interacting with society. Twenty-three applications were developed that addressed different themes. Self-valorization of the elderly as authors of their digital materials was observed during the course of the project, transforming them from passive to active students and including them in digital society.

Keywords: M-learning · Elderly learners · Inclusive pedagogy

1 Introduction

The number of elderly people is continually increasing all over the world, including in Brazil. According to the data presented by the IBGE [1], the number of people over 65 has risen significantly. In fact, in 2000 they made up 5.61% of the general population and in 2015 this number rose to 7.90%. It is expected that by 2030, the number will increase to 13.44%.

V.L. Uskov et al. (eds.), *Smart Education and e-Learning 2017*, Smart Innovation, Systems and Technologies 75, DOI 10.1007/978-3-319-59451-4_39

At the same time, a survey conducted by the Getulio Vargas Foundation University (FGV), in São Paulo, Brazil, revealed that in 2015 there were 306 million devices connected to the Internet in Brazil. Moreover, 154 million of these were smartphones [2]. Data clearly demonstrates both the increase of mobile devices and the elderly population, revealing new conceptions of the world, where transformations are directly influencing society and its cultures, both collectively and individually. It is therefore quite pertinent to investigate new pedagogical practices that can include older individuals in this constantly changing society.

Hence, this research aimed to identify possible pedagogical strategies for developing m-learning with the elderly in order to create inclusive pedagogical practices. M-learning incorporates the use of mobile technologies separately or in conjunction with other Information and Communication Technologies (ICT) [2]. In fact, "UNESCO believes that mobile technology can improve and enrich educational opportunities for students in different environments" [2].

There are various studies about the use of mobile devices for educational purposes, both in formal educational institutions (such as elementary schools, high schools, colleges, and universities) as well as in non-formal spaces (hospitals, digital inclusion, non-governmental organizations, etc.).

Thus, using pedagogical strategies supported in m-learning can help the elderly overcome the barriers of fear and insecurity related to the use of mobile devices and, consequently, the challenges they encounter in society in general. The development of smart pedagogy through pedagogical strategies aimed at the elderly can include this public in the contemporary highly connected and technological society. The next section of this article therefore presents two aspects related to aging and digital technologies with a focus on education. Next comes a discussion of the characteristics of m-learning, methodology, analysis, and presentation of the data. Lastly it provides final considerations.

2 Aging and Mobile Devices: An Educational Focus

The number of elderly people in Brazil will almost triple by 2050 [3]. This is also the trend in more developed countries, such as Spain and Portugal. The WHO stresses that improving and/or maintaining the quality of life of older people is becoming urgent and necessary, both for them to grow older healthier and increase their autonomy. Hence, the importance of permanent education for the elderly is highlighted as a way of promoting learning, sharing experiences, as well as aiding in their psychological, social, biological, cultural, and technical development.

Some authors such as Freire [4] and Delors [5] write about the importance of permanent education, or studying throughout one's entire life, since there is no specific age group designated for learning.

According to Doll, Ramos and Buaes [7] "educational activities aimed at the elderly should not be done without reflection on the pedagogical methods used and the professional training of the people involved". These strategies should be in accordance with the context of the elderly [7], providing creativity, self-knowledge, in a manner that adequately achieves the goals of educational gerontology.

At the same time, each year the number of mobile devices, such as tablets and smartphones, being produced and offered to the public increases. The "Brazilian Media Study" [8] observed that the cell phone was the most popular means of accessing the Internet (66%), followed by the tablet (7%). This demonstrates that Brazilians are using mobile phones for different purposes, including to access digital tools on the Internet. Unfortunately, this survey did not specify the data by age, making it difficult to conduct a detailed analysis of the use of the devices by the older public. However, has been shown previously that the increasing number of smartphones and tablets in homes has raised the elderly public's interest in using these devices.

Societal inclusion increasingly implies knowing about new technologies. In fact, it is a way of being socially integrated, because entertainment and communication permeate this area. Despite the insertion of digital technologies in daily life, a portion of the population still has difficulty or does not know how to handle them. The elderly are among the digitally excluded. They often do not use these technological resources because they do not master them. Possible factors could be lack of resources, not feeling safe to explore them on their own, or lack of confidence to seek help on the subject.

The elderly have been searching for spaces where that they can learn to use digital technologies, especially mobile devices. These spaces can be smart classrooms that allow the elderly to not only learn to use mobile devices, but also to reflect on their role as citizens in contemporary society. In education, research is still limited in the area of mobile learning and the elderly, as will be discussed below.

3 M-learning: Permanent Education Through Mobile Devices

M-learning refers to learning supported by the use of Information and Communication Technologies (ICT) [2]. Hence, the fundamental characteristic for learners is mobility. According to the same author, this process makes it possible to learn when individuals are far from each other and/or in informal educational spaces. In spite of distance and space, one can nonetheless be integrated with other learners, allowing for the formation of virtual and real networks of interaction.

Thus, according to UNESCO [2], m-learning enables mobility and connectivity of education. From this perspective, among the benefits considered by UNESCO is the flexibility that mobile learning provides. A student can "advance at his own pace and pursue his own interests, potentially increasing his motivation to pursue learning opportunities" [2]. Hence, it is possible to see the benefits of m-learning for the permanent education of the elderly.

Different studies are being conducted on mobile devices, especially smartphones, and the elderly due to the increasing demand for these technologies by the older population.

Zimer et al. [9] describe the development of an application using tablets, to improve memory in the elderly. This study aimed to maximize the levels of cognitive performance among the elderly.

Various authors have discussed which application interfaces should be used by the elderly. Strengers [10] addressed the most important features that should be considered for the use of smartphones by the elderly. The development of applications aimed at this audience should consider not only their cognitive and motor necessities, but also their social context and previous knowledge. Therefore, personal interests, curiosity, and memories can bring older people closer to digital technologies such as mobile devices [12–14]. In this context, creating applications with content that proposes critical reflections about themselves, or even the role of mobile devices within society, may also be an alternative.

Thus, developing applications (authorship) is a possible pedagogical strategy that can be used in m-learning with the elderly. Moreover, it is the methodology adopted in the research that will be presented below.

4 Methodology

This research uses a qualitative case study. The case study format was chosen in order to carry out research with different approaches and resources to collect data. The subjects were 23 elderly individuals, who were 60 years of age or over, who participated in a digital inclusion course at the Federal University of Rio Grande do Sul. To address the ethical questions of the study, all the participants signed an informed consent form.

Data was collected through various instruments: Participant observation, the elderly participants' technological production (applications), and an interview. The present research project consisted of five steps, as will be discussed below.

Step 1 - Development of the theoretical reference: This step was focused on the construction of the theoretical framework considering studies in the areas of Education, Aging, and Technologies (m-learning).

Step 2 - Courses to for the elderly to use mobile devices: Two courses called "DiMIApps - Mobile Devices for the Elderly: Apps" and "DiMIHobbie - Mobile Devices and Hobbies for the Elderly" were offered. The courses were designed in order to oofer smart classrooms for the elderly. They both took place in the second semester of 2015 and aimed to empower the elderly through the use of mobile devices and developing the authorship through the construction of applications.

Step 3 - Data collection and analysis: This step was focused on the analysis of the data collected in the previous stage. A qualitative analysis of the data was preformed through content analysis, including critical or hidden understandings of communication. To do so, we used the steps to conduct content analysis sugessted by Bardin [11].

Step 4 - Construction of pedagogical strategies: Based on the framework built in Step 1 and the data collected from the course in Steps 2 and 3 it was possible to map probable preliminary pedagogical strategies for m-learning for the elderly. These will then be re-evaluated by elderly individuals when they are applied in a course that will be offered in the first semester of 2017 with another group of elderly people at the same university. This will enable the last step to be developed.

Step 5 - Dissemination of results: Based on the data collected and its interpretation, the results will be published in articles, as well as presented at national and international conferences.

5 Analysis and Discussion of the Data Collected

In order to respond to the goal initially proposed, some notions regarding the use of mobile devices (smartphones and tablets) for the elderly were introduced, including settings, use of the applications, and forms of communication. Data were collected through the researchers' observations during the course, the applications produced by the elderly, and interviews with the participants which included questions about the use of mobile devices The individual construction of an application was proposed for the elderly students with the intention of instigating authorship, incorporating different resources such as sound, images, references for more information, etc. The tool used was AppyPie (http://www.appypie.com/).

A total of 23 elderly people, with a mean age of 70, participated in this study. Only three of the participants were male. Each elderly student chose a hobby that they wanted to explore further. Initially, everyone searched on the Internet for materials related to the subject that they had chosen and collected materials (photographs, information sites, videos, etc.). Next, there was an explanation of the functions and how to use the AppyPie tool. During the process, data collection instruments were applied, such as interviews and participant observation.

The applications that were developed dealt with different themes such as: Knitting, Volunteering, Brazilian Music, Dance for the Elderly, Digital Inclusion, Dog Health, Literature, Water Aerobics, Photography, Pilates, Growing Roses, Crafts, Gastronomy, and Tourism.

The students also stated that when building their application they verified that it was working properly. According to one of the elderly participants, "I tested everything, yep. I tested every link that I put in it. Then I went there in the application itself, pasted it, opened it, opened the right pages, the sites, everything (J)." The 23 elderly people also pointed out that they created texts in the form of abstracts in the application in order to be objective, short, and facilitate reading. They justified their choice as practical and the best way of addressing the subject in a more general manner.

All of the participants stated that they were careful about the size and color of the letters, background color, etc. They prioritized a set of pleasant and intuitive combinations. They began to construct materials based on their previous knowledge of the theme, and some then did more in-depth research. According to Doll, Machado and Cachioni [6], the elderly accumulate growing experience, which can make the application a rich resource in the process of teaching and learning. In fact, the elderly's previous experiences and knowledge may be contemplated in the construction process. The vast majority of the participants, 22, claimed to have chosen topics that interested them and that are or have been a part of their lives in some way. Satisfaction in learning and achieving personal goals are examples that have the greatest effect on external rewards and incentives [7].

The classes provided more in-depth analysis of this topic. It was observed that the choice of hobbies was a pleasurable experience, because it was within a context that the participants already knew. Thus it facilitated the research and presentation of the app. All applications were constructed in the Portuguese language, which is the mother tongue of the participants, as can be seen in Figs. 1 and 2. Some of the elderly people addressed hobbies from other times in their lives (such as the "My Brazilian Music Application") (Fig. 1), to reminisce. Others chose more current themes (such as the "Dog Health Application"), always trying to show how important and pleasant the hobby is (Fig. 2).

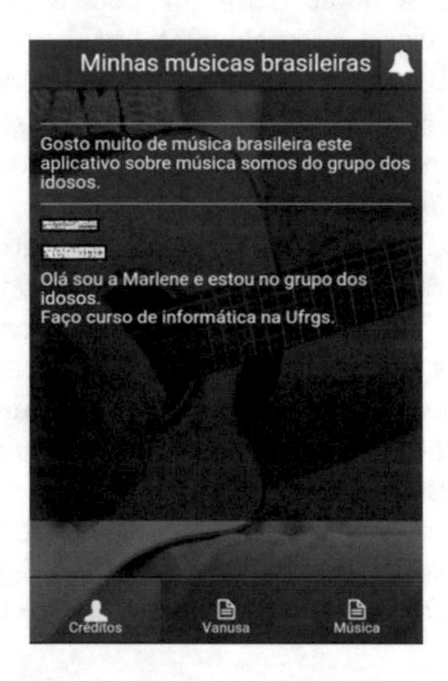

Fig. 1. "Credit screen" of "My Brazilian Music Application". Available at: http://ptsnappy.appypie.com/html5/minhas-músicas-brasileiras

On the final day of class, which focused on the completion of the apps, two older women were discussing the topics covered in their respective projects. One of the them started crying. She had become emotional because the recipe presented in the app reminded her of the good times she had with her mother in the kitchen. One elderly student (U) said, "In those days, on my birthday, I got those 'cookies'." Another participant then took one and put it in her mouth. Soon tears began to fall. She said, "Oh, I remember my mother making these cookies at Easter. And I also have this memory when I was 8 and 11 years old. We had a gigantic table and we covered it with cookies. I had a really big family. So you have this thing that you have made since you were a child (O)".

Based on the application of the data collection instruments, it was possible to map the pedagogical strategies considered relevant for m-learning with the elderly:

- Mobile devices: It is important to survey the mobile devices that the elderly access, especially the models and operating systems, because this will inform the organization of classes;

Fig. 2. "Credit screen" of "Dog Health Application". Available at: http://ptsnappy.appypie.com/html5/saÚde-dos-cÃes

- Applications: The choice of applications should be made based on a survey of the elderly themselves. Listing everything they wish to learn, starting by talking with them to instigate students them continue to use the devices. Applications should be free and not have advertisements that may make them difficult to use. The elderly prefer applications that are useful in their daily lives;
- Critical thinking: In addition to working with how to use mobile devices and their resources, it is also important to develop actions that encourage students to be critical of the tools used. After each application was explained, some questions were asked to evaluate the tool. These included not only the content and information provided in the application, but also layout, colors, language, etc. This activity assisted the elderly when it was time to develop their own materials (authorship);
- Authorship: Authorship should be instigated with older individuals. It can be authorship within applications like videos, music or photos, as well as building one's own apps. In order to achieve this goal, it is necessary to consider the limitations of the tools (accessibility and usability) for the public, since these difficulties will discourage students. It is also important to always plan and explain the objectives of the activity.

Therefore, the use of these pedagogical strategies can help the development of smart classrooms where the elderly can learn to use the mobile devices by interacting with these technologies in the classroom and in their daily lives.

The main difficulties in the use of mobile devices are the size and type of screen (touch screen) because many elderly people have visual limitations and lack of sensitivity in their fingers to operate the touch screens on the devices. These difficulties stem from human aging, but do not prevent the public from using such devices. Yet, they do require adaptation and the development of new approaches (pedagogical strategies) to enable the elderly to use mobile devices in spite of their limitations. For the most part, they pointed out that they will continue to use mobile devices because this type of technology provides mobility and enables instantaneous contemporary communication. This group of seniors will continue in courses on the use of mobile devices offered at the University in the year 2017.

The results collected point to the importance of working with mobile devices and the elderly, especially developing strategies that allow for authorship in m-learning.

6 Final Considerations

This study aimed to identify possible pedagogical strategies for the development of m-learning with the elderly in order to develop inclusive pedagogical practices.

Constructing an application was an important achievement for the elderly, because they were the authors of materials they developed for the device they were using. The applications were presented in an exhibition open to the public, making it possible to involve the students' family members and friends. Moreover, the elderly students were very happy and proud to show their work. Thus, in planning pedagogical strategies for the elderly to use mobile devices, one should consider their interests, expectations, skills, and difficulties, especially considering authorship. Developing inclusive pedagogical practices therefore requires planning strategies that consider questions and difficulties related to cognition, motor skills, vision, and hearing. This is particularly true for m-learning, because mobile devices differ from other technologies. Hence, the construction of applications can provide innumerable benefits for students, since it offers a range of materials such as texts, videos, images, presentations, stories, etc., helping to bring the elderly even closer to these technologies.

References

1. IBGE. Instituto Brasileiro de Geografia e Estatística. Projeções e estimativas da população do Brasil e das Unidades da Federação (2015). FGV. http://www.ibge.gov.br/apps/populacao/projecao/
2. UNESCO. Policy Guidelines for Mobile Learning. Publicado pela Organização das Nações Unidas para a Educação, a Ciência e a Cultura (UNESCO) (2013). http://unesdoc.unesco.org/images/0021/002196/219641e.pdf
3. OMS. Organização Mundial da Saúde. Resumo. Relatório Mundial de Envelhecimento e Saúde. Organização Pan-Americana da Saúde, Brasília (2015)
4. Freire, P.: Educação e Mudança. Paz e Terra, Rio de Janeiro (1993)
5. Delors, J.: A educação ou a utopia necessária. In: UNESCO. Educação: um tesouro a descobrir. Cortez, São Paulo (2010)
6. Doll, J., Cachioni, M., Machado, L.R.: As novas tecnologias e os idosos. In: Py, L. (ed.) Tratado de geriatria e gerontologia. GEN, Rio de Janeiro (2016)
7. Doll, J., Ramos, A.C., Buaes, A.S.: Educação e Envelhecimento. Educação Realidade **40**(1), 9–15 (2015)

8. Brasil. Presidência da República. Secretaria de Comunicação Social. Pesquisa brasileira de mídia: hábitos de consumo de mídia pela população brasileira. Secom. 4, Brasília (2014)
9. Zimer, M., et al.: Um aplicativo móvel para treino de memória em idosos: desenvolvimento e avaliação. Nuevas Ideas en Informática Educativa TISE. http://www.tise.cl/volumen9/TISE2013/715-718.pdf. Accessed 7 Jan 2017
10. Strengers, J.: Smartphone interface design requirements for seniors. http://dare.uva.nl/cgi/arno/show.cgi?fid=460020. Accessed 7 Jan 2017
11. Bardin, L.: Análise de conteúdo. Edições 70, Lisboa (2004)
12. Yang, M., Huang, H.: Research on interaction design of intelligent mobile phone for the elderly based on the user experience. In: Human Aspects of IT for the Aged Population: Design for Aging, vol. 9193(1), pp. 528–536 (2015)
13. Yu, Z., Liang, Y., Guo, B., Zhou, X., Ni, H.: Facilitating medication adherence in elderly care using ubiquitous sensors and mobile social networks. Comput. Commun. **65**(1), 1–9 (2015)
14. Mertens, A., Rasche, P., Theis, S., Wille, M., Schlick, C., Becker, S.: Influence of mobile ICT on the adherence of elderly people with chronic diseases. In: Zhou, J., Salvendy, G. (eds.) DUXU 2015. LNCS, vol. 9194, pp. 123–133. Springer, Cham (2015)

Developing a Web-Based Application as an Educational Support

Petra Poulova(✉) and Blanka Klimova

University of Hradec Kralove, Hradec Kralove, Czech Republic
{Petra.Poulova,Blanka.Klimova}@uhk.cz

Abstract. At present electronic education, the so-called eLearning, is quite widespread at all levels of educational institutions. There are many modalities of eLearning and web-based instruction is one of them. Therefore the purpose of this article is to discuss a procedure and pedagogical principles imposed on the development of any web-based application and technologies used in the creation of such web application. In addition, the authors of this article provide a specific example of the development of one smart web application which was designed for its use in the first grade of one basic school in the Czech Republic. The authors emphasize its benefits and list several improvements for its future use.

Keywords: Web · Application · Development · Education · Support · Learners

1 Introduction

At present electronic education, the so-called eLearning, is quite widespread at all levels of educational institutions. There are many definitions of eLearning. For example, the eLearning Action Plan defines it as the *use of new multimedia technologies and the Internet to improve the quality of learning by facilitating access to resources and services as well as remote exchanges and collaboration* [1]. Wagner defines it as *the educational process which uses information and communication technologies for designing courses, distributing the learning content, for teacher-learner and learner-learner communication and managing the whole process* [2]. Both definitions described above emphasize the enhancement of quality of education of eLearning. However, it was not always that case. Originally, the technological component was preferred (cf. [3] or [4]). Furthermore, there are many modalities of eLearning and web-based instruction is one of them. Thus, the purpose of this article is to discuss procedure pedagogical principles imposed on the creation of any web-based application and technologies used in the creation of such web application. In the end the authors of this article provide a specific example of such a web application which was designed for its use at a basic school.

2 Pedagogy

The procedure of the development of a web-based application is similar to the development of any eLearning course and includes the following steps:

V.L. Uskov et al. (eds.), *Smart Education and e-Learning 2017*, Smart Innovation, Systems and Technologies 75, DOI 10.1007/978-3-319-59451-4_40

- setting the aim, i.e., the purpose for which the application is developed;
- analysis of a target group, i.e., both teachers' and learners' age, IT skills, level of knowledge, and their existing experience with eLearning education;
- setting the study objectives, i.e., what the learner will be able to do after studying specific study material;
- choice of educational content, i.e., the choice of subject matter, its extent and organization [5].

In addition, the pedagogical principles of the creation of a web-based application are very important and they should involve the following aspects:

- Clarity of the application, i.e. the organization of the text, which should be structured into shorter paragraphs, and important information should be highlighted;
- Attractivity of the study material, i.e., the text should contain multimedia features such as animation, schemes, figures, or simulations, and there should be also practical examples;
- Motivational features, i.e., there should be a guide to this web-based learning in order not to discourage the learner already at the very beginning should they not know how to use this application, there should be self-tests, evaluation and communication tools;
- Feedback for the learners, provided by a tutor in order to stimulate learners in their studies [6, 7].

3 Technologies Used in the Development of a Web-Based Application

There is a number of technologies which can be used for the development of web-based applications. One of them is Spring Framework 4.0, which is an open-source application framework that facilitates and accelerates the development of J2EE applications for JAVA platform. The advantage of Spring Framework is its modularity and scintigraphic, which means that a developer can use only those modules s/he needs [8, 9].

Spring Framework is created by several modules which are grouped into bigger logical entities: Core Container, data Access/Integration, Object Oriented Programming, Aspects, Instrumentation, Messaging, Web and Test. Consult Fig. 1 below for more detailed information.

The basic foundation stones of Spring Framework are Dependency Injection and Aspects Oriented Programming. Dependency Injection is a general proposal model on the basis of which applications are developed by composing already used components with the purpose to reduce dependences among these components [9]. Object Oriented Programming (OOP) is a concept whose aim is to increase modularity of the developed application. OOP consists in setting aside portions of code that are used repeatedly, into the so-called aspects. Typical use of OOP is event logging, validation and transaction processing.

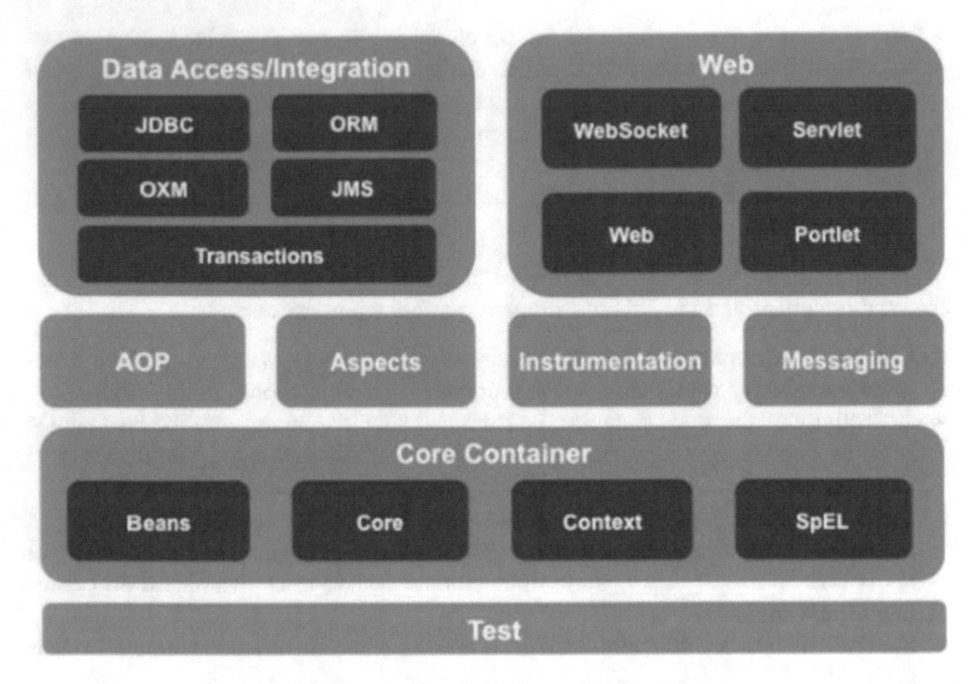

Fig. 1. Spring Framework modules [9]

4 Project of the Development of a Specific Application for the First Grade of a Basic School in the Czech Republic

The project assignment was made on the basis of consultations with teachers of the first grade of the basic school. The teachers were asked what they would expect from the application, how practicing of the study materials should be run, and what outcomes they would like to have. The assignment was divided into five basic parts: users' description in the application, course of practicing the study material, required output, motivational system for pupils and the application design.

In the application there are three types of users – pupil, teacher and teacher with administrator's rights. When entering the application, it is necessary for the user to use a login and a password. If the authentication is successful, the logged in user is redirected to his/her home page, the content of which varies according to which permission s/he has. The application enables keeping records of pupils and teachers. It is possible to add new pupils and teachers, adjust the present ones and if necessary, to delete them from the database. If the user is a pupil, s/he can see the statistics of his/her practicing next to each topic s/he has done. This statistics serves as feedback to a pupil so that s/he can see what subtopics are still difficult for him/her and should practice them more. By selecting the required subtopic s/he is straight redirected to practicing it. Each exercise consists of ten test questions which are selected on the pupil's past practicing. After completing the last question, the pupil can see an overview of all test questions with the right answers and feedback on his/her performance in the test.

Teachers have reports on pupils' practicing of individual topics. They have a possibility to obtain not only the summary reports of the whole class but also the statistics of practicing (number of completed exercises and number of correct and incorrect answers) for each pupil.

At the end of each practicing, the pupil can see the result of his/her performance. This report also contains verbal evaluation which differs according to how many points the student gained in his/her practicing so that in case of low score s/he could be stimulated for further practice. Pupils are also rewarded for a certain number of completed exercises by badges. Pupils of thc same class should be also motivated among one another, which can be achieved by the so-called ladders that can show them how well they perform in comparison with their schoolmates.

Figure 2 below presents the application in the form of a model of use cases which are based on the functional requirements analysis and a class diagram

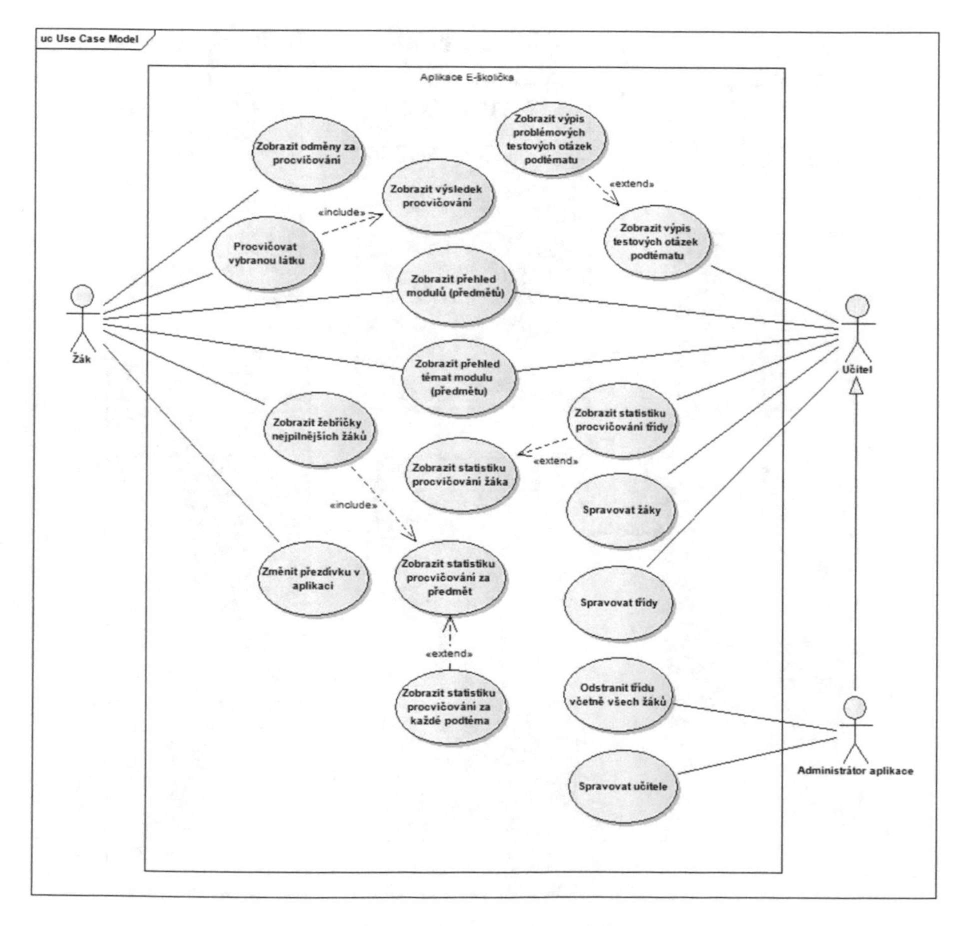

Fig. 2. Application model

The choice of the test questions for practicing is conducted on the basis of priority assignment which is maintained by TestItemPriority entity (Fig. 3).

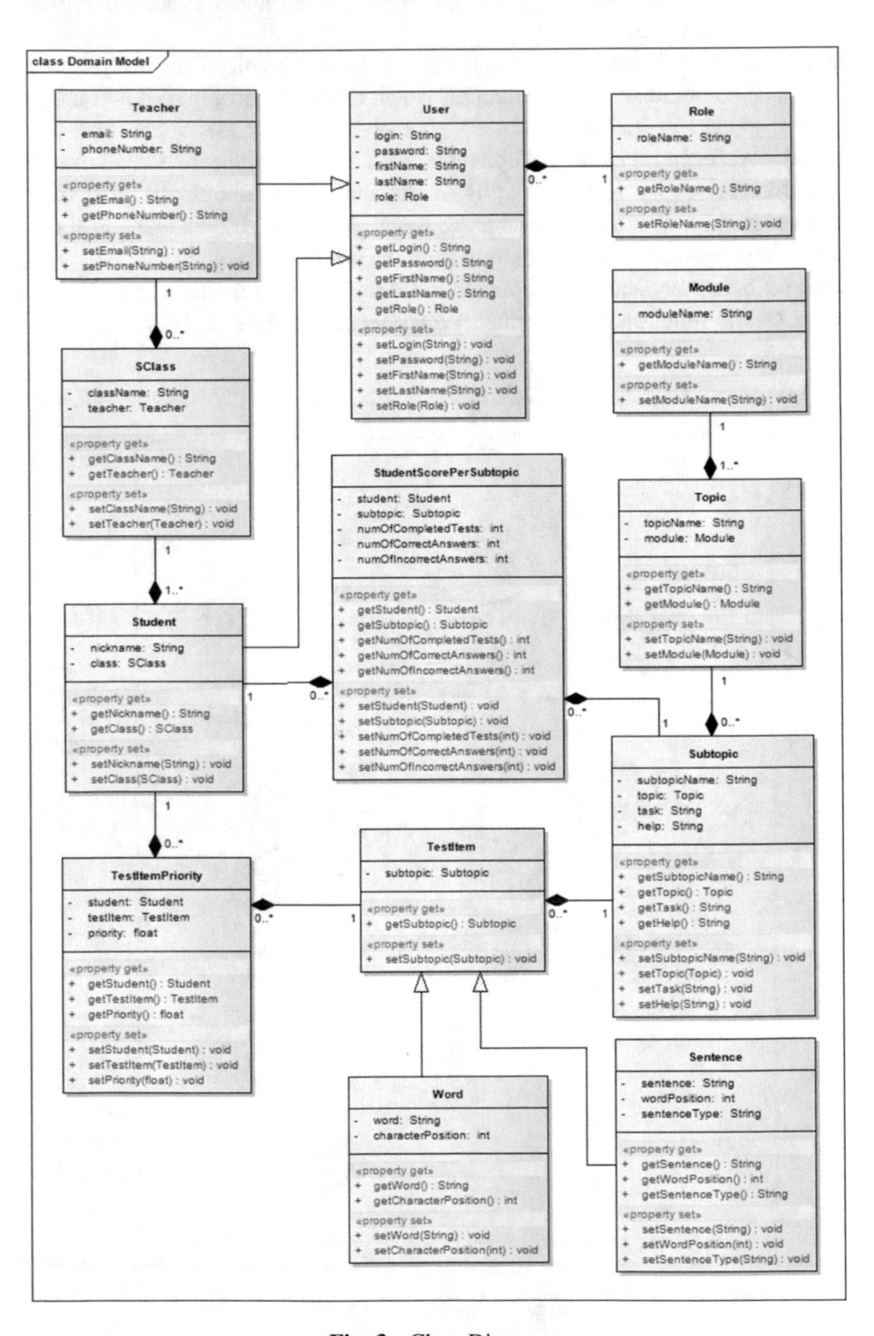

Fig. 3. Class Diagram

The value of priority of test question assignment ranges between 0 and 1. At the beginning all test questions are for a pupil priority 1, i.e., the highest. At the moment of evaluation of pupil's answer of practicing, the value of priority is automatically updated in the database as follows (Fig. 4):

```
@Entity
@Table(name = "test_item_priority")
public class TestItemPriority {

    @Id
    @GeneratedValue(strategy = GenerationType.AUTO)
    @Column(name = "priority_id")
    private Long id;

    @Column(name = "priority", nullable = false)
    private float priority;

    @ManyToOne(fetch = FetchType.LAZY)
    @JoinColumn(name = "student_id")
    private Student student;

    @ManyToOne(fetch = FetchType.LAZY)
    @JoinColumn(name = "id")
    private TestItem testItem;
```

Fig. 4. TestItemPriority entity

If the pupil answers correctly, the value of priority is lowered, otherwise increased. The aim is that the pupil will be gradually practicing those selected test questions that were the most difficult for him/her in the past (Fig. 5).

```
public static float calculate(float priority, boolean isCorrect) {
    if (isCorrect) {
        return decreasePriority(priority);
    } else {
        return increasePriority(priority);
    }
}
```

Fig. 5. Question priority

A few samples of the application for an illustration are provided below (Figs. 6, 7, 8 and 9).

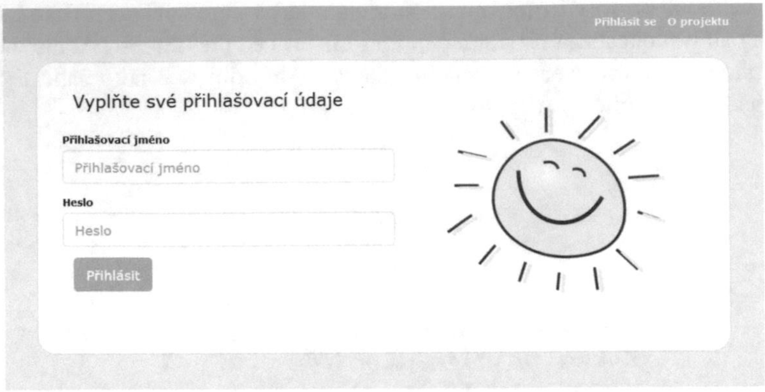

Přihlásit se O projektu
Vyplňte své přihlašovací údaje
Přihlašovací jméno
Přihlašovací jméno
Heslo
Heslo
Přihlásit

Fig. 6. Main screen

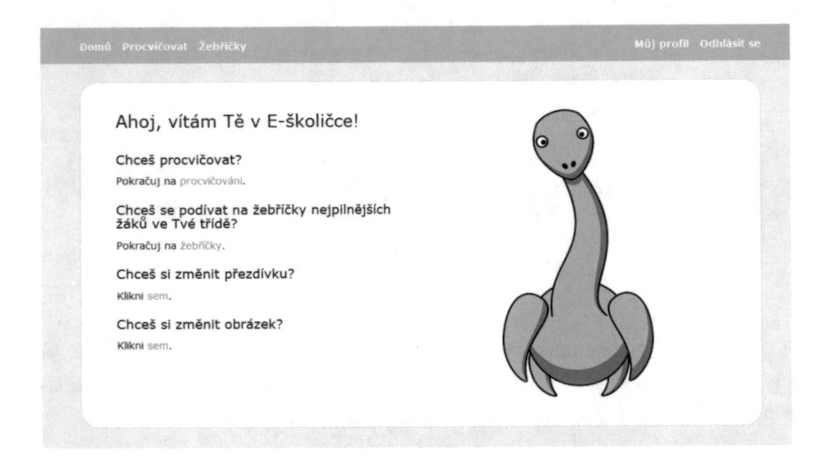

Fig. 7. Pupil's home page

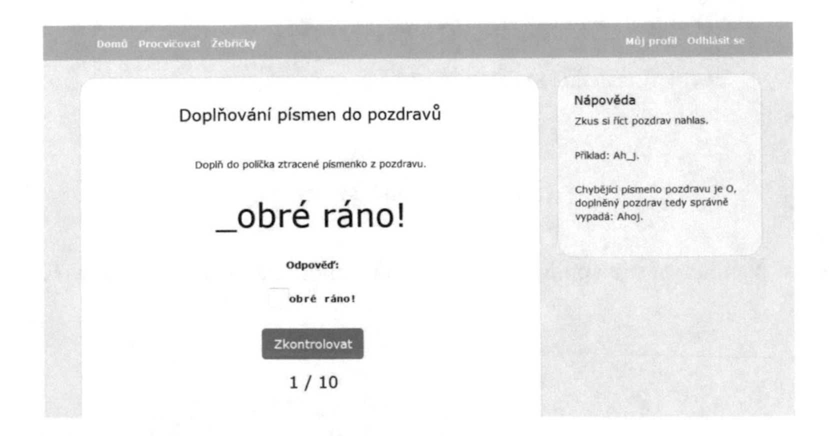

Fig. 8. Practicing

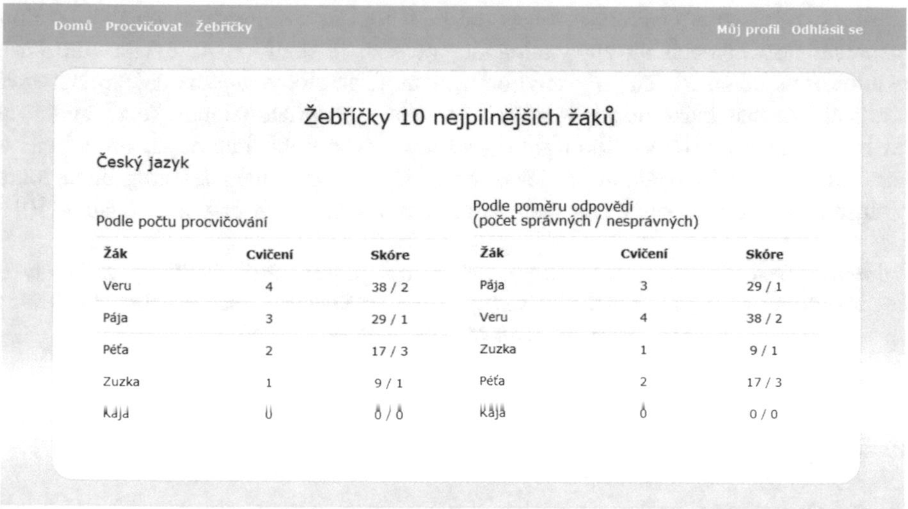

Domů Procvičovat Žebříčky Můj profil Odhlásit se

Žebříčky 10 nejpilnějších žáků

Český jazyk

Podle počtu procvičování

Žák	Cvičení	Skóre
Veru	4	38 / 2
Pája	3	29 / 1
Péťa	2	17 / 3
Zuzka	1	9 / 1
Kája	0	0 / 0

Podle poměru odpovědí (počet správných / nesprávných)

Žák	Cvičení	Skóre
Pája	3	29 / 1
Veru	4	38 / 2
Zuzka	1	9 / 1
Péťa	2	17 / 3
Kája	0	0 / 0

Fig. 9. Ladders of the most diligent students

5 Conclusion

The aim of the project was to design and implement the web application which can serve as a tool for the support of teaching at the first grade at the basic school. Therefore, the web application, called E-LittleSchool whose functionality meets the set requirements, was developed.

The application E-LittleSchool was launched in March 2016 at the basic school in Kamenne Zehrovice where it is used in the second year for practicing the Czech language. The pupils were instructed how to work with the application and where they can find individual parts: a list of subjects, topics and subtopics, practicing, ladders of the most diligent pupils and pupil's profile. The pupils of the second year were quick at orienteering in the application, which was also one of the requirements on the developed application.

On the basis of teacher's incentives, who directly works with the application, several improvements were defined for the future use. One of them is to stimulate pupils to practice and repeat even older topics on a regular basis. For this purpose, logics will be implemented into the application. The logics will monitor the elapsed time from the last practicing of each subtopic. If the delay is longer than the set limit, the pupil will be asked to start practicing. The second requirement includes comprehensive tests based on the revision of the whole topic and also a request to repeat these tests after some time. The third requirement is a functionality for generating tests and worksheets for printing. And the last requirement is a possibility to add pupils into the application in bulk.

Overall, the developed application seems to be positively perceived both by pupils and teachers; it is simple, easily controlled and well arranged. It enables pupils to practice the discussed subject matter and teachers receive immediate feedback on their pupil's performance. Overall, it meets the pedagogical requirements on the development and use of such an application in practice. In addition, it reflects the basic requirements of smart learning environment as technology-supported learning environment that can make adaptations and provide appropriate support (e.g., guidance, feedback, hints or tools) in the right places and at the right time based on individual learners' needs, which might be determined via analyzing their learning behaviours, performance and the online and real-world contexts in which they are situated [10].

Acknowledgements. The paper is supported by the project SPEV 2017 at the Faculty of Informatics and Management of the University of Hradec Kralove, Czech Republic. In addition, the authors thank Veronika Nemeckova for her help with the project.

References

1. The eLearning Action Plan (2001). http://eur-lex.europa.eu/LexUriServ/LexUriServ.do?uri=COM:2001:0172:FIN:EN:PDF (Cited 28 April 2016)
2. Wagner, J.: Don't be afraid eLearning. (Nebojte se eLearningu). Ceska skola, Praha (2005)
3. Korviny, P.: Moodle (not only) of OPF (Moodle (nejen) na OPF). OPF, Ostrava (2005)
4. Kveton, K.: Basics of distance and online education. (Zaklady distancniho a online vzdelavani). ČVUT, Praha (2003)
5. Egerova, D.: How to create study support for e-learning: methodological guide for authors study materials (Jak tvořit studijní opory pro e-learning: metodicka prirucka pro autory studijnich opor), 1st edn. WBU, Plzen (2011)
6. Vanecek, D.: Electronic Education (Elektronicke vzdelavani), 1st edn. CVUT, Praha (2011)
7. Vozenilek, V., Jilkova, J., Tolasz, R.: Climate change in e-learning teaching: starting, state, prototype, implementation (Klimaticka zmena v e-learningove vyuce: vychodiska, stav, prototyp, nasazeni), 1st edn. Univerzita Palackeho v Olomouci, Olomouc (2010)
8. Walls, C.A., Breidenbach, R.: Spring in Action. 2nd edn. Manning Publications Co., Greenwich (2008)
9. Spring Framework Reference Documentation. http://docs.spring.io/spring/docs/current/spring-framework-reference/htmlsingle/ (Cited 28 April 2016)
10. Klimova, B.: Assessment in smart learning environment – a case study approach. In: Uskov, V., Howlett, R.J., Jai, L.C. (eds.) Smart Innovation, Systems and Technologies, vol. 41, pp. 15–24 (2015)

The Use of LMS Blackboard Tools by Students in "Enterprise Accounting" Subject

Libuše Svobodová(✉) and Martina Hedvičáková

Department of Economics, Faculty of Informatics and Management,
The University of Hradec Králové, Rokitanského 62, 500 03 Hradec Králové,
Czech Republic
{Libuse.svobodova,martina.hedvicakova}@uhk.cz

Abstract. Education and the use of technology have seen a substantial positive trend over the last 17 years in the Czech Republic. Students have currently advanced technologies at their disposal, but they also have the ability to access online study materials at a vast majority of universities. The Faculty of Informatics and Management of the University of Hradec Králové is one of the major institutions that has influenced the Czech environment. This article analyzes the utilization of the Learning Management System Blackboard and related tools in connection with subject Enterprise Accounting as it is taught at the Faculty of Informatics and Management at the University of Hradec Králové. At the same time, the subject, conditions of study and basic information about students will be introduced. The main part of the article is focused on statistical data aimed on the use of individual tools within the subject by the students. The results shall be compared for full-time and part-time studies. The final part is focused on student feedback after completing the subject by blended learning. It has been established that full-time students use LMS tools more and they log in more often. Results shows that the time spent in LMS by students is very similar on average. Learning modules and assignments are the tools that are used the most.

Keywords: LMS blackboard · Statistics · Tools · Utilization

1 Introduction

The topic of financial accounting is of vital significance not only for accountants, but also for the managers and the owners of a company. Other connected groups include banks, investors, suppliers, customers, authorities and other institutions. The information obtained from financial accounting has several uses. For instance, it can be used for the purposes of managerial accounting, for managing the financial soundness of an enterprise or for evaluating and comparing companies.

Electronic education has been increasing, steadily developing and constantly changing over the last 20 years in the Czech Republic. The first full distance courses in the Faculty of Informatics and Management at the University of Hradec Králové have been intensively using e-learning courses that were formed in 1997 and were based on WWW pages. In the following years, the faculty started to use a virtual platform called LearningSpace in order to distribute e-learning courses. Later, the WebCT virtual

V.L. Uskov et al. (eds.), *Smart Education and e-Learning 2017*, Smart Innovation, Systems and Technologies 75, DOI 10.1007/978-3-319-59451-4_41

environment replaced it, and the LMS Blackboard is currently being used. In 2016, more than 100 courses are being taught on this platform. These courses are dedicated to full-time studies, part-time studies and to lifelong learning. Numerous teachers from the Faculty of Informatics and Management today use this virtual environment in order to support their lessons and provide blended learning. The learning environment provides tools that help teachers to ensure the quality of their preparation and instruction. In providing instructional support, there are tools that help to plan and manage the course of study, prepare and present study materials, facilitate communication between students and teachers, set and submit assignments, provide feedback via self-tests and present study results, including tracking. Subject is necessary to be run so that it motivates students. Motivation is one of the key conditions in any successful process of education. That is why study materials should engage students through constructivist elements.

Smart technologies can be used in the education process. They are defined as the technologies (includes physical and logical applications in all formats) that are capable to adapt automatically and modify behaviour to fit environment, senses things with technology sensors, this providing data to analyze and infer from, drawing conclusions from rules. It also is capable of learning that is using experience to improve performance, anticipating, thinking and reasoning about what to do next, with the ability to self-generate and self-sustain [12].

1.1 Literature Review – Motivation of Students

Teachers are still resolving how to motivate students and get them actively involved into the process of education. There is a wide range of ways how to motivate them. This paper is a small contribution into this current extensive topic.

Students can be motivated e.g. by the teacher's personality and interest in a specific area. Other factors affecting support of the learning process include the use of constructivism [1], social media, games and simulators. Modern elements can motivate students to develop deeper interest in an issue. The next important topic is learning styles. It is desirable to enable students to study in various ways, as different students may prefer different learning styles. This topic was addressed by the following authors, Frydrychová, Poulová and Jedlička [2], who answered the question: "Are students more successful if their learning style matches appropriate teaching style?" Poulová and Šimonová [3] focused on individual learning styles and university students.

Manochehr [4] presented an interesting study. For the instructor-based learning class (traditional), the learning style was irrelevant, but for the web-based learning class (e-learning), learning style was significantly important. The results showed that students with learning styles Assimilator (these learn best through lecture, papers and analogies) and Converger (these learn best through laboratories, fieldwork and observations) did better with the e-learning method. This mean that those learners that like to learn through thinking and watching and thinking and doing would learn better with e-learning. In addition, students with learning styles Accommodator (these learn best through simulations and case study) and Diverger (these learn best through brainstorming and logs) received better results with traditional instructor-based learning.

Interesting results were also obtained by Moravec et al. [5]. They compared two study groups. One without e-learning support and a second with the LMS Blackboard. Their research confirmed that providing an e-learning tool for students has a positive influence on their test results. Baris and Tosun [6] presented that e-portfolio supported education process influenced the success of students in positive way. Davies and Graff [7] were interested in the online participation and discussion. On the basis of their research, it may be concluded then that the reported beneficial effects of online participation and interaction do not necessarily translate into higher grades at the end of the year, with students who participated more frequently not being significantly awarded with higher grades. However, students who failed in one or more modules did interact less frequently than students who achieved passing grades. The educational process might also be supported by social media [8] or by playing games and utilizing simulators in the field of finance [9, 10].

2 Methods and Goals

In the article, the scientific questions will be evaluated on the basis of gained results.

- It is expected that part-time students who do not have so many face-to-face lessons will use the LMS Blackboard more than full-time students.
- It is expected that the most often used tools are learning modules and tests.

The main goal of the article is to analyze the utilization of LMS Blackboard tools in the educational process in "Enterprise Accounting" at the Faculty Informatics and Management at the University Hradec Králové. The article is focused on a comparison of study tools utilized in LMS Blackboard by selected groups of students. The research sample consisted of 70 students. Full-time and part-time students and their results obtained from the utilization of the LMS Blackboard in Enterprise Accounting for the 2015/2016 academic year are presented.

In order to evaluate the utilized tools, frequency distribution tables in .xls were used. The evaluation included all students who enrolled in the subject and who also enrolled to use the LMS Blackboard and this environment. The article is based on primary and secondary sources. The primary sources are represented by the statistics from the utilization of the LMS Blackboard and by the author's thoughts. Secondary sources include websites, technical literature, information gathered from professional journals, discussions and participation at professional seminars and conferences. It was then necessary to select, categorize and update available relevant information from the collected published material so that basic knowledge about the selected topic could be provided.

3 Learning Management System Blackboard

This part will provide a summary of the tools that are used in the virtual learning environment of the LMS Blackboard that is being used at the Faculty of Informatics and Management at the University of Hradec Králové. Teachers may choose which tools are substantial for them and they do not need to display the other ones on the home page - the main screen of the course. It serves as a guide-post. Basically, there are links and information: My announcements, My tasks, What´s new, My calendar, To do (with deadlines), Needs attention, Alert.

Tools that LMS Blackboard provide are:

- Syllabus
- Content, Information and Learning Modules
- Discussions
- Groups
- Announcements
- Assessments
- Assignments
- Calendar
- Chat
- Mail
- Media library
- Glossary
- Tools
- Web links
- My grades

4 "Enterprise Accounting" Subject

4.1 Introduction of the Subject

This subject is taken in the second year of studies at the Faculty of Informatics and Management at the University of Hradec Králové. A four-point grade system is used to evaluate the students' performance during their final examination. The sample of students was divided into two groups consisting of 42 full-time students and 28 part-time students. In the study group of part-time students, there were 33 students enrolled in the subject. Five students changed their minds and decided not to enrol in LMS and decided not to take the exam. Those students are not included in the evaluation.

Full-time students have two face-to-face classes weekly (2 h of lectures and 1 h of seminar) and can immediately discuss the topics with their instructor. Part-time students have only two six-hour meetings with their instructor. The subject is mandatory for Financial management students. For other study programs this subject is an elective. Ten students enrolled in this subject voluntarily. Six of them were students of Information management and four from Tourism management.

The course provides a more specific introduction to the nature of accounting of business entities, builds upon the basic principles of accounting, develops the knowledge that has already been gained and provides information about the accounting of more complicated economic operations. The content of the course is aimed at the current legislation on accounting for entrepreneurs in the Czech Republic.

In this sence, this accounting course is closely connected to other courses, in particular by managerial accounting, the tax system of the Czech Republic and the financial management of enterprises.

The main topics of the subject are:

- The legal and tax context of accounting
- Legal regulations in assets and liabilities appreciation
- Selected problems in the accounting of current assets
- Selected problems in inventory accounting
- Selected problems in the accounting of financial accounts
- Selected problems in accounting accruals of expenses and revenues
- Selected problems in accounting for equity
- Accounting for associations
- Recognition in bankruptcy
- Accounting for income and expenses, income tax payable and postponed
- Closing accounts, preparation of financial statements
- Consolidated Statements, methods of preparation, procedures
- Audit

A supporting e-course was created approximately 10 years ago in LMS WebCT and was run with both form of studies. The subject is based on the principles of constructivism. It encourages students' active participation, it works with previous experience gained by students and it is perceived as experiential learning and networking with practice. Furthermore, heuristic methods are used and it includes the principles of gradual support and individual work. Last but not least, cooperative forms of learning and thinking that are supported by students entering the problem are developed. Cross-curricular links are used, primary sources are supported and diverse perspectives are developed. An important role is also played by working with errors. All of the above-mentioned factors foster the educational development of students. Constructivism was also used in the subject entitled Business Theory [1].

4.2 Structure of the Course and Utilized Tools

Most of the study materials are located in training modules. These training modules include lectures (presentation in PowerPoint where fundamental terms and concepts are explained), extended texts in .html, links to interesting websites and exercises in MS Word converted to .html. The links are voluntarily. They can motivate students to gain a deeper understanding about individual topics related to the selected topic and provide examples, materials or articles from respected websites. Exercises are presented as assigned work and the answers are provided on the second page. Examples of interesting bachelor or diploma theses that are focused on the selected topics are also included in the environment. Figure 1 shows a screen from the e-course. It is dedicated to Enterprise accounting and includes lectures, examples and tasks and supplementary materials in the form of links.

The study materials and training modules come with tests that allow the students to auto-evaluate each topic. Students can revise their knowledge via tests that are aimed at

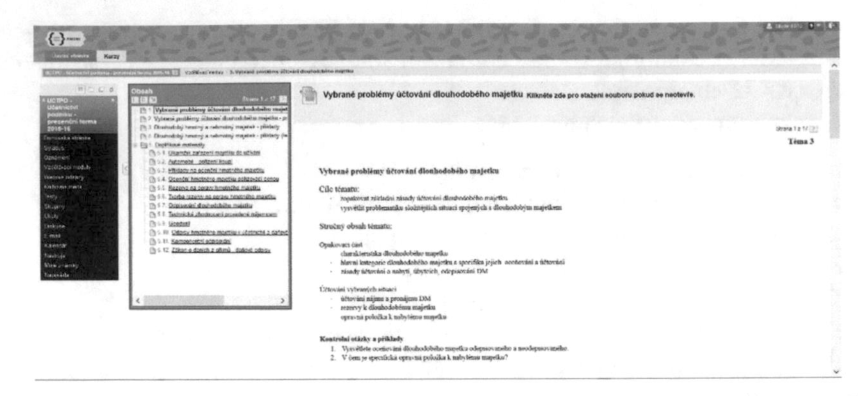

Fig. 1. An example of an education module focused on fixed assets [oliva.uhk.cz – course Enterprise accounting]

various topics from the financial accounting world. The tests serve as an additional control mechanism. The course includes 12 tests and one anonymous survey focused on evaluating the course study Students have the unlimited access.

Students also must do two mandatory assignments. The first is focused on a short literature search. Students look for 3 interesting articles focused on financial accounting in journals. The second one is connected with the utilization of the "Pohoda" accounting software. Students are given a step-by-step description of what to do in the software as shown in the screenshot. After the students do all the tasks in the accounting software, they uploaded their work to the LMS. Those two assignments were newly included in this academic year. Figure 2 presents study results by the students (grades of auto-tests, assignments etc.).

Úvodní stránka Kurzy

UCTPO - Účetnictví podniku - prezenční forma 2015-16 Přehled klasifikace

Domovská stránka
Sylabus
Oznámení
Vzdělávací moduly
Webové odkazy
Knihovna médií
Testy
Skupiny
Úkoly
Diskuse
E-mail
Kalendář
Nástroje
Chat
Obsah
Skupiny
Obsah kurzu
Moje známky
Nápověda

SPRÁVA KURZU
Ovládací panel

Zdroj UCTPO_KF | Místo určení UCTPO_PF_2015 is Kompletní. To access the detailed log, click here

Sloupec skryt. Může být zobrazen zpět ze stránky sloupce organizace.

Klasifikace : Přehled klasifikace

V Režimu čtení obrazovky tabulka je statická a známky mohou být vloženy na stránce detailu známky vybráním buňky v tabulce pro známku. V interaktivním módu v Přehledu klasifikace, známky mohou být napsány přímo do buňek. Použijte šipky nebo Tabulátor pro navigaci Přehledem klasifikace a Enter pro odeslání známky. Více informací

Vytvořit sloupec | Vytvoření výpočtového sloupce | Spravovat | Zprávy | Filtr | Pracovat offline

Přesunout na začátek | Email | Třídění sloupců podle: Pořadí rozvržení | Pořadí: Vzestupně

[illegible]	[illegible]	[illegible]	[illegible]	[illegible]	[illegible]	[illegible]	[illegible]	[illegible]	[illegible]	[illegible]	[illegible]	[illegible]
Uzdalová	Eva	bazalev1	11. leden 2016	K dispozici	994.00	994.00	115.00	--	80.00	90.00	230.00	[illegible]
Bednářová	Monika	bednamo1	12. únor 2016	K dispozici	56.00	56.00	--	✓	--	--	40.00	--
Burešová	Nikola	buresni1	18. únor 2016	K dispozici	1587.67	1587.67	100.00	✓	90.00	125.00	250.00	120.66667
Floriánová	Aneta	florian1	5. únor 2016	K dispozici	23.00	23.00	--	--	--	--	--	--
Fečíková	Lucie	fecikllu1	25. únor 2016	K dispozici	1713.00	1713.00	5.00	[illegible]	100.00	130.00	300.00	150.00
Grofová	Monika	grofomo1	18. únor 2016	K dispozici	231.00	231.00		[illegible]			210.00	
Hamtilová	Jana	hamtija1	11. únor 2016	K dispozici	121.00	121.00			100.00			
Hejduková	Martina	hejdumа1	4. leden 2016	K dispozici	1240.67	1240.67	135.00	--	100.00	110.00	250.00	116.66667
Homych	Vladimír	homyvl1	13. únor 2016	K dispozici	374.00	374.00	--	--	--	105.00	240.00	--
Imlaufová	Helena	imlaufhe1	9. únor 2016	K dispozici	29.00	29.00						
Jirková	Michaela	jirkomi2	7. únor 2016	K dispozici	191.00	191.00			100.00	70.00		

Vybrané řádky: 0

Přesunout na začátek | Email | Ikona legendy

Fig. 2. The classification of students [oliva.uhk.cz – course Enterprise accounting]

5 Tools Used by Students in the LMS Blackboard

The next part of the article provides the results of the utilization of tools in the "Enterprise accounting" course through the LMS Blackboard. Table 1 makes it clear that the expectations have been fulfilled. Students from both types of study most consulted Content areas, which are comprised of Assignments, Content, Learning modules, Media library, Syllabus, Assessments and Web links.

Table 1. An overall summary of user activity

	Full-time students		Part-time students	
	Hits	Percent	Hits	Percent
Announcements	191	3.61%	124	4.40%
Blogs			1	0.04%
Calendar	218	4.12%	9	0.32%
Content areas	4193	79.19%	2381	84.55%
Tools	31	0.59%	25	0.89%
Discussion	178	3.36%	67	2.38%
Groups	35	0.66%	65	2.31%
Email	99	1.87%	13	0.46%
My grades	346	6.53%	120	4.26%
Tasks	4	0.08%	11	0.39%
Total	5295		2816	

Part-time students consulted this content 5% more than full-time students. The Calendar was another major difference in the use of tools. It is evident that full-time students followed information about which topics would be lectured on during a particular week, or other important dates. Part-time students knew two session dates in advance and they were also told which topics would be lectured about on particular dates. Students usually search for exam dates in IS STAG, which is used for managing the study agenda (registering for subjects, entering the overall grades for all subjects, information about bachelor and master theses etc.) It may seem surprising that full-time students communicated by email the most. This was due to incorrectly sent tasks that the students were supposed to send via the Assignments tool.

When we focus on the usage of tools within content areas, we find out that the Learning modules are used the most. Full-time students used them almost 12% more than part-time students. Conversely, part-time students visit Assignments more often, by nearly 3%; the same holds true for visiting Assessments, by nearly 8%. Students in the combined type of study tried more self-tests voluntarily. The short credit test that was taken in written form by full-time students, but was taken online in LMS by part-time students might also have contributed to this (Table 2).

In statistics there were revealed interesting differences. The greatest number of visits of learning modules were for full-time students registered 229, 225 or 183

Table 2. All user activity inside content areas

	Full-time students		Part-time students	
	Hits	Percent	Hits	Percent
Assignments	666	15.88%	447	18.77%
Content			3	0.13%
Learning modules	2786	66.44%	1259	52.88%
Media library	33	0.79%	30	1. 26%
Syllabus	115	2.74%	107	4.49%
Assessments	551	13.14%	502	21.08%
Web links	42	1.00%	33	1.39%
Total	4193		2381	

approaches. In combined form, it was 160, 100 and 98 approaches. For daily tasks, students form the most subscribed 57 times, 29 times and 43. In the combined form was heading lower again. It was a 35, 34 or 28 approaches. Regarding tests for full-time, one student tested 135, 45 or by 41 tests. In the combined form of student results were affected with 157, 66 or 54 approaches. At full-time of study was accounted for the largest number of accesses to individual tools and materials with the number 230, 229, 225, 206, or 174. The largest total number of visits to individual instruments and materials in a part-time students has reached 372, 193, 163, 133, 126 or 125th.

In focusing on other statistics that can be obtained from the LMS Blackboard, we are able to state the number of times LMS was accessed and the time spent in LMS according to individual types of study. Table 3 makes it evident that even though students in the combined type of study spent as much time as full-time students in the environment on average, they had 9 fewer hits. Hits were added up for all users and consequently, the median statistic method was used. Lower values, for example the time – median for the combined type of study, could have been affected by students who only logged into LMS a few times and did not pursue their studies further.

Table 3. Hits and time spent in LMS sorted by type of study

	Full-time students	Part-time students
Hits	93	84
Time - mean	301	300
Time - median	243	222

If we focus on the statistics from the three tools that were used the most, it again was based on hits data. Both the average and the median were computed for individual tools and types of study. The data in Table 4 makes it clear that the number of times Assignments was accessed was nearly the same. However, there is a difference in the Assessments tool, where students in the combined type of study tested themselves

Table 4. Hits for the most used tools in LMS for different types of study

	Full-time students	Part-time students
Assignments - mean	16	16
Assignments - median	14	16
Assessments - mean	13	18
Assessments - median	5	8
Learning modules - mean	66	45
Learning modules - median	63	42

18 times on average. Full-time students used the testing tool 13 times on average. If we focus on the median statistical indicator, the testing was used much less; 8 times for those in the combined type of study and 5 times for full-time students. Overall, 14 tests were available for students in the environment. It is evident that most students did not try all of the tests voluntarily. As far as the most visited tool is concerned (Learning Modules), it was definitely used more by full-time students.

During a more profound analysis of access to the system, it was also established that full-time students logged in more regularly, typically a day before or on the day of a lecture or seminar. At the same time, the period spent in LMS increased the day before an exam or on the day of an exam. A different trend was observed for part-time students. The greatest web traffic was noted after the launch of the course, then prior to individual sessions and then again two days before an exam or on the day of the exam.

At the end of their studies, students were asked to fill in an anonymous poll. Overall, about 40% of the students responded. Students commended both on the content and the form of the study materials. They also had a positive view on the change of the form of seminar works as opposed to the preceding year. The students appreciated the fact that they could use and try out accounting software and fulfil tasks in it. Some of them worked with "Pohoda" software voluntarily, even though it was not set in the task. They found it beneficial that research seminar papers were published. However, they usually had no time to go through other students' papers. All of them agreed that they would definitely or probably take a similar course again during their learning. However, in view of the complexity of the subject matter, some of the part-time students would appreciate more face-to-face lessons in the connection with blended learning.

6 Discussion

During online learning, motivation to study is very important, as well as other factors. One of the major parts of the motivation is on the specific instructor that is teaching the students. A personal approach, helpful attitude, emphasis and enthusiasm about topics are most important. The most important factor for successful study is self-motivation. Well-prepared materials in the subject and a great instructor do not mean success.

Utilization of smart technologies and social networks in the process of the education is the next important issue in the nowadays turbulent time. The question should be: Why students do not use communication tools in the LMS instead of the other channels like Facebook, email, telephones, Skype, face-to-face etc.?

7 Conclusion

The article is dedicated to subject Enterprise accounting with respect to the use of tools in the LMS Blackboard by full-time students and part-time students. The results may be of benefit not only for accounting teachers, but also for teachers of other economic subjects or other fields. "Enterprise accounting" uses the LMS Blackboard as support for education. The course is built on the philosophy of constructivism and the methodology of adult education. Lectures, exercises, web links, assessments and assignments are added in the LMS. In the LMS is also used Calendar, Syllabus, Email and Discussion. After conducting an analysis of the data obtained from the LMS, we have to refute the first scientific question. Part-time students use the LMS less than full-time students. Based on the obtained data, it was established that part-time students usually download study materials and then they prepare themselves using these downloaded materials. The longest time spent in the LMS was at the beginning of its launch, then during the period before lectures or during lectures and then during the period before exams. Conversely, full-time students log into the LMS regularly. This tends to happen most often during lectures or exams. While full-time students study regularly, part-time students rather study at specific times.

Based on the data, we have to refute the second scientific question, too. Although Learning modules are the most used tool in the LMS, the second most used tool is not testing, but fulfilling tasks. This result was not expected. While there are 14 tests created in the environment, there are only 2 tasks.

Some authors mainly concentrate on study materials and revision. But other fields must also not be omitted. Strong importance is paid to the process of evaluation that is provided by students in questionnaires upon their completion of studying economic subjects. It describes their level of satisfaction with the course of study, i.e. their interest in e-learning, conditions for this way of study, how often and long they studied the course, whether the study materials suited them, whether they printed them or studied on-line, their instructor's approach, off-line version, etc. Students appreciated this supportive way of learning the subject so that they could check their knowledge.

Acknowledgements. This study is supported by internal research project no. 2103/2017 Investment under concept Industry 4.0 at Faculty of Informatics and Management, University of Hradec Kralove, Czech Republic. We would like to thank student Veronika Kadečková for cooperation in the processing of the article.

References

1. Rohlíková, L., Vejvodová, J., Černík, R.: Constructivism in the Practice of Universities. Západočeská univerzita v Plzni, Plzeň (2011)
2. Frydrychová Klímová, B., Poulová, P., Jedlička. M.S.: ICT and students' learning styles. AWER Procedia Inf. Technol. Comput. Sci. **3**, 142–147 (2013)
3. Poulová, P., Šimonová, I.: Individual learning styles and university students. In: IEEE Third Annual Global Engineering Education Conference on EDUCON 2012, pp. 128–133. IEEE, US Piscataway (2013)

4. Manochehr, N.-N.: The influence of learning styles of learners in e-learning environments: an empirical study. Information Systems Department, Qatar University, vol. 18, pp. 10–14. CHEER (2006)
5. Moravec, T., Štěpánek, P., Valenta, P.: The influence of using e-learning tools on the results of students at the tests. Procedia – Soc. Behav. Sci. **176**, 81–86 (2015)
6. Baris, F.M., Tosun, N.: Influence of e-portfolio supported education process to academic success of the students. Procedia – Soc. Behav. Sci. **103**, 492–499 (2013)
7. Davies, J., Graff, M.: Performance in e-learning: online participation and student grades. Br. J. Educ. Technol. **36**(4), 657–663 (2005)
8. Černá, M., Svobodová, L.: Current social media landscape. In: Proceedings of the Efficiency and Responsibility in Education 2013, pp. 80–86. ČZU, Prague (2013)
9. Mls, K. and Kol: Autonomous Decision Systems Handbook. BEN, Prague (2011)
10. Svobodová, L.: Computer games in economics education. J. Technol. Inf. Educ. **4**(2), 48–53 (2012)
11. http://oliva.uhk.cz/–courses Enterprise Accounting
12. http://www.igi-global.com/dictionary/smart-technology/38186

Teaching Students the Basics of Control Theory Using NI ELVIS II

Dmitrii Dobriborsci(✉), Dmitry Bazylev, and Alexey Margun

Department of Control Systems and Computer Science, ITMO University, Saint Petersburg, Russian Federation
dmitrii.dobriborsci@mail.ru
http://csi.ifmo.ru

Abstract. This paper describes a remote lab, currently available on the Department of Control Systems and Computer Science in ITMO University. The remote lab allows students with disabilities to gain knowledge in Electrical Engineering and Electronics, Control Theory, Systems Identification and etc. In this paper the developed laboratory setup for DC motor control is described. The paper shows the model of DC motor and designing the proportional-integral-differential controller tuned using Ziegler-Nichols and modal control approaches.

Keywords: DC motor · NI ELVIS · Remote lab · Ziegler-Nichols · Modal control · Education · Active learning

1 Introduction

One of some important changes in Russian educational system for the last decade of developments was an attempt to improve the access to higher education for disabled people for widening their participation in social life. However, this process is still at the initial stage and many problems in social and educational policies have not been addressed. Legislations concerning social protection and employment for disabled people are insufficient. Government does not provide enough support, for example financial and medical support for disabled people to be self-sufficient. There is very limited access to professional training and employment for them. However this is a problem of complex rehabilitation of a huge number of people. According to statistics, there are 11–15 million disabled people in Russia, that is more then 10% of entire population. Approximately 700 000 of them are children and young adults through the age of 18.

In Russia higher education provides an essential opportunity for independence and self-sufficiency for all people, including disabled people. Having higher education diploma is crucial in getting job in Russia. With respect to disabled

This paper was partially financially supported by the Government of Russian Federation, Grant 074-U01 and by the Ministry of Education and Science of Russian Federation (Project 14.Z50.31.0031) and by the Russian Federation President Grant 14.Y31.16.9281.

V.L. Uskov et al. (eds.), *Smart Education and e-Learning 2017*, Smart Innovation, Systems and Technologies 75, DOI 10.1007/978-3-319-59451-4_42

people, amongst those who are employed (13–15% of the disabled people population), sixty percent of them are having higher education diploma. Currently the participation of disabled in higher education is increasing. In 2005 there were about 20000 disabled students in 337 higher education institutions, and in 2008–2009 the number is more then 30 000. There are more than 1300 higher education institutions, and almost all of them train disabled students. However the percentage is lower compared to non-disabled people. It is the main motivation of the development remote laboratories in Russian Universities.

Nowadays electrical machines are widely used for generating electrical power, transforming voltage and robot control. That's why electrical engineering and electronics education becomes a very important problem. In [1] authors proposed system provides remote control of a DC motor by PWM control and remote measurement of a single phase alternator. The user interface of the system is designed without Web plug-ins so that users can use the systems with their smartphones or tablets as well as their personal computers. The paper [2] analyzes the effect of the use of VISIR in five different groups of students from two different academic years (2013–2014 and 2014–2015), with three teachers and at two educational levels. The empirical experience focuses on Ohm's Law. The results obtained are reported using a pretest and post-test design. The tests were carefully designed and analyzed, and their reliability and validity were assessed. The analysis of knowledge test question results shows that the post-test scores are higher that the pretest. The VISIR remote laboratory's positive effect on students' learning processes indicates that remote laboratories can produce a positive effect in students' learning if an appropriate activity is used.

The paper [3] describes the development of remotely accessible electrical drive sets for hands-on education in electrical machinery, electrical drive and control courses as well as certain aspects of motion control and power electronics. The remote access is provided via the developed client-server communication method using the TCP/IP protocol to run the Matlab/Simulink compatible electrical drive sets operating on the remote server side. The developed method allows access to the robotics and electrical drive test- beds, namely permanent magnet DC (PMDC), permanent magnet synchronous (PMSM), and induction motor (IM) motor-load sets, developed for both on-site and remote use at the Control Laboratory in the University of Alaska Fairbanks (UAF). The client-server communication is developed in C/C++ using wxWidgets to communicate with the Matlab/Simulink downloadable DS1104, which is the control unit used in all the electrical drive systems in the lab.

At ITMO University in the learning process is widely used the application of innovative mechatronic systems in automation and robotics learning. Smart E-learning method allows developing students interest in gaining practical skills during working with real robots remotely. As a result, the students' need for theoretical knowledge increases. The learning process represents "learning in project implementation" concept with ability to create own student project. [6]. In this paper authors present the last results on developing the remote lab based on the National Instruments hardware. Particularly, the control designing for a DC motor is described. This paper is an extension of the learning methods proposed in earlier papers [5,6]. The paper is organized as follows. After the "DC Speed Control" lab setup description in Sect. 2, the mathematical description of the DC drive in Sect. 3 is given. Section 4 presents the designing of a control laws using two different approaches. We conclude the paper with discussion on achieved results, application developed platform in research and education.

2 The Laboratory Setup Description

The laboratory setup includes the following components NI ELVIS II, prototiping board for NI ELVIS II, DC motor, motor driver, current sensor, ac/dc converter 220/24 V, personal computer.

Lets consider the main parts of the system components. The educational platform National Instruments Educational Laboratory Virtual Instrumentation Suite II (1) helps to form and to develop scientific and general professional competences such as: participation in organization and experiments conduction, processing of experimental studies using advanced information technologies and means and others [5]. The geared DC motor IG-42GM (2), its specifications are presented below in Table 1. Pololu High-Power Motor Driver 18v15 (3). It is a discrete MOSFET H-bridge motor driver enables bidirectional control of one high power DC brushed motor. The little board supports a wide 5.5 to 30 V voltage range and is efficient enough to deliver a continuous 15 A, without a heat sink. The Allegro ACS712 (4) provides economical and precise solutions for AC or DC current sensing in industrial, commercial and communication by the user. One of the typical applications include motor control. The system also consists of an AC/DC Converter 220/24 V (5) and personal computer (6) represented on Fig. 2.

Table 1. IG-42GM characteristics

Rated voltage	24 V
Rated current	≤2100 mA
Rated torque	9.5 kg-cm
Rated speed	240 rpm
Rated output	34.7 W
Weight	360 g

The experiments could be done remotely. The remote access to the laboratory is performed using LabView Remote Panels. Without any additional programming, a LabView program can be enabled for remote control through a common Web Browser. The user simply points the Web browser to the Web page associated with the application. Then, the user interface for the application shows up in the Web browser and is fully accessible by the remote user. The most exciting feature of remote panels is the ability to control LabVIEW remote VIs

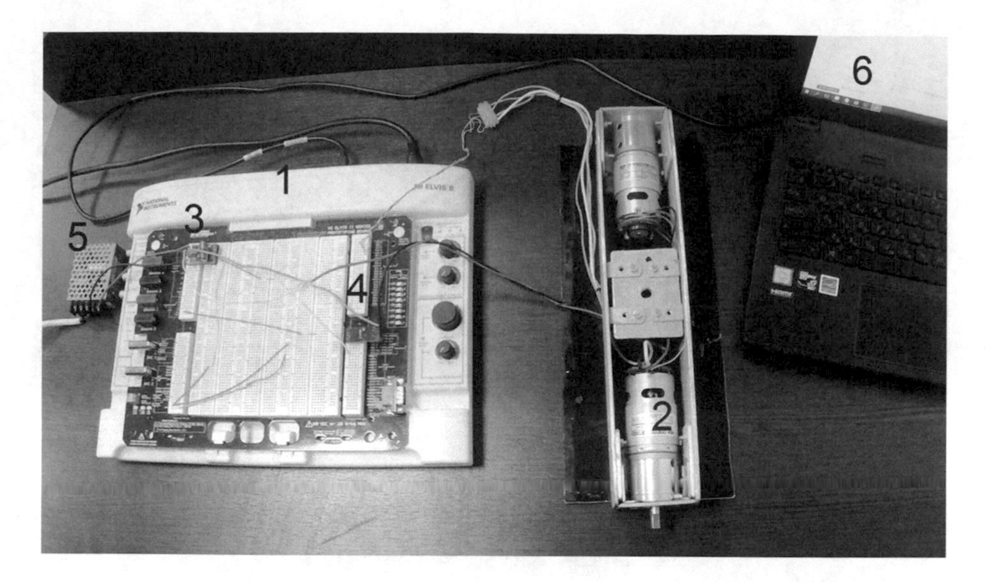

Fig. 1. The laboratory setup

from a web browser. To operate remote panels from a browser, you must first have the LabVIEW run-time engine and browser plug-in installed on the client machine. Next, you simply browse to the URL that comprises the server IP, network name, or domain name plus the HTML file name containing the embedded VI you want to operate. Within a LAN environment, the URL might appear as http://PcName:Port/ViName.html. If you are operating the VI from an internet connection, the URL might appear as http://lpAddr:Port/ViName.html. Such type of smart e-learning is used in ITMO University for disabled students (Fig. 1).

3 Mathematical Model of a DC Motor System

The theoretical material of the course is presented during classical lectures, combined with exercises for every student and online using the distance learning platform of ITMO University. The analyzing of this learning method was presented in earlier papers [7,8]. A DC motor system typically consists of a DC motor amplifier and sensors for position and current measurements. The interest here is to understand how to model this system for control purposes. Recall the dynamic equations of the DC motor given as [4]

$$\begin{aligned} L\frac{di}{dt} &= -Ri(t) - K_b\omega(t) + v(t) \\ J\frac{d\omega}{dt} &= -f\omega(t) + K_T i(t) - \tau_L(t) \\ \frac{d\theta}{dt} &= \omega(t) \end{aligned} \tag{1}$$

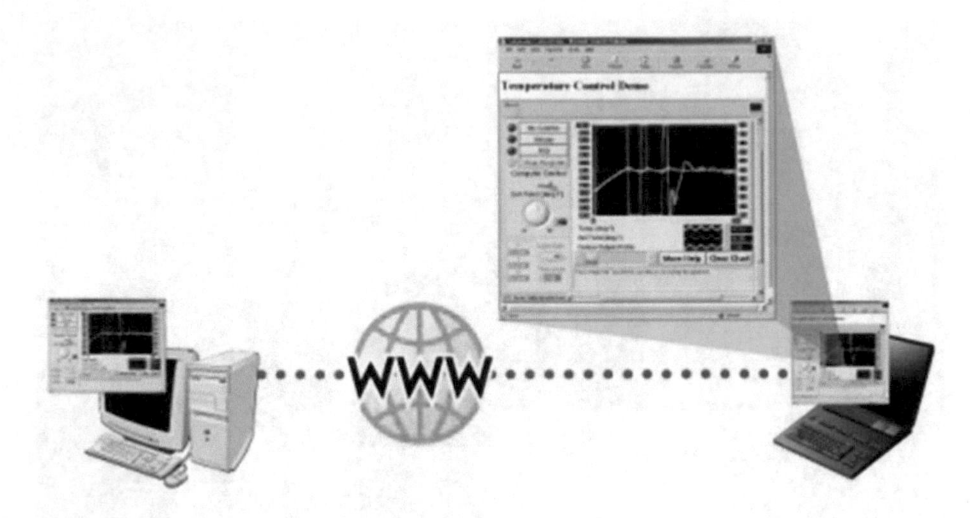

Fig. 2. The LabView remote panels

where, $L, R, \omega, v, J, f, i, \tau, \theta, K_b, K_T$ are electric inductance, electric resistance, angular velocity, voltage, moment of inertia of the rotor, viscous friction of the motor, current, rotor position, back electromotive force (EMF) constant, the torque constant, respectively. Also - $f\omega$ models the viscous friction torque on the motor due to both the bearings and to the brushes rubbing against th commutator. The voltage v is commanded to the motor through a power amplifier. The amplifier is limited in how much voltage it can actually put out. This limit is denoted by V_{max} in Fig. 3.

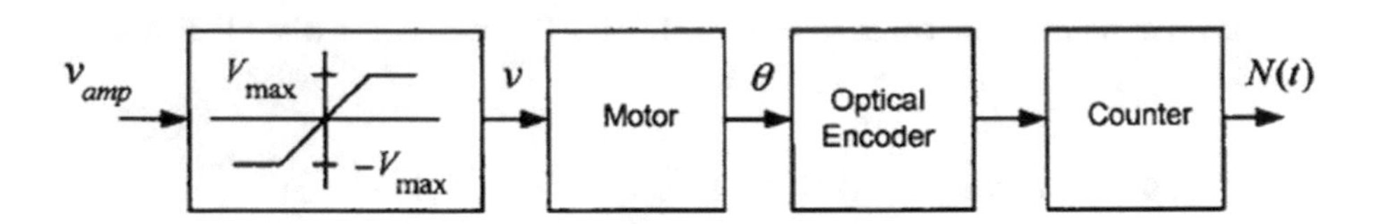

Fig. 3. Open-loop DC motor.

A common speed sensor used in industry is an optical encoder. The optical encoder consists of a set of windows spaced equally around a circular disk with a light source shining through the window when it is aligned with the source. A detector puts out a high voltage when there is light and a low voltage otherwise. For the current control case we use a Hall sensor.

4 Control Design for DC Motor

The objective is to design the speed control system for the DC motor. For the students various task options are offered (overshooting and transient time, desired speed value, steady-state error).

4.1 PID-Controller Tuned Using Classical Approaches

The parameters of the PID-controllers must be calculated using Ziegler-Nichols and modal control approaches.

4.2 Current Command

Typically, to get around the fact that the voltage is the input, one designs an inner current control loop that allows direct current command. That ism the voltage is forced by a controller to go to whatever value necessary to obtain the desired current. To understand how this is done, consider the Eq. (1) in the Laplace domain given by [4]

$$i(s) = \frac{-K_b\omega(s) + V(s)}{sL + R} \tag{2}$$

$$\omega(s) = \frac{K_T i(s) - \tau_L(s)}{sJ + f} \tag{3}$$

$$\theta(s) = \frac{-1}{s}\omega(s) \tag{4}$$

These algebraic relationships are illustrated in the block diagram shown in Fig. 4. Often the approximation that $L = 0$ is made to simplify the analysis of the system. However, a standard approach in industry is to put the amplifier into a current command mode which results in making the effect o L negligible [4].

To do so, the current is feedback (typically using analog electronics) through a proportional controller as shown in Fig. 5. Here, $i_r(s)$ is the Laplace transform of the reference current, $i(s)$ is the Laplace transform of the actual current in the motor, and $K_p > 0$ is a proportional gain.

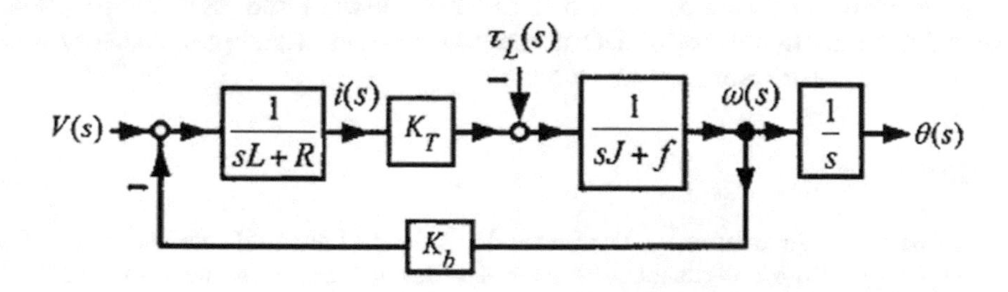

Fig. 4. Block diagram of a DC motor

4.3 Speed Controller

Using reduced-order model, it is straightforward to design a simple proportional speed controller as illustrated in Fig. 6. From the block diagram of Fig. 6, it follows that $\frac{\omega(s)}{\omega_{ref}(s)} = \frac{1}{\tau_m s+1}$. This is an example of the classical approach to control [4].

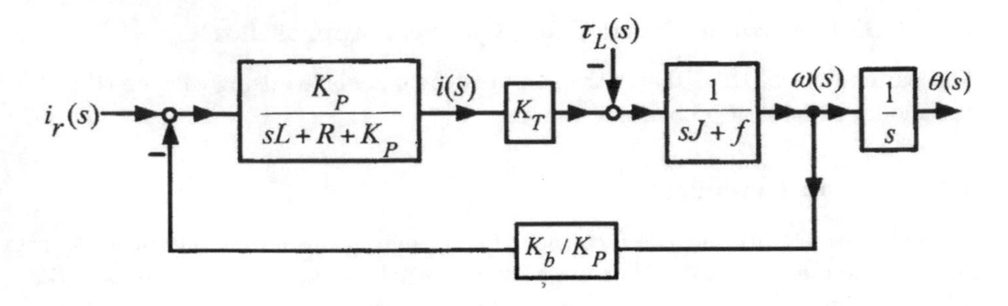

Fig. 5. DC motor inner current control loop

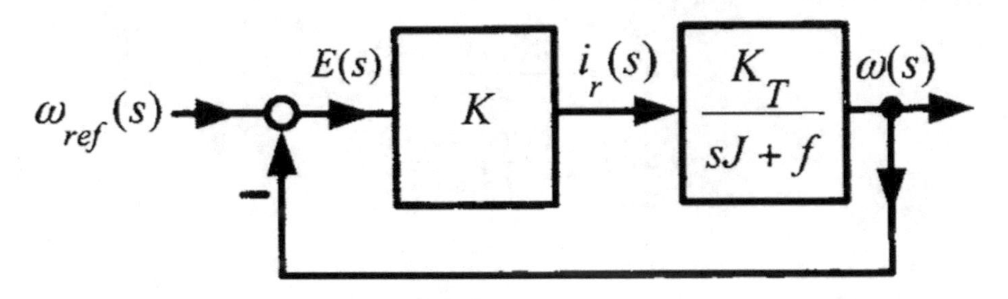

Fig. 6. Simple speed controller for a DC motor

5 Conclusion

The remote lab for electrical machines learning is presented in the paper. This laboratory is used at the ITMO University for training of disabled students. Hardware of described lab is based on National Instruments NI ELVIS II prototiping board. Software is based on LabView software and TCP/IP protocol. Lab allows remote control of DC motor via internet. Example of theory and tasks for laboratory work are described.

References

1. Ishibashi, M., Fukumoto, H., Furukawa, T., Itoh, H., Ohchi, M.: Development of a web-based remote experiment system for electrical machinery learners. In: The 2014 International Power Electronics Conference, pp. 724–729 (2014)
2. Garcia-Zubia, J., Cuadros, J., Romero, S., Hernandez-Jayo, U., Orduna, P., Guenaga, M., Gonzalez-Sabate, L., Gustavsson, I.: Empirical analysis of the use of the VISIR remote lab in teaching analog electronics. IEEE Trans. Educ. **PP**(99), 1–8 (2016)
3. Bogosyan, S., Gokasan, M., Turan, A., Wies, R.: Development of a remotely accessible Matlab/Simulink based electrical drive experiments. In: IEEE International Symposium on Industrial Electronics, ISIE (2007)
4. Chiasson, J.N.: Modeling and High Performance Control of Electric Machines. The Institute of Electrical and Electronics Engineers Inc., New York (2005)

5. Artemeva, M., Nikolaev, N., Dobriborsci, D., Nuyya, O., Slita, O.: NI ELVIS II in the concept of cognitive and active learning technologies. In: IDT 2016 - Proceedings of the International Conference on Information and Digital Technologies 2016, pp. 71–75 (2016)
6. Zimenko, K., Bazylev, D., Margun, A., Kremlev, A.: Application of innovative mechatronic systems in automation and robotics learning. In: Proceedings of the 2014 16th International Conference on Mechatronics (Mechatronika 2014) (2014)
7. Bazylev, D., Shchukin, A., Margun, A., Zimenko, K.: Applications of innovative "Active Learning" strategy in "Control Systems" curriculum. In: 3rd International KES Conference on Smart Education and e-Learning, vol. 59, pp. 485–494 (2016)
8. Kremlev, A., Bazylev, D., Margun, A., Zimenko, K.: Transition of the Russian federation to new educational standard: independent work of students as a factor in the quality of educational process. In: ERPA International Congresses on Education 2015, vol. 26 (2015). UNSP 01041

Use of Automatic Image Classification for Analysis of the Landscape, Case Study of Staré Jesenčany, the Czech Republic

Eva Trojovská, Pavel Sedlák(✉), Jitka Komárková, and Ivana Čermáková

Faculty of Economics and Administration, University of Pardubice, Pardubice, Czech Republic
eva.trojovska@gmail.com,
{pavel.sedlak, jitka.komarkova}@upce.cz,
ivana.cermakoval@student.upce.cz

Abstract. The landscape is changing, mainly due to man. The changes in the landscape are increasingly important, so it is important to monitor these changes, mainly due to the use of resources in the country in the future. The article deals with the use of automatic classification of data from remote sensing to analyse the development of the landscape on the case study in the village Staré Jesenčany, the Czech Republic.

Keywords: Image processing · Automatic classification · Comparison of classification · Landscape changes

1 Introduction

Studying changes of landscape is a very important issue in the modern society. It means that the society is interested in new findings, which can help to understand how nature and its particular elements exist and behave. Understanding functions, relationships and rules can support landscape management and further sustainable development e.g. prevention of devastating impacts of floods or drought. For example, studies [1, 2] are focused on landscape changes.

Remote sensing is very often used as a source of data for observation of landscape and terrain. There are many issues why this method is more reasonable than in situ observations, sampling and measurements and land surveying. The costs are lower (namely in a case when bigger areas are monitored), accuracy and spatial resolution is adequate to aims of studies and finally, data measured in various parts of electromagnetic spectrum are available. The last advantage is important for researches based on thermal imagery, various indices, etc. Satellites or aerial imagery is available for monitoring larger areas. Unmanned aerial vehicles (UAV) are increasingly used to monitor small areas. UAVs can often provide results faster and usually with higher spatial resolution.

Remotely acquired image material is either analogue or digital, and therefore the form determines also the possibility of further processing. Each image contains

V.L. Uskov et al. (eds.), *Smart Education and e-Learning 2017*, Smart Innovation, Systems and Technologies 75, DOI 10.1007/978-3-319-59451-4_43

information about topographic (geometric) properties of objects (e.g. size, the distance of objects and their relative position) and thematic information (e.g. type of surface soil moisture content). Processing of the digital image information, which is stored in digital form, together with the development of computer technology presents a series of processes to automate and accelerate some, particularly computational methods (transformation, creation of orthophoto, etc.). Other significant advantages of objectivity, accuracy and in a sense even lower processing costs. Besides these advantages the digital processing of remote sensing materials brings some disadvantages, and particularly in the area of automatic classification must also use knowledge from analogue interpretation. [3] Structure and sequence of methods of digital image processing are not completely uniform and closed.

2 Previous Researches

In the most general sense, the classification is a process in that each pixel is assigned a corresponding information. Its aim is to replace the value of the radiometric characteristics of the original image by values called informational class. The result of classification is then presented for example in the form of thematic maps [3].

The landscape as a term was emerged in the early nineties of the 20th century as one of the key words this time. The landscape is heterogeneous portion of the earth's surface, consisting of a set of interacting ecosystems, which is repeated in similar forms in that part of the surface [4, 5].

Monitoring changes in the landscape over time is generally based on monitoring changes in individual landscape components, their surface representation, dynamics and spatial configurations. To change the landscape type occurs when a different type of landscape elements becomes a landscape matrix, which either landscape component grows or fades. [6] Quantification of the structure of land cover always encouraged environmentalists to creating the most diverse indices or their acceptance of other disciplines [7–9].

At present, the theme of monitoring changes in the landscape and image classification goal of many projects. These themes are attractive for students in the processing of theses and become a commercial product of software companies in the field of geoinformatics also.

Detection of landscape changes is a very important part of landscape management. Remote sensing is used to collect up-to-date data in various parts of electromagnetic spectrum. Data are processed by many different methods.

Qindong and Shengyan [2] analysed landscape pattern changes in monitored period (1990–2013) and identified driving sources during that period. They used various landscape indices to calculate landscape changes.

Bortoleto et al. [1] focused on index of restoration in landscapes. They proposed a mathematical index named SIR that describes suitability of individual habitat patches for restoration within a landscape. A model based on the SIR used a map of distance classes among fragments and a map of habitat quality established according to each land cover category. SIR was obtained as a result of calculation.

At present, there are many works focusing on the automatic classification of images with high spatial resolution for the monitoring of change of landscape. They primarily use aerial photographs or images from the UAV. Automatic classification of images with high spatial resolution are solved, for example, in works [10–14]. However, the use of images from the archives of satellite images is often cheaper, a lot of works use this source [15–17].

3 Case Study: Classification of Image and Quantification of Areas

Methods of digital image processing with focus on automatic classification and monitoring changes in the landscape are applied to the analysis of the municipality Staré Jesenčany. The village is located near to city Pardubice in the Czech Republic. It is a small cadastral area of approximately 3.71 km^2. Typical for this landscape is a higher percentage of arable land, grass surface and construction of houses. There were analysed aerial photographs from firm Geodis, which were captured between 2003 and 2008. The classification was performed on the Landsat 7 satellite images from 2000 and 2010. The work was processed in software ArcGIS for Desktop from firm Esri.

A sequence of individual steps corresponds with the steps that are caught in a process map (Fig. 1), which was also created by the authors of this article [18].

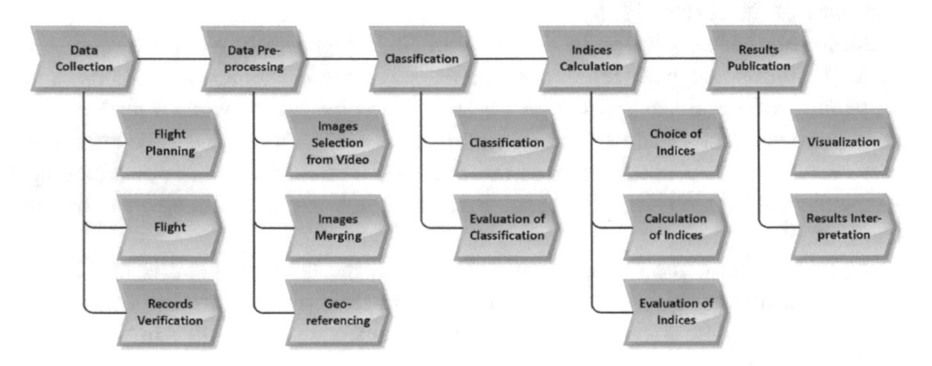

Fig. 1. The sequence of steps of image processing [18]

Algorithm ISODATA was applied on aerial photographs as the first method. At the beginning of this algorithm, number of clusters and the number of iterations is defined. Cluster, which becomes heterogeneous (based on the standard deviation values) is divided into two new clusters. Clusters, of that centers are closer than a predetermined value, are combined into one cluster. Clusters that contain fewer pixels than the pre-given number are cancelled and the pixels are assigned to the nearest clusters [3].

The results of this classification were particularly useful for obtaining pre-conditions of what we can expect from another classifier. For example, we can say that there is a great similarity between the areas of arable land and grass surface. The results were not used as input during the further processing.

The surface of the features creek or railway consists of a typical pixel structure and supervised classification will achieve better results here.

Maximum likelihood classification method was used as the basic method of aerial photographs. For each DN value, probability of its occurrence can be calculated. When it is presented it in the correlation field, each cluster with the same probability of a pixel can be connected to a certain value [3].

A training area for the following categories was created: building area, arable land, grass, communications, forest and pond. The big problem was to classify forested areas due to great similarities with the class grass, so they were only classified areas with shadows cast by trees. It is a major simplification, but it was assumed that there will be some compensation for the actual area of forest cover and the shadows of buildings. We can take this simplification into account if it is to monitor changes in the landscape percent per unit of the monitored area. The result of classification was full of isolated pixels and therefore postclassification adjustment was necessary. The difference between the original classification and adjusted image illustrates Fig. 2.

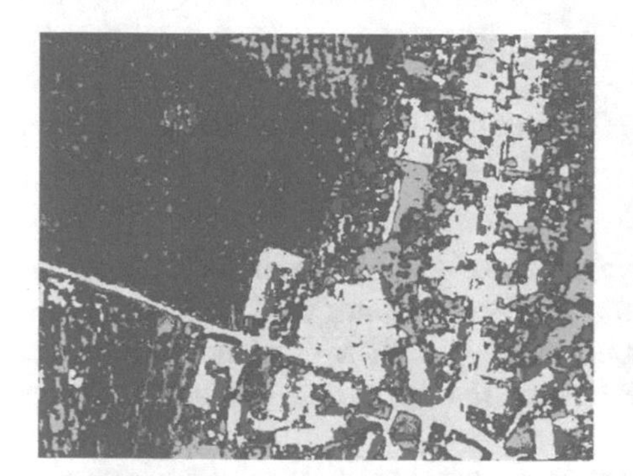

Fig. 2. The classification of image 2003 before and after adjustment (Source: authors)

The classification results for each time horizons were quantified in hectares and percentage of different classes in the study area.

An objective way to determine the accuracy of classification is to build the error matrix in which they are compared randomly selected reference points and classified points. Values of both error matrix were achieved through Raster Calculator tool, and function Combine in ArcGIS. From those matrices, it was determined that the average accuracy for 2003 is 0.56 and for 2008 is 0.63.

Another way to determine the accuracy of the classification was to compare the results of maximum likelihood classifier with the results of visual interpretation, which were available through the work of [19].

We can say that the automatic classification of aerial photographs gave generally satisfactory results. Since the classification of satellite images was to be expected that due to the small size of the territory and low resolution images of Landsat 7 (resolution 30 m), the results are less conclusive. Application of ISODATA classifier to satellite images from 2000 and 2010 provided very rough classification results that lead into three and four classes. Either classification using maximum likelihood algorithm did

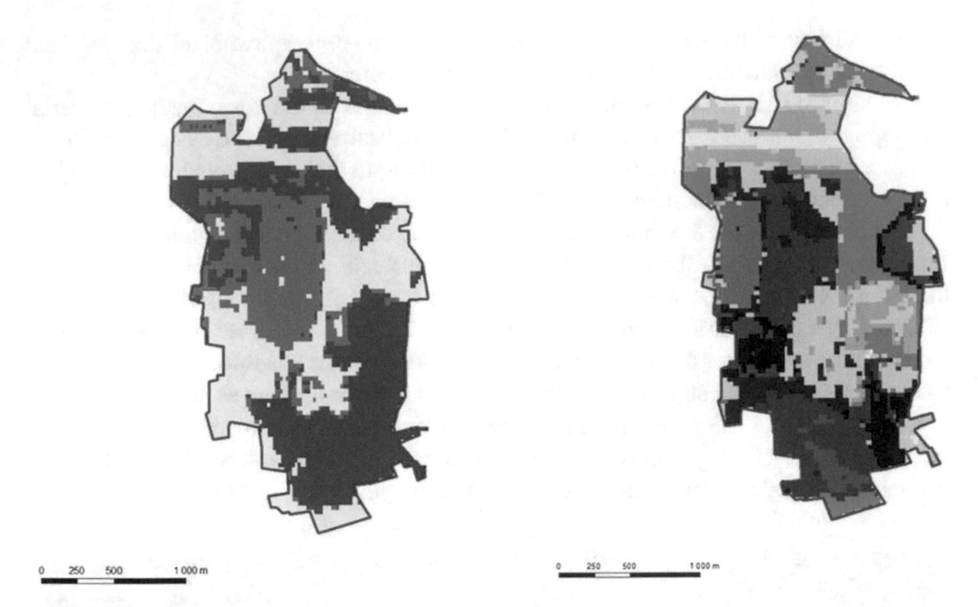

Fig. 3. Satellite images from 2000 and 2010, classified by maximum likelihood algorithm (Source: authors)

not provide a new information, it only provided already acquired knowledge about the area. We can see the general structure of the territory from this classification, but the details of it could not be obtained. Frequent erroneous classification of the pixels due to great similarities between some classes can be seen like in the case of classification of aerial photographs, see Fig. 3.

The visual interpretation was used to obtain the most accurate results of the classification of satellite images also. Although in this process, there is rather more emphasis put on interpretive skills of man than on manner of calculation program; and it is necessary to take into account the data itself. As already mentioned, the images are of a low resolution. Elements of territory in the image from 2010 are more recognizable than in the picture from 2000, where for example it is not possible to identify area of the pond. Visual interpretation of these images from the Landsat 7 shows Fig. 4.

For example, the very purpose of the study area and location of the pond, it is preferable to use a combination of bands of multispectral images in the order: red = 4, green = 5, blue = 3, which reveals the frontier land and water. The classification follows the displayed image by maximum likelihood classifier and visual interpretations. By means of Calculate Area, it was found that the area of a polygon obtained by visual interpretation is equal to 37 508 m^2, while the area resulting from supervised classification is 29 820 m^2, both in 2008. The area calculated by means of automatic classification is approximately 10 000 m^2 smaller. The surface of the pond during the years 2000–2010 is constant, so there is no change in the case of the second image.

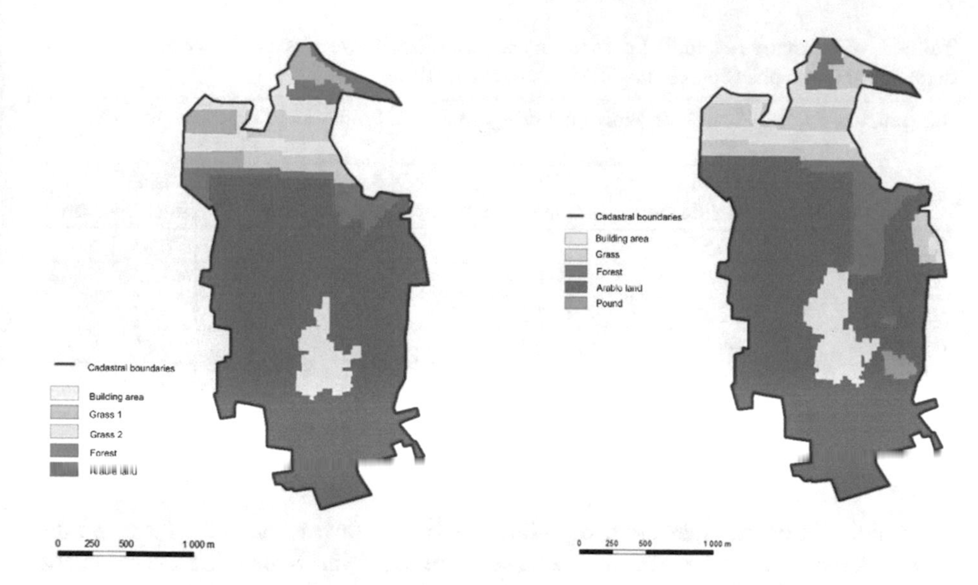

Fig. 4. Visual interpretation of satellite images from 2000 and 2010 (Source: authors)

4 Interpretation of Changes of Staré Jesenčany Municipality

In the case of the classification of aerial photographs, the difference between these two time horizons can be neatly summed up – see Fig. 5 and the following Table 1, where are listed quantitative differences from the visual interpretation of study area also.

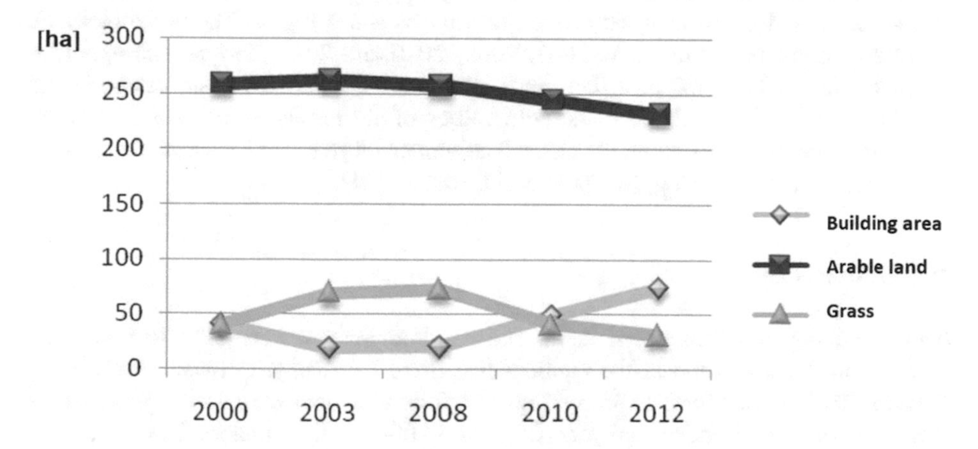

Fig. 5. The changes of basic categories in landscape (Source: authors)

Table 1. Quantitative capture of changes in each category between years 2008 and 2003, based on aerial photographs (Source: modified according [20])

Information class	Changes in individual categories [ha]		Proportional representation [%]	
	Automatic classification	Visual interpretation	Automatic classification	Visual interpretation
Building area	0.54	1.24	0.15	0.33
Arable land	−30.73	−5.21	−8.24	−1.40
Grass	32.87	3.39	8.90	0.91
Forest	−3.67	0.55	−0.99	0.15
Pond	0.65	0.20	0.17	0.06
Communication	0.04	0.02	0.01	0.01
Others	–	0.66	–	−0.05

Building area has increased, especially in the northwest part of village. Visual interpretation, however, reveals this increase more than twice higher than the automatic classification. The decline is typical for the area of arable land. In the case of forest monitoring for the result reflected a big difference between the visual interpretation and automatic classification of 2008. Although in 2008 were felled many trees and shrubs around the pond, on the whole territory, there is identifies a small increase in forest category (see visual interpretation results). The automatic classification gives bad results for this category. The calculation of result of the forest area in 2003 from automatic classification on the contrary is very close to reality. Category pond is a feature with an almost constant area. Growth of areas of communication category is small.

The graphic visualisation is a very good way to present the results of changes in the landscape, based on an analysis of satellite images – see Fig. 4. The horizontal axis represents particular years 2000, 2003, 2008, 2010 and 2012. The vertical axis represents areas in hectares. In a line graph, there are showed the basic categories of building area, arable land and grass, with values of the results from visual interpretation, therefore, the most accurate values. The values of year 2012 were obtained from Czech Office for Surveying, Mapping and Cadastre [21].

5 Conclusion

The automatic classification of aerial and satellite images were used to determine changes in the landscape in the municipality Staré Jesenčany. Software ArcGIS for Desktop 10 was used for data processing. The following data were used: aerial images from the company Geodis from years 2003 and 2008, satellite images from Landsat 7 ETM +, years 2000 and 2010. The supervised classification (Maximum Likelihood classifier) and unsupervised classification (method ISODATA) were used. The classification of satellite images was easier, but it did not provide good results due to the small size of the area and the low spatial resolution Landsat 7 images. Spectral image enhancement (composition change in RGB and vegetation indices) belongs to the significant benefits of multispectral images.

The classification results show that the proportion of arable land decreased due to new buildings. Also, there was a slight increase in area of forest and scrubland. The small increase can be observed in the category of communications as a result of new developments.

As far as spatial data are very important for the smart cities and regions sustainable development, suitable automatic classification, which is described in the paper, can be used as a part of smart technologies to support the development.

Acknowledgements. The paper has been completed with the kind support of SGS_2017_17 project of Faculty of economics and Administration, University of Pardubice.

References

1. Bortoleto, L.A.: Suitability index for restoration in landscapes: an alternative proposal for restoration projects. Ecol. Ind. **1**, 724–735 (2016)
2. Qindong, F., Shengyan, D.: Landscape pattern changes at a county scale: a case study in Fengqiu, Henan Province, China from 1990 to 2013. Catena **2**(2016), 152–160 (2016)
3. Lillesand, T.M., Kiefer, R.W., Chipman, J.W.: Remote Sensing and Image Interpretation, 6th edn. Wiley, Hoboken (2008). 756 pages
4. Forman, R.T.T.: Krajinná ekologie. 1. vyd. Praha: Academia, 583 pages (1993)
5. Hradecký, J., Buzek, L. Nauka o krajině: Učební texty Ostravské univerzity. Vyd. 1. Ostrava: Ostravská univerzita, 2001, 215 pages. https://katedry.osu.cz/kfg/user/hradecky/download/skripta.rar
6. Lipský, Z. Sledování změn v kulturní krajině: učební text pro cvičení z předmětu Krajinná ekologie. Kostelec nad Černými lesy: Lesnická práce, 71 pages (2000)
7. Guth, J., Kučera, T.: Monitorování změn krajinného pokryvu s využitím DPZ a GIS. Příroda **10**, 107–124 (1997). https://www.usbe.cas.cz/people/kucera/LE/TEXTY/landcov.pdf. Accessed 30 Jan 2012
8. Herzog, F., et al.: Landscape metrics for assessment of landscape destruction and rehabilitation. Environ. Manag. **27**(1), 91–107 (2001). Springer-Verlag
9. Pechanec, V., Brus, J., et al.: Decision support tool for the evaluation of landscapes. Ecol. Inform. **30**, 305–308 (2015)
10. Gonzalez, A., Rafael, G., et al.: Robus segmentation of aerial image data recorded for landscape ecology studies. In: Huang, F., Sugimoto, A. (eds.). LNCS, vol. 9555, pp. 61–72 (2016)
11. Movia, A., Benait, A., Crosilla, F.: Shadow detection and removal in RGB VHR images for land use unsupervised classification. Isprs J. Photogrammetry Remote Sens. **119**, 485–495 (2016)
12. Zhang, L., Li, A., et al.: Global and local saliency analysis for the extraction of residential areas in high-spatial-resolution remote sensing image. IEEE Trans. Geosci. Remote Sens. **58**, 3750–3763 (2016)
13. Caccetta, P., Collins, S., et al.: Monitoring land surface and cover in urban and peri-urban environments using digital aerial photography. Int. J. Digit. Earth **9**, 1–19 (2016)
14. Hung, C., Xu, Z., Sukkarieh, S.: Feature learning based approach for weed classification using high resolution aerial images from a digital camera mounted on a UAV. Remote Sens. **6**, 12037–12054 (2014)

15. Landsat Data Access. USGS. Landsat Missions 2015. http://landsat.usgs.gov/Landsat_Search_and_Download.php. Accessed 20 Sept 2015
16. Brunclík, T., Danquah, K.A.B.: Model of chlorophyll-a concentrations in inland water bodies based on Landsat data. In: Advances in Remote Sensing, Finite Differences and Information Security, pp. 215–218 (2012)
17. Pásler, M., Komárková, J., Sedlák, P.: Comparison of possibilities of UAV and Landsat in observation of small inland water bodies. In: Proceedings of International Conference on Information Society, pp. 45–49. Infonomic Society, Londýn (2015)
18. Čermáková, I., Komárková, J.: Modelling a process of UAV data collection and processing. In: Proceedings of International Conference on Information Society, pp. 161–164. Infonomics Society, Londýn (2016)
19. Hromádko, T.: Využití leteckých snímků pro analýzu vývoje krajiny v okolí obce Staré Jesenčany. Pardubice. Bachelor thesis. University of Pardubice (2009)
20. Trojovská, E.: Využití automatické klasifikace obrazu pro analýzu vývoje krajiny v obci Staré Jesenčany. Diploma thesis. University of Pardubice (2012)
21. Czech Office for Surveying, Mapping and Cadastre (Český úřad zeměměřičský a katastrální) K.ú.: 754412 - Staré Jesenčany - podrobné informace 2012. http://www.cuzk.cz/Dokument.aspx?PRARESKOD=10&MENUID=10016&AKCE=META:SESTAVA:MDR002_XSLT:WEBCUZK_ID:754412. Accessed 30 Jul 2012

Smart Education: National Initiatives and Approaches

Birth and Major Strategies of Smart Education Initiative in South Korea and Its Challenges

Seyeoung Chun(✉)

Department of Education, Chungnam National University, Daejeon, South Korea
sychun56@gmail.com

Abstract. Smart Education Initiative (SEI) was announced on the 29th June, 2011. Technology trend has been integrated into SMART concept as APPLE's iPhone and SAMSUNG's smart phone entered as the big giants into the world market. Korean society is very often called as the front line of smart society embedded with such highly digitized and connected technology. Korean education often cited as the miracle of the 20th century would be the pioneer to open the new millennial world of education by applying the smart technology. Although too many issues and concepts are not yet constructed as the solid ground for the daily guidelines for school teachers and students, the wider and deeper connection among them must be the key engine. This article touches on those issues and implications raised by Korean SEI, which is expected to contribute to build a meaningful step for the new millennial pedagogy.

Keywords: Smart education · Smart technology · Digital textbook · Education 3.0 · Korean education

1 Introduction

1.1 Background

Smart education has been a topic of conversation in all schools in South Korea (hereinafter referred to as "Korea") since June 29, 2011 when the Korean government announced its smart education initiative. Because of the social trends and students using smart devices, people think that it was the proper time to introduce smart education. But many others said that no specific qualities of smart education had been identified yet. As such, they wondered if smart education could be adopted in classrooms and implemented as expected.

Just after the pronouncement of smart education initiative, number of voices appeared for appealing for delaying the full implementation of SMART Education until scientific verification of its effectiveness and digital devices' impacts on health concern such as eyesight, mental health, and so on. And most teachers insisted that they needed some time until they would be required to apply it in the classrooms. In fact those concerns were considered by the government actions, although it was quite true not to deny some shortcomings of the initiative. Given the current situation, Korean government and society have been implementing by challenges and enthusiasm in order to reach to the full vision of smart education for the future, taking into account of those

V.L. Uskov et al. (eds.), *Smart Education and e-Learning 2017*, Smart Innovation, Systems and Technologies 75, DOI 10.1007/978-3-319-59451-4_44

conditions and opinions of schools, instead of rushing into the policy. Moreover, infrastructure needs, like supplying smart-devices and wireless-internet, costs a lot, and the voices of dissents about supply business of major companies are growing louder. There are many other aspects than these several issues if smart education would be well implemented and integrated into educational practices toward the future. Most of all the fundamental question would be its pedagogical ground, which might ask a great deal of research and development in this area. This paper has been prepared as one beginning effort in this sense.

1.2 Objectives of the Paper

This article has been prepared with the objective to review the policy background of smart education and its implementation process since its intake with a view to explore its pedagogical implication and future prospect.

For the purpose, this study reviewed the inception of smart education and its major strategies for implementation based on related policy documents. Some major academic research outcomes were also examined about the positive and negative effects of the smart education initiative in relation to the pedagogical implications of smart education, largely from the comparative perspective of orthodox educational and pedagogical history.

2 Policy Background of Smart Education Initiative (SEI) in Korea

2.1 ICT Utilization for Future Challenges in Korean Education

According to scores from the PISA 2009 assessment [1], among the 19 participating OECD countries, Korea ranked at the top in the digital reading assessments in addition to the high rank in other main subjects such as Math and Sciences. However, her education system leaned too much on academic assessments and test performance had also has been criticized for many years not to transform Korean society toward the future society and not to provide adequate education for students performing above and below average. Due to these complicated factors, demand for change in Korean education system had been increasing, and educating the new generation had become an urgent subject in order to keep pace in the age of a knowledge revolution that was rapidly changing in the 21st century.

The policy to integrate technology into education in Korea aimed at improving public education, at promoting economic growth, and at extending ICT accessibility. According to KERIS [2] annual report, Korea had been preparing and implementing the master plans for the use of ICT in education policy by periodically three phases to innovate education system since 1996. The first phase was about the foundation construction (1996 to 2000), the second phase was about proliferation and settlement (2001 to 2005), and the third and fourth phases were about acceleration (since 2006). In particular, Korea had established the new plan to improve education and science in the nation's HRD (Human Resources Development) in the fourth phase (2011-2015), emphasizing soft power of information more than hardware-based.

2.2 SMART Education Initiative

In 2011 the SMART Education has been initiated as an educational information policy that pushes ahead with the plans of the phase IV [3]. One of the most critical strategies for integrating the use of ICT into education had been to implement SMART Education (2011-2015). The goals of this new strategy were clearly stated in the title: SMART Education stands for Self-directed, Motivated, Adaptive, Resource enriched, and Technology embedded. By using ICT in the framework of education, science, and technology, SMART Education Initiative aimed to promote plans to enhance the agenda of schools and educational policies. SMART Education was therefore the first policy to attempt to overcome the limitations of existing ICT use in education. Since it is a policy to integrate ICT institutionally and culturally into education and curriculum, SMART Education promotes the use of ICT as a primary source of learning, rather than a supplementary one. Above all, SMART Education offers an education that enables learning through a teaching and learning system tailored to the needs of each individual student at any time and at any place. Through SMART Education Initiatives, the previous emphasis on the 3R's (Reading, aRithmetic, wRiting) will shift the focus in education to the system more suitable to the 21st century learner.

The strategy for implementing SMART Education is largely composed of five key tasks:

(1) Development and application of digital textbooks which overcome limits of printed textbooks, improve teaching at classroom, and implement customized learning;
(2) Institutionalization of online classes (which are admitted as regular classes) and introduction of online evaluation systems to guarantee learning options;
(3) Creation of an environment for the public use of educational content and reinforcement of information and communications ethics education to resolve adverse effects;
(4) Strengthening of teachers' competencies to put SMART Education into practice; and
(5) Practical use of the contents created by public agencies and individuals through a cloud computing-based environment in every school, and public use of learning resources and establishment of collaborative learning environments through collective intelligence and social learning [4, 5].

As a strategy for implementing the five key tasks, first of all, the relevant laws and systems to allow for changes in educational paradigm have been called for action. By extending the scope of learning from the educational place called "school" to any place, including homes and hospitals where learning happens, education can take place anywhere and at any time. Some options have been proposed to achieve these objectives; they are to use actively online classes, provide rich teaching and learning materials, and acknowledge online classes as equivalent to regular class hours, and be able to make free use of educational materials in class. The second task is to establish a virtuous cycle system that comprises of educational system, participants, teaching and learning methods, and educational environment. Although so many elements are involved in education, it is the most important one to enable individualized and customized learning through an accurate diagnosis and prescription of suitable lessons based on individual student's level. Therefore SMART education policy is strongly targeting to build necessary assessment systems and strengthen teachers' competencies as well. Third one is to eliminate information gaps and minimize malfunctions that may arise in the SMART Education environment. Applying SMART Education must take a keen consideration to the less privileged sectors of society in education, including low-income and multicultural families, and minimize the possible educational gap that may occur within the SMART Education environment. To prevent adverse effects such as Internet or game addiction, which may happen as learners are more exposed to ICT, SMART Education policy is designed to cover the related issues by utilizing necessary information about these problems and the prescriptions. The last task is focused on taking a care and prudence when introducing rapidly changing information technology to classrooms. Through practicing and researching by operating at pilot schools and phased expansion, the burden of SMART Education must be reduced, together with promoting societal understanding and positive perception of SMART Education through public campaigns. The promotion can be achieved by establishing and operating future education research centers equipped with theoretical research and on-site consulting functions.

2.3 Implementation Process SMART Education Initiative

After SEI was announced as a national agendum, Korean government made its action plan for 2011-2015 [6]. According to the plan, the initiative was divided into three phases: the first phase, research and preparation for the environment for 2011-12; the second phase, pilot implementation for 2013-2014; and the third and last phase, generalization after 2015.

During the first phase, legal and policy issues were reviewed and revised in order for smart education to take root. In this phase, the most important issue was to remove the copyright barrier for school teachers so that they could download and upload open educational content. Thanks to such a legal arrangement, there was a turning point in Korea's textbook policy, which authorized digital textbooks as an official textbook together with conventional paper textbooks. In 2013, the first year of the second phase, Korea selected three subjects, English, Science and Social Studies, for grades 3 and 4 (primary school), and grades 7 and 8 (secondary school), to be developed as the first pilot project for digital textbooks. These digital textbooks were expected to be distributed for classroom teaching beginning with the 2013 fall semester, although they might not be used on individual students' smart devices. By 2015 or afterwards, mobile and smart devices would not be the precondition for using digital textbooks. By then, digital textbooks were expected to be uploaded in the content cloud, from which every student and teacher can also download the content to mobile devices or PCs. If these pilot textbooks fit well into classroom teaching and learning, the society will build smart technology infrastructure, including Wifi network and individual devices, in schools in the third phase, beginning in 2015.

While implementing SEI in these phases, there are a lot of pros and cons about smart education. Proponents seem to have dreams and visions of smart education and society, where more people can enjoy learning and teaching to create a better world. Opponents seem to have worried about harms and dangers of a technology-dependent education and society where some children may be addicted to the internet and digital influence. First stream is about the effectiveness of SEI induced digital textbook on student's learning [7–11]. Particularly Son and Yoon's [12] research showed digital textbook utilization enhancing students' thinking ability and collaboration learning. Lee and Jang [13] recently supported also the positive effectiveness of digital media on learning. On the other hand, some different investigations have been also following up for more careful introduction of those smart and digital devices for children [14, 15] pointed out worries of elementary school teachers who are not fully prepared for using the high-technology mediated teaching material and Hong et al. [16] also the negative influences of technology induced classroom environment on emotional and psychological development of children. However quite many other researchers such as Jeong [17] are continuing their studies with a good prospect of the potentialities of smart education for the future paradigm building of education in Korean society, although there are at the same time critical approaches accompanying with.

What will be a better way to go ahead? There must be goods and bads, just like the history of human civilization. Technology must be smart enough to give us more freedom and prosperity, since pros and cons are our true partners to go with. For this partnership, this paper will review SEI in detail and discuss its pedagogical implications.

3 Pedagogical Implication of Korean Smart Education Initiative [18]

3.1 Legacies of Human Civilization: Brain and Collaboration with Tools

Human beings have been able to evolve and survive as the super-heroic species because they are different from others. Numerous explanations have been suggested. Three legacies among them must be reviewed thoroughly in order to go beyond the rigidity of outdated system and evolve into the creative future; they are brain, collaboration, and tool.

Brain is the root of human being that is creating and developing civilization. Although all the species have the brain, the size and capacity cannot be compared with human brain. Only one percent difference in DNA structure between homo-sapiens and chimpanzees creates countless differences between their brains and lives. However, human beings did not just stop there with one's own single brain but made further progress to enlarge it, big enough to create human civilization by collaborating and connecting their brains. This is a global or world brain in today's internet term. Human beings came to know that they could effectively cope with dangers of predators and catch more foodstuffs if they work together. Collaboration and connection thus became a fundamental aspect of human beings. They create a group and society beyond blood ties. Finally human beings made tools by utilizing brain and collaboration.

These three components of human civilization have been integrated into learning and teaching process from generation to generation as well as from here to there. This learning and teaching process has been conceptualized and evolved as a human system of education, which cannot be found anywhere except from human species. Pedagogy is a term for the knowledge system of this human education system. Just as human system and civilization have been developing and evolving, so has pedagogy. It has been known that the first educational institution in the history was the training school of teaching Sumerian scribe on the clay tablet for those who wanted to take the occupation of tablet scriber [19]. Sumerian cuneiform letter was the first tool and technology which human beings could store information by writing and remember it by reading. However it was such a secret code that ordinary people could not decode it easily. Clay tablet was the first type of media like a sheet of paper, developed from papyrus later. This has mediated between the story teller and the listener, teachers and students, father and son, friends and friends and so many relations on. Later over the years of history, numbers and mathematics, lyrics and arts, philosophy and science and finally religion and the Bible appeared and were disseminated all across the world. Pedagogy became a thick knowledge book which was far beyond ordinary people. And now teaching and learning job became a difficult mission for those who must master the pedagogic codes.

A new mission of pedagogy has been always called for mankind. King Sejong's invention of Hangeul, Korean script, was one of the miracles as Chun [20] explained. Finally, up to now digital and smart tools seem to have been good news focus. Smart education initiative is an answer to that mission.

3.2 Pedagogical Understanding of Smart Education

Smart education has been designed for the 21st century learners. Smart technology made smart education possible but smart education is something beyond technology. It will become an environmental setting and ecological system which digital natives will live in and prepare themselves for. Therefore, it is understood that teachers and seniors and parents should help the young generations to prepare for their future. However there may not exist any concrete ground for knowledge. The reason so many Korean students sleep in classrooms is that the present pedagogy does not fit with them. They may not want to listen to teachers' lecture any more but to experience multimedia connection with the real world. It is inevitable for them to think about eradicating the disconnection between teachers and students. Pedagogy is a set of knowledge about the connection and mediation between them. Those sleeping students must be woken up by a new pedagogy. However it is not an easy job to establish such a new pedagogy until we could get the higher understanding about the social and human change.

In fact, a society changes as its human environment changes. Then our mankind evolves to adjust and survive, which Darwin conceptualized as the evolution. History is the memory of our ancestors, how they have evolved so far, and the textbooks teach us to survive and succeed to them. This is an ever–changing story. Education 3.0 has a concept articulated on GETideas.org, an online resource and community for education leaders around the world to engage in a dialogue on the next generation education and learning systems. This site is a public service of the Cisco Global Education corporate social responsibility group [21]. The Education 3.0 concepts articulated on GETideas. org refer to a holistic and systemic approach to educational transformation as applied primarily to K-12 education, and identifies transformational qualities and strategies necessary across pedagogy and curriculum, leadership and governance, culture and infrastructure. SEI in this sense could be the Korean version of education 3.0. Smart education is apparently different from the previous e-Learning or ICT use in education in the sense that the former is using a comprehensive innovation strategy to change all the educational issues as the same as Education 3.0, while the latter used rather an optional or partial strategy to improve education by utilizing ICT. Now it is time to think about the more fundamental and radical aspects of smart education from the pedagogical viewpoint. There have been many concepts and terms used for such transformative innovations in education as e-Learning, u-Learning, distant learning, mobile learning and social learning. As those terms clearly show, all of them indicate 'learning' only, which implies that those concepts are targeting some changing features or aspects of learning by applying new learning methods or technological support. For instance, e-Learning indicates a certain set of learning style which can be differentiated from the conventional learning mainly based on paper-based media such as books, while e-Learning uses a variety of electronic tools such as PC and Internet. As in the case of e-Learning, others also indicate their unique learning styles mainly by the tools and methods they use.

However education 3.0 and smart education aim at the comprehensive and holistic approach integrating all aspects of education, namely, teaching and learning together. SEI takes this comprehensive approach and strategy for educational innovation in order to transform the present paradigm of education toward the 21st century learning by mobilizing all the policy tools. Within the framework of Education 3.0, we propose three elements in building a new pedagogy; pedagogy, technology and workforce. Smart education and education 3.0 are answers to the change brought by the new digital technology of computers and the Internet. Therefore it may be the first task to understand the characteristics of change occurred by digital and smart technologies, specifically in relation to its impact on teaching and learning process. As human civilization has been evolving, countless technologies have been discovered and accumulated. Education as the only system that human species have developed is a process to teach and learn that technology itself. One most striking difference between the conventional technology and the present smart technology is the different media connecting teachers and students. In the former system the medium was a book containing letters, while in the latter, digitized multimedia converging letters, sounds, motions, images and all kinds of human experience. Letters and numbers are such a complicated system that they are too difficult for ordinary people to master, except for so-called genius. Compared with letters and literacy, sounds and images are much easier to connect people together. Secret codes and abstract logics may not be necessary any longer. If teachers use motions and sounds, many more students will not fall asleep during class. If teachers do not instruct students to read books but do some activities, more students will be engaged in classroom activities. These activities must be integrated into a new pedagogical knowledge.

Thus, smart technology calls for a new pedagogy. The most important element in the new pedagogical setting is the workforce who participate in the process. We have called them teachers and students. By the way, in the conventional pedagogy this workforce was weighed heavier on teachers than students. It was rather a teacher-centered pedagogy. Now it is time to give the same weight between teachers and students. Workforce in this sense may be the proper term indicating the people who undertake their roles in educational process in smart education and education 3.0. SMART, the five initials for "smart" education initiative, stands for five characteristic goals of educational change or learning style: self-directed, motivated, adapted, resource-enriched and technology-embedded learning. We can reorganize them into three elements of education 3.0: technology, pedagogy and workforce.

Technology as the first element indicates that rapidly-changing features of the present ICT and smart devices have been bringing enormous changes into classroom activities. A resource-enriched learning environment will be implemented in the form of content cloud where teachers and students can freely and safely download and upload open educational resources and content in a so-called web 3.0 setting. The concept of "prosumer" will be a right term to grasp easily what it is in concrete. Students themselves participate in learning activities by searching, recreating, uploading and sharing information with their own stories. Teachers will help them with facilitating and protect them from harms and dangers. Conventionally immature students can do so because the media are no longer a secret code of letters, numbers and logics. Sophisticated and complicated lines of code written in the form of letters and numbers are no longer the hegemonic media with which our mankind has communicated and connected together in order to challenge for the uncertain external environment. Digital code and decode system is based on the simple symbols of 1 and 0, switch on and off. Everybody can decode the difference between 1 and 0, on and off, and sky and earth, yin and yang, male and female, and so on. Everybody can decode such sophisticated multimedia which can contain numerous stories they want to connect with their families and friends. Technology enables that. Smart technology makes it accessible to everybody whether a genius or fool. Once materials and codes of education become easy, this means that students can become self-directed and motivated learners. Eventually educational tasks and content will be adapted to every learner. This is the root of change smart technology will trigger to bring.

This change must be taken into account for the essential elements of educational innovation toward 3.0. In other words, we cannot effectively cope with the challenges of social and environmental changes unless we understand the realities of technological development. This change will eventually bring changes in pedagogy, That is, teaching and learning about how we human beings communicate and connect together with information for our group survival. In the end those pedagogical changes will challenge the conventional roles of teachers and students as the main actors or workforce in educational institutions. As we know pedagogy is the knowhow of teachers, only teachers can decode the secret code of letters and numbers, and they can maintain the power of knowledge to teach those students who are ignorant. Now students can easily decode them and even they can form a scrum of students to challenge teachers. Now it is the time teachers and students gathered together to become the same-level workforce in the learning space. This is why the ownership of learning should be returned to learners and students. Education 3.0 exactly tells us to do so, even though it would be hard to loosen the power enough.

With the development of teaching and learning technology to make learning easier, human civilization has evolved too. Media are a tool for ignorant human beings and learned adults as well, who can connect to the unknown world and extend the world of living. As McLuhan [22] said, media extend organs of human being vertically as well as horizontally. Human beings can connect to the future and to the past vertically and to the farthest corner of the globe horizontally. Teaching and learning technology has been established as pedagogy. Letters and literacy were the first revolution in the history of pedagogy. Until now, our mankind is struggling to gain literacy. However we know that the literacy for all is the mission impossible forever. Sleeping students are those strugglers with the task of literacy. We become sleepy when we are tired of heavy muscle work. Students become sleepy when they are tired of learning the impossible job.

The ultimate mission of pedagogy is therefore to find a way to transform the difficult job of literacy into an easy and fun job of learning. Throughout human history, letters have developed into a big knowledge to the point of creating a human creature like the clone sheep Dooley and to the extent of exploring the limitless space of the universe. Philosophy discovered by Socrates and modern sciences accelerated by Rene Descartes are also miracles. We are extending our bodies to the unknown universe and to the timeless eternity, although we can never reach the end. Gutenberg galaxy [23] led by the printing technology seems to be the most recent and powerful pedagogical world where teachers and students can teach and learn things much easier than ever. Teachers now distribute books with knowledge to students and save time and energy to prepare more things to teach in the next class. Students can now read and learn by themselves, unlike their ancestors such as Socrates and Plato who sat together and talked together for teaching and learning.

Schools and universities were formed to accomplish the universal literacy for all. Everybody goes to school, reads books and connects to the unknown world. It was not long ago, however, that we came to recognize that the mission of literacy for all is impossible. Life is never easy. The target is always moving forward and far over there when we think we are just close to it. We need again another pedagogy to go one step farther. When students get sleepy again, teachers must seek another pedagogy. The gaps between teachers and students are so wide that they cannot be connected together. Books and letters are too heavy and complicated to distinguish students from teachers. Books and letters became more and more complicated secret codes. Decoding technology is such a difficult skill for students to master.

4 Conclusion

Many Koreans have smart devices in their hands all the time. That is the phenomenon that we cannot deny. There is another phenomenon in school classrooms. Most teachers teach all things very hard, but, unfortunately, many students can't gain anything from their teaching. Therefore, students try to find out their own ways with their own tools. Two phenomena tell us that schools will be certainly challenged with the change of age, so that they should be changed and the key to the change can be found from the trend of smart.

In this paper, social and cultural changes were discussed, beginning with the inception of smart education initiative in Korea. Next, strategies for implementing smart education were made as a policy. The goals and visions of smart education initiative in Korea have been clearly explained in accordance with the five initials of smart; smart education refers to self-directed, motivated, adapted, resource-enriched and technology-embedded learning. In order to achieve this vision, three areas of policy tools were established. In the first area of system reform, three different tools are composed of digital textbook policy, on-line class and evaluation, and formation of an educational content market. The second area focuses on nurturing teachers' competencies, and the third area centers on the school infrastructure of smart technology.

The most important point and at the same time the conclusion of this paper are pedagogical implications of this smart education initiative. Three major implications are discussed and summarized at the end of this paper. First, smart education calls for a new pedagogy. That is, the new pedagogy should not just deal with letters and numbers but also address sounds and images. Second, teachers and students as workforce have the same importance in our classrooms. Third, a resource-enriched learning environment will be implemented in the form of content cloud where teachers and students can freely and safely upload and download open educational resources and content together.

References

1. Song, M., Rim, H., Choi, H., Park, H., Son, S.: OECD Program for International Students Assessment: Analyzing PISA 2012 Results, Korea Institute for Curriculum and Evaluation (2013)
2. KERIS (Korean Educational Research and Information Service). 2015 Annual Report of Educational Informatization in Korea (2015)
3. Chun, S., Kim, J., Kye, B., Jung, S., Jung, K.: Smart Education Revolution, 21st Books, Seoul (2013)
4. CIS (Council on Informatization Strategies) & MEST (Ministry of Education, Science and Technology). Smart Education Strategies. Unpublished documents (2011)
5. MEST (Ministry of Education, Science and Technology). Implementation Plans for SMART Education Strategies (2011)
6. Korea Education and Research Information Service. Formation of Plans for Smart Education Model Development and Implementation. KERIS, Seoul (2012)
7. Chun, S.Y., Jeon, M.A., Bang, I.J.: Analysis of the effects on using digital textbook in the classroom for smart education. J. Primary Educ. **27**(3), 137–161 (2014)
8. Jang, K.Y., Kim, H.J., Kye, B.Y.: A Time-Series Analysis of the Smart Education and Digital Textbook Effects. KERIS, Seoul (2014)
9. Kim, J.R., et al.: Development of Tools to Evaluate the Effectiveness of Smart Education and Digital Textbooks. KERIS, Seoul (2014)
10. Kim, Y.-A.: Classroom Revolution: Current Status of Smart Education and Its Prospects. KERIS, Seoul (2011)

11. Byun, H., Cho, W., Kim, N., Ryu, J., Lee, K.: Effects of e-Textbook (Research Report 2002-4). Korea Education & Research Information Service (KERIS), Seoul (2006)
12. Son, W.S., Yoon, M.S.: Study on effectiveness of digital thinking in the collaboration learning with digital textbook. J. Korean Assoc. Inf. Educ. **16**(1), 99–106 (2012)
13. Lee, O.H., Jang, S.S.: An analysis of junior high school students' perceptions with prior experience of digital textbooks for the use of digital textbooks. J. Educ. Inf. Media **22**(4), 755–776 (2016)
14. Ahn, S.H.: Analyzing the influence of digital textbook use for potential risk group of internet addiction and average group. J. Korean Assoc. Inf. Educ. **19**(4), 431–440 (2015)
15. Sol, M.K., Sohn, C.I.: A survey on teacher's perceptions about the current state of using smart learning in elementary schools. J. Korean Assoc. Inf. Educ. **16**(3), 309–318 (2012)
16. Hong, H.J., et al.: Teacher survey on the adoption of digital textbooks. J. Korean Educ. **40** (1), 109–130 (2013)
17. Jeong, Y.G.: Homo digitalis und der Smart-education-Plan in Korea. Theor. Pract. Educ. **20** (3), 145–165 (2015)
18. Chun, S.Y.: Korea's smart education initiative and its pedagogical implications. CNU J. Educ. Stud. **34**(2), 1–16 (2013). This part is partially quoted and modified from the reference
19. Cramer, S.N.: History Begins at Sumer. Philadelphia, Pennsylvania (1956)
20. Chun, S.Y.: Education and Freedom. Hakjusa, Seoul (2013)
21. CISCO. White Paper: 21st Century Learning. Unpublished documents (2010)
22. McLuhan, M.: Understanding Media: The Extensions of Man. Routledge, London (1964)
23. McLuhan, M.: The Gutenberg Galaxy: The Making of Typographic Man. University of Toronto Press, Toronto (1962)

Impact of International Accreditation for Education in the Russian Federation

Alexey Margun[1], Artem Kremlev[1], Dmitry Bazylev[1], and Ekaterina Zimenko[2](✉)

[1] Department of Control Systems and Informatics, ITMO University, 49 Kronverkskiy av., 197101 Saint Petersburg, Russia
alexeimargun@gmail.com, {kremlev_artem,bazylevd}@mail.ru
[2] Department of Economics and Strategic Management, ITMO University, 49 Kronverkskiy av., 197101 Saint Petersburg, Russia
zimenko.yekaterina@yandex.ru

Abstract. The paper describes the procedure of international accreditation and its impact for education in the Russian Federation. Namely, international accreditation of education programs is related primarily to the formation of quality assurance mechanisms in the field of education and adaptability of international and national standards of education within the Bologna process.

Keywords: International accreditation · Quality of educational process · Accreditation process

1 Introduction

In 2013 in accordance with the Decree of the Russian Federation President, the Government of the Russian Federation has developed and approved a plan of action for the development of Russian universities and enhance their competitiveness among the world's leading research and education centers. In 2020, at least five Russian universities should enter to the first hundred of leading universities according to international ratings as a result of the project implementation. Project of improving the competitiveness of the leading Russian universities among the world's leading research and education centers (Project 5–100) [1] aims to increase research capabilities of Russian universities and to strengthen their competitive position on the global education market. Work on the project, designed for 7 years, have been started in May 2013 in accordance the terms of the Decree 599 of the President of the Russian Federation "On Measures for Implementation of the state policy in the sphere of education and science". Since 2013, ITMO University is a member of the Project 5–100. In March 2015 the 5–100 International Council approved the submitted by University ITMO "Action Plan ("Roadmap") on the implementation of the program to improve the competitiveness of ITMO University among the world's leading research and education centers in the years 2013–2020. (Stage 2: 2015–2016)" [2].

V.L. Uskov et al. (eds.), *Smart Education and e-Learning 2017*, Smart Innovation, Systems and Technologies 75, DOI 10.1007/978-3-319-59451-4_45

According to the strategic initiative 2 "Global Education: personal growth and professional competitiveness" presented in the road map University ITMO, it was necessary to solve the problem 2.3, "ITMO - International quality management: international educational programs quality assessment and monitoring". This objective is aimed at regulatory methodological support and professional public expert evaluation of the effectiveness and compliance with international standards of high school educational programs and their international accreditation. The creation of international educational programs determines the need for total control of the quality of implemented educational services.

Such requirements are not caused by the general trend, but it caused by the fact, that international accreditation helps to assess the quality of ongoing education program and improve it by transition to SMART education concept. This is of great importance, because today the influence of human capital not enough for the development of modern education. It is necessary to change the educational environment: not only to increase the volume of education of labor resources, but also to change qualitatively the content of education, its methods, tools and environments for realization of universal transition to smart education. The introduction of new technologies into the sphere of education leads to the transition from the old scheme of the reproductive transfer of knowledge to the new creative form of education, where requirements for educational programs are increased especially to international ones. The development of the university's educational system in this direction is one of the key criteria for international accreditation.

To date, the assigned problem on international accreditation was fully implemented. The following sections describes the accreditation procedure, its criteria, experience of ITMO University and impact of accreditation on University education system.

2 Tools of Educational Services Quality Examination

Improving the quality of academic training programs in higher professional and postgraduate education is a priority of modern educational policy in Russia. One of the main tasks of modern education is the creation of a steady motivation for students' knowledge acquisition, the other is the search for new forms and tools for knowledge assimilation through creative solutions.

Modern society confronts universities with a new global task: the training of personnel with creative potential who are able to think and work in a new rapidly changing world. For this, it is necessary to teach students for new practical skills: to communicate in social networks, to select useful information, to work with electronic sources, to make personal knowledge bases, which requires a change in the nature of the educational process.

The state plays a central role in matters of influence assessment of transition to smart university concept, quality assurance, regulation of funding and monitoring implementation through accreditation, licensing and certification procedures. This approach is most common in the countries of the Bologna and the

Copenhagen agreements [3]. It should be noted, that in the Russia monitoring is carried out both by the Ministry of Education and Science of the Russian Federation through a system of regulatory authorities (state comprehensive evaluation of universities) and community organizations (Russian Engineering Education Association - REES, which also provides public and professional accreditation; Promotion Fund of the International Accreditation and Certification (IAC) in the field of education and high technologies, etc.). Therefore there is an urgent need to develop a uniform quality assurance system for training personnel of higher qualification. The urgency of this task is also determined by the need to develop criteria for comparability and mutual recognition of degrees and qualifications, mechanisms of international accreditation of academic training programs in the framework of the Bologna process, which is important for the competitiveness of Russian education. The criteria and procedure for professional and public accreditation should be developed in accordance with Regulation n. 6, Art. 96 of the Federal Law of December 29, 2012 no. 273-FZ "On Education in the Russian Federation", as well as taking into account the world experience in assessing the quality of technical and engineering education, and consistent with international standards EUR-ACE Framework Standards for Accreditation of Engineering Programmes and IEA Graduate Attributes and Professional Competencies.

Russian Engineering Education Association is a member of ENAEE and authorized for appropriation of the program on the basis of accreditation in accordance with these criteria, the mark EUR-ACE Label quality of engineering education programs to be entered in international registers ENAEE (European Network for Accreditation of Engineering Education) and FEANI (Federation of Professional Engineers). Graduates of educational programs accredited by RAEE with EUR-ACE Label, have an advantage in obtaining the title of "European Engineer" (EurIng) and European ENGCard. The main objective of the association is to ensure the functioning of the pan-European system for accreditation of engineering education programs. Since 2012 REEA is a member of the Washington Accord (an international agreement of national agencies, accrediting engineering programs). Because REEA is a member of international organizations for accreditation of engineering educational programs International Engineering Alliance and ENAEE, accreditation of educational programs carried out by REEA, is considered to be an international and recognized in all countries signatories of these agreements (http://www.ieagreements.org, http://www.enaee.eu). Educational program accredited by RAEE is a set of basic characteristics (planned learning outcomes, scope, content), organizational and pedagogical conditions and certification forms, which is presented in the form of curriculum, calendar academic schedule, the work programs of disciplines (modules, courses), and other components, as well as assessment tools and guidance materials needed to prepare for a specific professional development (technician, applied/academic bachelor, specialist and Master) in certain specialty/profile/field of study. REEA considering specialization and the profile in the direction of specialty training as an educational program.

Feasibility of public accreditation is caused by the need for an independent and objective assessment of the university in order to create healthy competitive environment and the need for tools of accurate measurement of educational services quality. Benefits of public accreditation lies in the fact that higher education institutions receive recommendations to management improvement. There are an effective staff training and strategic internal audit.

Passage of international accreditation of most Russian universities, including private, provides innovative approach in the field of education, as regularizing best foreign experience and a positive impact in the teaching methods and the quality of students knowledge.

Graduates of educational programs accredited by the criteria consistent with international standards may be registered in the International Registry International Engineering Technicians Register (graduates with a technician qualification), International Engineering Technologists Register (graduates with qualification Applied Bachelor), APEC Engineer Register and International Professional Engineers Register (graduates with qualification bachelor/specialist).

REEA criteria provide a common approach to professional public accreditation of educational programs at different levels, which stimulates the consistency and continuity of educational programs for the creation a common space of engineering education that meets world practice in Russia.

The criteria are designed to assess and validate the quality of training programs for graduates of educational institutions at the level of professional standards, requirements of engineering community and labor market, and international requirements for the competence of specialists in the field of engineering and technology. These criteria are focused on assessment of the achievement of educational objectives and planned learning outcomes. Learning outcomes are a set of universal (cultural) and professional competencies (knowledge, skills, experience), which are acquired by the graduates at the end of an educational program.

A necessary condition for the accreditation of the educational program are confirmation of expected results achievement by all graduates and their readiness for professional work in accordance with the objectives of the program.

The objectives of an educational program are formulated by educational organization implementing the program, and these objectives should be consistent with the organization's mission. The learning outcomes should be planned based on the goals of the educational program, and agreed with employers and other interested parties. For program accreditation the purposes and results must comply with the requirements of the Federal State Educational Standard (FSES) or Educational Organization Standard (EOS) and REEA requirements.

For professional public accreditation only licensed educational programs with state accreditation are allowed.

In accordance with the requirements of the FSES the content of educational programs of higher education is estimated by the credit units, namely, by credits of European Credit Transfer System (ECTS), recommended in the framework of the Bologna process. The educational program can be accredited by REEA only if it complies with all the criteria shown below:

- objectives of the program and the learning outcomes;
- the content of the program;
- the organization of educational process;
- teachers, teaching staff;
- preparation for professional activity;
- program resources;
- graduates.

3 Accreditation Procedure

Accreditation procedure takes place according to the following list.

1. Educational organization applies to the Director of the Accreditation Centre (AC) REEA for conducting of professional public accreditation of educational programs. Educational programs is profile direction or specialization specialty. If the educational organization is planning accreditation of several programs, the name and code of each profile/specialization are indicated.
2. The application is subject to review, if in the name of the graduate qualification there is engineering and technical terminology. The term of consideration of the application is 15 days from the date of receipt and registration in the Accreditation Center.
3. The application can be rejected for the following reasons:
   - improper filling of the application form;
   - lack of educational program in the current List;
   - the lack of engineering and technical terminology in name of the program;
   - lack of a license and (or) state accreditation of the program under consideration;
   - lack of information on the educational program on the website of the educational organization.
4. In case of disagreement with the decision of the Accreditation Center to reject the application the organization may submit a written statement to the Appeals Commission of REEA. The statement said the reason why the negative decision of the Accreditation Center was unlawful according to opinion of the educational organization.
5. In the case of positive decision the Accreditation Center and educational organization conduct a contract on professional public accreditation of educational programs.
6. Within 15 days after signing the contract the Accreditation Center directs to educational organization all criteria of professional public accreditation and guidance for self-examination of education program.
7. The educational organization up to 6 months conducts self-examination of the educational program in accordance with the requirements of the REEA and sends a report to the Accreditation Center.
8. Based on results of self-examination materials the Accreditation Center can take the following decisions:

- the continuation of accreditation procedures, namely, audit of the educational program directly in the educational institution;
- the need of self-examination materials improvements.

9. In the first case, the Accreditation Center agrees timing of the audit and work plan, informs about the composition of the expert committee for the audit. The educational institution shall inform in writing the Accreditation Center in the event of a reasoned removal of individual members of the committee of experts, and their replacement.
10. In the second case, the Accreditation Center agrees deadlines with the institution on improvement of self-examination report for reconsideration of the report by expert commission.
11. Accreditation Centre forms an expert commission to conduct an audit of the educational program. The expert commission consists of one representative of industry and at least 3 expert auditors who are experts in the evaluation of educational programs in technical fields and professions. Such expert commission may be formed for the accreditation of several educational programs in the organization. Each expert shall sign and submit to the Accreditation Center the statement about the absence of conflict of interest.
12. The visit of the expert committee in the educational organization lasts at least 3 days for the higher education programs and at least 2 days for secondary vocational education programs. At the end of the visit the chairman of the commission and head of the educational organization sign the Protocol on the audit conduction.
13. Based on a comprehensive analysis of audit results and self-examination materials the committee prepares a draft report on the evaluation of the educational program. This report is a detailed opinion on the compliance or non-compliance of the program to accreditation criteria and contains the dissenting opinions of the commission members, if they differ from the general conclusion.
14. The report is directed to the educational organization, not later than 3 weeks after completion of the audit. Within 2 weeks after receiving the report, educational organization may send to Accreditation Center reasoned objections on the report or comments about the violation of the audit procedure.
15. The Accreditation Center sends the report on the evaluation of the educational program, as well as the objections and observations of the educational organization, if any, to the REEA Accreditation Council.
16. The Accreditation Center approves the decision of the Accreditation Council. The educational organization receives the certificate of professional public accreditation of educational programs, signed by the President of the REEA, and the certificate on awarding by European quality label EUR-ACE Label, signed by the presidents of ENAEE and REEA. Accredited educational programs are recorded in the REEA and ENAEE registries, published in the media and on Internet sites of Accreditation Center and ENAEE. Data on accredited educational programs shall be reported to the Russian Ministry of Education and Science.

Schedule of educational programs accreditation procedure is presented in Fig. 1.

Stage of accreditation	Timing of the accreditation stages (in months)											
	1	2	3	4	5	6	7	8	9	10	11	12
Submission of application and its consideration by REEA AC												
Signing the contract on accreditation												
Direction of professional and public accreditation criteria to the educational organization												
Self-examination of educational organization and preparation of report												
Expertise of the self-examination report submitted by organization												
Formation of expert audit comission												
Audit of educational programs presented for accreditation												
Preparation of the report on the evaluation of the educational program and its sending to the educational organization. Sending of comments on reports by the educational organization												
Sending of evaluation report ti the REEA AC												
The decision on accreditation / non-accreditation of the educational program by Accreditation Council and approval by REEA management.												
Issuance of certificates and submitting of information on the program to the REEA and ENAEE registers												

Fig. 1. Schedule of educational programs accreditation procedure

In order to guarantee the rights of the educational organization the Accreditation Center forms special Appeals Commission. Members of the Commission must respect confidentiality and avoid conflicts of interest.

4 ITMO University Experience

Smart technologies are used in almost all educational programs that are being implemented at the ITMO University (see, for example [7–9], etc.). These technologies are not only in the instrumental technologies of the educational process (smart board, etc.), but in innovative curricula and disciplines. Smart technologies allow us to develop revolutionary teaching materials, as well as to form individual learning trajectories for students, realize personalization and adaptation of training, provide free access to necessary content around the world, etc.

The procedure for the accreditation of the University ITMO took place in several stages under the control of Russian Association for Engineering Education (REEA). REEA is a member of the European Network for Accreditation of Engineering Education and member of an international agreement Washington Accord.

In the first half of 2014 the procedure of "self-examination" of educational programs was conducted in accordance with international standards (the impact of transition to new educational standard is described, for example, in [4–6] etc.) Then there was an external examination of the data.

4–7 November 2014 ITMO University was visited by experts who assessed the quality of educational programs and their compliance with international standards. The delegation held meetings with the ITMO University management, faculty members and students, visits educational and scientific laboratories, computer classes during the attendance. Experts familiarized with the regulatory, organizational and methodological documents on educational programs, the state of the material and technical base of the university. The results of the examination have appeared positive and eight submitted programs have been accredited. The same procedure was carried out in 2015 year. As a result of another five programs have been accredited.

The following programs have been certified in 2014:

- Intelligent Control Systems of Technological Processes
- Biotechnology of Therapeutic, Special and Prophylactic Nutrition Products
- Applied Optics
- Overall Automation of Enterprise
- Embedded Computer Systems Design
- Integrated Analyzer Systems and Information Technologies of Fuel and Energy Complex Enterprises
- Quality Control of Products of Space Rocket Complexes
- Automation and Control in Educational Systems

in 2015:

- Methods of Diagnosis and Analysis in Bionanotechnology
- Devices for Research and Modification of Materials at the Micro- and Nanoscale Level
- Metamaterials
- Nanomaterials and Nanotechnologies for Photonics and Optoinformatics
- Optics of Nanostructures

Accreditation showed that educational programs presented by ITMO University has complex of characteristics necessary for compliance to international standards. The analysis of income students in 2015 of all 73 directions of training allows us to say that the programs with international accreditation, have a higher pass rate (an average of 10–15%) and more popular among high school seniors. Also it should be noted, that average mark and academic performance of students' from these areas is much better than in others directions of training.

5 Conclusion

The first results of the international accreditation in the Russian Federation allow us to speak about its absolute importance and necessity for the national higher education system. And it is related primarily to the formation of quality assurance mechanisms in the field of education and adaptability of international and national standards of education within the Bologna process. International accreditation and quality label EUR-ACE show recognition of the educational programs of the university at the European level. It helps to strengthen the university's position in international ratings, as well program graduates can get the professional title EUR ING, which will significantly improve their chances of further employment.

Acknowledgments. This work was supported by the Ministry of Education and Science of Russian Federation (Project 14.Z50.31.0031) and by Government of Russian Federation, Grant 074-U01.

References

1. Project 5-100. http://5top100.com/
2. Roadmap on the implementation of the program to improve the competitiveness of ITMO University among the world's leading research and education centers. http://5100.ifmo.ru/en/
3. Hernes, G., Martin, M.: Accreditation and the global higher education market. International Institute for Educational Planning, 284 p. (2008)
4. Kremlev, A.S., Bazylev, D.N., Margun, A.A., Zimenko, K.A.: Transition of the Russian Federation to new educational standard: independent work of students as a factor in the quality of educational process. In: Proceedings of ERPA International Congresses on Education. SHS Web of Conferences, vol. 26, pp. 1–9 (2016)
5. Abilkhamitkyzy, R., Aimukhambet, Z.A., Sarekenovac, K.K.: Organization of independent work of students on credit technology. Procedia - Soc. Behav. Sci. **143**, 274–278 (2014)
6. Hameed, S., Nileena, G.S.: IEEE student quality improvement program: to improve the employability rate of students. In: IEEE International Conference on MOOC, Innovation and Technology in Education (MITE), pp. 219–222 (2014)
7. Bazylev, D., Shchukin, A., Margun, A., Zimenko, K., Kremlev, A., Titov, A.: Applications of Innovative "Active Learning" Strategy in "Control Systems" Curriculum. In: Uskov V., Howlett R., Jain L. (eds.) Smart Education and e-Learning 2016. Smart Innovation, Systems and Technologies, vol. 59, pp. 485–494 (2016)
8. Bazylev, D., Margun, A., Zimenko, K., Kremlev, A., Rukujzha, E.: Participation in robotics competition as motivation for learning. Procedia - Soc. Behav. Sci. **152**, 835–840 (2014)
9. Zimenko, K., Bazylev, D., Margun, A., Kremlev, A.: Application of innovative mechatronic systems in automation and robotics learning. In: Proceedings of 16th International Conference on Mechatronics (Mechatronika 2014), pp. 437–441 (2014)

Algebraic Formalization of Sustainability in Smart University Ranking System

Natalia A. Serdyukova[1], Vladimir I. Serdyukov[2], Alexander V. Uskov[3](✉), Vladimir A. Slepov[1], and Colleen Heinemann[3]

[1] Plekhanov Russian University of Economics, Moscow, Russia
{nsns25, vlalslepov}@yandex.ru
[2] Institute of Management of Education of Russian Academy of Education, Moscow State Technical University n.a. N.E. Bauman, Moscow, Russia
W1S24@yandex.ru
[3] Department of Computer Science and Information Systems, Bradley University, Peoria, USA
uskov@bradley.edu

Abstract. The validity, truthfulness, and reliability of obtained results are of great value when it comes to theory and, specifically, when it comes to risk-free implementations of theoretical research outcomes, deliverables, and findings in practice. The reliability is particularly significant for people who are dependable on outcomes of innovative approaches and developments for human society, such as pedagogy, general theory of education, e-learning, economics, finance, etc. This paper presents the up-to-date outcomes of an on-going research and development project on analysis, design, and engineering of smart educational systems in general and, in particular, theoretical justification of the introduced *smart university* concept based on the algebraic formalization of smart systems. The obtained theoretical outcomes and findings were successfully validated by their application to an evaluation of sustainability of various universities' ranking systems.

Keywords: Smart system · Algebraic formalization · Sustainability · University rating systems

1 Introduction and Literature Review

In any field of science focused on the functioning of human society, there is a problem with the validity, truthfulness, and reliability of obtained data, outcomes, and findings. For this purpose, we will delve further into the introduced concept of algebraic formalization of smart systems [1–7]; we believe it will create a solid foundation for the justification of our current and future research statements, proposals, and obtained data.

In general, our proposed theoretical approach is based on (a) active use of various schemes and types of systems, and (b) a concept of system's suitability.

In this research project, we actively use various schemes and systems such as (1) the scheme reflecting the concept of formalization, including deductive Hilbert

V.L. Uskov et al. (eds.), *Smart Education and e-Learning 2017*, Smart Innovation, Systems and Technologies 75, DOI 10.1007/978-3-319-59451-4_46

system, (2) deductive Gentzen systems, (3) Boolean algebra, (4) Tarski's topological spaces with the consequences of joining an operator, (5) Lambek's category, (6) Kohn's universal algebras, (7) algebraic logic, (8) multi-sorted logics, and (9) Malt'sev's algebraic systems. Based on these schemes and systems, we plan to construct the algebraic hierarchy, which is a complete upper level of algebraic systems' semi-lattice. The premise is that systems' hierarchy constructed in such a way will allow us to construct a probabilistic algebraic formalization, or, in other words, a probabilistic algebraic logic. In its own turn, a probabilistic algebraic formalization will allow us to predict the functioning of smart systems and, in particular, smart systems in the theory of education and e-learning.

The other important topic for the common system theory, in general, and, particularly for the smart systems' theory, is a problem of systems' sustainability. The concept of sustainability is well-studied in terms of various quantitative parameters' availability that describe the dynamic behavior of a system [8]. We will consider discrete systems in a way as in [2–7]. Particularly, by the sustainability of a discrete system, we understand its ability to return to an equilibrium position by external factors after the end of any action; it is exactly as in the case of continuous time systems. The indices characterizing a quality of discrete systems were introduced [9]; they are designed (1) to evaluate the dynamic properties of a system, manifested in transient conditions, and (2) to determine the accuracy of the system characterized by errors in the steady state after a transition. Dynamic indicators of quality characterize the behavior of the transition process's free components in closed control systems or processes in an autonomous system; in this case, only stable systems are considered; however, convenient integrated indicators, which are a synthesis of qualitative and quantitative indicators of the phenomenon under study, are absent. We propose to use Cayley table of a group G_S of factors, determining the system S, to characterize the quality of dynamics of a closed associative smart system with feedback S. This makes it possible to regulate the behavior of the smart system, in some cases. The time factor is introduced in a construction of groups of factors G_S, determining the system S, for this purpose.

The probabilistic algebraic hierarchy allows one (1) to build an algorithm to determine the points (intervals) of sustainability loss by the smart system S, and (2) to determine the scenario of functioning of the smart system, S. We analyzed the sustainability of systems used for ranking of university activities as a practical example. The processes of decomposition and synthesis of systems are considered in this regard.

1.1 A Concept of Formalization in the Context of Algebraization of Logic: An Overview

The definitions below were originally introduced in [10, 11]. For the purpose of our research, the key point in the development of the notion of formalization is the analysis of relationships between logic and algebra. These days, the Boolean algebra methodology, stemming from the algebraic formalization of proposed logic, has become a very powerful and branched one [12]; for example, the closure operator with certain properties of a topological space, or a monoid, and a category came in sight. The main reasons for this were as follows: (1) a continual set of logical systems, (2) a study of

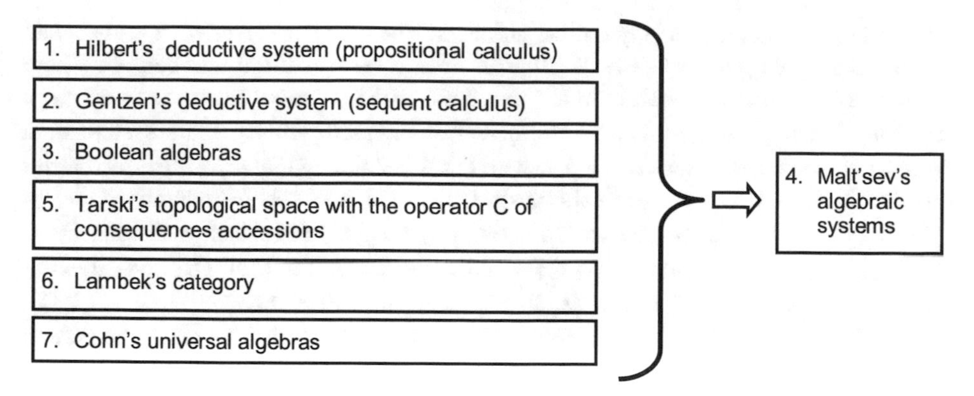

Fig. 1. The main contributors to the concept of formalization of algebraization logic

specific logical systems like some designs in the fashions of certain constructions, (3) the computer revolution, (4) the development of the information society, (5) the needs of insertional modeling, (6) the development of smart systems, etc. [11]. The main contributors to formalization of algebraization logic are presented in Fig. 1.

Hilbert's deductive systems and Gentzen's sequent calculus are the two directions of logic development [22]. Particularly, in the progressive development of the first trend, (1) the concept of formal language in a given alphabet, (2) the concept of semantics arose, (3) the concept of formal language syntax arose, and (4) the theorem about the deduction in the calculus of Hilbert type statements has been proven. The second trend, Gentzen's sequent calculus, has simplified the process of derivation and, therefore, has become of important mathematical value for Computer Sciences. It allows one to order the withdrawal process and avoid sorting through all the formulas during derivation.

Hilbert, Gentzen, Frege, Russell, Whitehead, Boole, Jevons, Pierce, and Shreder created the development of logic as a deductive system [12]. In fact, a different approach to logic was discovered; it (a) allows an investigation of basic properties of logical operations and (b) presents logic in the *Boolean algebra* form. The algebraization of logic then led to an introduction of Halmos's *algebraic logic* [12]. P. Cohn introduced the algebraic logic methods of universal algebra; then, Tarski built a deductive system as a topological space with the operator of accession of consequences, or the so-called Tarski's closure operator, and Lambek presented a deductive system as a category [12]. Particularly, in Lambek's representation, the objects of a category are formulas $A, B, \ldots$, and the morphisms $u : A \to B$ are proofs $A \vdash B$. A deductive system is a category if and only if $1_A\mu = u$, $v1_A = v$ and $w(vu) = (wv)u$. The categorical propositional logic and multi-level deductive systems are further discussed in [13]. The final stage of the formalization became from the concept of an algebraic system which was introduced by Malt'sev [10]. The concept of an algebraic system was then transformed into *the theory of models*. For the purpose of this research project, we need a definition of an algebraic system from [10].

Definition 1.1. Let α, β be any ordinal numbers. The type τ of an order (α, β) is called a pair of mappings $W(\propto) \rightarrow N, W(\beta) \rightarrow N$ into a set of natural numbers N, where $W(\propto)$ is a set of all ordinal numbers strictly less than α. Let type $\tau = \langle m_0, \ldots, m_\xi, \ldots; n_0, \ldots, n_\eta, \ldots \rangle$, where $\xi < \propto, \eta < \beta$. An algebraic system of a type τ is an object $\boldsymbol{A} = \langle A, \Omega_F, \Omega_P, \rangle$, where (1) A is a non-empty set (called the carrier or the basic set of an algebraic system $\boldsymbol{A}$), (2) the set of operations $\Omega_F = \langle F_0, \ldots, F_\xi, \ldots \rangle$ is defined on set A for every $\xi < \propto$, (3) the set of predicates $\Omega_P = \langle P_0, \ldots, P_\eta, \ldots \rangle$ is defined on a set A for every $\eta < \beta$. Let $\Omega = \Omega_F \cup \Omega_P$; one can write the system $\boldsymbol{A} = \langle A, \Omega_F, \Omega_P, \rangle \boldsymbol{A} = \langle A, \Omega_F, \Omega_P, \rangle$, or, in a more brief form, $\boldsymbol{A} = \langle A, \Omega \rangle$. An algebraic system $\boldsymbol{A} = \langle A, \Omega \rangle$ is finite if its basic set A is finite. An algebraic system $\boldsymbol{A} = \langle A, \Omega_F, \Omega_P, \rangle$ is an algebra if $\Omega_P = \emptyset$, and it is a model if $\Omega_F = \emptyset$.

Definition 1.2. Under the formalization, we will understand the record or a presentation of any phenomena, process, or practical result denoted by P and expressed verbally by the model FP in a special language $\delta = <M, C_M>$ with its own semantics and syntax, or the language of formalization. The model FP is presented in the form of symbols or constructions of this language.

1.2 Formalization. Expressive Properties. A Hierarchy of Formalizations

Multi-sorted logics of the first order, infinite logics, ω - logic, R - logic, weak second–order logic, etc. were invented later; they improved the expressive properties of formalization. The idea of multi-sorted logics is used in insertion modeling [23]. Insertion modeling is the technology of system design founded on the theory of agents and environments that have been applied to verification problems in various subject areas. One of the main applications of this technology is the Verification of Requirement Specification (VRS) system. The VRS software package provides means of checking the requirements for hardware and software systems by (a) automated theorem proving and (b) using symbolic and deductive techniques in model validation. This raises the following question: How should a set of formalizations hierarchy be defined? Actually, this is the question of how to construct a lattice of logics. Let us show that it can be done with the help of Malt'sev's algebraic systems taking into account the fact that any logic is defined by Malt'sev's algebraic system.

Let's define the lattice of logics, or the hierarchy of formalizations of the lattice of algebraic systems, as $\boldsymbol{N} = \langle L, \cup, \cap \rangle$, and the binary relation ρ on the set of all algebraic systems $\{\boldsymbol{A}_\propto = \langle A_\propto | \Omega_F^\propto, \Omega_P^\propto \rangle, \propto \epsilon \Lambda\}$ in the following way: $\boldsymbol{A}_\propto = \langle A_\propto | \Omega_F^\propto, \Omega_P^\propto \rangle \rho \boldsymbol{A}_{\boldsymbol{\beta}} = \left\langle A_\beta | \Omega_F^\beta, \Omega_P^\beta \right\rangle$ is true if and only if $A_\propto \subseteq A_\beta, \Omega_F^\propto \subseteq \Omega_F^\beta, \Omega_P^\propto \subseteq \Omega_P^\beta$. Then, ρ is a reflexive, transitive, and antisymmetric binary relation on the set of all algebraic systems, so the binary relation of partial order ρ is given on the set of all algebraic systems. The relation ρ defines a lattice of all algebraic systems. As a result, a set of all algebraic systems with a binary relation of a partial order ρ is a complete upper semi-lattice. Let $N = \{\boldsymbol{A}_\gamma | \gamma \epsilon \Gamma\} \subset M$. An algebraic system with the basic set $\left\langle \bigcup_{\gamma \epsilon \Gamma} A_\gamma \right\rangle$ of the signature $\left\langle \bigcup_{\gamma \epsilon \Gamma} \Omega_F^\gamma, \bigcup_{\gamma \epsilon \Gamma} \Omega_P^\gamma \right\rangle$ is the upper bound of set N if

$\left\langle\bigcup_{\gamma\epsilon\Gamma} A_\gamma\right\rangle$ is $\bigcup_{\gamma\epsilon\Gamma} A_\gamma$ joined with the results of all operations from $\bigcup_{\gamma\epsilon\Gamma} \Omega_F^\gamma$, and is applied to elements of $\bigcup_{\gamma\epsilon\Gamma} A_\gamma$. So, we have an upper complete semi-lattice of algebraic systems. An algebraic system $\left\langle\bigcap_{\gamma\epsilon\Gamma} A_\gamma, \bigcap_{\gamma\epsilon\Gamma} \Omega_{F_\gamma}^\gamma, \bigcap_{\gamma\epsilon\Gamma} \Omega_{P_\gamma}^\gamma\right\rangle$ is the exact lower bound of the set N, which differs from $\emptyset$ if $\bigcap_{\gamma\epsilon\Gamma} A_\gamma \neq \emptyset$. If (a) we add an empty set $\emptyset$ to the set of all algebraic systems and (b) consider an empty algebraic system as an empty set with an empty signature, or, in other words, consider $\langle A, \emptyset, \emptyset,\rangle$ where A can be empty, as an algebraic system, then we obtain a complete lattice of all algebraic systems.

For the purpose of this research project, we need to use the theorem of Birkhoff [11]: Every distributive lattice is isomorphic to a lattice of sets.

1.3 Probabilistic Algebraic Formalization

Let $\boldsymbol{N} = \langle N, \bigvee\wedge\rangle$, where $N = \{\boldsymbol{A}_\gamma = \langle A_\gamma, \Omega_F^\gamma, \Omega_P^\gamma\rangle | \gamma\epsilon\Gamma\}$ is a complete lattice of algebraic systems and $\widehat{\mathrm{E}} = \left\langle\left\langle\bigcup_{\gamma\epsilon\Gamma} A_\gamma\right\rangle, \bigcup_{\gamma\epsilon\Gamma} \Omega_F^\gamma, \bigcup_{\gamma\epsilon\Gamma} \Omega_P^\gamma\right\rangle$. Let's define an unary operation $'$ in the following way: $\boldsymbol{A}' = \langle\langle\hat{\mathrm{E}} \backslash A_\gamma\rangle, \bigcup_{\gamma\epsilon\Gamma, \propto\neq\gamma} \Omega_F^\gamma, \bigcup_{\gamma\epsilon\Gamma, \propto\neq\gamma} \Omega_P^\gamma,$ where $\widehat{\mathrm{E}} = \left\langle\left\langle\bigcup_{\gamma\epsilon\Gamma} A_\gamma\right\rangle, \bigcup_{\gamma\epsilon\Gamma} \Omega_F^\gamma, \bigcup_{\gamma\epsilon\Gamma} \Omega_P^\gamma\right\rangle$, $\hat{0} = \hat{\mathrm{E}}'$. If for any $\boldsymbol{A}_\gamma = \langle A_\gamma, \Omega_F^\gamma, \Omega_P^\gamma\rangle \epsilon \boldsymbol{N}$ there exists the only $\boldsymbol{A}_\gamma'$, which satisfies the conditions $\boldsymbol{A}_\gamma \vee \boldsymbol{A}_\gamma' = \hat{\mathrm{E}}$ and $\boldsymbol{A}_\gamma \wedge \boldsymbol{A}_\gamma' = \hat{0}$, then $\boldsymbol{N} = \langle N, \vee\wedge, ' \rangle$ is a complete distributive lattice with supplements.

Definition 1.3. Let the set of algebraic systems N be closed under countable intersections of basic sets A_γ of algebraic systems from N and let N also be closed under countable intersections of algebraic systems signatures Ω_F^γ from N. In this case, $\boldsymbol{N} = \langle N, \vee\wedge, ' \rangle$ is a σ - algebra and one can define the function of a probability measure $p: N \to [0; 1]$, which satisfies the following conditions:

(1) $p(\vee_\propto(\boldsymbol{A}_\propto = \langle A_\propto | \Omega_F^\propto, \Omega_P^\propto\rangle)) = \sum_\propto p(\boldsymbol{A}_\propto = \langle A_\propto | \Omega_F^\propto, \Omega_P^\propto\rangle)$, where the set of indices $\{\propto | \propto \in \omega\}$ is countable, $\langle A_\propto | \Omega_F^\propto, \Omega_P^\propto\rangle \wedge \left\langle A_\beta | \Omega_F^\beta, \Omega_P^\beta\right\rangle = \hat{\mathrm{O}}$, where $\hat{\mathrm{O}}$ is the minimal element of a lattice $\boldsymbol{N} = \langle N, \vee\wedge, ' \rangle$ for any $\propto, \beta \in \omega, \propto \neq \beta$.

(2) $p\left(\widehat{\mathrm{E}}\right) = 1$, where $\widehat{\mathrm{E}}$ is the maximal element of a lattice $\boldsymbol{N} = \langle N, \vee\wedge, ' \rangle$.

Thus, we get an algebraic probabilistic logic or probabilistic algebraic formalization, where the algebraic system $\boldsymbol{A}_\propto = \langle A_\propto | \Omega_F^\propto, \Omega_P^\propto\rangle$ describes an abstract system S more precisely than the algebraic system $\left\langle A_\beta | \Omega_F^\beta, \Omega_P^\beta\right\rangle$ if $\boldsymbol{p}(\boldsymbol{A}_\propto = \langle A_\propto | \Omega_F^\propto, \Omega_P^\propto\rangle) \geq \boldsymbol{p}\left(\boldsymbol{A}_\beta = \left\langle A_\beta | \Omega_F^\beta, \Omega_P^\beta\right\rangle\right)$.

Definition 1.4. Let $\mathrm{S} = \langle N, \bigvee\wedge\rangle$, where $N = \{\boldsymbol{A}_\gamma = \langle A_\gamma, \Omega_F^\gamma, \Omega_P^\gamma\rangle | \gamma\epsilon\Gamma\}$, be a complete lattice of algebraic systems. The set of atomics elements $\Omega_N = \{\boldsymbol{A}_\tau | \tau\epsilon T\}$ of this lattice is called a space of elementary algebraic formalizations.

The sense of the introduced Definition 1.4 is clarified by the Stone's theorem [11]: "Every finite Boolean algebra is isomorphic to the algebra of all subsets of its atoms[1]." By the symbol $\boldsymbol{A} \models \varphi$, we will denote the fact that the formula φ is identically true on the algebraic system $\boldsymbol{A}$.

Let us introduce the notion of a random formula in the following way.

Definition 1.5. A formula φ of a signature Ω_β, where $\beta \in \Gamma, \Omega_\beta = \Omega_F^\beta \cup \Omega_P^\beta$, is called a random one if, for an algebraic system $\langle \boldsymbol{A}_\gamma | \boldsymbol{A}_\gamma \models \varphi,$ where φ is a formula of a signature $\Omega_\beta, \gamma\rho\beta\rangle$, generated by algebraic systems $\boldsymbol{A}_\gamma$ in such a way that $\boldsymbol{A}_\gamma \models \varphi$, where φ is a formula of a signature Ω_β and $\gamma\rho\beta$, the inclusion $\langle \boldsymbol{A}_\beta | \boldsymbol{A}_\beta \models \varphi,$ where φ is a formula of a signature $\Omega_\beta\rangle \epsilon N$ is true. In this case, $\varphi = \varphi(\boldsymbol{x})$, where $\boldsymbol{x} = (x_1, \ldots, x_n)$, or φ does not contain any variables.

Let us define a distribution function for a random formula as follows.

Definition 1.6. The probability of a random formula φ of a signature Ω_β is called the probability $p(\varphi)$, defined as follows: $\boldsymbol{p}(\boldsymbol{\varphi}) = \boldsymbol{p}(\langle \boldsymbol{A}_\delta \,|\, \boldsymbol{A}_\delta \models \boldsymbol{\varphi}, \textit{\textbf{where}}\, \boldsymbol{A}_\delta \boldsymbol{\rho} \boldsymbol{A}_\beta \rangle)$. The value $\boldsymbol{p}(\boldsymbol{\varphi}) = \boldsymbol{p}(\langle \boldsymbol{A}_\delta | \boldsymbol{A}_\delta \models \boldsymbol{\varphi}, \textit{\textbf{where}}\, \boldsymbol{A}_\delta \boldsymbol{\rho} \boldsymbol{A}_\beta \rangle)$ is defined because $\langle \boldsymbol{A}_\delta | \boldsymbol{A}_\beta \models \boldsymbol{\varphi}, \textit{\textbf{where}}\ \boldsymbol{A}_\beta \boldsymbol{\rho} \boldsymbol{A}_\beta \rangle \in \boldsymbol{N}$.

The simplest property of the function p runs as follows: $p(\boldsymbol{A}_\propto \bigvee \boldsymbol{A}_\beta) = p(\boldsymbol{A}_\propto) + p(\boldsymbol{A}_\beta)$, if $\boldsymbol{A}_\propto \wedge \boldsymbol{A}_\beta = \widehat{\mathrm{O}}$.

Definition 1.7. We define the distribution function $F(\varphi)$ of the random formula φ as follows: $F_\varphi(A_\gamma) = p(\langle \boldsymbol{A}_\delta | \boldsymbol{A}_\delta \models \varphi, \textit{where}\, \delta\rho\gamma\rangle)$.

Let's consider the simplest property of a distribution function of a random formula. Let $\boldsymbol{A}_\propto$ be minimal with respect to relation ρ, an algebraic system of the lattice $\boldsymbol{N}$, such that $\boldsymbol{A}_\beta \models \varphi$. Consequently, $\boldsymbol{A}_\propto \rho \boldsymbol{A}_\beta$, and $p(A_\propto) \le F(\varphi) < p(\boldsymbol{A}_\beta)$.

1.4 Probability Distribution of the Complete Countable Distributive Lattice of Algebraic Systems

Let us give several additional definitions to understand the probability distribution of the complete countable distributive lattice of algebraic systems.

Definition 1.8. The function $\xi : N \to R$ that is given on a complete lattice of algebraic systems $\langle \boldsymbol{N} = N, \bigvee, \wedge, \rangle$, where $N = \{\boldsymbol{A}_\gamma = \langle A_\gamma, \Omega_F^\gamma, \Omega_P^\gamma \rangle | \gamma \epsilon \Gamma\}$, $\boldsymbol{A}_\beta \in N, \xi(\boldsymbol{A}_\beta) \in R$, and taking values in the set R of real numbers, is called a random function of algebraic formalizations $\boldsymbol{N}$ if the complete inverse image of $\xi(N) \subseteq R$ under the map ξ is an element of N. In other words, $\xi^{-1}(\xi(N)) \epsilon N$.

[1] Recall that the element a is an atom of Boolean algebra if and only if it is the minimal non-zero element with respect to the induced partial order.

Algebraic formalization	A_1	A_2	...	A_k	...
Probability	p_1	p_2	...	p_k	...

Fig. 2. The probability distribution of the complete countable distributive lattice $\boldsymbol{N}$

Definition 1.9. The probability distribution of the complete countable distributive lattice $\boldsymbol{N} = \langle N, \bigvee, \wedge, \rangle$, where $N = \{\boldsymbol{A}_i = \langle A_i, \Omega_F^i, \Omega_P^i \rangle | i = 1, 2, \ldots.\}$, $\boldsymbol{A}_i \in N$, $i = 1, 2, \ldots$, of algebraic systems is called a table (given in Fig. 2) in which $\sum_{i=1}^{\infty} p_i = 1$, if the lattice $\boldsymbol{N}$ is infinite. (A note: The top line of this table is ordered by relation ρ)

Definition 1.10. If the lattice $\boldsymbol{N} = \langle N, \vee \wedge \rangle$ is finite, then $N = \{\boldsymbol{A}_i = \langle A_i, \Omega_F^i, \Omega_P^i \rangle | i = 1, 2, \ldots .k\}$ and the probability distribution of the complete distributive finite lattice $\boldsymbol{N} = \langle N, \vee \wedge \rangle$, where $N = \{\boldsymbol{A}_i = \langle A_i, \Omega_F^i, \Omega_P^i \rangle | i = 1, 2, \ldots k\}$, $\boldsymbol{A}_i \in N$, $i = 1, 2, \ldots k$, is called the table (given in Fig. 3) in which $\sum_{i=1}^{k} p_i = 1$. (A note: The top line of this table is ordered by relation ρ).

Algebraic formalization	A_1	A_2	...	A_k
Probability	p_1	p_2	...	p_k

Fig. 3. The probability distribution of the complete distributive finite lattice $\boldsymbol{N}$

Definition 1.11. The distribution function of a random function $\xi : N \to R$ of an algebraic formalization lattice $\boldsymbol{N} = \langle N, \vee \wedge \rangle$, where $N = \{\boldsymbol{A}_\gamma = \langle A_\gamma, \Omega_F^\gamma, \Omega_P^\gamma \rangle | \gamma \epsilon \Gamma\}$, $\boldsymbol{A}_\beta \in N$, $\xi(\boldsymbol{A}_\beta) \in R$, is called a function $F(x)$ whose value at the point $x = \boldsymbol{A}_\gamma$ is equal to the probability that a random function $\xi : N \to R$ of the lattice of formalizations $\boldsymbol{N} = \langle N, \vee \wedge \rangle$ will take a value less then x.

Probabilistic algebraic formalizations are used while considering different systems' scenarios of development because they randomly change during the process of developing a system in general and, particularly, a smart system.

We have also constructed an interpretation of the classic laws of parametric statistics to model the development of a system for a finite number of scenarios. Through this, we found the relationships between the concept of final sustainability and the classical Lyapunov's notion of stability.

Now let us consider two examples of applications of the described hierarchy of algebraic formalizations - in the area of smart learning course and in the field of Abelian groups.

1.5 Hierarchy of Algebraic Formalizations: Examples of Application

The first example of the usage of the hierarchy of algebraic formalizations is implicitly presented in [14]; it actually describes a formalization of a course of high algebra and the theory of numbers as a course for smart learning.

The second example is as follows. In one area of mathematics, namely, in the theory of Abelian groups, the problem of the description of torsion free abelian groups of finite rank was a well-known problem; however, it was solved independently by A. G. Kurosh, A.I. Malt'sev, and Derry in 1937 [21]. The proposed solution used one of the possible algebraic formalizations, namely, a language of the theory of matrices. Corner [20] has introduced the notion of a quasi–isomorphism, which is a generalization of a notion of isomorphism; he arrived to the solution by using the hierarchy of logic languages suitable for solving the problem of abelian torsion free groups. To solve the same problem, Kulikov and Fomin [24] have used p-adic numbers and, ultimately, provided a more powerful and expressive logic language to describe the properties of torsion free abelian groups of finite rank. In other words, they also used the hierarchy of logic languages suitable to solve this problem.

2 The Proposed Approach and Algorithm for Sustainability of Systems

The introduced probabilistic hierarchy of algebraic formalizations allows us to study systems' sustainability in general and, in particular, the sustainability of smart systems.

For the purpose of this research project, we will need a notion for a final sustainability of a system that is analogous to the classical concept of stable equilibrium.

Definition 2.1. Let $\boldsymbol{G_S} = \langle G_S, \circ, ^{-1}, e \rangle$ be a group of factors representing the system S. So, S is a closed associative system with feedback. Let G_S be finite, consisting of n elements, i.e. $|G_S| = n$. Let $G_{1S}, G_{2S}, \ldots, G_{mS}$ be all pairwise non-isomorphic groups of n elements. Groups $\boldsymbol{G_{1S}}, \boldsymbol{G_{2S}}, \ldots, \boldsymbol{G_{mS}}$ are called final states of system S. The system S is called a final sustainable one if it has only one final state.

Definition 2.2. Groups $G_{1S}, G_{2S}, \ldots, G_{mS}$ are also called the scenarios of development of system S.

Consequence 2.3. Based on Definitions 2.1 and 2.2, system S, which is represented by a simple groups of factors $\boldsymbol{G_S}$, is final sustainable.

A system S, which is represented by a groups of factors $\boldsymbol{G_S}$, consisting of a finite prime number of factors, is final sustainable.

The next problem (that arises in this context) regards how the time factor can be entered into a design of a group of factors that represent the associative closed system with a feedback.

2.1 The Time Structure of Algebraic Formalization

Let us introduce the time factor into the construction representing an associative closed system with a feedback S.

The time factor is represented by a linearly ordered set $\langle T, \leq \rangle$ in [15]. Let us interpret time a little bit differently, namely, by considering an abelian linearly ordered group $\boldsymbol{T} = \langle T, +, -, 0 \leq \rangle$. In this interpretation, we assume the datum point from "the modern period *0*". From this, we can review past time periods with the help of the operation '-'.

Let $\boldsymbol{G_S} = \langle G_S, ^\circ, ^{-1}, e \rangle$ be a group of factors representing the system S. Let us consider the Cartesian product of sets $\boldsymbol{G_S} \times \boldsymbol{T}$. We will use the following notations:

(1) an ordered pair $\langle a, t \rangle$ will be denoted as $a(t)$; that is, $a(t) \rightleftharpoons \langle a, t \rangle$;
(2) the connections between factors, which represent the system S and the factor time, are defined by binary operations $\oplus$ and $*$ on $G_S \times T$ in a way that satisfies the conditions:

(2.1) $a(x) \oplus a(y) = a(x+y)$, for every $a \in G_S$, for every $x, y \in T$,
(2.2) $a(x) * b(x) = (a^\circ b)(x)$, for every $a, b \in G_S$, for every $x \in T$.

As a result, every $a \in G_S$ defines the homomorphism $\breve{a} : T \to G_S \times T$. such that, for any $x \in T$, we have $x \xrightarrow{\breve{a}} a(x)$.

For a complete inverse image of $M \subseteq G_S \times T$ under the map $\breve{a}$, we have the inclusion $(\breve{a})^{-1}(M) \subseteq T$. Every $x \in T$ defines a homomorphism $\hat{x} : G_S \to G_S \times T$ such that, for any $a \in G_S$ we have $a \xrightarrow{\hat{x}} a(x)$. For a complete inverse image of $M \subseteq G_S \times T$ under the map $\hat{x}$, the inclusion $(\hat{x})^{-1}(M) \subseteq G_S$ takes place.

Now let $T(t_1; t_2)$ be the minimal subgroup on the relation of inclusion containing all $x \epsilon T$ such that $t_1 \leq x \leq t_2$. Let $G_S \times T(t_1; t_2) \subseteq M \subseteq G_S \times T$ and $x \in T$. Then we have the inclusion $(\hat{x}^{-1})(G_S \times T(t_1; t_2)) \subseteq (\hat{x}^{-1})(M) \subseteq G_S$. A right action ρ of a group G_S on $(\hat{x}^{-1})(M) \subseteq G_S$ is a map $\rho : G_S \times (\hat{x}^{-1})(M) \to (\hat{x}^{-1})(M)$, such that $(a, g) \mapsto ag$ and $ae = a, ag_1 g_2 = (ag_1)g_2$, where $a \epsilon G_S, g, g_1, g_2 \in (\hat{x}^{-1})(M)$. A right action ρ of a group G_S on the set $(\hat{x}^{-1})(M) \subseteq G_S$ defines the homomorphism : $G_S \to Symm((\hat{x}^{-1})(M))$ for which $a \mapsto ag$, where $a \in G_S, g \in (\hat{x}^{-1})(M)$. $Symm((\hat{x}^{-1})(M))$ is a group of all permutations on the set $((\hat{x}^{-1})(M)) \subseteq G_S$ and is a subgroup of the group $SymmG_S$.

Let $i : G_S \to SymmG_S \cong S_n$, where S_n is a symmetric group of all permutations of degree n. Let's designate an isomorphism $SymmG_S \cong S_n$ by $\propto$. Let A_n be an alternating group of permutations of degree n. Let us choose a set M in such a way that the inclusion $i^{-1} \propto^{-1} (A_n) \subseteq M$ takes place; then, $(\hat{x})^{-1}(i^{-1} \propto^{-1} (A_n)) \subseteq (\hat{x}^{-1})(M)$ и $(\hat{x}^{-1})(M)$ is a simple group. The last statement is derived from the diagram in Fig. 4.

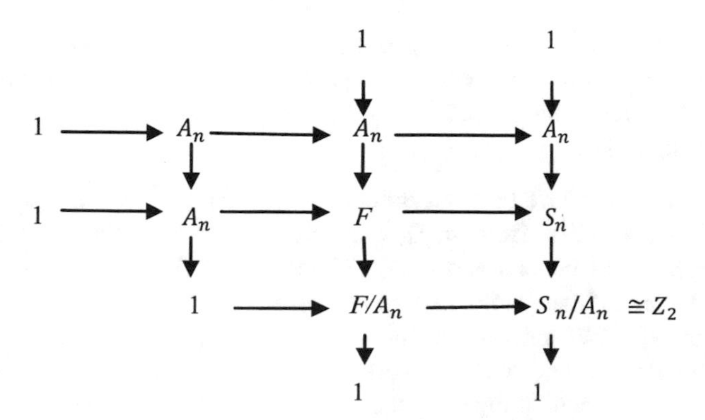

Fig. 4. Development of $(\hat{x})^{-1}(i^{-1} \propto^{-1} (A_n)) \subseteq (\hat{x}^{-1})(M)$ и $(\hat{x}^{-1})(M)$ as a simple group

We have $F/A_n \cong 1$ or $F/A_n \cong Z_2$, so $F \cong A_n$ or $F \cong S_n$; hence, we obtain the following statement:

Theorem 2.4. A right action of the group of factors G_S, which determines the system S on the set $(\hat{x}^{-1})(M)$, where $i^{-1} \propto^{-1} (A_n) \subseteq M$, defines a single scenario of a development of the system S for any instant of time x.

If at any point of time t_0 the set of factors acts on the system S, which contains a subgroups of factors isomorphic to A_n in a specific way, namely as $i^{-1} \propto^{-1} (A_n)$, then the system S is developed by the unique scenario at the moments of time $t \geq t_0$.

Theorem 2.5. To make it possible to adjust a development scenario of system S, it is required to have an existence of the instant time $x \in T$ in such a way that, for any $M \subseteq G_S \times T$, the condition $i^{-1} \propto^{-1} (A_n) \subsetneq M$ is performed.

The last condition can be rewritten as: $(\exists x \in T)(\forall M \subseteq G_S \times T)(\exists y)((y \in i^{-1}\alpha^{-1} (A_n)) \wedge (y \notin M))$.

Definition 2.6. An element $x \in T$ is called the point of no return of the system S if the following conditions are true: (1) $i^{-1} \propto^{-1} (A_n) \subseteq M$ and (2) x is the minimal one by the $\leq$ relation. (A remark: the inclusion $i^{-1} \propto^{-1} (A_n) \subseteq M$ depends on the parameter x).

Definition 2.7. The system S is sustainable for the period of time $t_1 \leq t \leq t_2$ if S does not change the scenario of development in this interval.

2.2 System's Development Scenarios and Intervals of Sustainability Loss: A Proposed Algorithm

The following problem was identified over the course of this research project: (1) How should we determine the points and intervals of system's loss of sustainability? In order to answer this important question, we developed the algorithm to determine the development scenarios of system **S** and points and time intervals of loss of suitability by system **S**.

Let $G_S = \{a_i | i = 1, \ldots, n\}$, $|G_S| = n$, and $G_{1S}, G_{2S}, \ldots, G_{mS}$ be pairwise non-isomorphic groups of n elements. Let $T_{1S}, T_{2S}, \ldots, T_{mS}$ be Cayley tables for groups $G_{1S}, G_{2S}, \ldots, G_{mS}$, respectively. Let the equality $a_i^{\circ} a_j = a_{rk}$, where $\in \{1, \ldots, n\}$, be true in the Cayley table T_{kS}.

Assumption 1. Assume we have all indices $a_i(t), a_j(t), a_{rk}(t)$ for $t_1 \leq t \leq t_2$. We will assume that every index from $a_i(t), a_j(t), a_{rk}(t)$ does not change the nature of monotony in the time interval $t_1 \leq t \leq t_2$. Then, two of these indices have the same nature of monotony. If these are a_i and a_j, and a_{rk} has another nature of monotony, then the Caley table T_{kS} does not work in the time interval $t_1 \leq t \leq t_2$. This means that the algebraic formalization G_{kS} does not work anymore. This means we choose a way of modeling the development of the system S among those Cayley tables $T_{1S}, T_{2S}, \ldots, T_{mS}$ in which a_i, a_j,a_{rk} have the same nature of monotony.

Remembering lemma about compressing segments and appropriately diminishing the interval $t_1 \leq t \leq t_2$ we get the point of loss of sustainability or the interval of loss of sustainability by system S. This depends on the quality of a system's statistic data.

Assumption 2. Let the scenario of development G_{kS} take place in the time interval $t_0 \leq t \leq t_1$. Then the ratio of the lengths $q = \frac{|[t_1;t_2]|}{|[t_0;t_2]|}$ of the time intervals shows the probability of the default scenario of development G_{kS} of system S. To show that the ratio of the lengths of the time intervals defines a probability measure, let $X = \{G_{1S}, \ldots, G_{mS}\}$ be a set of all development scenarios of system S, which is described by a finite set of factors; then, a set X is finite.

Assumption 3. Let $P(X)$ be a set of all subsets of the set X; in this case, $|X| = 2^m$. As a result, the algebra $\langle P(X), \cup, \cap, ' \rangle$ is a finite σ-algebra.

The proposed algorithm can be described as follows:

Step 1: In accordance with above-mentioned Assumption 1, we choose all of the scenarios of the development $\{\mathrm{G_{is}} | 1 \leq \mathrm{i} \leq \mathrm{m}, \mathrm{i} \in \mathrm{I_k}\}$ for the system S such that, for each, the probability of its default in the time interval $\mathrm{t_1} \leq \mathrm{t} \leq \mathrm{t_2}$ is greater than or equal q.

Step 2: Repeat Step 1 for all development scenarios for system S from the set $\{\mathrm{G_{1S}}, \ldots, \mathrm{G_{mS}}\} \backslash \{\mathrm{G_{is}} | 1 \leq \mathrm{i} \leq \mathrm{m}, \mathrm{i} \in \mathrm{I_k}\}$.

Step 3: Since the number of all possible development scenarios of system S is finite, at some finite step j, the process will be terminated.

Step 4: On the penultimate step $j - 1$, we get the development scenario of the $G_{k_{j-1}S}$ for system S in the time interval $t_1 \leq t \leq t_2$.

The nature of monitoring indices of factor, which determine if the behavior or functioning of system S is discrete; as a result, in order to predict the behavior of system S to determine its development scenario, we need the most well-known laws of random value distribution. (A remark: It is easy to see that this algorithm defines points of no return of the system S).

3 A Validation of Obtained Research Outcomes (An Application of Algebraic Formalization to Sustainability of University Ranking Systems)

We validated the obtained research outcomes by formalizing a system synthesis and the use of the technique of the theory of extensions of Abelian groups. Particularly, we used the proposed algorithm to examine the sustainability of several well-known world university ranking systems that are focused on the evaluation of the effectiveness of universities all over the world. Those systems include, but are not limited to, (1) THE World University Rankings system [16], (2) QS World University Rankings [17], (3) Academic Ranking of World Universities [18], and (4) Global Research University Profiles (GRUP) [19]. It is a well-known fact that the first three systems use "a decomposition into subsystems" technique during the process of university evaluation process. (A note: In this research project we use the following definition: “A system is a sustainable one if after its withdrawing from the state of equilibrium (rest) by the external effects or forces, it returns back to the state of equilibrium after termination of external influences”. From the algebraic formalization’s point of view, this means that there are restrictions on the number of final states of the system [8, 9].)

3.1 Evaluation of The World University Rankings System

This system uses 13 parameters (or evaluation criteria) with weights, which are expressed in percentages from the total score on the following categories of evaluation criteria:
Category I: “Teaching and learning environment” - 30% upon parameters:

1. The survey of the scientific university staff’s environments - 15%;
2. The ratio of the number of professors and teaching staff to the number of students in high school - 4.5%;
3. The ratio of the number of masters who completed a PhD to the number of bachelors enrolled in masters - 2.25%;
4. The ratio of the number of professors and teaching staff with a PhD to the total number of professors and teaching staff - 6%;
5. The ratio of an income of an educational institution from the scientific research to the number of scientific staff - 2.25%;

Category II: “Research – volume, income, reputation” - 30% upon parameters:

6. Research reputation among similar universities - 18%;
7. Income from research activities correlated with the number of employees and normalized for purchasing power parity - 6%;
8. Assessment of the environment for scientific research calculated as the ratio of the numbers of papers indexed in Thomson Reuters as scientific to the number of professors and teaching staff and then this ratio normalized;
9. Assessment of the environment for scientific development - 6%

Category III: “Citation, influence, authority” - 30% upon parameters:

10. The category consists from one indicator an index of citing. The ranking does not include universities that published less than 200 papers per year.
11. Data are normalized to reflect the difference in the various scientific fields.

Category IV: “Income from production activities: innovations” - 2.5% upon parameter:

12. Evaluation of knowledge transfer - innovations, consulting, invention - 2.5%.

Category V: “International image” - 7.5% upon parameters:

13. The share of foreign undergraduate and graduate students in relation to local students and postgraduates - 2.5%;

14. The share of foreign citizens among the professors and teaching staff in relation to the total number of professors and teaching staff - 2.5%;
15. The share of scientific publications, which have at least one foreign co-author relative to the total number of publications of the University - 2.5%.

There are five categories (or blocks) in this ranking system; as a result, in accordance with Consequence 2.3 above, this ranking is a rather sustainable one.

3.2 Evaluation of the QS World University Rankings System

The QS system contains the following 6 indicators (or evaluation criteria):

1. Expert survey in the scientific community - 40%;
2. Employers survey - 10%;
3. The ratio of the number of students to the number of professors and staff - 20%;
4. The number of citations per staff member of the teaching staff - 20%;
5. The share of foreign students - 5%;
6. The share of foreign employees among professors and teaching staff - 5%.

In accordance with Consequence 2.3, an introduction of Indicator 6 makes this system less sustainable.

3.3 Evaluation of the ARWU World University Rankings System

The ARWU system uses the following indicators (evaluation criteria) combined into five groups:

1. Alumni (notably, holders of the Nobel Prize or Fields Medal) - 10%;
2. Awards (a total number of employees of the University that received the Nobel Prize or Fields Medal while working at the University) - 20%;
3. HiCi (the number of scientists most highly cited in 21 subject areas) - 20%;
4. N&S (the number of scientific articles published in the journals in Nature and Science areas from 2008 to 2016) - 20%;
5. PUB (a total number of articles indexed in the databases Science Citation Index-Expanded and Social Science Citation) - 20%;
6. PCP (a calculated parameter that is based on five points of the preceding indicators divided into an equivalent number of academic staff working full time) - 10%. (A note: this indicator is derived from the previous indicators).

There are five basic blocks of evaluation criteria in this ranking; as a result, in accordance with Consequence 2.3 above, this ranking system is a rather sustainable one.

3.4 Evaluation of the GRUP World University Rankings System

Shanghai Jiaotong University established the global research profiles of universities known as the so-called Global Research University Profile (GRUP) [19]. It uses the following indicators (or evaluation criteria): (1) 14 indicators for students, (2) 9 for professors and teaching staff, (3) 13 to describe the infrastructure and economy of the university, and (4) 5 for research, totaling a total number of indicators is 41, which is a simple number. Based on Consequence 2.3, this ranking is rather sustainable.

Let's consider a system $\boldsymbol{S}$ which represents THE World University Rankings. A decomposition of this system gives 5 subsystems, namely S_1, S_2, S_3, S_4, S_5. These correspond to each of the five categories mentioned above. Let G_S be a group of factors that represent the system S. Let B_1, B_2, B_3, B_4, B_5 be groups of factors, which represent subsystems S_1, S_2, S_3, S_4, S_5, respectively. We may apply the additional restriction on system S and subsystems S_1, S_2, S_3, S_4, S_5. The operation of the factors composition is a commutative one. Under this restriction, a synthesis of system S is described by the following theorem.

Theorem 2.8. Let the operation of the factors' composition, which represent the closed associative system with a feedback, be a commutative one. The synthesis of the systems S_1, S_2 is described by the group of factors $Ext(B_2, B_1)$,[2] the synthesis of the systems S_1, S_2, S_3 is described by the group of factors $Ext(B_3, Ext(B_2, B_1))$, the synthesis of the systems S_1, S_2, S_3, S_4 is described by the group of factors $Ext(B_4, Ext(B_3, Ext(B_2, B_1)))$, and the synthesis of the systems B_1, B_2, B_3, B_4, B_5 is described by the group of factors $Ext(B_5, Ext(B_4, Ext(B_3, Ext(B_2, B_1))))$. Because the numbers of factors (that represent a close associative system S with a feedback with commutative operation of composition of factors) is finite, in this case, the following existing theorem can simplify the synthesis process of the system S.

Theorem 2.9 [20]. If A and C are finite groups, then the group of extensions $Ext(C, A)$ is isomorphic to the group of all homomorphisms $Hom(C, A)$. That is, $Ext(C, A) \cong Hom(C, A)$. For example, if $\boldsymbol{G_S} \cong \boldsymbol{Z_m}$ is a cyclic group of order m, then $Hom(\boldsymbol{Z_m}, \boldsymbol{Z_m}) \cong \boldsymbol{Z_m}$. $Hom(\boldsymbol{Z_m}, \boldsymbol{Z_m}) \cong \boldsymbol{Z_m}$. The synthesis process gives us exactly m different variants of the system's synthesizing. Hence, using (a) structures of analyzed university ranking systems (such as THE, QS, ARWU and GRUP systems), (b) a description of the group *Hom (C, A)*, (c) Theorem 2.9, and (d) the final stability of blocks that make up a ranking system, we obtain estimates of sustainability of these systems. These show a rather conservative nature of these systems. It is important to note here that, by using Theorem 2.9, one can build a new ranking system with a higher sustainability in comparison with the systems discussed above.

[2] The group of extensions of an abelian group B_1 by the abelian group B_2 [20].

4 Conclusions

Conclusion 1. The obtained research outcomes and described findings enabled us to formulate the following recommendations to ensure the highest possible effectiveness of a smart university. It is strongly recommended to (1) develop an optimization model for each group of indicators (or, evaluation criteria) of smart university system's efficiency and effectiveness, (2) if needed, synchronize this optimization model and indicators with those in the world leading university ranking systems (such as THE, QS, ARWU, GRUP, etc.), (3) create an overall optimization model for the reliable planning and development of a smart university with an emphasis on (a) sustainable and reliable design and development on SmU components – SmU stakeholders, software, hardware, technology, smart pedagogy, etc., and (b) continuity of existing scientific (research) schools. The main features of a smart university – adaptation, sensing, inferring, anticipation, self learning, and self-optimization – provide any university with great potential for significant improvement of university's ranking in the world.

Conclusion 2. In order to improve a consistency between world/international and national university ranking systems we need:

1. A clarification of the evaluation criteria (indicators) for the national university ranking system, and a clear and transparent linkage (correspondence) between evaluation criteria of national and well-known world university ranking evaluation systems. Additionally, it is necessary to establish (a) a continuous monitoring system for university ranking evaluation criteria using various Web-based information technologies and (b) a reliable calculation of indicators for evaluation criteria, the indicators and/or groups of indicators that are determined and actively used in the world university ranking systems.
2. A development of optimization models for each group of evaluation criteria' indicators. Based on to-be-developed overall optimization models, we will be able to plan a progressive development of reliable higher education systems to ensure an entry and permanent presence of leading universities in the world's key ranking systems. The priority in those optimization models should be given to sustainable development and continuity of both existing and new scientific schools.

References

1. Serdyukova, N.A.: On generalizations of purities. Algebra Logic **30**(4), 432–456 (1991)
2. Serdyukova, N.A.: Optimization of tax system of Russia, Parts I and II, Budget and Treasury Academy, Rostov State Economic University (2002). in Russian
3. Serdyukova, N.A., Serdyukov, V.I.: The new scheme of a formalization of an expert system in teaching. In: Proceedings of ICEE/ICIT 2014 Conference, Paper 032, Riga (2014)
4. Serdyukova, N.A., Serdyukov, V.I., Slepov, V.A.: Formalization of knowledge systems on the basis of system approach. In: Smart Education and Smart e – Learning, Smart Innovation, Systems and Technologies, SEEL 2015, vol. 41, pp. 371–380. Springer (2015)

5. Serdyukova, N.A., Serdyukov, V.I.: Modeling, simulations and optimization based on algebraic formalization of the system. In: 19th International Conference on Engineering Education, 20–24 July 2015, Zagreb, Zadar (Croatia), Proceedings of ICEE2015 New Technologies and Innovation in Education for Global Business, pp. 576–582 (2015)
6. Uskov, A.V., Serdyukova, N.A., Serdyukov, V.I., Byerly, A., Heinemann, C.: Optimal design of IPSEC-based mobile virtual private networks for secure transfer of multimedia data. In: Pietro, G., Gallo, L., Howlett, R.J., Jain, L.C. (eds.) Intelligent Interactive Multimedia Systems and Services 2016. SIST, vol. 55, pp. 51–62. Springer, Cham (2016)
7. Serdyukova, N.A., Serdyukov, V.I., Uskov, V.L., Ilyin, V.V., Slepov, V.A.: A formal algebraic approach to modeling smart university as an efficient and innovative system. In: Uskov, V.L., Howlett, R.J., Jain, L.C. (eds.) Smart Education and e-Learning 2016. SIST, vol. 59, pp. 83–96. Springer, Cham (2016). doi:10.1007/978-3-319-39690-3_8
8. Demidovich, B.P.: Lectures on Mathematical Theory of Sustainability, p. 472. Nauka, Moscow (1967). in Russian
9. Bratus, A.S., Novojilov, A.S., Rodina, E.V.: Discrete dynamical systems and models in ecology, p. 135, Moscow State University of Railway Engineering, Moscow (2005). in Russian
10. Maltcev, A.I.: Algebraic Systems, p. 392. Nauka, Moscow (1970). in Russian
11. Rasiowa, H., Sikorski, R.: The Mathematics of Metamathematics. Polska Akademia Nauk, Monografie Matematyczne **41**, 593 (1963)
12. Karpenko, A.S.: The Logic at the Turn of Millennium, in Russian. logic.iph.ras.ru
13. Vasyukov, V.L.: Categorical Logic, ANO Institute of Logics, p. 208 (2005). in Russian
14. Kulikov, L.Y.: Algebra and Number Theory, p. 559. Vysshaya Shkola, Moscow (1979)
15. Mesarovich, M., Takahara, Y.: General System Theory: Mathematical Foundations. Mathematics in Science and Engineering, vol. 113. Academic Press, New York (1975)
16. The World University Rankings. https://www.timeshighereducation.com/world-university-rankings
17. QS World University Rankings. https://www.topuniversities.com/qs-world-university-rankings
18. Academic Ranking of World Universities (ARWU). http://www.shanghairanking.com/
19. Global Research University Profiles (GRUP). http://www.shanghairanking.com/grup/
20. Fucks, L.: Infinite Abelian Groups, vol. 1, p. 335. Academic Press, New York and London (1970). And vol. 2, p. 416. Academic Press, New York and London (1973)
21. Kurosh, A.G.: Theory of Groups, p. 648. Nauka, Moscow (1967). in Russian
22. Hilbert, D., Bernays, P.: Foundations of Mathematics, Logical Calculus and Formalization of Arithmetic, p. 551. Nauka, Moscow (1979). in Russian
23. Nikitchenko, N.S., Timofeev, V.G.: On the application of composition – nominative logics in insertion modelling. Int. J. Control Syst. Mach. **6**(242), 57–63 (2012)
24. Kulikov, L.Y., Fomin, A.A.: Ninth all-union symposium on group theory. Uspekhi Mat. Nauk **40**(6(246)), 167–171 (1985)

Government Enterprise Architecture for Big and Open Linked Data Analytics in a Smart City Ecosystem

Martin Lnenicka, Renata Machova, Jitka Komarkova(✉), and Miroslav Pasler

Faculty of Economics and Administration, University of Pardubice, Pardubice, Czech Republic
martin.lnenicka@gmail.com, {renata.machova, jitka.komarkova, miroslav.pasler}@upce.cz

Abstract. Opening up government data introduces a change about how governments operate and how stakeholders use data. At the same time, the large amounts of data need to be collected, processed, transmitted, and published with the aim of enhancing the quality of urban life. Now, these trends are shaping smart cities efforts in the public sector. This paper aimed to clarify the concepts of enterprise architecture, big and open linked data analytics, and smart city and how they are related to each other. A conceptual framework was proposed to analyze the requirements of BOLD analytics in a smart city ecosystem. This was contextualized by decomposing selected smart city architectures that served as a basis for understanding the interrelationships between components. This framework can be used as a guide to help developers and designers in creating government enterprise architectures for BOLD analytics in smart cities.

Keywords: Government enterprise architecture · Big and open linked data · Data analytics · Smart city ecosystem · Conceptual framework

1 Introduction

The diffusion and use of Information and Communication Technologies (ICT) among citizens and businesses is increasing rapidly and governments need to be prepared for these challenges. The literature indicates that by using ICT in an innovative way, governments can improve their performance [1–3]. There can be also seen a progress of ICT goals from cost-saving strategies and improving the quality and effectiveness of internal operations and public services delivery to dealing with transparency, openness and accountability through engagement of stakeholders in policy and decision making processes [4–6]. At the same time, the growth of available data and the ease with which these can be accessed using different devices are changing the way of interacting with stakeholders, whereas open linked data focus on the publishing, sharing and combining of these data. Big data analytics then facilitates faster analysis and better utilization of resources and improved personalization [7, 8]. Taken together, the concept of Big and

V.L. Uskov et al. (eds.), *Smart Education and e-Learning 2017*, Smart Innovation, Systems and Technologies 75, DOI 10.1007/978-3-319-59451-4_47

Open Linked Data (BOLD) analytics has gained considerable popularity in the public debate [5].

Since there is increasing effort worldwide to transform cities into smart cities, any smart city has to address two challenges. On the one hand, the enormous amount of data and their analytics that can help all stakeholders to provide and accept better decision processes. On the other hand, the need for these data to be available to the same stakeholders in order to provide added value to these data sources and to reach the required level of sustainability and improve the living standards through this openness [4, 6, 7, 9]. Similarly, Batty et al. [1] stated that more recent attempts to make cities smarter are focusing on improving the management and sustainability of a city through monitoring and analyzing the relevant data, with the intention of improving the quality of life if its citizens. According to Al Nuaimi et al. [4], the recent technology that has a huge potential to enhance smart city services is big data analytics. BOLD can contribute to the smartness of cities by linking and combining data, obtaining valuable insights from a large amount of data collected through various sources, and employing data or predictive analytics to improve better use of resources [5, 7]. However, in BOLD analytics settings, it becomes difficult to manage data using traditional applications [6]. In addition, it is crucial to establish the main requirements that should be fulfilled by the system and the chosen enterprise architecture in supporting the success of the smart city project [8, 10–13].

Thus, this paper aims to provide a conceptual framework for a data-centric government enterprise architecture in relation to the requirements for BOLD analytics in the smart city ecosystem. Smart city ecosystem assumptions are derived from Abu-Matar [14] and further extended based on the literature review. The term requirement follows the definition provided by Kyriazopoulou [12] as "*an official declaration of some functionality that aims to achieve the existing or the upcoming needs of the city*". In this regard, government enterprise architecture defines the structure, interrelationship and functionalities of all potential components, which deliver all expected smart city services to stakeholders. Further, a stakeholder is a person, group or organization that has an interest in, or is potentially impacted by, the operations of the public sector and its agencies or institutions. The main contribution then lies in the integration of two strands of research in e-government literature, i.e. BOLD analytics and the smart city ecosystem, and by investigating what are their key requirements, components and interrelationships for government enterprise architecture. In contrast to previous research, this paper clearly discusses and defines the importance of data analytics lifecycle in the smart city ecosystem.

This paper is organized as follows. First, the key concepts of this paper are defined. Subsequently, the most important requirements are identified and conceptualized in an ecosystem. Then, the key components and interrelationships in a conceptual framework are identified and followed by a discussion of the challenges and limitations for BOLD analytics in smart cities. Finally, conclusions are made by addressing the contributions of the paper.

2 Theoretical Background

2.1 Towards Government Enterprise Architecture for Smart City Ecosystems

While there are various definitions of a smart city, Caragliu et al. [2] defined a city to be smart when "*investments in human and social capital and traditional (transport) and modern (ICT) communication infrastructure fuel sustainable economic growth and a high quality of life, with a wise management of natural resources, through participatory governance*". A typology of smart city functions was presented by Batty et al. [1] and includes smart economy, people, governance, mobility, environment, and living. Since each city is shaped by various aspects; a smart city can be considered as a contextualized interplay among technological, managerial and organizational, and policy innovations [15] or as an ecosystem of various layers or views [14].

The achievement of these goals is feasible through the implementation of architectures that can incorporate together the various components of the smart city and make them interact in an effective way [1, 12, 16]. More precisely, Mamkaitis et al. [13] argued that the need for this architecture is determined by the existence of a collection of different initiatives and organizations, common goals, and competition between cities. Government enterprise architecture, business process modelling, and software development are a way to technological, organizational and managerial innovations that change traditional bureaucracy in regard to the needs of citizens [15, 17]. In this context, the architecture framework has to align and fully support the government's transformation objectives for smart city initiatives as well as interest of other stakeholders. This topic was studied by Vilajosana et al. [8] who suggested a model that will help to scale up smart cities deployments, generating a feedback model involving citizens and developers as well as utilities and operators. McGinley and Nakata [17] proposed a community architecture framework based on the Zachman framework and development methodology that can support diverse stakeholder requirements. They defined this architecture as "*a representation of the relationship of the community stakeholders' perspectives to the processes and data that support them*".

A conceptual architectural framework for smart cities addressing important quality and functional requirements was introduced by Kakarontzas et al. [18]. Regarding the type of data storage they recommended to use local data infrastructure and external data infrastructure (e.g. cloud storage). The access should be provided based on the users' role. Mamkaitis et al. [13] introduced the concept of an urban enterprise, which enables to view the smart city from the enterprise perspective. Al-Hader et al. [16] debated smart city infrastructure architecture development framework and surveyed positional accuracy of locating the assets as a base of the smart city development architecture. A reference architecture for smart city projects that contains architectural building blocks, best practices, and patterns was introduced by Abu-Matar [14]. The meta-model identified eight views, their corresponding stakeholders, concerns, and model type, together with the relationship among the views. Anthopoulos and Fitsilis [10] concluded to a common enterprise architecture, which identified the blue prints for urban information based development.

Some other authors, such as Abu-Matar [14], Da Silva et al. [11], Kyriazopoulou [12] or Vilajosana et al. [8], discussed the architectures that have already been proposed in the literature in order to find the most appropriate ones to be combined and applied to the development of smart cities. Their results showed that most of architectures focused on a technological solution, but smart city is social rather than technological issue. Thus, it is important to make involved stakeholders be a part of the solution and include them in the smartening process.

2.2 Phases and Activities of Big and Open Linked Data Analytics

The smart concept is represented in transmitting and receiving various data using communication protocols from and to the network element [16]. Effective analysis and utilization of these data is a key factor for success in many service domains. Hashem et al. [7] discussed the visions of big data analytics to support smart cities by focusing on how big data can fundamentally change urban populations at different levels. As reported by Al Nuaimi et al. [4], the big data platforms, tools and services will store, process, and mine smart cities applications information in an efficient manner to produce information to enhance different smart city services. On the other hand, Vilajosana et al. [8] reported that the need for policy changes, limited capital availability, and funding structures are preventing investment in smart cities.

The importance of open data for smart city initiatives was explored by Pereira et al. [3] who revealed evidences that open data initiatives contribute to enhance the delivery of public value in smart city contexts. Attard et al. [9] defined the data value network as: "*a set of linked activities having the aim of adding value to data in order to exploit it as a product*", where different stakeholders can participate by executing one or more activities and each activity can consist of a number of value creation techniques. Within a smart city, this network can have major impacts on its stakeholders, especially where application and services built on open data are used in a decision making process [3, 9]. Janssen et al. [5] derived a new taxonomy for forms of collecting and opening data to create smart cities. This taxonomy can be used by initiatives aimed at opening data to determine which ways of opening data are appropriate. A review of applications for BOLD analytics was provided by Al Nuaimi et al. [4].

According to Da Silva et al. [11], management of services in smart cities requires a constant monitoring, equipped with mechanisms for data collection. Data must be processed and analyzed, returning as response some kind of action to ensure the provision of services at satisfactory levels of quality and effectiveness [7]. The active participation of stakeholders in providing information will help in enhancing the quality of collected data and the performance of the applications [4]. Thus, all these data must be published to receive feedback. The key phases and activities are in Fig. 1.

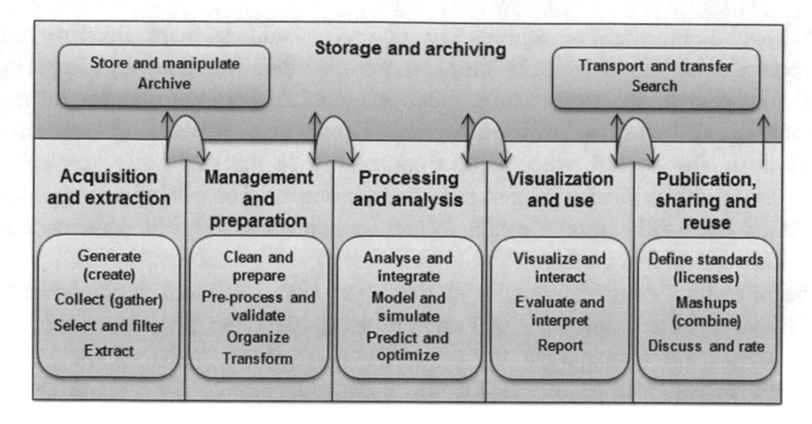

Fig. 1. Phases and their activities of the BOLD analytics lifecycle

3 Classification of Requirements for BOLD Analytics

Government enterprise architecture is able to handle the complexity of a smart city in which large amounts of data in various formats and from many domains are collected, processed, transmitted, and published with the aim of enhancing the quality of life of all citizens and other stakeholders. Thus, a first step is to classify the basic requirements that are essential to understanding the components and interrelationships. From the survey of architectures made in the previous section, a set of key requirements that must be addressed when developing this architecture is presented in Table 1.

This overview highlights the importance of smart network infrastructure and its availability, which provides computing resources, data transport, and necessary storage capacity for data streaming and processing platforms. The process of data management in smart cities is managed by the BOLD analytics lifecycle, its phases and associated

Table 1. Overview of most important requirements in a smart city ecosystem

Requirement	Literature sources
Smart network infrastructure and its availability	[4, 7, 10–12, 18]
Data streaming and processing platforms	[4, 7, 11, 12]
Advanced algorithms	[4, 7, 10, 12]
Data lifecycle and workflow	[4, 6, 7, 10, 19]
Open standard technology and interoperability	[4, 6, 10–12, 15, 18, 19]
Security and privacy	[4–7, 11, 12, 18]
Evaluation and real-time monitoring	[10–12]
Economy and cost savings	[4, 7, 10, 12]
Government role, management and organization	[4–6, 10]
Participation, collaboration and cooperation	[5, 10, 15]
Stakeholders awareness and skills	[4, 6, 11, 15]
Stakeholders satisfaction and usability	[10, 12, 18]
Sustainability and adaptability	[7, 10–12]

activities, while advanced algorithms are required to address them. Security and privacy requirements as well as evaluation and real-time monitoring must be implemented across all layers of the architecture. Open standard technology and interoperability requirements must be solved as well, because these are closely related to economy and cost savings. The overall structure and boundaries of the smart city ecosystem are determined by the existing legal and policy environment. The role of governments and their ability and willingness to establish and implement sustainable policies and programs is crucial in this context.

The most important requirement in the smart city ecosystem is characterized by involved stakeholders and their different roles shaping the processes and services provided. However, almost none of the previous research papers have investigated this topic more thoroughly. As the sustainability and adaptability were identified as key goals, this gap may negatively affect their fulfilment. This paper is aware of this issue and provides a list of possible roles. Among them is the role of domain expert the one that indicates the merit of providing services in the smart city. Although there are various providers and users of services or data, domain expert knows what data must be collected and are required for the execution and efficient operation of the service and is thus needed to transfer this knowledge to developers and designers of government enterprise architecture. Another problem arises when stakeholders do not have an

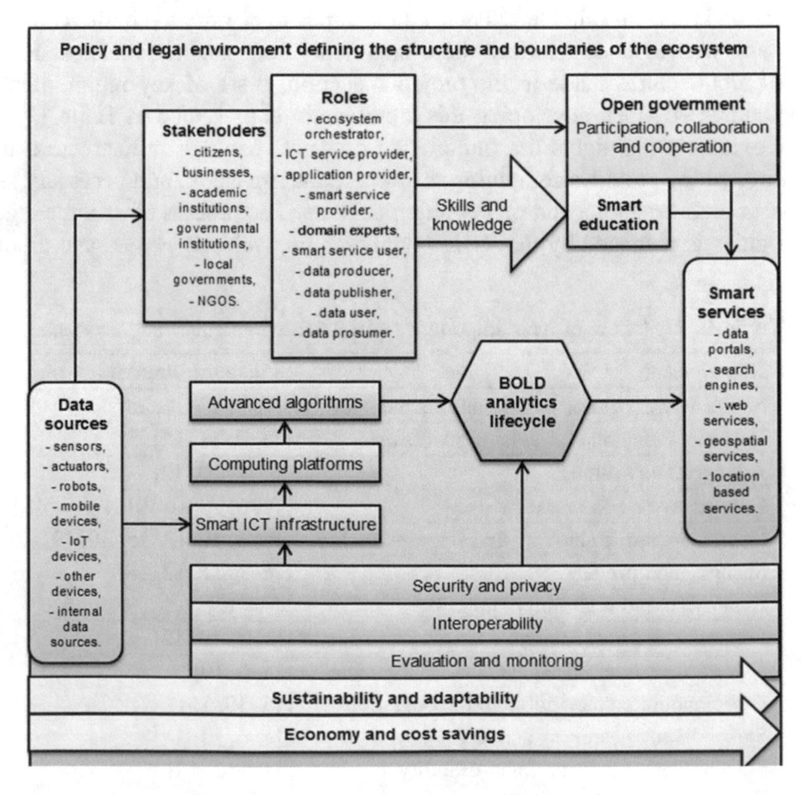

Fig. 2. Representation of the requirements and their relationships in the smart city ecosystem

adequate level of skills. This requirement is solved in the context of smart education, which encompasses participation, collaboration and cooperation of stakeholders as well as measuring their satisfaction of the experience of the services provided. The ecosystem is shown in Fig. 2.

4 Defining a Conceptual Framework

Generally, all the existing platforms offer similar characteristics. Vilajosana et al. [8] reported that they offer seamless interconnection with monitoring systems at the infrastructure level. On top, there emerge big data structures allowing for storing and analyzing the generated data, which are offered to stakeholders as services through standardized interfaces in an open data fashion [5, 9]. Thus, the conceptual framework needs to be debated starting from the smart ICT infrastructure preparation stage. This should be discussed while doing the concept of design of all infrastructure networks [7, 16]. These networks of data sources form a smart environment that incorporates various sources of data, which need to be managed. These phases are covered by the BOLD analytics lifecycle (see Fig. 1). In order to satisfy these requirements, ICT infrastructure has to be designed to support data-intensive computing and advanced algorithms as well as demands of various stakeholders who need to use it.

To facilitate such applications and services large distributed computational and storage facilities are needed [4, 7]. Thus, the second step is focused on building up the proper data storage and archiving architecture that would reflect the proposed infrastructure networks. Data architecture has to reflect the completeness of the network assets as well as the consistency, fault tolerance, data integrity and scalability [7, 16]. Al Nuaimi et al. [4] argued that this architecture should support both offline applications (historical data) as well as low latency processing to serve effectively in real-time applications. In the above context, cloud computing promises solutions to such challenges by facilitating data storage and delivering the capacity to support other phase of BOLD analytics [7, 19]. There are two ways data can be retrieved from data repositories. First, using Application Programming Interface (API)s or second, suitable extractors can be implemented to extract data from these distributed repositories [19]. Programming models for BOLD analytics are the last component for data architecture. There should be provided platforms, tools and services for both batch processing and stream processing, such as Apache Hadoop or Apache Storm. The smart analytics can use scalable machine learning algorithms or other advanced data mining algorithms to provide extraction of patterns from these data [7].

Smart application services represent the communication channel, in which people and machines directly interact with each other to make smart decisions. According to Nam and Pardo [15], a smart city is not system-driven but service-oriented. Al-Hader et al. [16] argued that these services are accessible through wireless mobile devices and are enabled by services oriented enterprise architecture including Web services, the Extensible Markup Language (XML), mobilized software applications, real-time and streaming interfaces, and standardized open data APIs. This provides the interactivity and interfaces that stakeholders need to browse the information based on their topics of interest, submit queries and get access to unified resources [8, 19]. With respect to open government movement, related processes of participation, collaboration and cooperation must be taken into account when governments set out policies. In this regard, feedback from the various stakeholders is constantly collected and monitored. These activities aim to facilitate the publication, management and sustainability of the open data publication and reuse process. Especially, the requirement of specific skills and knowledge needed to support these processes must be specified.

Therefore, this component can be viewed as the last piece of an architectural puzzle, since it provides the missing link between the delivery of smart services to stakeholders and the strategic goals of governments represented by e-government and governance architecture. This solution overcomes the problem found by Kakarontzas et al. [18], who mentioned that many smart initiatives do not have a business plan and just followed innovation spirit or resources savings as their potential business goals.

Further, data security is considered to be one of the most serious requirements in the extent that it is mandatory for a system to satisfy it [12]. Developers and designers must include security and privacy policies and procedures as an integral part of the design and implementation, e.g. data anonymization techniques or authorization based on roles. It is strongly recommended that all these components should be based on open standards and open formats, increasing the interoperability of ICT systems. Using open standards and open formats also means that technology can be easier shared and replicated within governments, which results in saving resources. Evaluation and monitoring component must be employed to obtain detailed feedback. These components support

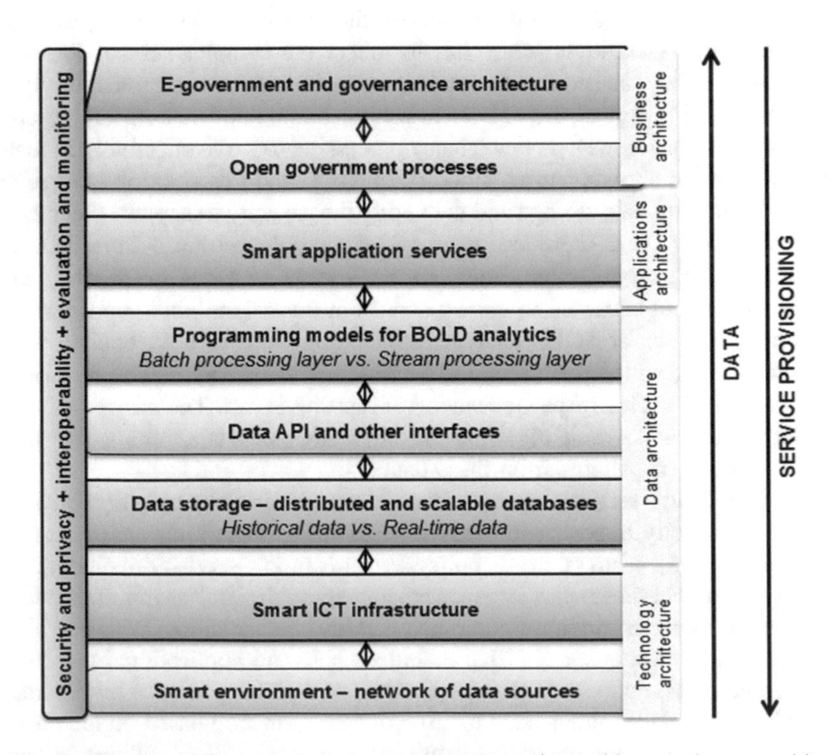

Fig. 3. Conceptual framework for government enterprise architecture in smart cities

the other components. Finally, the use of The Open Group Architecture Framework (TOGAF) and its Architecture Development Method (ADM) as a general approach to develop architecture with sufficient details can be recommended [13]. In this way, developers and designers could clearly define a problem, target stakeholders, goals and processes to achieve the goals.

The conceptual framework is shown in Fig. 3. Each layer represents the component that is needed to meet the requirements of BOLD analytics for smart cities.

5 Discussion and Limitations

Conceptualizing a smart city ecosystem as government enterprise architecture for BOLD analytics can provide easier understanding and interpretation of required components and interrelationships between them. However, there are also several challenges and limitations to this approach. In general, governments around the world are concerned about the cost and benefits of implementing a smart city [4]. Considering that cities usually have different characteristics, it is natural to imagine that the same architecture does not apply to all of them, mainly because one cannot generalize local, financial, social and environmental restrictions [11]. Another limitation is the effort to enable participation, collaboration and cooperation processes without adequate classification of stakeholders' roles [9]. These challenges were addressed in this paper.

There are also obstacles related to ICT infrastructure and various devices that span private and public networks and may cause security and integrity problems. Capabilities, skills, and attitudes of stakeholders must shape all the architectural decisions. As noted by Al Nuaimi et al. [4], the need for highly educated well qualified people to design, develop, deploy and operate smart city infrastructures, platforms and applications is growing rapidly. Thus, specialized education and training programs in these fields need to be developed and offered to stakeholders [6, 11, 15]. Janssen et al. [5] reported that a major challenge is to deal with data distributed over various data sources and how to align these data push with the stakeholders' needs. In this regard, data about their activities must be collected, processed and feedback ensured to meet the requirements of transparency and sustainability. On the other hand, Molinari et al. [6] argued that there is a clear tension between transparency and privacy. Therefore, there are still a lot of challenges to be solved.

6 Conclusions

This paper contributes to a preliminary understanding of the basic components and their interrelationships that are required to successfully develop, manage, and implement the new government enterprise architecture for BOLD analytics in the smart city ecosystem. Relying on existing research, a conceptual framework is proposed to illustrate this contribution to enhance the delivery of smart public services.

At first, authors identify key requirements and describe the ecosystem. Especially the importance of involved stakeholders and their roles in the ecosystem is emphasized since it was realized that this topic is not adequately addressed in the previous literature.

Then, the conceptual framework is designed and the most important components are discussed to clarify exact meaning of the requirements. Its main contribution is the relation between business and applications architecture in the public sector, which is realized using open government processes. The framework can be used as a starting point for actual smart city architectures. It should also provide a mechanism to stimulate value generation from BOLD analytics in smart cities.

Acknowledgement. This research is supported by University of Pardubice, SGS_2017_17 project.

References

1. Batty, M., et al.: Smart cities of the future. Eur. Phys. J. Spec. Top. **214**(1), 481–518 (2012)
2. Caragliu, A., Del Bo, C., Nijkamp, P.: Smart cities in Europe. J. Urban Technol. **18**(2), 65–82 (2011)
3. Pereira, G.V., Macadar, M.A., Luciano, E.M., Testa, M.G.: Delivering public value through open government data initiatives in a smart city context. Inf. Syst. Front. **19**, 1–17 (2016)
4. Al Nuaimi, E., Al Neyadi, H., Mohamed, N., Al-Jaroodi, J.: Applications of big data to smart cities. J. Internet Serv. Appl. **6**(1), 1–15 (2015)
5. Janssen, M., Matheus, R., Zuiderwijk, A.: Big and open linked data (BOLD) to create smart cities and citizens: insights from smart energy and mobility cases. In: Tambouris, E., et al. (eds.) EGOV 2015. LNCS, vol. 9248, pp. 79–90. Springer International Publishing, Cham (2015)
6. Molinari, A., et al.: Big data and open data for a smart city. In: IEEE Smart Cities Inaugural Workshop, December 2014 in Trento, Italy, pp. 1–8. IEEE (2014)
7. Hashem, I.A.T., et al.: The role of big data in smart city. Int. J. Inf. Manage. **36**(5), 748–758 (2016)
8. Vilajosana, I., et al.: Bootstrapping smart cities through a self-sustainable model based on big data flows. IEEE Commun. Mag. **51**(6), 128–134 (2013)
9. Attard, J., Orlandi, F., Auer, S.: Data driven governments: creating value through open government data. In: Hameurlain, A., et al. (eds.) Transactions on Large-Scale Data-and Knowledge-Centered Systems XXVII. LNCS, vol. 9860, pp. 84–110. Springer, Heidelberg (2016)
10. Anthopoulos, L., Fitsilis, P.: From digital to ubiquitous cities: defining a common architecture for Urban development. In: 2010 Sixth International Conference on Intelligent Environments, pp. 301–306. IEEE (2010)
11. Da Silva, W.M., et al.: Smart cities software architectures: a survey. In: Proceedings of the 28th Annual ACM Symposium on Applied Computing, pp. 1722–1727. ACM, New York (2013)
12. Kyriazopoulou, C.: Architectures and requirements for the development of smart cities: a literature study. In: Helfert, M., et al. (eds.) Smartgreens 2015 and Vehits 2015. CCIS, vol. 579, pp. 75–103. Springer International Publishing, Cham (2015)
13. Mamkaitis, A., Bezbradica, M., Helfert, M.: Urban enterprise: a review of smart city frameworks from an Enterprise Architecture perspective. In: 2016 IEEE International Smart Cities Conference (ISC2), pp. 1–5. IEEE (2016)
14. Abu-Matar, M.: Towards a software defined reference architecture for smart city ecosystems. In: 2016 IEEE International Smart Cities Conference (ISC2), pp. 1–6. IEEE (2016)

15. Nam, T., Pardo, T.A.: Smart city as urban innovation: focusing on management, policy, and context. In: Proceedings of the 5th International Conference on Theory and Practice of Electronic Governance, pp. 185–194. ACM, New York (2011)
16. Al-Hader, M., Rodzi, A., Sharif, A.R., Ahmad, N.: Smart city components architecture. In: 2009 International Conference on Computational Intelligence, Modelling and Simulation, pp. 93–97. IEEE (2009)
17. McGinley, T., Nakata, K.: A community architecture framework for smart cities. In: Foth, M., Brynskov, M., Ojala, T. (eds.) Citizen's Right to the Digital City, pp. 231–252. Springer, Singapore (2015)
18. Kakarontzas, G., Anthopoulos, L., Chatzakou, D., Vakali, A.: A conceptual enterprise architecture framework for smart cities: a survey based approach. In: 2014 11th International Conference on e-Business (ICE-B), pp. 47–54. IEEE (2014)
19. Khan, Z., Anjum, A., Kiani, S.L.: Cloud based big data analytics for smart future cities. In: Proceedings of the 2013 IEEE/ACM 6th International Conference on Utility and Cloud Computing, pp. 381–388. IEEE (2013)

User Identification in a Variety of Social Networks by the Analysis of User's Social Connections and Profile Attributes

Marina V. Lapenok[1(✉)], Anna V. Tsygankova[1], Nataliya G. Tagiltseva[1], Lada V. Matveyeva[1], Olga M. Patrusheva[2], and Natalya V. Gerova[3]

[1] Ural State Pedagogical University, Ekaterinburg, Russia
lapyonok@uspu.me, {anya_tsygankova, musis52nt}@mail.ru, lada-matveeva@yandex.ru

[2] Ural Bank for Reconstruction and Development, Ekaterinburg, Russia
podsnejnik1993@gmail.com

[3] Ryazan State University named for S. Yesenin, Ryazan, Russia
nat.gerova@gmail.com

Abstract. In the process of e-learning students are often organized in different groups in social networks for remote educational interaction. Social networks of modern Smart society are used by unscrupulous people as a means of preparing and committing various types of criminal activity. The article demonstrates the results of the development of the algorithm of identification of the user of social networks based on the analysis of text attributes and social ties which can be included in the complex of measures to combat illegal activities in social networks by law enforcement agencies. The algorithms on the similar subject developed by other authors are analyzed. The authors propose the implementation of the algorithm and experimental results on its use with specific percentages.

Keywords: Social networks in modern Smart society · User identification · Social graph · Graph database · Social connections

1 Introduction

In the process of e-learning, students are organized into various groups in social networks for remote interaction with a view to the implementation of collective training projects, consulting with teachers and classmates. Social networks in today's Smart society are frequently used by unscrupulous people as a means of preparing and committing various criminal activities. In the modern world there is a dangerous tendency of the penetration of criminal threats in the Internet due to the minimal legal regulations of internal relations, lack of knowledge by most users of the basic concepts of information security measures, and the widespread network. In particular, modern social and cultural phenomenon – social networks have a great potential as a medium for preparing and committing crimes.

A social network is a platform, online service or web site that is designed to build reflection and the organization of social relationships.

V.L. Uskov et al. (eds.), *Smart Education and e-Learning 2017*, Smart Innovation, Systems and Technologies 75, DOI 10.1007/978-3-319-59451-4_48

The service must have the basic features under which it can be considered a social network:

- each user has a personal public or partially public webpage – a profile containing some information about the user;
- each user has an opportunity to create and maintain a list of other users with whom they have any relationship in the framework of the social network (e.g., friendships, business contacts, etc.);
- browsing and bypassing links between users within the service (for example, a user can see the "friends" of "friends");
- there is a possibility to join users in online communities in accordance with their interests (groups) [1].

It should be emphasized that in social networks, generally with no restriction on purpose of grouping users, it is possible to create a group of "positive" users with shared ideological views or professional interests, and groups with extremist views, and groups to promote narcotic drugs and psychotropic substances, to provide commercial sexual services.

We are aware of cases [2, 3], when social networking was used to commit various criminal acts or actions inciting to commit the latter, such as:

- search and communication of supporters of the extreme nationalist ideas, terrorism, supporters of drug use and other individuals with deviant views;
- distribution of pathogenic information that could cause harm, especially to an underage person, including materials of extremist nature, promoting narcotic drugs, psychotropic substances and their analogues, advertising the services of an intimate nature. The propagation speed of destructive content is large and often not inferior to real communication, and the indication of personal information in social networks causes the purposeful distribution of materials to provide maximum impact;
- influence on the formation of certain opinions and views as individuals and segments of social networks;
- various types of fraud, including using virtual currency and credit cards. The review of unauthorized transfers of funds in 2014 from the Central Bank of the Russian Federation [4] states that the total amount of unauthorised transactions with payment cards issued in the territory of the Russian Federation, in 2014 amounted to 1.58 billion rubles;
- use of social engineering – the section of social psychology, intended to introduce a certain model of behavior into human consciousness and thereby to manipulate the actions using the "vulnerability" of people – fear, compassion, subordination or negligence. Long before the growing prevalence of social networking seeking personal or confidential information required a lot of time and effort, the social network greatly simplifies this task;
- collection of information about the object before the burglary. Burglaries on the tip have become especially widespread, that cause the victims themselves spreading photos of their houses and summer cottages with a description of their internal structure, location and valuables in social networks, sharing their plans about

vacations, with specific dates through the services, and thereby attracting the attention of offenders;
- collection of information of interest before kidnapping and human trafficking;
- organization and coordination of actions aimed at direct confrontation with the legally elected government, terrorism;
- finding out incriminating facts about users for later blackmail.

Thus, social networks in today's Smart society have a criminogenic potential opportunity of use of interaction within the Internet as for the preparation and commission of a crime, and for the formation of criminal behaviour of the user or thinking of the victim.

However, it should be noted that the fight against criminal actions in social networks of modern Smart society is not developed and is typically limited to a theoretical justification of the need for such. The practical implementation of ideas is very rare. Law enforcement agencies in rare cases use the traces of criminal activities, and other available social networking information for the prevention, detection and investigation of crimes. Scientific methods of monitoring of information space of social networks of modern Smart society are not developed, technical means are not used; there are no mechanisms to assess threats.

2 Theoretical Part

One of the most popular tasks, the solution of which has the prospect of the use in the operational - investigation activities to detect computer-savvy criminals, is the problem, most known as entity resolution (the problem of entity resolution – identification of a variety of database records that are related to the same real-world object), applied to profiles in social networks, i.e. the problem of identifying the same user in different social networks[1].

In this work, an algorithm solving this problem for users of social networking and its implementation are submitted.
The main problems of the information obtained from different social networks are:

- distribution among many different services;
- failure, due to the possible anonymity and secrecy of individuals and greatly complicating the process of establishing the profile of the corresponding user.

In solving the above problems, the social information of the user and the concept of social similarity provide substantial assistance. Since the user himself forms his social connection within the social network, the latter unmask him well. It should be noted that even in the case when the user for some reason does not put (or deletes) data in the profile:

- with high probability the user will maintain social ties, that is, he will have a specific list of contacts ("friends");

[1] The research is financially supported by the Russian Scientific Foundation, project №16-18-02102.

– users whom an anonymous user has in his contact list may not hide information about themselves in the profile. Thus, it is possible to use information of individuals in the contact list ("friends") of the user to define his profile in another social network.

Let's look at Fig. 1, which presents profiles of the same user in two different social networks. The identification of profiles for a single user is possible due to the presence of a certain area of intersection of the contact list ("friends") of the user who also has profiles in these networks. The size of the area of intersection depends on the number of pairs of profiles that were found to be similar according to the results of calculating the textual similarity (based on text data).

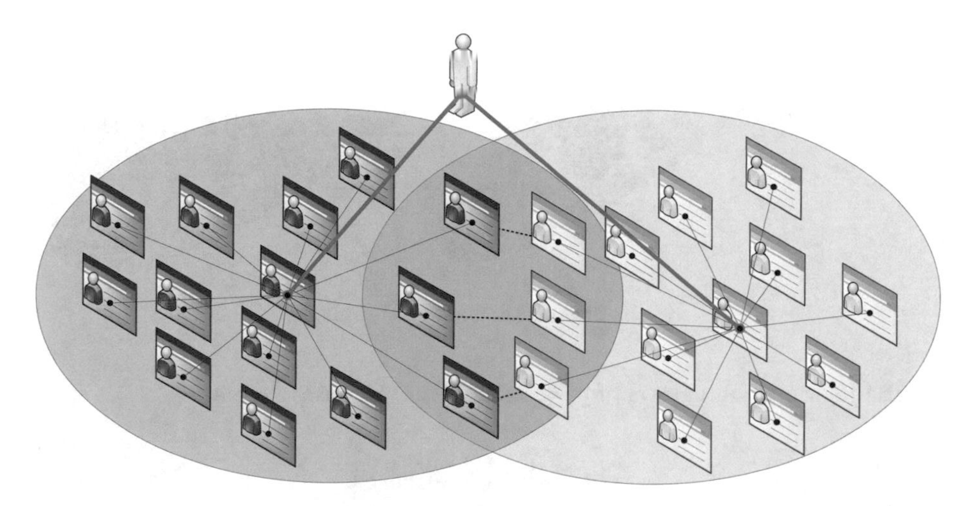

Fig. 1. Determination of similarity through the structure of social relations

The result of the comparison may be a situation where one profile in a social network matches multiple profiles from different social network with equal values of the text similarities and there is a problem of choosing between them.

Figure 2 shows two social graphs <A, B>. We mean by social graph the graph whose nodes are represented by profiles with different attributes, and the edges are social relationships between accounts (e.g. "friendship"). Connections in a graph can be directed or undirected depending on the semantics of relations that they reflect.

The result of the algorithm to find profiles belonging to the same real person in different social networks in modern Smart society, is a well-defined couple of profiles (x, y) such that $x \in A$, $y \in B$. We denote the mapped profile to profile $x \in A$ as $pr(x) \in B$ - projection of x profile $\in$ A in B.

To solve the problem of the choice social information of all projections is used, for which the index of text similarity exceeds some threshold value. The definition of social similarity is as follows. We determine all adjacent vertices to the desired node $x \in A$. We denote the set of these vertices as $N(x) = \{n(x)1, n(x)2, \ldots, n(x)m\}$. For each neighboring vertex n(x)i all projections in the graph B {pr(n(x)i)1, pr(n(x)i)2, …, pr(n(x)i)o} $\in$ PR(n(x)i) for which textual similarity exceeds some threshold value.

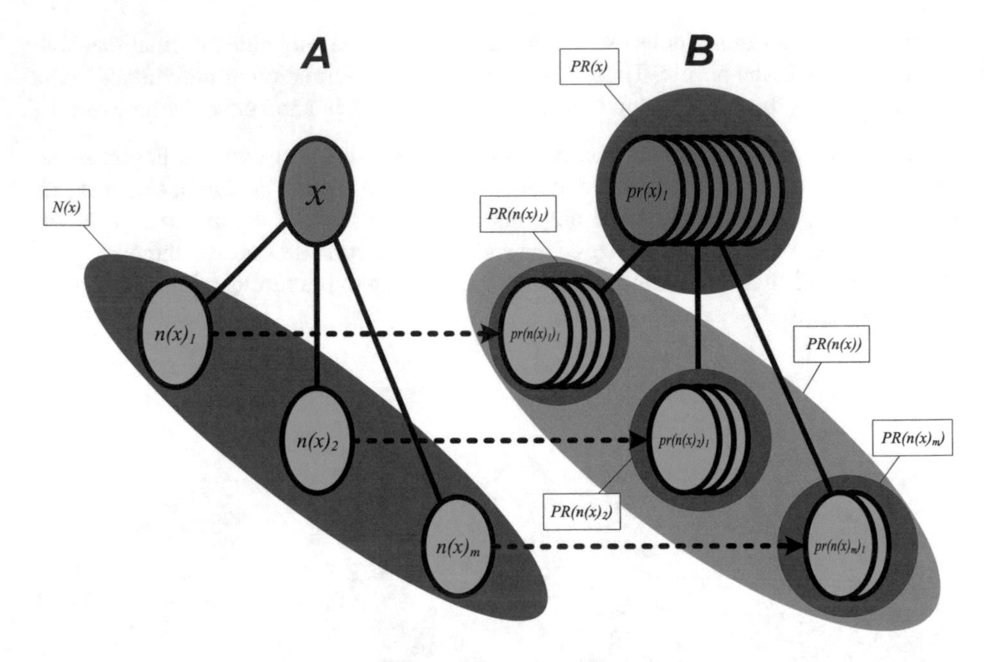

Fig. 2. Determination of social similarity

The joining of all the sets of projections PR(n(x)i) forms the set

$$\mathrm{PR(n(x))} = \bigcup_{i=1}^{m} PR(n(x)i)$$

For each of such projections pr(n(x)i)j neighboring vertices are determined, the joining of which forms a set of projections of the required vertices.

x ∈ A : PR(x) = {pr(x)1, pr(x)2, …, pr(x)o}. For each projection the set of its adjacent vertices N(pr(x)k) is determined in graph B. For each projection pr(x)k indicator of social affinity is calculated by the formula (1) [5].

$$\mathrm{SMatch} = \frac{|\mathrm{N(pr(x)_k)} \cap \mathrm{PR(n(x))}|}{|\mathrm{N(x)}|} \tag{1}$$

3 Implementation

In the development and implementation of algorithm for solving the problem of identification of the user in various social networks a lot of research was conducted. When studying the results of the conducted research the following common assumptions have been identified:

- simplification of the model of comparison of textual account attributes;
- insufficient use of social information of accounts;

- when calculating the similarity of the structure of social relations each user of a social network was put in line the only user of other social network based on the comparison results of the text fields of the account.

After studying the works of other authors, in the course of the study, an emphasis was made to eliminate the above assumptions. In the developed approach social relations are actively used by reviewing and analyzing the list of contacts ("friends") of the user, and the information filled by the person in a social network profile is used. The algorithm consists of the following steps:

(1) collection of baseline data and their presentation in a common format;
(2) consistent application of functions to compute similarity to the corresponding fields of the two accounts, and deciding on a text similarity based on a binary classifier (with the calculation of the numerical index);
(3) comparison of the structure of social ties of accounts and calculation of index of social similarity;
(4) calculation of the resulting similarity index of accounts.

The initial data for the algorithm are the following:

- the attributes filled by the user in a social network profile. Examples of such data include: surname, name, patronymic, date of birth, school, higher education institution, work place, interests, user message (also known as "wall profile");
- social links: the user's contact list ("friends"), user's participation in various communities;
- reference data, such as links to other users in the texts, reference information on their profiles in other social networks, etc.;
- other data.

In the first stage of the algorithm the following steps are performed.

1. The collection of information of interest of users of social networks in modern Smart society using the API - methods of the social networking services is carried out.
2. The development of a unified format of representation of information from different social networks and bringing it to this format are implemented. The need to bring data to a common format is caused by the problem of the heterogeneity of the data formats in which the social networks return information about users. The heterogeneity lies in the different names of the fields and the grouping of data. To avoid the problem it is necessary to form a compliance scheme which allows the source data to put to a common structure.
3. The collection of baseline data, which will be the calculation of similarity for different attributes and calculation of indicators of text similarity is produced. The criteria for data selection are following:
   - availability and accessibility of data, in relation to which the search of profiles of one person occurs. On the one hand, specific or inaccessible data should be excluded. The unavailability of certain data for extraction may be due to limitations of the user in the security settings. On the other hand, the data that is

most often filled in by users in different social networks are included in the basic data. These are such attributes as name, surname, gender, date of birth, city of residence, school, institution of higher education, the place of work;
- informative data for the algorithm. An example of uninformative data can serve as such a parameter as the presence of the user on the site at the moment, the availability of foreign currency in the user's profile account of the social network and its quantity, etc.;
- location of the original data in the graph database Neo4j. The graph database Neo4j was selected to store the user's data collected in the process of implementation of the algorithm, because the used data have a graph structure, the main elements of which are nodes and links between them (edges). According to some authors [6], for problems with the natural graph structure graph management systems of database can significantly outperform relational performance, and also have the advantages of data visualization and making changes.

4. Within this database there is the possibility of adding parameters to each node or edge, defined by the attributes of the user's account.

In the second stage of the algorithm there is the calculation of an index of textual similarity for all data attributes that were selected in the first stage. Taking into account the fact that different data need different comparison functions, they are divided into groups of attributes:

- attributes requiring an exact match;
- attributes that allow a partial match.

For the first group of attributes functions of matches use the elementary comparison operation to test the equality of fields and present the result as a value of logical type. Examples of attributes of the first category are the name, gender, date of birth.

Spelling errors, different word order, abbreviations tend to the second category of attributes. Therefore, the similarity of two sequences of data can be measured by algorithms of search of subline in line, fuzzy comparison, which will get the number of similarities. The second group of attributes includes the city of residence, school, institution of higher education.

The third stage is the calculation of the indicators of social similarity.
The fourth step is the calculation of the result of the similarity index.
The calculation of similarity index is produced by Formula 2:

$$\mathrm{Match} = \mathrm{w_T} \cdot \mathrm{TMatch} + \mathrm{w_S} \cdot \mathrm{SMatch} \tag{2}$$

where w_T and w_S are the weight values for indicators of text and social similarities, accordingly, at that, $w_T + w_S = 1$.

The similarity of two profiles in different social networks is evaluated due to the action of the algorithm.

The implementation of the algorithm includes the following steps:
Step 1. Original data: the sample profile in social network 1. Using a set of ready-made methods called API social network API, or, provided that in API social network there are no methods that return the required data using GET requests to a specific page we receive the following attributes and social connections of profile:

- a list of "friends" of a profile in social network 1;
- attributes of the desired profile and all its contacts ("friends") from the list of contacts ("friends"): name, surname, date of birth, hometown, country, education (high school, college, etc.) place of work;
- key words, the formation of which comes from the text placed in the profile of the author (the so-called "wall") in accordance with the developed algorithm in the study. An algorithm for extracting keywords includes the following steps:

  1. Exclusion of stop words and punctuation from the text. Stop words include pronouns, introductory words, interjections, adverbs, particles, exclamations, numerals.
  2. Reduction of words to normal form: nouns - singular, nominative case, verbs are infinitives, adjectives are given to nouns from which were formed.
  3. Counting of the frequency of words.
  4. Calculation of the frequency of threshold, depending on the amount of text.
  5. Formation of the final array of keywords from the words frequency use of which is higher than the threshold value.

For the implementation of the algorithm pymorphy2 module of Python was used, which is a morphological analyzer. With it, the words were listed in normal form, and grammatical information about the words, in particular, the part of speech was obtained.
Step 2. Conducting a sequential search for each profile from the list of contacts ("friends") from step 1 and displaying it in social network 2. Read more:

- **Primary filtration**, the purpose of which is the formation of an array of profiles, similar to the profiles from social network 1for attribute values. For the primary filter, the following algorithm is used: by the profile attributes from step 1 using the search function provided by the API that maps each profile from the list of "friends" with all users of social network 2 (the standard search function can be used correctly for all users of social network 2). The result of the primary filtration is a number of "similar" attribute values of the profiles from social network 2. They are the comparison group, for each profile from the list of "friends". The attributes, the list of "friends" of each profile from the comparison group (the attributes obtained using the API methods of social network 2 and the key words of the "wall" for each of the comparison group) are an additional result. The result is placed into the database Neo4j.
- **Extension of the comparison group** for each of the contacts ("friends"), the purpose of which is to increase the number of profiles of social network 2, consisting in the comparison group for each of the "friends". The algorithm of extension of the comparison group foresees the search of profiles that were not found by direct search using the API and adding these profiles to the comparison

group. The initial data for the implementation of such a search are dictionaries of names, universities or schools. The internal structure of the dictionaries used at this stage, seems similar to a dictionary of synonyms of the Russian language. Different names of proper names are presented as synonyms, for example, Ekaterinburg – "Eburg" – "Ekb" – Ekat or Lyceum №180 – Lyceum "Poliforum" – 180 Lyceum. The result: an increased number of profiles, consisting in the comparison group.

– **Filtering the expanded list,** the purpose of which is the formation of values of text similarity of profile attributes and text similarity of the "wall" of the comparison group and setting appropriate thresholds. In this case, if the value of the text similarity of attribute profiles or the wall for a profile of the comparison group is below the threshold, then the profile is deleted from the comparison group, the algorithm is used for group comparisons of each of the contacts ("friends").

Additionally, we compare each profile from each comparison group using a variety of techniques with the appropriate contact ("friend") from the original list of contacts ("friends") of stage 1. Methods of comparison used when filtering:

1. comparison – the result of which is a Boolean value – whether there is a match or not.
2. application of fuzzy search by attributes of the profiles, using the algorithm for calculating the Levenshtein distance, i.e., counting the minimum number of operations of inserting a single symbol, deleting a single symbol and replacing one symbol with another, needed to transform one line into the other. The result is a numerical coefficient of text similarity for each profile from the comparison group. All numerical coefficients are combined into the general table of the numerical coefficients of text similarity. It turns out a separate table for each contact ("friend") from the list of contacts ("friends") of the original profile.
3. search for a subline in the line. A comparison of the attributes is as follows:
   – the name is checked for an exact match or search for a subline in the line;
   – the surname is checked for an exact match, search for a subline in the line or in the absence of the latter the Levenshtein distance and its threshold value is calculated;
   – date of birth is checked for an exact match (excluding the year of birth);
   – city is checked for an exact match or search for a subline in the line using the dictionary;
   – country is checked for an exact match;
   – school, college, university, career are checked for an exact match or search for a subline in the line;
   – key words from the wall are checked for an exact match and subline search in the line.

The filtering of the extended list enables you to identify a comparison group profile for each contact ("friend"), that is, those profiles in social network 2, which may belong to the same contact ("friend") of the desired profile from social network 1. The result is a list of all contacts ("friends") in social network 2 for each object of the comparison group.

Step 3. We record the list of all contacts ("friends") of social network 2 in graph database Neo4j. In the resulting graph it is necessary to calculate the degree of each vertex – they are the candidates for the desired profile of the person in social network 2. The desired profile is the top most. Step 3 is illustrated in Fig. 3.

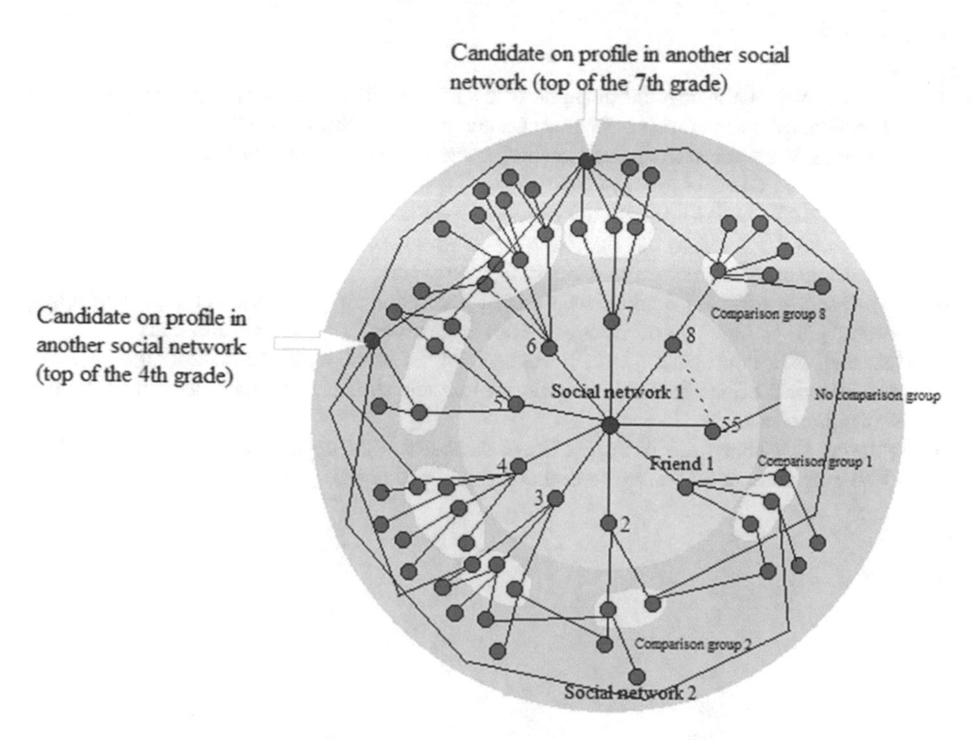

Fig. 3. Searching the required profile in social graph

This algorithm was tested in the social networks "Classmates" and "InContact" in today's Russian-speaking part of the Internet. When run on 1000 random users who have profiles in social networks "Classmates" and "InContact" having not less than 15 contacts ("friends") in each of the social networks, in 84% of cases among the results provided by the algorithm was the Profile of the user in another social network.

4 Conclusion

In the study, the algorithm of identification of one user in different social networks was developed, described, implemented and tested. The result can be considered from a mathematical point of view (a numerical index of similarity) and from a visual point of view (graph vertex).

The algorithm was tested on the social networks in today's Russian-speaking part of the Internet "Classmates" and "InContact". Using an algorithm will allow to counteract the penetration of criminal threats to social networks Smart modern society, where in the process of e-learning students communicate for the purpose of implementation of educational training projects, consulting with teachers and classmates.

References

1. Solovyov, V.S.: Criminogenic potential of social networks: a methodology for assessment and neutralization measures. Saratov Center for the Study of Organized Crime and Corruption, Voronezh (2013). sartraccc.ru/Explore/soloviev1.docx. (in Russian)
2. Levshits, N.G.: Cases of fraud in social networks while raising funds for medical treatment, Moscow (2015). http://echo.msk.ru/blog/nlevshits/1519362-echo/. (in Russian)
3. Sokolova, A.A.: Average damage from fraud in social networks, Moscow (2015). http://rusbase.com/news/internet-fraud-statistics/. (in Russian)
4. Overview of the Central Bank of the Russian Federation on unauthorized transfers of funds, Moscow (2014). http://www.cbr.ru/psystem/P-sys/survey_2014.pdf. (in Russian)
5. Sinadskiy, N.I., Patrusheva, O.M., Bezuglaya, M.V., Sushkov, P.: Identification of users of social networks on the basis of similarity of text and structure of social networks, Tyumen (2015). (in Russian)
6. Robinson, I., Webber, J., Eifrem, E.: Graph databases. New opportunities for connected data (2015). http://info.neo4j.com/rs/neotechnology/. (in Russian)

Author Index

V.L. Uskov et al. (eds.), *Smart Education and e-Learning 2017*, Smart Innovation, Systems and Technologies 75, DOI 10.1007/978-3-319-59451-4

Printed in the United States
By Bookmasters